METHODS OF ELECTRONIC-STRUCTURE CALCULATIONS

Wiley Series in Theoretical Chemistry

Series Editors

D. Clary, University College London, London, UK
A. Hinchliffe, UMIST, Manchester, UK
D. S. Urch, Queen Mary and Westfield College, London, UK
M. Springborg, Universität Konstanz, Germany

Thermodynamics of Irreversible Processes: Applications to Diffusion and Rheology
Gerard D. C. Kuiken
Published 1994, ISBN 0 471 94844 6

Modeling Molecular Structures
Alan Hinchliffe
Published 1995, ISBN 0 471 95921 9 (cloth), ISBN 0 471 95923 5 (paper)
Published 1996, ISBN 0 471 96491 3 (disk)

Molecular Interactions
Edited by **Steve Scheiner**
Published 1997, ISBN 0 471 97154 5

Density-Functional Methods in Chemistry and Materials Science
Edited by **Michael Springborg**
Published 1997, ISBN 0 471 96759 9

Theoretical Treatments of Hydrogen Bonding
Dušan Hadži
Published 1997, ISBN 0 471 97395 5

The Liquid State — Applications of Molecular Simulations
David M. Heyes
Published 1997, ISBN 0 471 97716 0

Quantum-Chemical Methods in Main-Group Chemistry
Thomas M. Klapötke and Axel Schulz
Published 1998, ISBN 0 471 97242 8

Metal Clusters
Edited by **Walter Ekardt**
Published 1999, ISBN 0 471 98783 2

METHODS OF ELECTRONIC-STRUCTURE CALCULATIONS

From Molecules to Solids

Michael Springborg
Department of Chemistry, University of Konstanz,
Germany

JOHN WILEY & SONS, LTD
Chichester • New York • Weinheim • Brisbane • Singapore • Toronto

National   01243 779777
International (+44) 1243 779777
e-mail (for orders and customer service enquiries): cs-books@wiley.co.uk
Visit our Home Page on http://www.wiley.co.uk
or http://www.wiley.com

Other Wiley Editorial Offices

John Wiley & Sons, Inc., 605 Third Avenue,
New York, NY 10158-0012, USA

WILEY-VCH Verlag GmbH, Pappelallee 3,
D-69469 Weinheim, Germany

Jacaranda Wiley Ltd, 33 Park Road, Milton,
Queensland 4064, Australia

John Wiley & Sons (Asia) Pte Ltd, Clementi Loop #02-01,
Jin Xing Distripark, Singapore 129809

John Wiley & Sons (Canada) Ltd, 22 Worcester Road,
Rexdale, Ontario M9W 1L1, Canada

Library of Congress Cataloging-in-Publication Data

Springborg, Michael.
    Methods of electronic-structure calculations : from molecules to solids / Michael Springborg.
        p.   cm. — (Wiley series in theoretical chemistry)
    Includes bibliographical references and index.
    ISBN 0-471-97975-9 (alk. paper) — ISBN 0-471-97976-7 (pbk. : alk. paper)
    1. Quantum chemistry. 2. Electronic structure. I. Title.   II. Series.
QD462.S695   2000
541.2′8 — dc21   99–058318

British Library Cataloguing in Publication Data

A catalogue record for this book is available from the British Library

ISBN 0 471 97975 9 hardback
ISBN 0 471 97976 7 paperback

Typeset in 10/12pt Times by Laser Words, Madras, India

This book is printed on acid-free paper responsibly manufactured from sustainable forestry,
in which at least two trees are planted for each one used for paper production

Contents

Series Preface

Theoretical chemistry is one of the most rapidly advancing and exciting fields in the natural sciences today. This series is designed to show how the results of theoretical chemistry permeate and enlighten the whole of chemistry together with the multifarious applications of chemistry in modern technology. This is a series designed for those who are engaged in practical research. It will provide the foundation for all subjects which have their roots in the field of theoretical chemistry.

How does the materials scientist interpret the properties of the novel doped-fullerene superconductor or a solid-state semiconductor? How do we model a peptide and understand how it docks? How does an astrophysicist explain the components of the interstellar medium? Where does the industrial chemist turn when he wants to understand the catalytic properties of a zeolite or a surface layer? What is the meaning of 'far-from-equilibrium' and what is its significance in chemistry and in natural systems? How can we design the reaction pathway leading to the synthesis of a pharmaceutical compound? How does our modeling of intermolecular forces and potential energy surfaces yield a powerful understanding of natural systems at the molecular and ionic level? All these questions will be answered within our series which covers the broad range of endeavour referred to as 'theoreitcal chemistry'

The aim of the series is to present the latest fundamental material for research chemists, lecturers and students across the breadth of the subject, reaching into the various applications of theoretical techniques and modeling. The series concentrates on teaching the fundamentals of chemical structure, symmetry, bonding, reactivity, reaction mechanism, solid-state chemistry and applications in molecular modeling. It will emphasize the transfer of theoretical ideas and results to practical situations so as to demonstrate the role of theory in the solution of chemical problems in the laboratory and in industry.

D. Clary, A. Hinchliffe, D. S. Urch and M. Springborg
June 1994

Preface

This book is based on courses held for students of either physics or chemistry at the University of Konstanz, Germany. The idea behind the courses is to present the basic principles behind currently applied methods for the calculation of electronic and structural properties of specific materials. In contrast to most other textbooks I have attempted to cover areas both from chemistry and from physics so that the students of each field also obtain a feeling for the other field.

The student is expected to possess an introductory knowledge of quantum theory. Nevertheless, the first part of the book, Preliminaries, contains a description of fundamental quantum-theoretical principles and methods that are of ultimate importance for the remaining parts. Part II, Basic Methods, contains a detailed description of Hartree–Fock-based and density-functional-based methods that form the basis for almost all current electronic-structure methods. Also this part is of fundamental importance. On the other hand, the last two parts, Special Properties and Special Systems, contain less detailed descriptions of various properties and systems that can be studied with the methods of Part II. The idea here is that the students can choose those subjects that are of particular interest to them and obtain a feeling for how these are specifically treated without being burdened by all formalistic and mathematical details.

Without mentioning anyone in particular, I would like to stress that this book could not have been made without support, encouragement, and help from a number of students, friends, colleagues, and coworkers, as well as from all those scientists and publishers that have permitted me to use their work here. Any error is, however, solely my responsibility.

Michael Springborg
Konstanz, 1999

Part I
PRELIMINARIES

This introductory part contains a short overview of some of those methods of quantum theory that will be of ultimate importance for the remaining parts of this book. Most of it may be well known but the presentation can then serve as a repetition as well as an introduction to the notation used throughout the book.

1 Introduction

One of the central issues of basic and applied chemistry and physics is to understand the relations between the properties of a given material on the one side and its composition and structure on the other. To this end very many studies of single materials as well as of classes of related materials are carried through and by proper interpretation of the results so obtained the above relations are sought. The methods that are applied may be either experimental or theoretical, but ultimately only through a combination of both approaches can a detailed understanding be achieved.

The class of materials that is studied is very broad and includes isolated atoms, isolated molecules, molecules in solutions, weakly interacting molecules of molecular crystals and glasses, polymers, macromolecules, clusters, colloids, perfectly periodic crystals, crystals with impurities or defects, surfaces of crystals, interfaces between crystals, adsorbents on surfaces, superlattices, liquids, amorphous solids, alloys, grain boundaries, etc. Similarly, very many properties are studied including structure (i.e., relative positions of the nuclei), mechanical properties, stability, reactivity, reactions, thermal properties, phase transitions, magnetic properties, interactions, excitation energies, linear and non-linear responses on external perturbations, phase diagrams, etc.

The field of theoretical calculations of the properties of specific materials lies between experimental studies and phenomenological theoretical studies. As in experiment, specific materials are studied, and precise values for the properties of interest are sought. On occasion, these are in turn used as input for models that are derived at a more abstract phenomenological level and that can be used in providing further information on the material of interest. On the other hand, the field is purely theoretical as in the more phenomenological studies.

The advantage of the theoretical studies in contrast to the experimental studies is that the systems under study are well defined. Therefore, the results are not obscured by unwanted effects like those due to impurities, inhomogeneities, etc. Furthermore, the theoretical studies often allow the study of properties that are beyond the reach of experiment. On the other hand, the danger of those studies is that one has almost in all cases to study idealized systems, which then may have only very little connection with reality. Moreover, in some situations the above effects are exactly those that are of interest, but due to their complexity not treatable by theoretical methods. Therefore, an interplay between theoretical and experimental studies is very important, and very often the combination of both is more than the sum of each study separately.

In principle, all these materials and properties can be studied theoretically by solving the (time-independent) Schrödinger equation (neglecting for the moment relativistic effects)

$$\hat{H}\Psi = E\Psi. \tag{1.1}$$

Here, Ψ is the wave function of all participating particles and $\hat{H}$ is the Hamilton operator. For the systems of interest the particles are the electrons and the nuclei, and we shall for simplicity base our discussion on a position-space representation. Thus, for a given material with M nuclei and N electrons, Ψ is a function of $3N$ position coordinates for the N electrons, N spin coordinates for the N electrons (whereas the position coordinates each can take any value between $-\infty$ and $+\infty$, the spin variables for the electrons can only take one of the two values $-\frac{1}{2}$ and $+\frac{1}{2}$), $3M$ position coordinates for the M nuclei, and M spin coordinates for the M nuclei. In total, Ψ is accordingly a function of $3(M+N)$ continuous variables and $M+N$ discrete variables.

Solving Eq. (1.1) is in principle the perfect approach for obtaining any desired information on a given material. The problem is, however, that only in very special cases (that are only rarely met in reality) is it possible to solve this equation. In order not to completely abandon the theoretical studies of the materials one has therefore to invoke various approximations. Thereby, one has to be guided by two counteracting principles: the approximations should lead to simplifications that make the resulting approximations of Eq. (1.1) solvable. On the other hand, they should not be so severe that the results are strongly obscured by them. It is thus most often a delicate and far from trivial problem to solve Eq. (1.1) approximately, although it is often done on an apparently routine basis. It is, however, very important to be aware of the consequences of any approximation applied in solving Eq. (1.1) as well as understanding when a given approximation is good and when not. It is the purpose of the present manuscript to give some background information on standard approaches for solving Eq. (1.1) approximately and thereby also discuss when the different approximations work and when not.

We will concentrate on describing the principles behind the different computational techniques instead of giving a detailed discussion of their applications. Typically, one is interested in understanding the properties of a given (class of) material(s). From a set of (calculated or measured) observations one attempts to extract some information on the properties of the material corresponding to following the path (1) $\rightarrow$ (4) below.

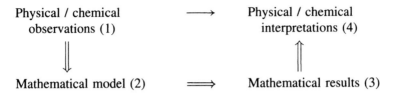

However, almost always one reformulates the problem of interpreting the physical or chemical observations as a mathematical problem, e.g., that of solving the Schrödinger equation for the system of interest. Subsequently one solves this mathematical problem more or less accurately, and, finally, interprets the mathematical solution in terms of physical or chemical properties of the material of interest. This corresponds to following the path $(1) \rightarrow (2) \rightarrow (3) \rightarrow (4)$ above. In this manuscript we are mainly concerned with the step $(2) \rightarrow (3)$ for the special case in which the mathematical problem is that of solving the Schrödinger equation.

It will be assumed that the basic principles of quantum theory are well known, but we will in Chapter 2 nevertheless introduce a few basic concepts and results that will be used throughout the manuscript. We will concentrate on describing the principles by deriving the mathematical formulas behind them. Thereby exact general mathematical proofs will be less important, but the philosophy will rather be that of a practitioner: there may be mathematically relevant cases where the approaches do not work or are invalid, but for the practical applications these are irrelevant and will therefore not be considered.

We shall concentrate on time-independent properties and accordingly seek solutions to the time-independent Schrödinger equation. Only at some few places shall we include the discussion of time-dependent properties. Similarly, we will at most places neglect relativistic effects and only briefly describe them at selected places.

There exist many textbooks on the principles of calculating electronic properties of materials using more or less accurate methods, and we shall not attempt here to give a list of these. Most often, however, they concentrate on either small (e.g., molecular) systems or on large (e.g., crystalline) materials and others are treated only marginally. Furthermore, the applied methods are either so-called wavefunction-based ones or density-based ones, and once again others are often treated only at a secondary level. In addition, often smaller systems are treated by wavefunction-based methods, and larger ones by density-based methods. Accordingly, manuscripts that cover the broader class of materials and that consider both wavefunction-based methods and density-based methods on an equal footing are scarce. This manuscript is an attempt to fill this gap.

2 Operators

2.1 WHAT IS AN OPERATOR?

Ultimately we attempt to solve the time-independent Schrödinger equation

$$\hat{H}\Psi = E\Psi \tag{2.1}$$

in a more or less approximate way. As mentioned above, Eq. (2.1) cannot be solved exactly (except for some very few cases) and approximations are therefore sought. It will then be important to be able to estimate how accurate the approximations are as well as to be able to improve the approximate solutions in a systematic way. In order to avoid simply arbitrarily guessing, it is extremely useful to make use of the properties of the Hamilton operator $\hat{H}$. Therefore, we shall start out here by recalling some of the properties of this operator. We shall discuss a general operator $\hat{A}$ (note, we will mark an operator by putting a caret on it), but for later applications we shall almost exclusively let this operator be the Hamilton operator. It is, however, important to stress that the discussion is more general than only applying to the Hamilton operator. Finally, we shall at the moment not bother about the precise form of the operator.

An operator acts on a function. It may have very many different forms, of which some simple examples are

$$\hat{A}_1 f_1(x) = \frac{d f_1(x)}{dx} \tag{2.2}$$

(i.e., the operator $\hat{A}_1$ acts on a function of one variable and returns the derivative of this function),

$$\hat{A}_2 f_2(x) = \cos[f_2(x)] + \pi + 4[f_2(x)]^2, \tag{2.3}$$

$$\hat{A}_3 f_3(x) = \int_{-\infty}^{x} f_3(y)\,dy, \tag{2.4}$$

$$\hat{A}_4 f_4(x, y) = \frac{\partial f_4(x, y)}{\partial x} + f_4(0, y^2), \tag{2.5}$$

$$\hat{A}_5 f_5(x) = (\cos^2 f_5(x), \exp[f_5(x) + 2]), \tag{2.6}$$

$$\hat{A}_6 f_6(x) = \begin{cases} 4 & \text{if } f_6(x) = 1 \\ 0 & \text{if } f_6(x) = 0, \end{cases} \tag{2.7}$$

$$\hat{A}_7 f_7(x, y) = f_7(y, x). \tag{2.8}$$

We shall use these examples in discussing some basic principles.

First, it is important that the operator and the function on which it acts are mutually consistent. This means, e.g., that the operator $\hat{A}_1$ can only act on functions that depend on exactly one variable, and an expression like $\hat{A}_4[x^2 + yz]$ does not make sense, whereas $\hat{A}_4[x^2 + y]$ does. Furthermore, the operator $\hat{A}_6$ can only act on functions that take only the two values 0 and 1. (This therefore resembles a spin variable. In that case one could imagine that $\hat{A}_6$ is an operator that acts on the z-component of the spin of one electron, where it is well known that this component takes only the values $-\frac{1}{2}$ and $+\frac{1}{2}$.)

Second, it is possible to perform more operations subsequently, if the intermediates 'fit'. Thus,

$$\hat{A}_1 \hat{A}_2 (x^2) = \hat{A}_1[\hat{A}_2(x^2)]$$
$$= \hat{A}_1[\cos(x^2) + \pi + 4x^4]$$
$$= \frac{d}{dx}[\cos(x^2) + \pi + 4x^4]$$
$$= -2x \sin(x^2) + 16x^3, \tag{2.9}$$

whereas an expression like $\hat{A}_4 \hat{A}_1 f(x)$ does not make sense. The expression $\hat{A}_6 \hat{A}_2(x^2)$ is meaningful only if the x values have been restricted so that $\hat{A}_2(x^2)$ is either 0 or 1.

Third, in the general case the order of the operators is important. Interchanging the two operators in Eq. (2.9) yields

$$\hat{A}_2 \hat{A}_1 (x^2) = \hat{A}_2[\hat{A}_1(x^2)]$$
$$= \hat{A}_2[2x]$$
$$= \cos(2x) + \pi + 16x^2, \tag{2.10}$$

which clearly is different from the expression of Eq. (2.9). If, however, for two operators $\hat{A}$ and $\hat{B}$ we have that for *all* functions f that they can act on

$$\hat{A}\hat{B}f = \hat{B}\hat{A}f, \tag{2.11}$$

then the two operators are said to *commute*. Eq. (2.11) may also be formulated with the help of the commutator

$$[\hat{A}, \hat{B}] \equiv \hat{A}\hat{B} - \hat{B}\hat{A} \tag{2.12}$$

as

$$[\hat{A}, \hat{B}] = 0. \tag{2.13}$$

Fourth, the operator $\hat{A}_7$ is an example of a so-called permutation operator that later will become important. It interchanges two arguments, and in the case of Eq. (2.8) we have as an example

$$\hat{A}_7(x^2 + y) = y^2 + x. \tag{2.14}$$

The operators act on functions that in turn depend on one or more variables. If the function is the wavefunction of a single electron in position representation, it depends on three position coordinates and one spin coordinate. Often each position coordinate may take any value between $-\infty$ and $+\infty$, whereas the spin coordinate can only take the two values $-\frac{1}{2}$ and $+\frac{1}{2}$.

2.2   EXPECTATION VALUES

It will very often be useful to consider integrals over all the possible values of the different variables or coordinates. We shall denote this by

$$\int \cdots d\vec{x}, \tag{2.15}$$

Here, $\vec{x}$ is a combined parameter containing all variables, i.e., for the single electron above it includes the three position coordinates and the single spin coordinate. The integral sign is to be interpreted in its most general case, and the integral is to be taken over all possible values of the variables. For the single electron the integration is accordingly over a three-dimensional position space plus a summation over the two values of the spin variable.

For a two-electron system the parameter $\vec{x}$ will contain three position coordinates for the first electron, three position coordinates for the second electron, one spin coordinate for the first electron, and one spin coordinate for the second electron. In some cases it will be useful to show this dependence explicitly and we shall in those cases denote the integrals by one of the two equivalent expressions

$$\int \cdots d\vec{x} = \int \int \cdots d\vec{x}_1 \, d\vec{x}_2. \tag{2.16}$$

Integrals of the type

$$\int f_1^*(\vec{x}) f_2(\vec{x}) \, d\vec{x} \equiv \langle f_1 | f_2 \rangle \tag{2.17}$$

will be called overlap matrix elements and have above been written with the usual bracket notation of Dirac. If this integral equals 0, the two functions f_1 and f_2 are said to be *orthogonal*.

In addition we will need integrals of the type

$$\int f_1^*(\vec{x})\hat{A}f_2(\vec{x})\,\mathrm{d}\vec{x} \equiv \langle f_1|\hat{A}|f_2\rangle, \tag{2.18}$$

that are matrix elements for the operator $\hat{A}$. We shall here implicitly assume that all quantities 'fit together'. Thus, expressions where f_1 is a function of a two-electron system (i.e., having six position coordinates and two spin coordinates), whereas f_2 is a function of an eight-electron system, are not valid, and are not considered here at all!

The operators may obey two further properties. First, an operator $\hat{A}$ is Hermitian *if f*

$$\langle f_1|\hat{A}|f_2\rangle = \langle f_2|\hat{A}|f_1\rangle^* \tag{2.19}$$

is valid for *any* two functions f_1 and f_2 [for which Eq. (2.19) makes sense!].

Second, $\hat{A}$ is a linear operator, *if f* for *any* two constants c_1 and c_2 and any two functions f_1 and f_2,

$$\hat{A}(c_1f_1 + c_2f_2) = c_1\hat{A}f_1 + c_2\hat{A}f_2. \tag{2.20}$$

Almost all physical relevant operators are hermitian, and many of those are also linear. The Hamilton operator, as a special case, is both.

2.3 AN EXAMPLE

Let us close this section by considering an example. As a physical realization of the example one may think of particles confined to a potential well in one dimension.

We consider functions of one continuous variable $f(x)$, and we will assume that the only relevant functions are those that vanish at $x = a$ and $x = b$, i.e.,

$$f(a) = f(b) = 0. \tag{2.21}$$

The potential well is thus assumed to span the interval $[a; b]$. We consider the operator $\hat{A}_1$ of Eq. (2.2). Since

$$\hat{A}_1(c_1f_1 + c_2f_2) = \frac{\mathrm{d}}{\mathrm{d}x}(c_1f_1 + c_2f_2)$$

$$= c_1\frac{\mathrm{d}}{\mathrm{d}x}f_1 + c_2\frac{\mathrm{d}}{\mathrm{d}x}f_2$$

$$= c_1\hat{A}_1f_1 + c_2\hat{A}_1f_2, \tag{2.22}$$

the operator is obvious linear.

On the other hand,

$$
\begin{aligned}
\langle f_1 | \hat{A}_1 | f_2 \rangle &= \int_a^b f_1^*(x) \frac{\mathrm{d}}{\mathrm{d}x} f_2(x) \, \mathrm{d}x \\
&= [f_1^*(x) f_2(x)]_a^b - \int_a^b \left[\frac{\mathrm{d}}{\mathrm{d}x} f_1^*(x) \right] f_2(x) \, \mathrm{d}x \\
&= -\int_a^b f_2(x) \frac{\mathrm{d}}{\mathrm{d}x} f_1^*(x) \, \mathrm{d}x \\
&= -\left[\int_a^b f_2^*(x) \frac{\mathrm{d}}{\mathrm{d}x} f_1(x) \, \mathrm{d}x \right]^* \\
&= -\langle f_2 | \hat{A}_1 | f_1 \rangle^*.
\end{aligned}
\tag{2.23}
$$

Hence, due to the minus sign in the last equation, the operator $\hat{A}_1$ is not Hermitian. We add that in deriving Eq. (2.23) we have in the second identity used the standard expression for an integral of a product of one function and the derivative of another ($\int f \cdot g' \, \mathrm{d}x = f \cdot g - \int g \cdot f' \, \mathrm{d}x$), and in the third identity used Eq. (2.21).

Upon passing we add that one may introduce the concept of adjoint operators. Thus, for a given operator $\hat{A}$ the adjoint operator $\hat{A}^\dagger$ obeys

$$
\langle f_1 | \hat{A} | f_2 \rangle = \langle f_2 | \hat{A}^\dagger | f_1 \rangle^*.
\tag{2.24}
$$

For a Hermitian operator we have obviously

$$
\hat{A}^\dagger = \hat{A},
\tag{2.25}
$$

whereas we for the special case of $\hat{A}_1$ have

$$
\hat{A}_1^\dagger = -\hat{A}_1.
\tag{2.26}
$$

3 Eigenvalues and Eigenfunctions

3.1 GENERAL PROPERTIES

Solving the time-independent Schrödinger equation

$$\hat{H}\Psi = E\Psi \tag{3.1}$$

involves determining an eigenvalue E and an eigenfunction Ψ. This is accordingly a special case of the general problem of finding eigenvalues and eigenfunctions for a given (Hermitian and linear) operator $\hat{A}$. We shall here, accordingly, consider the general problem of solving

$$\hat{A}f_i = a_i f_i, \tag{3.2}$$

where a_i is an eigenvalue and f_i the corresponding eigenfunction. i is used in labelling the different solutions to Eq. (3.2).

Let us consider two examples. If $\hat{A}$ is the Hamilton operator for a harmonic oscillator in one dimension, the eigenvalues are the well-known $(n + \frac{1}{2})\hbar\omega$ with n being any non-negative integer, and the eigenfunctions are Gaussian functions times Hermite polynomials of the position coordinate. In this case the eigenvalues and eigenfunctions are labelled by the integer n and are numerable.

For the electron of a hydrogen atom we have eigenfunctions characterized by the three integers n, l, m describing the position-space dependence of the wavefunction, plus a single half-integer describing the spin dependence. These functions are all bounded, i.e., they decay exponentially far away from the nucleus, and can be counted (i.e., they are numerable). Their eigenvalues are negative as a consequence of the functions being bounded. There is, however, also a set of so-called continuum states with positive energies. These states are not bounded and the eigenvalues are not discrete but form a continuum. Also these are characterized by three numbers describing the position-space dependence and one that describes the spin-dependence. Their precise form, properties, etc., are not important for the present discussion, but it will merely be stressed that there exist many solutions to Eq. (3.2), and that the eigenvalues may be either discrete or form a (finite or infinite) continuum or even, as for the hydrogen atom, a combination of the two.

3.2 HERMITIAN OPERATORS

Let us now study the properties of the eigenfunctions and eigenvalues when the operator $\hat{A}$ is Hermitian. To this end we consider two eigenfunctions,

$$\hat{A}f_n = a_n f_n$$
$$\hat{A}f_m = a_m f_m \tag{3.3}$$

and construct

$$\langle f_m|\hat{A}|f_n\rangle = \langle f_m|\hat{A}f_n\rangle = \langle f_m|a_n f_n\rangle = a_n\langle f_m|f_n\rangle. \tag{3.4}$$

Simultaneously,

$$\langle f_m|\hat{A}|f_n\rangle = \langle f_n|\hat{A}|f_m\rangle^* = \langle f_n|\hat{A}f_m\rangle^*$$
$$= \langle f_n|a_m f_m\rangle^* = a_m^*\langle f_n|f_m\rangle^* = a_m^*\langle f_m|f_n\rangle. \tag{3.5}$$

Combining Eqs. (3.4) and (3.5) now yields

$$(a_n - a_m^*)\langle f_m|f_n\rangle = 0. \tag{3.6}$$

For $n = m$ we obtain accordingly that a_n is real. This result does make a lot of sense: any physical observable is described quantum-theoretically by a Hermitian operator, and the fact that the eigenvalues of these are real means that measuring such an observable leads to a real (in contrast to complex) result.

Equation (3.6) shows subsequently (since the eigenvalues are real) that eigenfunctions belonging to different eigenvalues are orthogonal,

$$\langle f_m|f_n\rangle = 0 \qquad \text{for} \qquad a_m \neq a_n. \tag{3.7}$$

There may exist more different eigenfunctions belonging to the same eigenvalue. A well-known example is the electronic eigenfunctions for the hydrogen atom when the operator $\hat{A}$ equals the Hamilton operator. Here, e.g., the six $2p$ and two $2s$ functions (including spin) all have the same energy. In the general case, any linear combination of those will also be an eigenfunction. Thus, from the above $2s$ and $2p$ functions for the hydrogen atom, also the set $2s\sigma$, $2p_x\sigma$, $(2p_x + 2p_y)\sigma$, and $(2p_x + 2p_y + 2p_z)\sigma$ (with σ being either the α or β spin-function) forms a set of eight functions that are eigenfunctions to the Hamilton operator. It is clear that these functions are not orthogonal. It is, however, always possible to construct a new set (new linear combinations) such that these are orthogonal. Out of the eight $2s$ and $2p$ functions for the hydrogen atom we obtain thereby for instance the $2s\sigma$, $2p_x\sigma$, $2p_y\sigma$, and $2p_z\sigma$ functions or the $2s\sigma$, $2p_{-1}\sigma$, $2p_0\sigma$, and $2p_1\sigma$ functions. There are different ways of obtaining a set of orthogonal functions from one that is not (for instance, one may apply the Schmidt procedure), but here it is only important to stress that it is possible, not to describe exactly how.

Through this procedure we may accordingly replace Eq. (3.7) by

$$\langle f_m | f_n \rangle = 0 \qquad \text{for} \qquad m \neq n. \tag{3.8}$$

Finally, if any function f_n obeys Eq. (3.2), so does a constant times this function. Since, moreover, $\langle f_n | f_n \rangle$ is a real and positive number, we may assume that the functions f_n have been multiplied by constants, so that $\langle f_n | f_n \rangle = 1$. In total we end up with a set of eigenfunctions obeying

$$\langle f_m | f_n \rangle = \delta_{n,m}. \tag{3.9}$$

We stress that compared with Eq. (3.7) we have done nothing more than rescale the eigenfunctions to normalize them and form new linear combinations to make them orthogonal. In total, the eigenfunctions are said to be orthonormal. Below we shall almost exclusively assume that we have constructed such a set of orthonormal functions.

The fact that the eigenvalues a_n are real means that they can be placed on a real axis (cf. Fig. 3.1).

Furthermore, for almost all cases of our interest there is always a smallest eigenvalue, so that we can sort the eigenvalues according to increasing size. Denoting the smallest one a_0 we have in that case

$$a_0 \leq a_1 \leq a_2 \leq \cdots \leq a_{n-1} \leq a_n \leq a_{n+1} \leq \cdots. \tag{3.10}$$

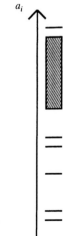

Figure 3.1 A schematic representation of the eigenvalues of a given Hermitian operator containing both discrete and continuum parts

Without proof we shall give one further important property of the eigenfunctions: They form a complete set. This means that any function g can be expanded in the set $\{f_n\}$,

$$g = \sum_n c_n f_n, \tag{3.11}$$

where the c_n are some constants. Of course, g and f_n have to depend on the same type of variables. This means that the hydrogen-atom eigenfunctions can be used to expand functions that depend on three continuous variables that may take any value between $-\infty$ and $+\infty$ and one variable that only can take the two values $-\frac{1}{2}$ and $+\frac{1}{2}$.

Multiplying Eq. (3.11) by f_m^* and integrating over all variables, we obtain

$$\langle f_m | g \rangle = \sum_n c_n \langle f_m | f_n \rangle. \tag{3.12}$$

Finally, the orthonormality of the functions f_n [Eq. (3.9)] gives

$$c_m = \langle f_m | g \rangle, \tag{3.13}$$

valid for any m.

3.3   COMMUTING OPERATORS

That two operators commute (cf. Section 2.1) can be used to advantage, as we shall see in detail later in this manuscript. Here, we shall establish the mathematical foundations for the applications later. Before this, however, it will be useful to recognize that the concept of commuting operators is often used without being explicitly discussed.

One example is the case where one of the two commuting operators is the Hamilton operator of the electrons of a given system and the other is a symmetry operator. Consider, e.g., the benzene molecule of Fig. 3.2.

It is clear that the benzene molecule has a number of symmetry elements, including a mirror plane (that of all nuclei), a six-fold axis, six two-fold axes,

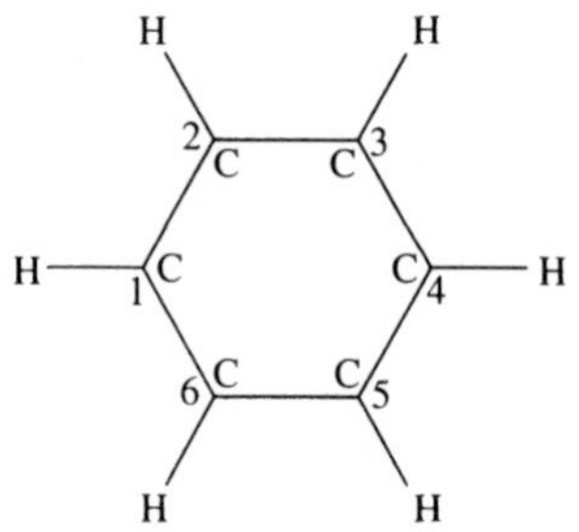

Figure 3.2   The benzene molecule C_6H_6. The numbers indicate the different CH units

etc. We shall here only consider one of the symmetry operations, i.e., a rotation of 60° about the six-fold axis. Such a rotation will bring CH unit no. 1 onto no. 2, no. 2 onto no. 3, etc. This means that the nuclear coordinates will be interchanged as just specified. But due to the symmetry of the system we do not have to modify the electronic coordinates equivalently, i.e., the symmetry operator and the Hamilton operator for the electrons commute. If we replaced one of the CH units by an N atom, this would no longer be the case.

The symmetry operation is used in attributing to any eigenfunction of the Hamilton operator not only an energy but also a label describing how it transforms according to the symmetry operator. It is simplest when the symmetry operator is a mirror operation or an inversion. In that case the labels g and u are often used in describing whether the function is symmetric (g) or antisymmetric (u) with respect to this symmetry operation. Alternatively formulated, g and u tells whether the eigenvalue o of

$$\hat{O}\Psi = o\Psi \qquad (3.14)$$

(with $\hat{O}$ being the symmetry operator) equals $+1$ or -1. For other symmetry operations (e.g., the rotation of 60° mentioned above), the value o will be different, although for the symmetry operations we always have

$$|o| = 1. \qquad (3.15)$$

Notice that $\hat{O}$ for the symmetry operations is often non-Hermitian, but in most other cases where we make use of the fact that two operators commute they are both Hermitian. Actually, the case of symmetry operators is most conveniently treated with group theory, to be discussed in Chapter 7.

Another place where commuting operators is used is that of hydrogen-atom-like eigenfunctions, where the quantum number n describes the Hamilton operator eigenvalue, whereas l and m describes those for the operators of the square of the angular momentum and of its z component, respectively. Finally, s and m_s describe the eigenvalues with respect to the spin operators $\hat{s}^2$ and $\hat{s}_z$, respectively.

In all cases we have that two or more operators commute. Here, we shall only consider the case of two commuting operators, but add that generalization to more is straightforward. We consider accordingly two Hermitian, linear operators $\hat{A}$ and $\hat{B}$ obeying

$$[\hat{A}, \hat{B}] = 0. \qquad (3.16)$$

The important point is that it is possible to obtain a complete set of functions that simultaneously are eigenfunctions to both operators, i.e., a set of functions obeying

$$\hat{A}f_n = a_n f_n$$
$$\hat{B}f_n = b_n f_n. \qquad (3.17)$$

Repeating the example from above this means that when the one operator equals the Hamilton operator for the electrons of some system and the other, e.g., an inversion-symmetry operator, the wavefunctions can be classified not only with respect to their energies but also according to whether they are of g or u symmetry.

In order to prove Eq. (3.17) we define first the two operators

$$\hat{C}_+ = \hat{A} + i\hat{B}$$
$$\hat{C}_- = \hat{A} - i\hat{B}. \tag{3.18}$$

These two operators are each others adjoint operator since

$$
\begin{aligned}
\langle f_1 | \hat{C}_+ | f_2 \rangle &= \langle f_1 | \hat{A} + i\hat{B} | f_2 \rangle = \langle f_1 | \hat{A} | f_2 \rangle + i \langle f_1 | \hat{B} | f_2 \rangle \\
&= \langle f_2 | \hat{A} | f_1 \rangle^* + i \langle f_2 | \hat{B} | f_1 \rangle^* = \langle f_2 | \hat{A} | f_1 \rangle^* - \langle f_1 | i\hat{B} | f_2 \rangle^* \\
&= \langle f_2 | \hat{A} - i\hat{B} | f_1 \rangle^* = \langle f_2 | \hat{C}_- | f_1 \rangle^*,
\end{aligned}
\tag{3.19}
$$

where we have used the fact that $\hat{A}$ and $\hat{B}$ are Hermitian.

Furthermore, $\hat{C}_+$ and $\hat{C}_-$ commute, since

$$
\begin{aligned}
[\hat{C}_+, \hat{C}_-] &= [\hat{A} + i\hat{B}, \hat{A} - i\hat{B}] = (\hat{A} + i\hat{B})(\hat{A} - i\hat{B}) - (\hat{A} - i\hat{B})(\hat{A} + i\hat{B}) \\
&= \hat{A}\hat{A} + \hat{B}\hat{B} + i\hat{B}\hat{A} - i\hat{A}\hat{B} - \hat{A}\hat{A} - \hat{B}\hat{B} + i\hat{B}\hat{A} - i\hat{A}\hat{B} \\
&= 2i[\hat{B}, \hat{A}] = 0,
\end{aligned}
\tag{3.20}
$$

since $\hat{A}$ and $\hat{B}$ commute.

Without proof we state that the eigenfunctions to any of the two operators $\hat{C}_+$ or $\hat{C}_-$ form a *complete set*. We consider one of the eigenfunctions to $\hat{C}_+$,

$$\hat{C}_+ g_n = c_n g_n. \tag{3.21}$$

Since c_n is a constant, Eq. (3.19) gives immediately

$$
\begin{aligned}
\langle f_2 | \hat{C}_- - c_n^* | f_1 \rangle^* &= \langle f_1 | \hat{C}_+ - c_n | f_2 \rangle = \langle f_1 | (\hat{C}_+ - c_n) f_2 \rangle \\
&= \langle (\hat{C}_- - c_n^*) f_2 | f_1 \rangle^*.
\end{aligned}
\tag{3.22}
$$

In the first two expressions we insert

$$f_1 = g_n$$
$$f_2 = (\hat{C}_- - c_n^*) g_n. \tag{3.23}$$

Please notice that this choice is made because it works; not because it is physically meaningful!

Then

$$
\begin{aligned}
\langle (\hat{C}_- - c_n^*) g_n | \hat{C}_- - c_n^* | g_n \rangle^* &= \langle g_n | (\hat{C}_+ - c_n) | (\hat{C}_- - c_n^*) g_n \rangle \\
&= \langle g_n | (\hat{C}_+ - c_n)(\hat{C}_- - c_n^*) | g_n \rangle.
\end{aligned}
\tag{3.24}
$$

$\hat{C}_-$ and $\hat{C}_+$ commute, and, therefore, so do $\hat{C}_+ - c_n$ and $\hat{C}_- - c_n^*$, too. This gives for the expression in Eq. (3.24)

$$\langle g_n|(\hat{C}_- - c_n^*)(\hat{C}_+ - c_n)|g_n\rangle. \tag{3.25}$$

Due to Eq. (3.21) this vanishes, whereby the first expression in Eq. (3.24) vanishes. This, on the other hand, is an integral of the absolute square of $(\hat{C}_- - c_n^*)g_n$ and can therefore only vanishes if the integrand is identical zero, i.e., if

$$\hat{C}_-g_n = c_n^*g_n. \tag{3.26}$$

In total, $\hat{C}_-$ and $\hat{C}_+$ have the same eigenfunctions, but the eigenvalues are each others' complex conjugates.

For the operators of Eq. (3.19) this leads to

$$\hat{C}_+g_n = (\hat{A} + i\hat{B})g_n = c_ng_n$$
$$\hat{C}_-g_n = (\hat{A} - i\hat{B})g_n = c_n^*g_n. \tag{3.27}$$

The sum and the difference of these two equations give immediately

$$\hat{A}g_n = \frac{1}{2}[c_n + c_n^*]g_n$$
$$\hat{B}g_n = \frac{1}{2i}[c_n - c_n^*]g_n, \tag{3.28}$$

which completes the proof that there exists a common complete set of eigenfunctions for both operators. The (real) eigenvalues are given by Eq. (3.28).

4 Factorization; Time and Spin Dependence

4.1 TIME DEPENDENCE

We have begun each section by recalling that the ultimate goal is to solve the time-independent Schrödinger equation

$$\hat{H}\Psi = E\Psi. \tag{4.1}$$

Actually, this equation is a special case of the more general time-dependent Schrödinger equation

$$\hat{H}\Psi_0 = i\hbar\frac{\partial\Psi_0}{\partial t}. \tag{4.2}$$

For most of the cases of our interest, $\hat{H}$ does not contain any explicit time dependence. On the other hand, Ψ_0 depends both on time (t) and on all other (position and spin) coordinates ($\vec{x}$). It is now useful to make the assumption that Ψ_0 is a product of two factors, one depending only on $\vec{x}$ and the other only on t:

$$\Psi_0(\vec{x}, t) = \Psi(\vec{x})A(t). \tag{4.3}$$

We insert this into Eq. (4.2),

$$\hat{H}[\Psi(\vec{x})A(t)] = i\hbar\frac{\partial\Psi(\vec{x})A(t)}{\partial t}. \tag{4.4}$$

By using that $\hat{H}$ is a linear operator and does not depend on t we find

$$A(t)\hat{H}\Psi(\vec{x}) = i\hbar\Psi(\vec{x})\frac{\partial A(t)}{\partial t}. \tag{4.5}$$

Subsequently, we divide by $A \cdot \Psi$ [simply ignoring the $(\vec{x}, t)$ points where this function vanishes] giving

$$\frac{\hat{H} \Psi(\vec{x})}{\Psi(\vec{x})} = i\hbar \frac{\frac{\partial A(t)}{\partial t}}{A(t)}. \tag{4.6}$$

The crucial step is now to realize that the left-hand side contains no dependence on t, whereas the right-hand side contains none on $\vec{x}$. Nevertheless, the two sides are identical for *all* $\vec{x}$ and t, which only can be the case when they equal a constant, say E.

We end therefore up with two equations, of which the one is

$$i\hbar \frac{\partial A(t)}{\partial t} = E \cdot A(t), \tag{4.7}$$

which gives

$$A(t) = \exp\left[-i\frac{Et}{\hbar}\right], \tag{4.8}$$

where we have set a constant prefactor equal to 1.

The other equation is

$$\hat{H}\Psi = E\Psi, \tag{4.9}$$

i.e., Eq. (4.1).

The procedure in Eq. (4.3) is an example of factorization, where we — by intelligent inspection! — observe that by writing the sought function (which depends on more variables) as a product of two functions, each depending on fewer variables, we can reduce the problem of calculating the function we seek. The factorization in Eq. (4.3) was somewhat atypical in that the one factor depends only on one coordinate (t) whereas the other contains the dependence on all other coordinates. We stress, however, that the factorization (4.3) is only an assumption: there are cases where it does not work, but these will not be identified through this procedure. For instance, the differential equation

$$\frac{\partial y}{\partial x} = a\frac{\partial y}{\partial t} \tag{4.10}$$

has

$$y = \cos(ax + t) \tag{4.11}$$

as one of its solutions, which cannot be factorized as $y_x(x) \cdot y_t(t)$.

4.2   SPIN DEPENDENCE

A further example will illustrate the approach once more. In the non-relativistic case, $\hat{H}$ contains no explicit spin dependence. Thus, when the Schrödinger equation (4.9) is that for an N-particle system, Ψ contains a dependence on $3N$ position coordinates and N spin coordinates,

$$\Psi(\vec{x}) = \Psi(\vec{r}_1, \vec{r}_2, \vec{r}_3, \ldots, \vec{r}_N, \sigma_1, \sigma_2, \sigma_3, \ldots, \sigma_N) \equiv \Psi(\vec{r}, \vec{\sigma}). \qquad (4.12)$$

In this case we may assume that Ψ can be written as a product of a position-dependent part times a spin-dependent one,

$$\Psi(\vec{r}, \vec{\sigma}) = \psi(\vec{r})\Theta(\vec{\sigma}), \qquad (4.13)$$

but upon insertion in Eq. (4.9) we obtain no information on the spin-dependent part, but only

$$\hat{H}\psi(\vec{r}) = E\psi(\vec{r}). \qquad (4.14)$$

It should finally be stressed once more that the factorization is only an assumption. Through this one may obtain some of the solutions to the Schrödinger equation, but there may be other solutions that cannot be written as a simple product of two factors that depend on different coordinates.

5 Variational Principle; Lagrange Multipliers

5.1 VARIATIONAL PRINCIPLE

In this section we shall describe the variational principle. This principle is the foundation for all the approximate methods that we shall discuss later. We therefore repeat that our ultimate goal is to solve the time-independent Schrödinger equation

$$\hat{H}\Psi = E\Psi, \tag{5.1}$$

but realize that due to the complexity of this equation, it can only in very few cases be solved exactly. It would, therefore, be desirable to be able to approximate the solutions to Eq. (5.1),

$$\Phi \simeq \Psi. \tag{5.2}$$

The fundamental question is, how do we determine Φ, how accurate is Φ, and how do we compare different approximate Φ?

To address this question we start somewhat more generally. We consider the general eigenvalue problem corresponding to Eq. (5.1),

$$\hat{A}f_n = a_n f_n, \tag{5.3}$$

where $\hat{A}$, as $\hat{H}$, is a Hermitian and linear operator.

Since $\hat{A}$ is Hermitian, all the eigenvalues a_n are real, and we shall assume that there exists a smallest one, so that we can sort the eigenvalues according to

$$a_0 \leq a_1 \leq a_2 \leq a_3 \leq \cdots \leq a_{n-1} \leq a_n \leq a_{n+1} \leq \cdots. \tag{5.4}$$

This procedure can obviously be carried through when $\hat{A}$ equals the Hamilton operator. Actually, it is one of the basic principles of the (non-relativistic) quantum theory that the system seeks — at least at the absolute temperature zero — that state that corresponds to the lowest energy, i.e., the eigenvalue of our interest is a_0. On the other hand, when, e.g., $\hat{A}$ equals the position operator, $\hat{x}$, there is no smallest eigenvalue. But in almost all cases where we seek approximate solutions to Eq. (5.3), the operator $\hat{A}$ is the Hamilton operator, and the variational principle works.

We need to repeat one further property of the solutions to Eq. (5.3), i.e., the eigenfunctions can per construction be made orthonormal, and we shall therefore assume (notice, this is no lack of generality) that

$$\langle f_n | f_m \rangle = \delta_{n,m}. \tag{5.5}$$

We stress that we do not explicitly need the eigenfunctions f_n or the eigenvalues a_n. We just need to know that they exist and that the eigenfunctions form a complete set in order to carry the arguments through that lead to the variational principle.

A further fundamental principle of quantum theory is that expectation values of experimental observables for a given system in a given state are calculated as matrix elements of (normalized) eigenfunctions. Therefore, if we knew that the system of our interest occupied the state f_n, then the expectation value for the operator $\hat{O}$ is calculated as

$$\frac{\langle f_n | \hat{O} | f_n \rangle}{\langle f_n | f_n \rangle}. \tag{5.6}$$

A special case is the trivial case that $\hat{O}$ equals $\hat{A}$ and that $n = 0$. Then Eq. (5.6) takes the form

$$\frac{\langle f_n | \hat{O} | f_n \rangle}{\langle f_n | f_n \rangle} = \frac{\langle f_0 | \hat{A} | f_0 \rangle}{\langle f_0 | f_0 \rangle} = \frac{\langle f_0 | a_0 | f_0 \rangle}{\langle f_0 | f_0 \rangle} = \frac{a_0 \langle f_0 | f_0 \rangle}{\langle f_0 | f_0 \rangle} = a_0. \tag{5.7}$$

We shall now ask ourselves what happens with the above expectation value (5.7) when we replace f_0 by an approximate function

$$f_0 \simeq \phi. \tag{5.8}$$

The idea is that we can choose ϕ absolutely freely, whereby we also can choose ϕ so that the integrals entering

$$\frac{\langle \phi | \hat{A} | \phi \rangle}{\langle \phi | \phi \rangle} \tag{5.9}$$

can be evaluated. Furthermore, we stress that by evaluating Eq. (5.9) we do not need to know anything about the exact function f_0. However, we immediately arrive at the question: if two persons suggest two different approximate functions ϕ, how do we distinguish between them, and, in particular, how do we determine which one is the better (whatever that may mean). To answer this question partly we use the completeness of the eigenfunctions f_n so that also any test function ϕ of Eq. (5.8) can be expanded as

$$\phi = \sum_n c_n f_n. \tag{5.10}$$

We only need to know that this expansion in principle is possible; not its precise form. We insert it into Eq. (5.9) and use the fact that the operator $\hat{A}$ is a linear operator. Then

$$
\frac{\langle \phi | \hat{A} | \phi \rangle}{\langle \phi | \phi \rangle} = \frac{\left\langle \sum_{n_1} c_{n_1} f_{n_1} \left| \hat{A} \right| \sum_{n_2} c_{n_2} f_{n_2} \right\rangle}{\left\langle \sum_{n_1} c_{n_1} f_{n_1} \middle| \sum_{n_2} c_{n_2} f_{n_2} \right\rangle}
$$

$$
= \frac{\left\langle \sum_{n_1} c_{n_1} f_{n_1} \middle| \sum_{n_2} c_{n_2} \hat{A} f_{n_2} \right\rangle}{\left\langle \sum_{n_1} c_{n_1} f_{n_1} \middle| \sum_{n_2} c_{n_2} f_{n_2} \right\rangle}
$$

$$
= \frac{\left\langle \sum_{n_1} c_{n_1} f_{n_1} \middle| \sum_{n_2} c_{n_2} a_{n_2} f_{n_2} \right\rangle}{\left\langle \sum_{n_1} c_{n_1} f_{n_1} \middle| \sum_{n_2} c_{n_2} f_{n_2} \right\rangle}
$$

$$
= \frac{\sum_{n_1,n_2} \langle c_{n_1} f_{n_1} | c_{n_2} a_{n_2} f_{n_2} \rangle}{\sum_{n_1,n_2} \langle c_{n_1} f_{n_1} | c_{n_2} f_{n_2} \rangle}
$$

$$
= \frac{\sum_{n_1,n_2} c_{n_1}^* c_{n_2} a_{n_2} \langle f_{n_1} | f_{n_2} \rangle}{\sum_{n_1,n_2} c_{n_1}^* c_{n_2} \langle f_{n_1} | f_{n_2} \rangle}
$$

$$
= \frac{\sum_{n_1,n_2} c_{n_1}^* c_{n_2} a_{n_2} \delta_{n_1,n_2}}{\sum_{n_1,n_2} c_{n_1}^* c_{n_2} \delta_{n_1,n_2}}
$$

$$= \frac{\sum_n c_n^* c_n a_n}{\sum_n c_n^* c_n} = \frac{\sum_n |c_n|^2 a_n}{\sum_n |c_n|^2}$$

$$\geq \frac{\sum_n |c_n|^2 a_0}{\sum_n |c_n|^2} = \frac{a_0 \sum_n |c_n|^2}{\sum_n |c_n|^2} = a_0. \qquad (5.11)$$

At the first step we have inserted the expansion of Eq. (5.10) into Eq. (5.9) and thereby used two different summation variables n_1 and n_2 for clarity. Subsequently, we have used that $\hat{A}$ is a linear operator so that the outcome of letting it act on a linear combination of functions equals the linear combination of the operator acting on the individual functions. Then we use that the functions f_n are eigenfunctions of $\hat{A}$. Afterwards, we move the summations inside the bras and kets to the outside, together with the constant coefficients c_{n_1} and c_{n_2} and the eigenvalues a_{n_2}. By using the orthonormality of the eigenfunctions [Eq. (5.5)], we can reduce the double summations to single summations (and simultaneously change the summation index to n). The crucial step at the inequality is due to the assumption (5.4). This step is valid since every a_n is multiplied by a non-negative real number. Having replaced all a_n by a_0, we can place this constant outside the summation, whereby we easily end up with the final result of Eq. (5.11).

Equation (5.11) tells us that no matter what approximate function ϕ of Eq. (5.8) we suggest, the expectation value (5.9) will always be higher than or identical to the lowest eigenvalue a_0. The identity will only occur when ϕ is an exact eigenfunction for $\hat{A}$ for the lowest eigenvalue a_0.

A simple example will illustrate the importance of this theorem. Let us assume that $\hat{A}$ is the Hamilton operator and we seek the lowest energy for some system. We choose one ϕ, calculate the expectation value (5.9), and obtain the number -90 in some (unimportant) units. By choosing another ϕ we obtain -92. We know, from Eq. (5.11), that both results are above or identical to the lowest exact energy. Therefore, the lowest exact energy must be -92 or below. This is what the theorem states. We will now *assume* that the ϕ function that led to the result -92 also is the best approximation to the true eigenfunction of the lowest energy. This is an assumption, although it is in most cases also a good approximation. But it raises also another question: what is meant by a good approximation? Before addressing this through an example in Section 5.4 we shall extend the variational principle slightly.

5.2 AN EXAMPLE

The importance of the variational principle is due the possibility of introducing approximate eigenfunctions to, e.g., the Hamilton operator and to compare different approximate eigenfunctions. We demonstrated this above for a simple example where two functions were compared. But one may also study a whole class of functions and out of this obtain the 'best' function. Let us illustrate the approach through a simple example.

We study a one-dimensional case with a single particle moving in a potential well. The profile of this well is shown in Fig. 5.1 and is given through

$$V(x) = \begin{cases} -V_0 & \text{for } |x| \le \ell \\ 0 & \text{for } |x| > \ell. \end{cases} \tag{5.12}$$

The Schrödinger equation for this particle is given by

$$\left[-\frac{\hbar^2}{2m}\frac{d^2}{dx^2} + V(x) \right] \Psi(x) = E\Psi(x), \tag{5.13}$$

with $V(x)$ defined above. The Hamilton operator is obviously

$$\hat{H} = -\frac{\hbar^2}{2m}\frac{d^2}{dx^2} + V(x). \tag{5.14}$$

The ground state is the state of the lowest total energy. By looking at the potential of Fig. 5.1 it seems intuitively correct that the particle occupies the potential well as much as possible, i.e., that the wavefunction has the largest amplitude in the region $|x| \le \ell$. A Gaussian centred at the origin is such a function and we therefore propose the approximate function

$$\phi = \exp(-\alpha x^2). \tag{5.15}$$

For very small α this function is broad and extends well outside the region of the well, whereby the potential energy goes up. On the other hand, in that case the kinetic energy [the first term on the right-hand side of Eq. (5.14)] is small. For large α the opposite is true: the potential energy is low, but the kinetic energy

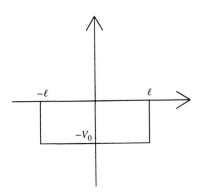

Figure 5.1   The one-dimensional potential well with finite depth

goes up. Thus, the 'best' α will most likely lie somewhere in between. However, we can simply determine it through variation. Thus, the quantity

$$\frac{\langle\phi|\hat{H}|\phi\rangle}{\langle\phi|\phi\rangle} \tag{5.16}$$

becomes a function of α,

$$\frac{\langle\phi|\hat{H}|\phi\rangle}{\langle\phi|\phi\rangle} \equiv \tilde{E}(\alpha). \tag{5.17}$$

For any α, $\tilde{E}(\alpha) \geq E_0$ (with E_0 being the ground-state energy), and therefore that α that leads to the lowest $\tilde{E}(\alpha)$ leads to the best estimate for E_0. Accordingly, we determine α by requiring

$$\frac{\partial}{\partial\alpha}\tilde{E}(\alpha) = 0. \tag{5.18}$$

Simultaneously, we will then assume that the resulting function ϕ is the best approximation to the exact ground-state eigenfunction.

We have not carried these calculations through here since they are not important for our general arguments, but hope that the basic ideas are clear.

5.3 VARIATION

In the general case we seek an approximate solution to the lowest eigenvalue and to the corresponding eigenfunction for the eigenvalue equation

$$\hat{A}f_0 = a_0 f_0. \tag{5.19}$$

The eigenfunction (f_0) depends on the coordinates of the system of our interest (in the example above, of the one-dimensional coordinate x), i.e.

$$f_0 = f_0(\vec{x}). \tag{5.20}$$

Equivalently, an approximate solution ϕ will depend on $\vec{x}$. But in addition we may let it depend on further parameters, $p_1, p_2, \ldots, p_{N_p}$ (in the case above of the previous subsection there was only one such parameter, α). Accordingly we write

$$\phi = \phi(p_1, p_2, \ldots, p_{N_p}; \vec{x}). \tag{5.21}$$

As in Eq. (5.17) the expectation value (5.9) becomes a function of these extra parameters,

$$\frac{\langle\phi|\hat{A}|\phi\rangle}{\langle\phi|\phi\rangle} \equiv \tilde{a}(p_1, p_2, \ldots, p_{N_p}), \tag{5.22}$$

and we will determine these parameters by minimizing this function, i.e., by requiring

$$\frac{\partial a(p_1, p_2, \ldots, p_{N_p})}{\partial p_1} = \frac{\partial a(p_1, p_2, \ldots, p_{N_p})}{\partial p_2} = \cdots = \frac{\partial a(p_1, p_2, \ldots, p_{N_p})}{\partial p_{N_p}} = 0.$$

(5.23)

By introducing more and more parameters it should be possible to obtain very accurate approximations to the exact ground-state eigenfunctions, but, unfortunately, the equations (5.22) very easily become too complicated to be solvable. There is, however, one important exception, the so-called linear variation, which we shall discuss further below. But before this we shall discuss a little further the quality of the approximate eigenfunctions obtained through this variation, by looking at a very simple example.

5.4 THE HYDROGEN ATOM

The Hamilton operator for the single electron of a hydrogen atom is

$$\hat{H} = -\frac{\hbar^2}{2m} \nabla^2 - \frac{e^2}{4\pi\varepsilon_0} \frac{1}{r},$$

(5.24)

where we have assumed that the nucleus is placed at the origin.

In order to avoid having to remember all the fundamental constants, it is useful to work with so-called atomic units. We shall later describe them in more detail but here we just list them. We set

$$\hbar = m = |e| = 4\pi\varepsilon_0 = 1.$$

(5.25)

Energies are then given in so-called hartrees (1 hartree $\simeq$ 27.21 eV) and lengths in bohr (1 bohr $\simeq$ 0.5292 Å). With these, the Hamilton operator of Eq. (5.24) becomes

$$\hat{H} = -\frac{1}{2}\nabla^2 - \frac{1}{r}.$$

(5.26)

The eigenvalue problem

$$\hat{H}\psi_n = e_n \psi_n$$

(5.27)

has as its lowest eigenvalue

$$e_0 = -\tfrac{1}{2}$$

(5.28)

and the corresponding eigenfunction is

$$\psi = \frac{1}{\sqrt{\pi}} e^{-r},$$

(5.29)

(this function is normalized) as can be found in most introductory textbooks on quantum theory.

Since we in this case know the exact solution we can directly compare it with results of calculations using the variational method. This is in almost all other cases not possible.

We will apply the variational method for the test function

$$\phi = \left(\frac{2\alpha}{\pi}\right)^{3/4} e^{-\alpha r^2}, \tag{5.30}$$

i.e., a Gaussian (that furthermore is normalized). In Fig. 5.2 we compare the exact wavefunction [Eq. (5.29)] with that of Eq. (5.30) for different values of α.

Since ϕ of eq. (5.4.7) is normalized, the expectation value of Eq. (5.22) becomes

$$\tilde{E}(\alpha) = \langle \phi | \hat{H} | \phi \rangle, \tag{5.31}$$

which can be calculated to yield

$$\tilde{E}(\alpha) = \frac{3\alpha}{2} - 2 \left(\frac{2\alpha}{\pi}\right)^{1/2}. \tag{5.32}$$

This function is shown in Fig. 5.3.

Taking the derivative of this with respect to α gives the value of α at the minimum,

$$\alpha = \frac{8}{9\pi} \simeq 0.283, \tag{5.33}$$

and the expectation value for this value of α is

$$\tilde{E}\left(\alpha = \frac{8}{9\pi}\right) = -\frac{4}{3\pi} \simeq -0.424. \tag{5.34}$$

By looking at Fig. 5.2 we see that this value of α does appear to give a reasonable (but no more!) description of the exact wavefunction although the agreement between the exact lowest energy [Eq. (5.28)] and the approximate one [Eq. (5.32)] may suggest a better agreement.

However, the application of the variational principle is not restricted to the Hamilton operator. We therefore construct another operator that happens to have the same eigenfunction of the lowest eigenvalue as the Hamilton operator of the hydrogen atom. Please notice that this construction does not result from physical reasons but is used since it may illustrate some few aspects of the variational method.

The operator we construct is

$$\hat{G} = \left(\hat{H} + \tfrac{1}{2}\right)^2, \tag{5.35}$$

where $\hat{H}$ is the operator of Eq. (5.26). This operator has exactly the same eigenfunctions as $\hat{H}$ of Eq. (5.26) (this can be shown by using the completeness of the latter), and the lowest eigenvalue is 0 for the function of Eq. (5.29).

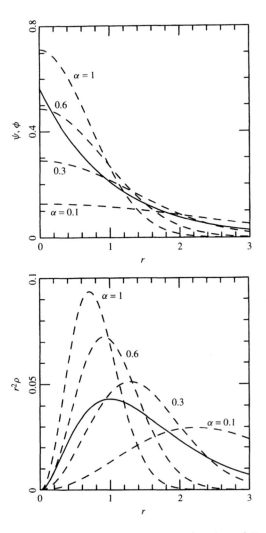

Figure 5.2   The solid line shows the exact hydrogen function of Eq. (5.29), whereas the dashed curves show Gaussians of Eq. (5.30) with $\alpha = 0.1$, 0.3, 0.6, and 1.0. The most compact function is that of the largest α. The upper panel shows the functions themselves, whereas the lower panel shows the functions squared times r^2 (i.e., a radial electron density)

Here we calculate

$$\tilde{G}(\alpha) = \langle \phi | \hat{G} | \phi \rangle, \tag{5.36}$$

which can be calculated to yield

$$\tilde{G}(\alpha) = \frac{1}{8} + \frac{9}{4}\alpha + \frac{19}{32}\alpha^2 - 2\alpha\sqrt{\frac{2\alpha}{\pi}} - \sqrt{\frac{2\alpha}{\pi}}. \tag{5.37}$$

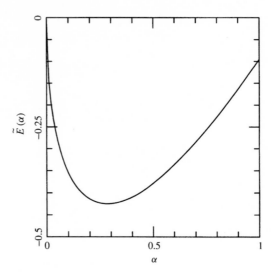

Figure 5.3   The function $\tilde{E}(\alpha)$ of Eq. (5.32)

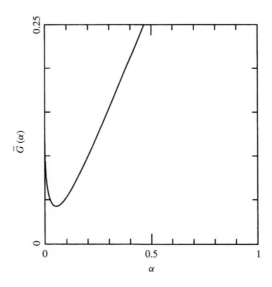

Figure 5.4   The function $\tilde{G}(\alpha)$ of Eq. (5.37)

This function is shown in Fig. 5.4.

The minimum of $\tilde{G}(\alpha)$ is found for

$$\alpha \simeq 0.0508, \tag{5.38}$$

i.e., a significantly smaller value than that of Eq. (5.33). By inspecting Fig. 5.2 we see that the larger value of α [Eq. (5.33)] leads to a function that more resembles

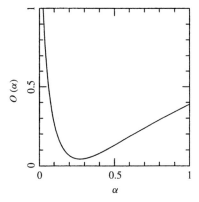

Figure 5.5 The function $O(\alpha)$ of Eq. (5.39)

the true ground-state function, which lends supports to the usual approach of using the Hamilton operator in variationally giving an approximate wavefunction. It does, however, also show that one may occasionally obtain results that are less trustworthy, although the formulas have been applied correctly.

Finally, we may also directly measure the quality of a given approximate wavefunction in the present example by calculating

$$O(\alpha) = \langle \phi - \psi | \phi - \psi \rangle. \tag{5.39}$$

This function is smallest when the difference between the exact and the approximate function is smallest (using the usual norm). $O(\alpha)$ is shown in Fig. 5.5.

The minimum in Fig. 5.5 is found for

$$\alpha \simeq 0.27, \tag{5.40}$$

i.e., for a value close to that of Eq. (5.33).

5.5 LINEAR VARIATION AND LAGRANGE MULTIPLIERS

As mentioned above, the problem of solving the equations (5.23) becomes very easily very complicated also when the number of parameters is not very large. There is, however, one — very important — case, where the equations, although in a slightly modified form, can be solved straightforwardly.

We want to minimize the expectation value

$$\frac{\langle \phi | \hat{A} | \phi \rangle}{\langle \phi | \phi \rangle}. \tag{5.41}$$

We shall now assume that the minimization is done by varying the *coefficients* to some *fixed* so-called basis functions. This means that we will write ϕ as a linear combination of fixed basis functions,

$$\phi(\vec{x}) = \sum_{i=1}^{N_b} c_i \chi_i(\vec{x}), \tag{5.42}$$

where N_b is the number of basis functions $\{\chi_i\}$, and where the variation is done by varying the coefficients $\{c_i\}$ but keeping the basis functions $\{\chi_i\}$ fixed.

Equation (5.23) takes then the form

$$\frac{\partial}{\partial c_k} \frac{\langle \phi | \hat{A} | \phi \rangle}{\langle \phi | \phi \rangle} = 0 \tag{5.43}$$

for all $k = 1, 2, \ldots, N_b$, or by taking the complex conjugate of this equation (this is not really a necessary step, but the equations look simpler!),

$$\frac{\partial}{\partial c_k^*} \frac{\langle \phi | \hat{A} | \phi \rangle}{\langle \phi | \phi \rangle} = 0. \tag{5.44}$$

We may restate this as requiring that

$$\frac{\partial}{\partial c_k^*} \langle \phi | \hat{A} | \phi \rangle = 0, \qquad k = 1, \ldots, N_b, \tag{5.45}$$

when we simultaneously require that the function ϕ is normalized:

$$\langle \phi | \phi \rangle = 1. \tag{5.46}$$

This means that the single equation (5.44) can be replaced by the two equations (5.45) and (5.46).

We shall combine the two equations (5.45) and (5.46) into one by introducing one extra parameter. This is done by considering the quantity

$$K = \langle \phi | \hat{A} | \phi \rangle - \lambda[\langle \phi | \phi \rangle - 1]. \tag{5.47}$$

The extra parameter λ is a so-called *Lagrange multiplier*, and Eqs. (5.45) and (5.46) then take the form

$$\frac{\partial K}{\partial c_k^*} = \frac{\partial K}{\partial \lambda} = 0. \tag{5.48}$$

The Lagrange multiplier is a mathematical trick that allows one to incorporate one (or more) extra boundary condition(s) [in this case, that of Eq. (5.46)] into a variation. It is not obvious whether the multiplier has any meaning (although this often is the case), but it is first of all introduced due to its convenience.

By inserting K of Eq. (5.47) into the last identity of Eq. (5.48) we recover immediately the boundary condition (5.46),

$$\langle \phi | \phi \rangle - 1 = 0. \tag{5.49}$$

From the first equations of Eq. (5.48) we obtain, upon inserting Eq. (5.42),

$$\frac{\partial}{\partial c_k^*}\left[\left\langle \sum_i c_i \chi_i \left|\hat{A}\right| \sum_j c_j \chi_j \right\rangle - \lambda\left(\left\langle \sum_i c_i \chi_i \left| \sum_j c_j \chi_j \right.\right\rangle - 1\right)\right]$$

$$= \frac{\partial}{\partial c_k^*}\left(\left\langle \sum_i c_i \chi_i \left|\hat{A}\right| \sum_j c_j \chi_j \right\rangle - \lambda\left\langle \sum_i c_i \chi_i \left| \sum_j c_j \chi_j \right.\right\rangle\right)$$

$$= \frac{\partial}{\partial c_k^*}\sum_{i,j} c_i^* c_j [\langle \chi_i|\hat{A}|\chi_j\rangle - \lambda\langle \chi_i|\chi_j\rangle]$$

$$= \sum_j c_j [\langle \chi_k|\hat{A}|\chi_j\rangle - \lambda\langle \chi_k|\chi_j\rangle] \equiv 0. \tag{5.50}$$

This gives immediately the following eigenvalue equation,

$$\sum_j c_j \langle \chi_k|\hat{A}|\chi_j\rangle = \lambda \sum_j c_j \langle \chi_k|\chi_j\rangle, \tag{5.51}$$

where we have as many equations ($k = 1, 2, \ldots, N_b$) as we have basis functions, i.e.

$$k = 1, 2, \ldots, N_b. \tag{5.52}$$

In Eq. (5.50) we recognize matrix elements of the type $\langle \chi_k|\hat{A}|\chi_j\rangle$ and of the type $\langle \chi_k|\chi_j\rangle$. It is useful to define two matrices containing these, i.e.,

$$\underline{\underline{A}}_{kj} = \langle \chi_k|\hat{A}|\chi_j\rangle \tag{5.53}$$

and

$$\underline{\underline{O}}_{kj} = \langle \chi_k|\chi_j\rangle. \tag{5.54}$$

The equations (5.51) are then easily rewritten as the generalized matrix eigenvalue equation

$$\underline{\underline{A}} \cdot \underline{c} = \lambda \cdot \underline{\underline{O}} \cdot \underline{c}, \tag{5.55}$$

where $\underline{c}$ is a column vector containing the sought coefficients $\{c_j\}$. Eq. (5.55) is called the secular equation.

That Eqs. (5.55) and (5.51) are equivalent can easily be shown by writing out the matrix equations (5.55) and inserting the definitions of Eqs. (5.53) and (5.54).

The *very* important point is that the matrix eigenvalue equation (5.55) easily can be solved using standard computer routines. The results will contain the coefficients c_j and the eigenvalue λ. However, the coefficients c_j may have to be multiplied by a common constant in order to satisfy the normalization condition (5.46). Actually, there will not be only one solution, but N_b solutions. One question is thus, which one do we choose.

Once the coefficients are known, the sought expectation value

$$\frac{\langle\phi|\hat{A}|\phi\rangle}{\langle\phi|\phi\rangle} = \frac{\sum\limits_{i,j} c_i^* c_j \langle\chi_i|\hat{A}|\chi_j\rangle}{\sum\limits_{i,j} c_i^* c_j \langle\chi_i|\chi_j\rangle} \tag{5.56}$$

can be calculated. But in our case the Lagrange multiplier has a very simple interpretation that makes this step unnecessary and that furthermore allows us to choose the correct eigenvalue and eigenfunction immediately.

To realize this we multiply Eq. (5.51) by c_k^* and sum all the different equations for the different k, which gives

$$\sum\limits_{j,k} c_k^* c_j \langle\chi_k|\hat{A}|\chi_j\rangle = \lambda \sum\limits_{j,k} c_k^* c_j \langle\chi_k|\chi_j\rangle, \tag{5.57}$$

or

$$\lambda = \frac{\sum\limits_{j,k} c_k^* c_j \langle\chi_k|\hat{A}|\chi_j\rangle}{\sum\limits_{j,k} c_k^* c_j \langle\chi_k|\chi_j\rangle}, \tag{5.58}$$

i.e., the eigenvalue (or Lagrange multiplier) *is* the sought expectation value! Furthermore, out of the N_b eigenvalues and eigenfunctions, we shall choose that corresponding to the smallest eigenvalue.

5.6   AN EXAMPLE

As an example we consider the potential of Fig. 5.6, given by

$$V(x) = (x^2 - 1)^2, \tag{5.59}$$

which is shown in Fig. 5.

The corresponding one-dimensional Schrödinger equation,

$$\left[-\frac{\hbar^2}{2m}\frac{d^2}{dx^2} + V(x) \right] \psi(x) \equiv \hat{H}\psi = E \cdot \psi(x), \tag{5.60}$$

cannot be solved directly and instead we seek an approximate solution. To this end we *choose* some fixed basis functions $\{\chi_j\}$ for which we can calculate the matrix elements $\langle\chi_k|\hat{H}|\chi_l\rangle$ and $\langle\chi_k|\chi_l\rangle$. These functions could, e.g., be

$$\chi_1(x) = e^{-(x-1)^2}$$

$$\chi_2(x) = e^{-(x+1)^2}$$

$$\chi_3(x) = (x-1)e^{-(x-1)^2}$$

$$\chi_4(x) = (x+1)e^{-(x+1)^2} \tag{5.61}$$

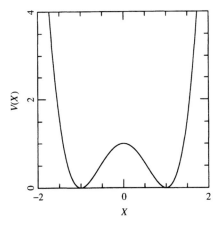

Figure 5.6 The potential $V(x)$ of Eq. (5.59)

and, consequently, we approximate the ground-state eigenfunction to Eq. (5.60) as

$$\psi(x) \simeq \phi(x) = c_1\chi_1(x) + c_2\chi_2(x) + c_3\chi_3(x) + c_4\chi_4(x), \qquad (5.62)$$

where the coefficients c_i are sought.

Applying the method of the previous subsection, these coefficients can be found through the following 4×4 generalized eigenvalue problem (the secular equation),

$$\begin{pmatrix} \langle\chi_1|\hat{H}|\chi_1\rangle & \langle\chi_1|\hat{H}|\chi_2\rangle & \langle\chi_1|\hat{H}|\chi_3\rangle & \langle\chi_1|\hat{H}|\chi_4\rangle \\ \langle\chi_2|\hat{H}|\chi_1\rangle & \langle\chi_2|\hat{H}|\chi_2\rangle & \langle\chi_2|\hat{H}|\chi_3\rangle & \langle\chi_2|\hat{H}|\chi_4\rangle \\ \langle\chi_3|\hat{H}|\chi_1\rangle & \langle\chi_3|\hat{H}|\chi_2\rangle & \langle\chi_3|\hat{H}|\chi_3\rangle & \langle\chi_3|\hat{H}|\chi_4\rangle \\ \langle\chi_4|\hat{H}|\chi_1\rangle & \langle\chi_4|\hat{H}|\chi_2\rangle & \langle\chi_4|\hat{H}|\chi_3\rangle & \langle\chi_4|\hat{H}|\chi_4\rangle \end{pmatrix} \cdot \begin{pmatrix} c_1 \\ c_2 \\ c_3 \\ c_4 \end{pmatrix}$$

$$= E \cdot \begin{pmatrix} \langle\chi_1|\chi_1\rangle & \langle\chi_1|\chi_2\rangle & \langle\chi_1|\chi_3\rangle & \langle\chi_1|\chi_4\rangle \\ \langle\chi_2|\chi_1\rangle & \langle\chi_2|\chi_2\rangle & \langle\chi_2|\chi_3\rangle & \langle\chi_2|\chi_4\rangle \\ \langle\chi_3|\chi_1\rangle & \langle\chi_3|\chi_2\rangle & \langle\chi_3|\chi_3\rangle & \langle\chi_3|\chi_4\rangle \\ \langle\chi_4|\chi_1\rangle & \langle\chi_4|\chi_2\rangle & \langle\chi_4|\chi_3\rangle & \langle\chi_4|\chi_4\rangle \end{pmatrix} \cdot \begin{pmatrix} c_1 \\ c_2 \\ c_3 \\ c_4 \end{pmatrix}. \qquad (5.63)$$

The solution sought is the one that corresponds to the lowest eigenvalue E.

6 Perturbation Theory

6.1 THE NON-DEGENERATE CASE

As the name indicates, perturbation theory is applied when the effects of perturbations are included. The idea behind it is that we know the wavefunctions and eigenvalues for some given system (most often only approximately through variational methods as described in the preceding section) and want to include the effects of some extra interaction whose effects are supposed to be small. Before we start with the basic equations, two simple well-known examples will illustrate the principles.

We consider two non-interacting molecules, e.g., a water molecule, H_2O, and a methanol molecule, CH_3OH. The electrons of each of those will occupy different orbitals, and in the leftmost and rightmost parts of Fig. 6.1 we show the energies of one of the orbitals for each molecule. We assume subsequently that the two molecules start interacting, but only weakly, e.g., through a hydrogen bond for the two molecules mentioned above. That the interactions are weak will mean that the molecular orbitals will change only little. As is well known, this will lead to two new orbitals that differ only little from the original one; cf. the middle part of Fig. 6.1. This means that the energetically lowest orbital, ϕ_b, which is the bonding linear combination, will have an energy only slightly different from that of the ϕ_B orbital of the pure B system and that also the orbital itself will be very similar to the ϕ_B orbital. Similarly, the other hybrid orbital ϕ_a will be very similar to the original ϕ_A orbital of the pure A system. Here, we have assumed — as in Fig. 6.1 — that the energy of the ϕ_A orbital is higher than that of the ϕ_B orbital.

In carrying these arguments through we have assumed that the effects of the interactions are small, so that it is a good starting point to consider first the non-interacting systems and then, subsequently, include the effects of the interactions as something that changes the original orbitals only little. That is, the interactions are included perturbatively.

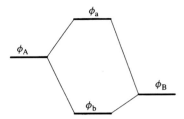

Figure 6.1 Schematic representation of how the energy levels of two non-interacting systems A and B change when weak interactions between them are turned on

As the other example we consider an isolated transition-metal atom from the $3d$ series (e.g., Ti, V, Fe, Cr, ...). Neglecting spin we have five $3d$ orbitals that are energetically degenerate as shown in the left part of Fig. 6.2. The $3d$ orbitals are furthermore very compact in the sense that they do not extend far away from the nucleus. When the atom is placed inside a crystal it may occupy many different positions, but we shall here assume that it sits at a position of very high symmetry (in order to simplify the discussion), i.e., at a position of cubic symmetry. The orbitals will feel the presence of the surrounding host, and actually the symmetry around the atom of our interest will no longer be spherical (as it is for the isolated atom) but will be lowered to cubic symmetry. However, the $3d$ orbitals are well localized close to the nucleus and will therefore feel this symmetry lowering as only a smaller effect. The orbitals will then split as shown in the right part of Fig. 6.2, i.e. there will be one group of three orbitals and another at another energy containing two orbitals, but the splitting is only small.

Once again, these arguments work when it is justified that the effects of the extra interactions that are included are small compared with the original effects. Furthermore, in both cases we have assumed that we somehow have obtained the orbitals and the energy levels for the system before the extra interactions are included, and that we can use these as basis for discussing the effects of the extra interactions.

In the general case we assume accordingly that the total Hamilton operator for the complete system can be written as

$$\hat{H} = \hat{H}_0 + \Delta\hat{H}, \tag{6.1}$$

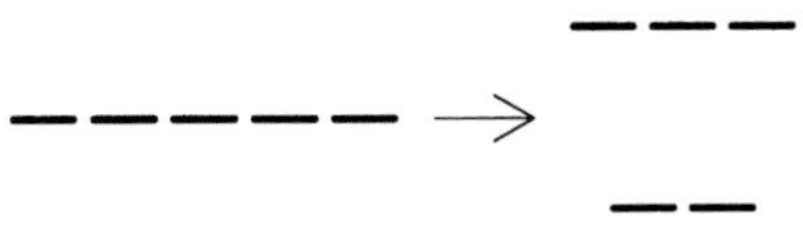

Figure 6.2 Schematic representation of how the energy levels of the $3d$ electrons of a transition-metal atom change when the atom is placed inside a crystal with cubic symmetry around the transition-metal atom

where $\hat{H}_0$ is that of the system without the extra interaction and $\Delta\hat{H}$ is that of the extra interaction. For the transition-metal atom discussed above, $\hat{H}_0$ will be the Hamilton operator for the isolated atom and $\Delta\hat{H}$ the extra potential generated from the crystalline host. For the two weakly interacting molecules $\hat{H}_0$ will be the sum of the two Hamilton operators for the isolated A and B molecules where it is assumed that the electrons of the A molecule do not experience any effects from the presence of the B molecule, and vice versa. $\Delta\hat{H}$ is then these extra effects.

In Eq. (6.1) we will furthermore assume that

$$\Delta\hat{H} \ll \hat{H}_0, \tag{6.2}$$

although we stress that this equation can be interpreted only with difficulty (for instance, is d/dx smaller or larger than d^2/dx^2?). The meaning of Eq. (6.2) is to be as follows: We will assume that we are interested in some orbital ϕ (e.g., the $3d$ orbitals of the transition-metal atom, or the orbitals ϕ_A and ϕ_B of the example of Fig. 6.1). Then, Eq. (6.2) is to be interpreted as meaning

$$\langle\phi|\Delta\hat{H}|\phi\rangle \ll \langle\phi|\hat{H}_0|\phi\rangle, \tag{6.3}$$

although there may be other orbitals (in which we are *not* interested) where this is not fulfilled.

We shall rewrite Eqs. (6.1) and (6.2) as

$$\hat{H} = \hat{H}_0 + \lambda\hat{H}_1, \tag{6.4}$$

with

$$\Delta\hat{H} = \lambda\hat{H}_1 \tag{6.5}$$

and

$$|\lambda| \ll 1. \tag{6.6}$$

As is clear from the discussion above, we shall assume that we know the solutions to the Schrödinger equation without the perturbation ($\Delta\hat{H}$), i.e., we have solved

$$\hat{H}_0\psi_i^{(0)} = E_i^{(0)}\psi_i^{(0)}, \tag{6.7}$$

where the upper indices '0' indicate that these are the solutions for the Schrödinger equation for $\lambda = 0$.

The eigenfunctions in Eq. (6.7) can without loss of generality be assumed to be orthonormal,

$$\langle\psi_i^{(0)}|\psi_j^{(0)}\rangle = \delta_{i,j}, \tag{6.8}$$

and so they form a complete set of orthonormal functions.

We will be interested in a single one of those orbitals and in particular how this changes when the perturbation is turned on. Compared with the examples of

Figs. 6.1 and 6.2 this means that the solutions of Eq. (6.7) are *all* orbitals, not only those shown in the figures, but we will be interested only in those of the figures. We will thus concentrate on one single orbital $\psi_k^{(0)}$ and its corresponding energy $E_k^{(0)}$. We will first treat the case where (before the perturbation is turned on) there is no other orbital with the same energy than that one, i.e., it will be applicable to the example of Fig. 6.1 but not to that of Fig. 6.2.

Accordingly, we focus on the function satisfying

$$\hat{H}_0\psi_k^{(0)} = E_k^{(0)}\psi_k^{(0)}, \tag{6.9}$$

where we assume that

$$E_i^{(0)} \neq E_k^{(0)} \qquad \text{for} \qquad i \neq k. \tag{6.10}$$

We seek the solution to the equation that includes the perturbation,

$$\hat{H}\psi_k = (\hat{H}_0 + \lambda\hat{H}_1)\psi_k = E_k\psi_k. \tag{6.11}$$

Since the functions $\psi_i^{(0)}$ form a complete set, also the solution to Eq. (6.11) can be written as a linear combination of those. Thus, we know that it is possible to write

$$\psi_k = \sum_i c_i\psi_i^{(0)}, \tag{6.12}$$

although we do not know the coefficients c_i.

The idea is now to write all unknown quantities as series in λ. As long as λ is small (and this is exactly the basic assumption) this is possible. The unknown quantities are the energy eigenvalue E_k and the expansion coefficients c_i. We write accordingly

$$E_k = E_k^{(0)} + \lambda E_k^{(1)} + \lambda^2 E_k^{(2)} + \cdots \tag{6.13}$$

for the energy and

$$c_i = c_i^{(0)} + \lambda c_i^{(1)} + \lambda^2 c_i^{(2)} + \cdots \tag{6.14}$$

for the expansion coefficients.

For $\lambda = 0$ the energy should equal that of the unperturbed system. Therefore, the first term on the right-hand side of Eq. (6.13) is $E_k^{(0)}$. Furthermore, for $\lambda = 0$ we know also that $\psi_k = \psi_k^{(0)}$, which gives that

$$c_i^{(0)} = \delta_{i,k}. \tag{6.15}$$

We insert Eqs. (6.13) and (6.14) into Eq. (6.11) with ψ_k expressed as in Eq. (6.12) and obtain then

$$(\hat{H}_0 + \lambda\hat{H}_1)\sum_i [c_i^{(0)} + \lambda c_i^{(1)} + \lambda^2 c_i^{(2)} + \cdots]\psi_i^{(0)}$$

$$= [E_k^{(0)} + \lambda E_k^{(1)} + \lambda^2 E_k^{(2)} + \cdots]\sum_i [c_i^{(0)} + \lambda c_i^{(1)} + \lambda^2 c_i^{(2)} + \cdots]\psi_i^{(0)}. \tag{6.16}$$

Each of the Hamilton operators $\hat{H}_0$ and $\hat{H}_1$ is a linear operator, so that the left-hand side of Eq. (6.16) can be rewritten as

$$\hat{H}_0 \sum_i [c_i^{(0)} + \lambda c_i^{(1)} + \lambda^2 c_i^{(2)} + \cdots] \psi_i^{(0)}$$

$$+ \lambda \hat{H}_1 \sum_i [c_i^{(0)} + \lambda c_i^{(1)} + \lambda^2 c_i^{(2)} + \cdots] \psi_i^{(0)}$$

$$= \sum_i [c_i^{(0)} + \lambda c_i^{(1)} + \lambda^2 c_i^{(2)} + \cdots] \hat{H}_0 \psi_i^{(0)}$$

$$+ \lambda \hat{H}_1 \sum_i [c_i^{(0)} + \lambda c_i^{(1)} + \lambda^2 c_i^{(2)} + \cdots] \psi_i^{(0)}$$

$$= \sum_i [c_i^{(0)} + \lambda c_i^{(1)} + \lambda^2 c_i^{(2)} + \cdots] E_i^{(0)} \psi_i^{(0)}$$

$$+ \lambda \hat{H}_1 \sum_i [c_i^{(0)} + \lambda c_i^{(1)} + \lambda^2 c_i^{(2)} + \cdots] \psi_i^{(0)}, \qquad (6.17)$$

where, for the last identity, we have used Eq. (6.7).

We substitute Eq. (6.17) into Eq. (6.16) and multiply subsequently with $[\psi_l^{(0)}]^*$, where l specifies any of the orbitals of Eq. (6.7). Finally, we integrate over all space, which yields

$$\sum_i [c_i^{(0)} + \lambda c_i^{(1)} + \lambda^2 c_i^{(2)} + \cdots] E_i^{(0)} \langle \psi_l^{(0)} | \psi_i^{(0)} \rangle$$

$$+ \lambda \left\langle \psi_l^{(0)} \middle| \hat{H}_1 \middle| \sum_i [c_i^{(0)} + \lambda c_i^{(1)} + \lambda^2 c_i^{(2)} + \cdots] \psi_i^{(0)} \right\rangle$$

$$= [E_k^{(0)} + \lambda E_k^{(1)} + \lambda^2 E_k^{(2)} + \cdots]$$

$$\times \sum_i [c_i^{(0)} + \lambda c_i^{(1)} + \lambda^2 c_i^{(2)} + \cdots] \langle \psi_l^{(0)} | \psi_i^{(0)} \rangle. \qquad (6.18)$$

Since the functions $\psi_i^{(0)}$ are orthonormal this gives

$$[c_l^{(0)} + \lambda c_l^{(1)} + \lambda^2 c_l^{(2)} + \cdots] E_l^{(0)}$$

$$+ \lambda \sum_i [c_i^{(0)} + \lambda c_i^{(1)} + \lambda^2 c_i^{(2)} + \cdots] \langle \psi_l^{(0)} | \hat{H}_1 | \psi_i^{(0)} \rangle$$

$$= [E_k^{(0)} + \lambda E_k^{(1)} + \lambda^2 E_k^{(2)} + \cdots][c_l^{(0)} + \lambda c_l^{(1)} + \lambda^2 c_l^{(2)} + \cdots], \quad (6.19)$$

where we also have used that $\hat{H}_1$ is a linear operator.

Both sides of Eq. (6.19) are Taylor series in λ although appearing somewhat mixed up. Nevertheless, the two series can only be identical *iff* the factors to each power λ^n are identical on the two sides. Thus, Eq. (6.19) can be written in the general form

$$a^{(n)}\lambda^n = b^{(n)}\lambda^n, \tag{6.20}$$

where the coefficients $a^{(n)}$ and $b^{(n)}$ depend on the quantities of Eq. (6.19). The requirement

$$a^{(n)} = b^{(n)} \tag{6.21}$$

gives for $n = 0$:

$$E_k^{(0)} = E_k^{(0)}, \tag{6.22}$$

which is trivially fulfilled. For $n = 1$ we find

$$c_l^{(1)}E_l^{(0)} + \sum_i c_i^{(0)}\langle\psi_l^{(0)}|\hat{H}_1|\psi_i^{(0)}\rangle = E_k^{(0)}c_l^{(1)} + E_k^{(1)}c_l^{(0)}, \tag{6.23}$$

whereas $n = 2$ gives

$$c_l^{(2)}E_l^{(0)} + \sum_i c_i^{(1)}\langle\psi_l^{(0)}|\hat{H}_1|\psi_i^{(0)}\rangle = E_k^{(0)}c_l^{(2)} + E_k^{(1)}c_l^{(1)} + E_k^{(2)}c_l^{(0)}. \tag{6.24}$$

With the help of these equations we shall now determine the lowest-order terms in Eqs. (6.13) and (6.14).

First we consider Eq. (6.23) for the special case $l = k$ (since l can be chosen freely, this is allowed). Then

$$c_k^{(1)}E_k^{(0)} + \langle\psi_k^{(0)}|\hat{H}_1|\psi_k^{(0)}\rangle = E_k^{(0)}c_k^{(1)} + E_k^{(1)}, \tag{6.25}$$

where we have used Eq. (6.15). From Eq. (6.25) we immediately obtain

$$E_k^{(1)} = \langle\psi_k^{(0)}|\hat{H}_1|\psi_k^{(0)}\rangle. \tag{6.26}$$

Already at this point we have accordingly the first effects of the inclusion of the perturbation. Equation (6.26) tells how the energy changes to lowest order in the perturbation [cf. Eq. (6.13)], i.e.,

$$\begin{aligned}
E_k &= E_k^{(0)} + \lambda E_k^{(1)} + \cdots \\
&= E_k^{(0)} + \lambda\langle\psi_k^{(0)}|\hat{H}_1|\psi_k^{(0)}\rangle + \cdots \\
&= E_k^{(0)} + \langle\psi_k^{(0)}|\lambda\hat{H}_1|\psi_k^{(0)}\rangle + \cdots \\
&= E_k^{(0)} + \langle\psi_k^{(0)}|\Delta\hat{H}|\psi_k^{(0)}\rangle + \cdots.
\end{aligned} \tag{6.27}$$

Equation (6.23) gives for the case $l \neq k$,

$$c_l^{(1)}E_l^{(0)} + \langle\psi_l^{(0)}|\hat{H}_1|\psi_k^{(0)}\rangle = E_k^{(0)}c_l^{(1)} \tag{6.28}$$

giving

$$c_l^{(1)} = \frac{\langle\psi_l^{(0)}|\hat{H}_1|\psi_k^{(0)}\rangle}{E_k^{(0)} - E_l^{(0)}}. \tag{6.29}$$

This equation is valid for $l \neq k$ and due to assumption (6.10) we do not divide by 0. This equation does not however, tell what $c_k^{(1)}$ is (i.e., we have to exclude the case $k = l$) and in order to determine this latter quantity we must proceed somewhat differently.

Sofar we have determined the coefficients up to λ^1 and no further. This means that we have implicitly assumed that everything of higher order in λ is so small that it can be ignored. Let us therefore be internally consistent and normalize the modified eigenfunction, but only keeping terms up to first order in λ:

$$
\begin{aligned}
\langle \psi_k | \psi_k \rangle &\simeq \left\langle \sum_i (c_i^{(0)} + \lambda c_i^{(1)}) \psi_i^{(0)} \,\middle|\, \sum_j (c_j^{(0)} + \lambda c_j^{(1)}) \psi_j^{(0)} \right\rangle \\
&= \sum_{i,j} (c_i^{(0)} + \lambda c_i^{(1)})^* (c_j^{(0)} + \lambda c_j^{(1)}) \langle \psi_i^{(0)} | \psi_j^{(0)} \rangle \\
&= \sum_i (c_i^{(0)} + \lambda c_i^{(1)})^* (c_i^{(0)} + \lambda c_i^{(1)}) \\
&= (1 + \lambda c_k^{(1)})^* (1 + \lambda c_k^{(1)}) + \sum_{i \neq k} \lambda^2 |c_i^{(1)}|^2.
\end{aligned}
\tag{6.30}
$$

When only including terms up to λ^1, this function is only normalized if

$$
c_k^{(1)} = 0. \tag{6.31}
$$

In deriving Eq. (6.30) we have used the orthonormality of the functions $\psi_i^{(0)}$.

The second-order correction to the energy can be obtained from Eq. (6.24) for $l = k$,

$$
c_k^{(2)} E_k^{(0)} + \sum_i c_i^{(1)} \langle \psi_k^{(0)} | \hat{H}_1 | \psi_i^{(0)} \rangle = c_k^{(2)} E_k^{(0)} + c_k^{(1)} E_k^{(1)} + E_k^{(2)} \tag{6.32}
$$

or

$$
\begin{aligned}
E_k^{(2)} &= \sum_i c_i^{(1)} \langle \psi_k^{(0)} | \hat{H}_1 | \psi_i^{(0)} \rangle - c_k^{(1)} E_k^{(1)} \\
&= \sum_{i \neq k} c_i^{(1)} \langle \psi_k^{(0)} | \hat{H}_1 | \psi_i^{(0)} \rangle \\
&= \sum_{i \neq k} \frac{\langle \psi_i^{(0)} | \hat{H}_1 | \psi_k^{(0)} \rangle}{E_k^{(0)} - E_i^{(0)}} \langle \psi_k^{(0)} | \hat{H}_1 | \psi_i^{(0)} \rangle \\
&= \sum_{i \neq k} \frac{|\langle \psi_i^{(0)} | \hat{H}_1 | \psi_k^{(0)} \rangle|^2}{E_k^{(0)} - E_i^{(0)}}.
\end{aligned}
\tag{6.33}
$$

We have here used Eqs. (6.29) and (6.31) as well as the fact that $\hat{H}_1$ is Hermitian.

To second order in the perturbation we thus have, analogously to Eq. (6.27),

$$E_k = E_k^{(0)} + \langle \psi_k^{(0)} | \Delta \hat{H} | \psi_k^{(0)} \rangle + \sum_{i \neq k} \frac{|\langle \psi_i^{(0)} | \Delta \hat{H} | \psi_k^{(0)} \rangle|^2}{E_k^{(0)} - E_i^{(0)}} \tag{6.34}$$

The eigenfunction is, to first order in the perturbation,

$$\psi_k = \psi_k^{(0)} + \sum_{i \neq k} \frac{\langle \psi_i^{(0)} | \Delta \hat{H} | \psi_k^{(0)} \rangle}{E_k^{(0)} - E_i^{(0)}} \psi_i^{(0)}. \tag{6.35}$$

Higher-order terms can be calculated in an equivalent way, but we shall not do so here.

Finally, if the effects of the perturbation are to be studied for more eigenfunctions, each obeying Eq. (6.10) (e.g., both ϕ_A and ϕ_B for the two weakly interacting molecules in Fig. 6.1), the above procedure has to be repeated for each of those.

6.2 AN EXAMPLE

We consider a very simple case, i.e., that of a single particle moving in a one-dimensional potential well. The Schrödinger equation

$$\left[-\frac{\hbar^2}{2m} \frac{d^2}{dx^2} + V(x) \right] \psi_n(x) \equiv \hat{H} \psi_n(x) = \varepsilon_n \psi_n(x) \tag{6.36}$$

with

$$V(x) = \begin{cases} 0 & \text{for } 0 \leq x \leq a \\ \infty & \text{otherwise} \end{cases} \tag{6.37}$$

has the (normalized) solutions

$$\psi_n(x) = \sqrt{\frac{2}{a}} \sin\left(\frac{n\pi}{a} x\right) \tag{6.38}$$

and the energy eigenvalues

$$\varepsilon_n = \frac{\hbar^2 \pi^2}{2ma^2} n^2. \tag{6.39}$$

Inside the potential well (i.e., for $0 \leq x \leq a$) we now add a small perturbation, $\Delta V(x)$. To first order in the perturbation the energies change then by

$$\Delta \varepsilon_n = \langle \psi_n | \Delta V | \psi_n \rangle = \frac{2}{a} \int_0^a \Delta V(x) \sin^2\left(\frac{n\pi}{a} x\right) dx, \tag{6.40}$$

which, in principle, can be calculated once the precise form of the perturbation is known. [It could, e.g., be $\Delta V(x) = c \cdot x$ or $\Delta V(x) = d \cdot x^2$ with c and d some constants].

Similarly, the nth eigenfunction is, to lowest order in the perturbation, changed by

$$\Delta \psi_n(x) = \sum_{m \neq n} \frac{\langle \psi_m | \Delta V | \psi_n \rangle}{\varepsilon_n - \varepsilon_m} \psi_m(x) = \frac{2ma^2}{\hbar^2 \pi^2} \sum_{m \neq n} \frac{\langle \psi_m | \Delta V | \psi_n \rangle}{n^2 - m^2} \psi_m(x). \quad (6.41)$$

Also this expression can in principle be evaluated.

In contrast to this example, however, in most cases one does not know the complete set of eigenvalues and eigenfunctions, which means that one then has to truncate the infinite series in Eq. (6.2.6) to only the known eigenvalues and eigenfunctions. This is also the case for higher-order corrections to the eigenvalues.

6.3 THE DEGENERATE CASE

The formulas of Section 6.1 were explicitly developed under the assumption that the eigenfunction that was studied was non-degenerate, i.e., that there was no other orbital with the same energy for the unperturbed case. Already the formulas [for instance Eqs. (6.34) and (6.35)] show that we will encounter problems when trying to use the formulas when we have more orbitals that are energetically degenerate with that of interest. This means also that we cannot use the results above in studying the transition-metal atom of Fig. 6.2.

At a more fundamental level our approach is based on the idea that the perturbation leads to only small changes in eigenfunctions and eigenvalues. We shall use this in discussing the problems associated with the degenerate case a little more.

In order to outline the problems we first discuss a simple example. We consider an isolated atom and assume that we have one electron that can occupy three p orbitals. We place a coordinate system with the origin at the nucleus and describe the orbitals as being the p_x, the p_y, and the p_z orbitals. It is clear that the energies of these three orbitals are identical, but also that their definition is arbitrary. We may, e.g., define any other three orthogonal p orbitals and obtain absolutely identical results for any physical observable. One can suggest using three p orbitals along the $(1,1,1)$, $(-1,1,0)$, and the $(-1,-1,2)$ directions. We shall call these 'alternative p functions'.

We now consider first, however, the original p_x, p_y, and p_z orbitals. Subsequently, we add a perturbation. We assume that the electron is exposed to a weak field along the z direction and that this gives rise to an extra term to the Hamilton operator,

$$\Delta \hat{H} = \text{cst} \cdot z^2. \quad (6.42)$$

where cst is a constant. To lowest order in this perturbation, it will only affect the p_z orbital but not the p_x and p_y orbitals. Furthermore, the changes of the p_z orbital will most likely only be modest. Thus, we will expect that out of the three p orbitals one will change a little and the two others will be left unperturbed.

But, if we would base our discussion of the effects on the three p electrons on the three alternative p functions suggested above the picture will apparently be different. However, ultimately the same physical results will appear, so that the extra Hamilton operator of Eq. (6.42) will make the three alternative p orbitals change into one along the field direction (i.e., along the z axis) and the other two perpendicular to this one. This will actually happen independently of the strength of the field, so that also an infinitesimally small perturbation will lead to changes in the eigenfunctions that are not infinitesimally small.

The solution to this apparent paradox is that if we have degenerate eigenfunctions without the perturbation, we have to form such linear combinations that the changes induced by the perturbation in fact are small. Accordingly, had we based the calculations on the alternative p orbitals we would have to form linear combinations of those so that one of them becomes the p_z orbital. The main problem is how to choose these linear combinations.

These ideas will now be described in the general case, but in contrast to Section 6.1 we shall only determine the relevant linear combinations as well as the first-order changes in the energy. Once these have been chosen, one may determine higher-order changes by applying the strategy of Section 6.1.

We assume that we have a set of degenerate orbitals satisfying the Schrödinger equation for the unperturbed case,

$$\hat{H}_0 \psi_k^{(0)} = E_k^{(0)}. \tag{6.43}$$

From those that have the same eigenvalue we seek new linear combinations

$$\tilde{\psi}_k^{(0)} = \sum_j{}' a_j \psi_j^{(0)}, \tag{6.44}$$

where the prime indicates that we sum only over those orbitals for which

$$E_j^{(0)} = E_k^{(0)}. \tag{6.45}$$

In the preceding subsection we wrote the eigenfunctions for the perturbed system as

$$\psi_k = \sum_i [c_i^{(0)} + \lambda c_i^{(1)} + \lambda^2 c_i^{(2)} + \cdots] \psi_i^{(0)}. \tag{6.46}$$

The theory was so constructed that for vanishing perturbation ($\lambda = 0$), there should be no changes in the eigenfunctions, which led to $c_i^{(0)} = \delta_{i,k}$. Here, however, this is not the case. Here, we construct new linear combinations through Eq. (6.44), so that these functions in Eq. (6.45) describe the functions for the unperturbed system. Accordingly, in the present case we have

$$a_j = c_j^{(0)} \tag{6.47}$$

but only for those j that satisfy Eq. (6.45). Accordingly, Eq. (6.15) is no longer valid.

Equation (6.19) is, however, still valid, and we shall therefore start at that point. Collecting the coefficients to λ^n for $n = 1$ on both sides gives then, replacing Eq. (6.23),

$$\sum_i c_i^{(1)} E_i^{(0)} \langle \psi_l^{(0)} | \psi_i^{(0)} \rangle + \sum_i c_i^{(0)} \langle \psi_l^{(0)} | \hat{H}_1 | \psi_i^{(0)} \rangle = E_k^{(0)} c_l^{(1)} + E_k^{(1)} c_l^{(0)}. \quad (6.48)$$

Due to the orthonormality of the functions $\psi_i^{(0)}$ the first summation on the left-hand side reduces to a single term. Furthermore, since the coefficients $c_j^{(0)}$ are non-zero only for those j that satisfy Eq. (6.45), the second summation on the left-hand side reduces to one over exactly those functions. For the example above this would mean that the summation is only over the three p functions and not over other p functions or over s or d functions, etc.

Equation (6.48) takes then the form

$$c_l^{(1)} E_l^{(0)} + {\sum_i}' c_i^{(0)} \langle \psi_l^{(0)} | \hat{H}_1 | \psi_i^{(0)} \rangle = E_k^{(0)} c_l^{(1)} + E_k^{(1)} c_l^{(0)}. \quad (6.49)$$

The function $\psi_l^{(0)}$ was arbitrarily chosen, and we can choose it as we want. Therefore, we let it be one of the functions for which

$$E_l^{(0)} = E_k^{(0)}. \quad (6.50)$$

This casts Eq. (6.49) into

$${\sum_i}' c_i^{(0)} \langle \psi_l^{(0)} | \hat{H}_1 | \psi_i^{(0)} \rangle = E_k^{(1)} c_l^{(0)} \quad (6.51)$$

with l being any of the indices for which Eq. (6.50) is satisfied.

In total, Eq. (6.51) is a matrix eigenvalue problem of the form

$$\underline{\underline{H}}_1 \cdot \underline{c}^{(0)} = E_k^{(1)} \cdot \underline{c}^{(0)}. \quad (6.52)$$

The size of the matrix $\underline{\underline{H}}_1$ equals the number of functions for which Eq. (6.50) [(6.45)] is satisfied. The eigenvalues, $E_k^{(1)}$, define the first-order changes in the energies,

$$E_k = E_k^{(0)} + \lambda E_k^{(1)} + \cdots, \quad (6.53)$$

and the corresponding eigenvectors $\underline{c}^{(0)}$ define the linear combinations of the original eigenfunctions, so that the perturbation induces only a small change in the eigenfunction,

$$\tilde{\psi}_k^{(0)} = {\sum_i}' c_i^{(0)} \psi_i^{(0)}. \quad (6.54)$$

If higher-order corrections are required — either in the energies or in the eigen-functions — one can apply the procedure of the preceeding subsection with the starting eigenfunctions given by Eq. (6.54).

6.4 AN EXAMPLE

We consider an extension of the example above, i.e., we consider a hydrogen atom in the presence of an external electrostatic field along the z axis. Then

$$\Delta \hat{H} = \text{cst} \cdot z. \tag{6.55}$$

Moreover, we study the $n = 2$ orbitals, i.e., the $2s$, $2p_x$, $2p_y$, and $2p_z$ orbitals. In this case, the matrix $\underline{\underline{H}}_1$ of Eq. (6.52) becomes

$$\underline{\underline{H}}_1 = \begin{pmatrix} \langle 2s|\Delta\hat{H}|2s\rangle & \langle 2s|\Delta\hat{H}|2p_x\rangle & \langle 2s|\Delta\hat{H}|2p_y\rangle & \langle 2s|\Delta\hat{H}|2p_z\rangle \\ \langle 2p_x|\Delta\hat{H}|2s\rangle & \langle 2p_x|\Delta\hat{H}|2p_x\rangle & \langle 2p_x|\Delta\hat{H}|2p_y\rangle & \langle 2p_x|\Delta\hat{H}|2p_z\rangle \\ \langle 2p_y|\Delta\hat{H}|2s\rangle & \langle 2p_y|\Delta\hat{H}|2p_x\rangle & \langle 2p_y|\Delta\hat{H}|2p_y\rangle & \langle 2p_y|\Delta\hat{H}|2p_z\rangle \\ \langle 2p_z|\Delta\hat{H}|2s\rangle & \langle 2p_z|\Delta\hat{H}|2p_x\rangle & \langle 2p_z|\Delta\hat{H}|2p_y\rangle & \langle 2p_z|\Delta\hat{H}|2p_z\rangle \end{pmatrix}. \tag{6.56}$$

Due to symmetry, a number of these matrix elements vanish, so that $\underline{\underline{H}}_1$ reduces to

$$\underline{\underline{H}}_1 = \begin{pmatrix} 0 & 0 & 0 & \langle 2s|\Delta\hat{H}|2p_z\rangle \\ 0 & 0 & 0 & 0 \\ 0 & 0 & 0 & 0 \\ \langle 2p_z|\Delta\hat{H}|2s\rangle & 0 & 0 & 0 \end{pmatrix}. \tag{6.57}$$

Thus, the perturbation leads to lowest order to unmodified $2p_x$ and $2p_y$ orbitals, whereas the $2s$ and $2p_z$ orbitals become mixed. That is, as new basis functions we shall use $(1/\sqrt{2})(2s + 2p_z)$, $(1/\sqrt{2})(2s - 2p_z)$, $2p_x$, and $2p_z$, and the first-order shifts of the eigenvalues are $\pm|\langle 2p_z|\Delta\hat{H}|2s\rangle|$ and 0.

6.5 TIME-DEPENDENT PERTURBATION THEORY

The time-dependent Schrödinger equation is

$$\hat{H}\psi(\vec{x}, t) = i\hbar \frac{\partial}{\partial t}\psi(\vec{x}, t), \tag{6.58}$$

where $\vec{x}$ includes all coordinates (position coordinates and spin) that are not the time (t).

We shall assume that $\hat{H}$ contains a dominating time-independent term and a small perturbation that is time-dependent,

$$\hat{H} = \hat{H}_0 + \Delta\hat{H}(t). \tag{6.59}$$

As above, we shall also assume that we have solved the time-independent Schrödinger equation without the perturbation,

$$\hat{H}_0 \tilde{\psi}_k^{(0)}(\vec{x}) = E_k^{(0)} \tilde{\psi}_k^{(0)}(\vec{x}). \tag{6.60}$$

According to the discussion of Section 4.1, the time-dependent solutions corresponding to those of Eq. (6.60) are given by

$$\psi_k^{(0)}(\vec{x}, t) = \tilde{\psi}_k^{(0)}(\vec{x}) \exp\left(-\frac{iE_k^{(0)}t}{\hbar}\right). \tag{6.61}$$

In order to distinguish between the functions of Eq. (6.60) and those of Eq. (6.61) we have put tildes on the former.

With the inclusion of the time-dependent perturbation we shall write the time-dependent wavefunctions as

$$\psi_k(\vec{x}, t) = \sum_i c_i(t) \psi_i^{(0)}(\vec{x}, t), \tag{6.62}$$

where we use the fact that the functions of Eq. (6.60) form a complete set. We shall furthermore assume (this is, as always, no restriction!) that the functions of Eq. (6.60) are orthonormal.

We insert Eq. (6.62) into the time-dependent Schrödinger equation and obtain

$$[\hat{H}_0 + \Delta\hat{H}(t)] \sum_i c_i(t) \psi_i^{(0)}(\vec{x}, t) = i\hbar \frac{\partial}{\partial t} \sum_i c_i(t) \psi_i^{(0)}(\vec{x}, t), \tag{6.63}$$

or by using that $\hat{H}_0$ is time-independent as well as the definitions of $\psi_k^{(0)}(\vec{x}, t)$,

$$\sum_i c_i(t) E_i^{(0)} \psi_i^{(0)}(\vec{x}, t) + \Delta\hat{H}(t) \sum_i c_i(t) \psi_i^{(0)}(\vec{x}, t)$$

$$= i\hbar \sum_i \left[\frac{\partial c_i(t)}{\partial t} - \frac{iE_i^{(0)}}{\hbar} c_i(t)\right] \psi_i^{(0)}(\vec{x}, t), \tag{6.64}$$

or

$$\Delta\hat{H}(t) \sum_i c_i(t) \psi_i^{(0)}(\vec{x}, t) = i\hbar \sum_i \frac{\partial c_i(t)}{\partial t} \psi_i^{(0)}(\vec{x}, t). \tag{6.65}$$

We multiply this equation by $[\psi_j^{(0)}(\vec{x}, t)]^*$ (j being arbitrarily chosen) and integrate over all $\vec{x}$. This gives then, due to the orthonormality of the functions of Eq. (6.60),

$$i\hbar \frac{dc_j(t)}{dt} = \sum_i \langle \tilde{\psi}_j^{(0)} | \Delta\hat{H}(t) | \tilde{\psi}_i^{(0)} \rangle \exp(i\omega_{ji}t) c_i(t), \tag{6.66}$$

where the matrix element is time-dependent (i.e., the integration is only over $\vec{x}$), and where

$$\omega_{ji} = \frac{E_j^{(0)} - E_i^{(0)}}{\hbar}. \tag{6.67}$$

As above we substitute

$$\Delta \hat{H}(t) = \lambda \hat{H}_1(t), \tag{6.68}$$

and assume — as in Section 6.1 — that λ is small. Furthermore, we expand the coefficients $c_i(t)$ in powers of λ,

$$c_i(t) = c_i^{(0)}(t) + \lambda c_i^{(1)}(t) + \lambda^2 c_i^{(2)}(t) + \cdots. \tag{6.69}$$

Inserting this into Eq. (6.66), using that both the left-hand side and the right-hand side is a Taylor series in λ (this is exactly the same procedure as in Section 6.1, but actually slightly simpler), identifying the coefficients to the same powers of λ (i.e., to λ^n) on both sides, leads for $n = 0$ to

$$i\hbar \frac{dc_j^{(0)}(t)}{dt} = 0 \tag{6.70}$$

and for $n > 0$ to

$$i\hbar \frac{dc_j^{(n)}(t)}{dt} = \sum_i \langle \tilde{\psi}_j^{(0)} | \Delta \hat{H}(t) | \tilde{\psi}_i^{(0)} \rangle \exp(i\omega_{ji}t) c_i^{(n-1)}(t). \tag{6.71}$$

An important case is the one where the system is in a fixed state until a certain time t_0 where the perturbation is turned on. That is, we assume that

$$c_j(t) = \delta_{j,k} \qquad \text{for} \qquad t < t_0, \tag{6.72}$$

where the state of the system before the perturbation is turned on is labelled k. Then, from Eq. (6.70),

$$c_j^{(0)}(t) = \delta_{j,k} \tag{6.73}$$

and

$$c_j^{(1)}(t) = \frac{1}{i\hbar} \int_{t_0}^t \langle \tilde{\psi}_j^{(0)} | \Delta \hat{H}(t') | \tilde{\psi}_k^{(0)} \rangle \exp(i\omega_{jk}t') \, dt' \tag{6.74}$$

and related expressions for the higher-order corrections.

Here, it may furthermore be assumed that the duration of the perturbation is very short, so that its time dependence can be approximated by a δ-function. Alternatively, it may be considered very long, so that t above is best replaced by ∞. But here we shall not go into further discussions about any of those limits. It should, however, be obvious that due to such external, time-dependent perturbations, orbitals that were unoccupied before the perturbation was turned on may become occupied, and also that the populations may become time-dependent, including oscillating.

7 Symmetry and Group Theory

7.1 SYMMETRY

Many systems possess symmetries that can be exploited both in simplifying the problem of solving the Schrödinger equation (in some cases even change it from an impossible one into a manageable one) and in classifying the solutions according to their behaviour under the symmetry operations. With this symmetry information it is often possible to give much more precise estimates on experimental observables than would have been possible without it (e.g., the symmetry classifications can often be used in predicting which transitions or excitations are allowed and which are not). We shall here put the application of symmetry on a little more formal basis so that it to some extent can be applied as (powerful) machinery. First, however, we shall describe symmetries in some detail through some examples.

The ammonia molecule, NH_3, is shown in Fig. 7.1. It contains three hydrogen atoms that define an equilateral triangle as well as a nitrogen atom placed above (or below) the centre of that triangle. We shall here neglect the famous umbrella vibrations of this molecule according to which the nitrogen atom flips back and forth between being above and below the triangle, but instead assume that the nuclei stay fixed as in Fig. 7.1.

NH_3 contains a number of symmetry elements. First of all we have the three-fold axis through the nitrogen atom and the centre of the H_3 triangle. This defines two symmetry operations: a rotation of $120°$ (which brings hydrogen atom 1 onto No. 2, No. 2 onto No. 3, and No. 3 onto No. 1) and a rotation of $240°$ (which brings hydrogen atom 1 onto No. 3, No. 2 onto No. 1, and No. 3 onto No. 2). In addition there are the three mirror planes each passing through the nitrogen atom, one hydrogen atom, and the centre of the line joining the other two hydrogen atoms. Each of the corresponding symmetry operations maps one hydrogen

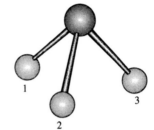

Figure 7.1 The ammonia molecule NH₃. In order to
clarify the discussion the three hydrogen atoms have been
given different numbers

atom onto itself and interchanges the other two. Finally, there is the trivial
symmetry operation, the so-called identity operation, that maps the complete
system unchanged on itself. In total, we see that we have six symmetry operations.

Due to the symmetry the Hamilton operator will look exactly the same when
performing one of the symmetry operations. This means that interchanging the
nuclei as described above will not change the electronic distribution or the solutions to the Schrödinger equation. In the language of Section 3.3 the Hamilton
operator and any of the symmetry operators $\hat{O}_i$ commute,

$$[\hat{H}, \hat{O}_i] = 0. \tag{7.1}$$

Another example is the CO molecule. It is (has to be) linear and contains as
only symmetry element a C_∞ axis passing through both nuclei. This means that
any rotation (of the angle θ, $0 \le \theta < 2\pi$) maps the system onto itself. Here the
special case $\theta = 0$ corresponds to the identity operation.

The N_2 molecule resembles the CO molecule but has as additional symmetry
element an inversion centre at the midpoint between the two nuclei. This means
that we in addition have the symmetry operations consisting of a combined
rotation of θ plus an inversion. Both for the N_2 and the CO molecule we have
an infinite number of symmetry operations.

Polyvinylchloride (Fig. 7.2) is a polymer where each chain (macromolecule)
is very long; i.e., so long that we shall assume that it is infinite. In that case the
material consists of an infinite sequence of repeated equivalent units, each unit
containing two of the carbon atoms of the backbone. The symmetry operations
consist of all the translations along the polymer axis of an integer times the
unit cell length, $n \cdot a$, $n = \cdots, -2, -1, 0, 1, 2, \ldots$. In that case $n = 0$ for the
identity.

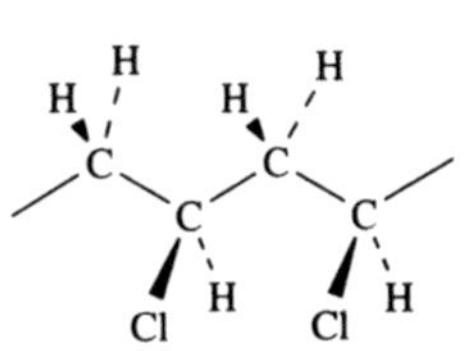

Figure 7.2 Polyvinylchloride (PVC)

Also polycarbonitrile, Fig. 7.3, can be considered as an infinite, periodic chain. Each unit cell contains one CHN unit, and the symmetry operations are those of polyvinylchloride plus all combined operations consisting of one such translation plus a reflection in the plane of all nuclei.

Also the hypothetic helical polymer of Fig. 7.4 will be assumed to be infinite and periodic, although, in this case, the symmetry operations are slightly more complicated: any of them can be described as an integral of a screw-axis operation, which in turn consists of a translation (of h) and rotation (of v), where h and v are shown in Fig. 7.4.

Finally, the NaCl crystal of Fig. 7.5 contains a number of symmetry elements. First of all, there is the translations along any of the three main axes. In addition, there are various mirror planes, symmetry axes, inversion centres, etc. The complete class of symmetry operations is large and will therefore not be described in detail here. It does, however, contain all the translations along any of the three main axes (plus combinations thereof) as well as the combined operations consisting of these translations and any of the reflections, rotations, inversions, etc.

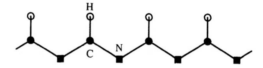

Figure 7.3   Polycarbonitrile

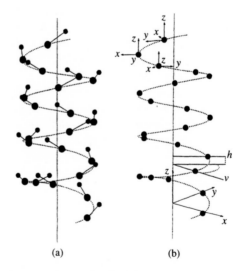

Figure 7.4   A hypothetic helical polymer. (a) shows the polymer with two atoms per repeated unit, whereas (b) shows only one of these atoms together with the parameters h and v describing the primitive symmetry operation

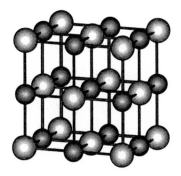

Figure 7.5 A part of a NaCl crystal

In all cases we have a (finite or infinite) set of symmetry operations that all obey Eq. (7.1). Group theory offers a powerful tool with which these symmetry operations can be used in simplifying a number of calculations. Group theory is not restricted to symmetries, and the results to be derived in the next subsection can also be applied — with only smaller modifications — to other cases where group theory is found useful. Here, we shall nevertheless base our discussion on the symmetry operations.

7.2 GROUP THEORY

For each of the systems discussed above we have identified a set of symmetry operations, $\{\hat{O}_i\}$. This set forms what mathematically is called a group, since it fulfills the following requirements:

1. For any two symmetry operations $\hat{O}_i$ and $\hat{O}_j$ the combined operation $\hat{O}_i\hat{O}_j$ (meaning *first* performing $\hat{O}_j$ and then $\hat{O}_i$) is also a member of the set,

$$\hat{O}_i\hat{O}_j = \hat{O}_k. \tag{7.2}$$

For polycarbonitrile, e.g., of Fig. 7.3 $\hat{O}_j$ may be the translation of n_j units combined with the reflection in the mirror plane, whereas $\hat{O}_i$ is the translation of n_i units. The combined operation $\hat{O}_i\hat{O}_j$ becomes then a translation of $n_i + n_j$ units combined with the reflection in the mirror plane, which obviously also is one of the symmetry operations for this system. For the NH_3 molecule of Fig. 7.1 we could, e.g., let $\hat{O}_i$ be the reflection in the plane passing through the nitrogen atom, the hydrogen atom no. 1, and the midpoint between hydrogen atoms no. 2 and 3, and $\hat{O}_j$ be the rotation about the C_3 axis of 120°. The combined operation brings first the three hydrogen atom (1,2,3) into the set (2,3,1) (here the first number means that hydrogen atom no. 2 occupies the original site of no. 1, etc.). Subsequently, symmetry operation $\hat{O}_j$ brings this set onto the set (2,1,3). In total this is seen to be equivalent to the reflection in the plane passing through the nitrogen atom, hydrogen atom no. 3, and the midpoint between hydrogen atoms no. 1 and 2, i.e., to another symmetry operation.

2. For any three symmetry operations $\hat{O}_i$, $\hat{O}_j$, and $\hat{O}_k$ we have

$$\hat{O}_i(\hat{O}_j\hat{O}_k) = (\hat{O}_i\hat{O}_j)\hat{O}_k. \tag{7.3}$$

This means that one may combine either the first two operations first or the last two first. We shall not prove this in any way, and the interested reader may convince him-/herself about its validity.

3. There exists one operation $\hat{O}_0$ so that

$$\hat{O}_i\hat{O}_0 = \hat{O}_0\hat{O}_i = \hat{O}_i \tag{7.4}$$

for all operations $\hat{O}_i$. This element is the identity whose existence we have stressed at certain places above.

4. For each operation $\hat{O}_i$ there exists an inverse $\hat{O}_i^{-1}$ so that

$$\hat{O}_i\hat{O}_i^{-1} = \hat{O}_i^{-1}\hat{O}_i = \hat{O}_0. \tag{7.5}$$

For polycarbonitrile the inverse of any translation by n units is the translation by $-n$ units. If the translation is accompanied by the reflection in the mirror plane, so is its inverse. For NH_3, the inverse of any reflection is the operation itself, whereas the inverse of the rotation of $120°$ is that of $240°$.

When all these four criteria are satisfied the set of operations form a group and we can apply all the mathematical theorems of group theory. This is the case for the symmetry operations of interest.

In some cases we have in addition that the symmetry operations commute, but we stress that this is often *not* the case.

Let us now return to the discussion of the ammonia molecule of Fig. 7.1. In order to extend the discussion we introduce a coordinate system as shown in Fig. 7.6. Above we only considered what happened when we let the symmetry operations act on the nuclei, but now we shall see what happens when they act on functions in space.

We will assume that all equivalent atoms are described equivalently. This means that in Fig. 7.6 we will assume that we have identical $1s$ functions on all hydrogen atoms, whereas on the nitrogen atom we have $1s$, $2s$, and $2p$ functions. When we let any of the symmetry operations $\hat{O}_i$ act on any one of these functions, let us call it f_j, the result can be written exactly as a linear combination of the original functions,

$$\hat{O}_i f_j = \sum_k a^i_{jk} f_k, \tag{7.6}$$

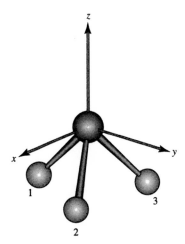

Figure 7.6 The ammonia molecule NH_3. In order to clarify the discussion the three hydrogen atoms have been given different numbers. Furthermore, a global coordinate system has been introduced compared with Fig. 7.1

i.e., the functions will transform among each other. This is exactly the fundamental aspect of the symmetry operations.

As a very simple example we consider a one-dimensional case. We assume furthermore that we have inversion symmetry about $x = 0$. In this case we have two symmetry operations, the identity $\hat{I}$ and the inversion $\hat{O}$. Moreover, from *any* function $f(x)$, we construct the two functions

$$f_+(x) = f(x)$$
$$f_-(x) = f(-x). \tag{7.7}$$

This corresponds to making sure that the set of functions that we are considering satisfies the symmetry of the system, i.e., f_+ and f_- are the two functions we shall study.

Then,

$$\hat{I}f_+ = f_+$$
$$\hat{I}f_- = f_-$$
$$\hat{O}f_+ = f_-$$
$$\hat{O}f_- = f_+. \tag{7.8}$$

For the NH_3 molecule Eq. (7.6) states that the hydrogen $1s$ functions will be mapped onto each other by the symmetry operations, whereas the nitrogen $1s$, $2s$, and $2p_z$ functions will be mapped onto themselves by the symmetry operations, and the nitrogen $2p_x$ and $2p_y$ functions will be mapped onto linear combinations of these two functions.

One of the basic theorems of group theory is that for any group it is possible to construct new functions g_j from the functions f_j, i.e.,

$$g_j = \sum_i c_{ji} f_i \qquad (7.9)$$

(we shall below discuss how this is done in practice) that behave particularly nicely. Also these new functions transform according to an equation like Eq. (7.6),

$$\hat{O}_i g_j = \sum_k d^i_{jk} g_k, \qquad (7.10)$$

but the important point is that the k summation will run only over a subset of the g_k functions. Furthermore, different subsets of functions will not mix. We have indicated this above for the NH_3 molecule. There we found that the hydrogen $1s$ functions will transform among each other but will not mix with any of the other functions, and that out of the nitrogen $2p$ functions, the $2p_x$ and $2p_y$ functions mix but the $2p_z$ function does not mix with any other function under any symmetry operation.

For the simple one-dimensional case above, the new linear combinations are

$$g_1 = f_+ + f_-$$
$$g_2 = f_+ - f_-. \qquad (7.11)$$

For these we have

$$\hat{I} g_1 = g_1$$
$$\hat{I} g_2 = g_2$$
$$\hat{O} g_1 = g_1$$
$$\hat{O} g_2 = -g_2, \qquad (7.12)$$

whereby the summations in Eq. (7.10) are limited to one single term. Thus, when we compare this equation with Eq. (7.8) we see that in Eq. (7.8) the two functions f_+ and f_- mix (i.e., depending on which symmetry operation we apply on any of the two functions we may get either f_+ or f_-), whereas in Eq. (7.12) the two function g_1 and g_2 do not mix (i.e., starting with, e.g., g_1 no symmetry operation leads to any contribution from g_2). This is an example of the statement above that the summation in Eq. (7.10) runs over only a smaller set of functions compared with that in Eq. (7.6).

Equation (7.10) states that it is possible to construct certain functions that can be split into different subsets. Then any function belongs to only one subset and *independent* of which symmetry operation we consider, the transformation of any of those functions can be written as a linear combination of functions belonging to *only* the same subset. These subsets are called *irreducible representations*, and the functions of any of the subsets is said to belong to the corresponding irreducible representation or to transform according to it.

The constants d^i_{jk} of Eq. (7.10) are unique and do not depend on the initial functions f_i. Independently of which set of functions we start out with we will end up with the same set of constants d^i_{jk} (as long as the set of functions is not so small that we 'miss' some of them). On the other hand, these constants are specific for the group that contains the symmetry operations. For a given irreducible representation the constants d^i_{jk} define a set of square matrices (one for each symmetry operation labelled i) and they have the same size for all i. Different sets of functions that transform according to the same set of matrices transform according to the same irreducible representation. For any group there is only a well-known set of different irreducible representations.

Furthermore, apparently different groups may be equivalent. Thus, the group of symmetry operations for polyvinylchloride of Fig. 7.2 and that of the helical polymer of Fig. 7.4 lead to the same set of matrices and these groups are therefore said to be *isomorphic*, and from a group-theoretical point of view they are absolutely equivalent.

As mentioned above, the constants d^i_{jk} define square matrices whose size is independent of i. The size is determined by how many functions mix for this specific irreducible representation, and the size is accordingly called the *dimension* of the irreducible representation.

A very important point is that *any* function $F(\vec{r})$ (since we here use group theory in discussing symmetries in three-dimensional space, we shall let the function depend on a three-dimensional position coordinate, although the result is not limited to this case but is generally valid within group theory) can be written as a linear combination of the form

$$F(\vec{r}) = \sum_{i=1}^{N_{\text{i.r.}}} \sum_{k=1}^{n_i} C_{ik} G_{ik}(\vec{r}), \qquad (7.13)$$

where $N_{\text{i.r.}}$ is the number of different irreducible representations and n_i the dimension of the ith irreducible representation. Furthermore, for a given irreducible representation i the functions G_{ik} transform according to that representation, although their precise form depends on the function $F(\vec{r})$. In Eq. (7.13) $F(\vec{r})$ is said to be decomposed after the irreducible representation, and in some sense it equals that of expanding any function in a complete set of basis functions.

The set of different groups (i.e., groups that are not isomorphic) is not very large so it can be, and has been, classified. Therefore, all the information about any group is available in the literature (in particular for the groups that are relevant for the symmetry operations in three-dimensional position space), which is one reason why group theory is so useful.

So far we have learnt that it is possible to take any set of original functions (like those of the NH_3 molecule of Fig. 7.6) and construct new linear combinations from those, so that the new linear combinations belong to different irreducible representations. Moreover, we have learnt that any function may be written as a linear combination of functions belonging to different irreducible representations. The overall relevant question is, how this information can be of any use in electronic-structure calculations for a given system once we know its symmetry properties.

First of all, it turns out that it is extremely useful to construct so-called symmetry-adapted functions. These are the new linear combinations that transform according to the irreducible representations constructed from the original functions. We shall illustrate the approach on the NH_3 molecule of Fig. 7.6. To this end we need the concept of character tables. These are defined from matrices containing the constants d^i_{jk}. It turns out that these matrices have characteristic traces (the trace of a square matrix is the sum of the diagonal elements) for the different irreducible representations. By looking at the definition of these constants [Eq. (7.10)] we see that the traces are actually the number of functions that for a given symmetry operation are mapped onto themselves (including prefactors, as, e.g., can occur if the functions change sign due to the operation).

Table 7.1 shows the character table for the group called C_{3v}, which is the name for the group of the symmetry operations for the NH_3 molecule. There are three irreducible representations that by tradition are called A_1, A_2, and E, and we have three different *types* of symmetry operations, i.e., the identity $(\hat{I})$, the two rotations $(\hat{C}_3)$ of either $120°$ or $240°$ about the z axis, and the three reflections $(\hat{\sigma})$ in planes containing the nitrogen atom, one hydrogen atom, and the midpoint between the other two hydrogen atoms.

The character tables are all we need for our purpose, and these are the ones that are found in the literature. It is useful to study the characters for the identity operation first. These give immediately the dimension of the different irreducible representations (since the identity leaves everything unchanged, all functions of a given representation are mapped exactly onto themselves). In Table 7.1 we see therefore that we have only one- and two-dimensional irreducible representations. This means that, although we found that all the three hydrogen $1s$ functions mix under the symmetry operations, it is possible to form new linear combinations that do not mix.

In order to do so, we define so-called symmetry projection operators. For each irreducible representation (each row in Table 7.1) we construct

$$\hat{P}_k = \sum_i \chi_k(\hat{O}_i)\hat{O}_i, \qquad (7.14)$$

Table 7.1 The character table for the C_{3v} group that describes the symmetry operations of the NH_3 molecule of Fig. 7.6

Irreducible representation	$\hat{I}$	$\hat{C}_3$ (two rotations)	$\hat{\sigma}$ (three reflections)
A_1	1	1	1
A_2	1	1	-1
E	2	-1	0

where $\chi_k(\hat{O}_i)$ is the character for the kth irreducible representation and the symmetry operation $\hat{O}_i$. Group theory tells us why this works, but here we shall not go into that but instead simply apply it. The idea is that when applying $\hat{P}_k$ on any function the result is a new function that transforms according to the irreducible representation k (if it is not identically vanishing).

For our one-dimensional example of Eqs. (7.7), (7.8), (7.11), and (7.12), we find very easily (we show the character table for this group in Table 7.2),

$$\hat{P}_g = \hat{I} + \hat{O}$$
$$\hat{P}_u = \hat{I} - \hat{O}. \qquad (7.15)$$

Then,

$$\hat{P}_g f_+ = f_+ + f_-$$
$$\hat{P}_g f_- = f_- + f_+$$
$$\hat{P}_u f_+ = f_+ - f_-$$
$$\hat{P}_u f_- = f_- - f_+. \qquad (7.16)$$

The first two equations show that $f_+ + f_-$ transforms according to the g representation, and the last two equations show that $f_+ - f_-$ transforms according to the u representation. This was actually what we above already used in Eqs. (7.11) and (7.12).

For the NH_3 example we find the following three symmetry projection operators:

$$\hat{P}_{A_1} = \hat{I} + \hat{C}_3(120°) + \hat{C}_3(240°) + \hat{\sigma}_1 + \hat{\sigma}_2 + \hat{\sigma}_3$$
$$\hat{P}_{A_2} = \hat{I} + \hat{C}_3(120°) + \hat{C}_3(240°) - \hat{\sigma}_1 - \hat{\sigma}_2 - \hat{\sigma}_3$$
$$\hat{P}_E = 2\hat{I} - \hat{C}_3(120°) - \hat{C}_3(240°). \qquad (7.17)$$

We have here distinguished between the three reflection operations through the indices 1, 2, and 3, whereas the two rotations are specified through the angle of rotation.

Subsequently, we simply apply these operators on the different functions. For our example let us apply them on the $1s$ functions of the hydrogen atoms. We

Table 7.2 The character table for the group consisting of the identity and the inversion operator

Irreducible representation	$\hat{I}$	$\hat{O}$
g	1	1
u	1	-1

will let $1s(m)$ denote the $1s$ function of hydrogen atom no. m and find then

$$\hat{P}_{A_1} 1s(1) = 1s(1) + 1s(2) + 1s(3) + 1s(1) + 1s(2) + 1s(3). \tag{7.18}$$

Constant prefactors are not important, so we end up with the result that $1s(1) + 1s(2) + 1s(3)$ belongs to the irreducible representation A_1.

Repeating this for the two other $1s$ functions will not bring anything new; i.e., we end up with the same linear combination. Furthermore,

$$\hat{P}_{A_2} 1s(m) = 0 \tag{7.19}$$

independently of m, so that the hydrogen $1s$ functions do not lead to functions of the A_2 representation.

Finally,

$$\hat{P}_E 1s(1) = 2 \cdot 1s(1) - 1s(2) - 1s(3)$$

$$\hat{P}_E 1s(2) = 2 \cdot 1s(2) - 1s(1) - 1s(3)$$

$$\hat{P}_E 1s(3) = 2 \cdot 1s(3) - 1s(1) - 1s(2). \tag{7.20}$$

Only two of those are linearly independent (e.g., the third expression can be obtained as the negative of the sum of the first two), and we may thus choose any two as defining a pair of functions (since the E representation is two-dimensional, we need two functions) that transform according to the E irreducible representation.

In total, we have now constructed three new linear combinations of the original three hydrogen $1s$ functions. The only difference is that these new functions have well-defined symmetry properties, i.e., are symmetry-adapted.

The procedure may be repeated for the functions on the nitrogen atom. The result is that each of the $1s$, $2s$, and $2p_z$ functions belong to the A_1 irreducible representation. Furthermore, the two functions p_x and p_y form another pair that transforms according to the E irreducible representation.

As another example we shall consider polyvinylchloride of Fig. 7.2 or the helical polymer of Fig. 7.4. As mentioned above, the two groups of symmetry operators for the two systems are isomorphic, and therefore the two systems are, from a group-theoretic point of view, identical. We shall just consider one function per unit cell, i.e., an s function on one of the atoms. An arbitrary symmetry operation (e.g., a translation of n unit cells for polyvinylchloride) will map the function from unit m onto that of unit $m + n$.

Table 7.3 shows the character table for the relevant group. Here, we have an infinite set of symmetry operations (corresponding to all the different n that describe the translations for polyvinylchloride) and, equivalently, we have an infinite set of (one-dimensional) irreducible representations. Nevertheless, we can collect all the information in a simple table as shown in Table 7.3. The parameter k can take all values in the interval

$$0 \leq k < 2\pi. \tag{7.21}$$

Table 7.3 The character table for the group that describes the symmetry operations for polyvinylchloride of Fig. 7.2 or the helical polymer of Fig. 7.4

Irreducible representation	Operation
k	e^{ikn}

By denoting the above-mentioned s function of the mth unit $\phi_s(m)$, and the operator that translates the system n units $\hat{O}(n)$, we can construct the symmetry-adapted functions for the irreducible representation k by applying the operator of Eq. (7.14) on (for instance) the function of the 0th unit,

$$\hat{P}_k \phi_s(0) = \sum_n e^{ikn} \hat{O}(n)\phi_s(0)$$

$$= \sum_n e^{ikn} \phi_s(n). \tag{7.22}$$

Finally, we shall consider polycarbonitrile of Fig. 7.3. For this we have, in addition to the translation operations, also the combined symmetry operations consisting of a translation and a reflection in the plane of all nuclei. We shall here consider two different types of functions per unit of which one is an s function as above, whereas the other is a p function perpendicular to the plane of the nuclei. The two functions of the mth unit will be denoted $\phi_s(m)$ and $\phi_p(m)$, respectively.

Table 7.4 shows the character table for the relevant group, which is very similar to that of Table 7.3, except that it is doubled. Equivalent to Eq. (7.22) we consider

$$\hat{P}_{k+}\phi_s(0) = \sum_n e^{ikn} \hat{O}(n)\phi_s(0) + \sum_n e^{ikn} \hat{O}(n)\phi_s(0) = 2\sum_n e^{ikn}\phi_s(n)$$

$$\hat{P}_{k-}\phi_s(0) = \sum_n e^{ikn} \hat{O}(n)\phi_s(0) - \sum_n e^{ikn} \hat{O}(n)\phi_s(0) = 0$$

$$\hat{P}_{k+}\phi_p(0) = \sum_n e^{ikn} \hat{O}(n)\phi_p(0) - \sum_n e^{ikn} \hat{O}(n)\phi_p(0) = 0$$

$$\hat{P}_{k-}\phi_p(0) = \sum_n e^{ikn} \hat{O}(n)\phi_p(0) + \sum_n e^{ikn} \hat{O}(n)\phi_p(0)$$

$$= 2\sum_n e^{ikn}\phi_p(n), \tag{7.23}$$

where we have used that the s functions are symmetric and the p functions are antisymmetric with respect to reflection in the plane of the nuclei. Moreover, the first sum in each expression above is the result of applying the translations, whereas the second sum is that of the translation + reflection operations.

Table 7.4   The character table for the group that describes the symmetry operations for polycarbonitrile of Fig. 7.3

Irreducible representation	Translation	Translation + reflection
$k+$	e^{ikn}	e^{ikn}
$k-$	e^{ikn}	$-e^{ikn}$

Equation (7.22) shows that the function

$$\sum_n e^{ikn} \phi_s(n) \tag{7.24}$$

belongs to the $k+$ irreducible representation, whereas the function

$$\sum_n e^{ikn} \phi_p(n) \tag{7.25}$$

belongs to the $k-$ representation.

The importance of the irreducible representations becomes clear when we calculate matrix elements. As the simplest ones we calculate first overlap matrix elements between two functions ϕ_1 and ϕ_2,

$$\langle \phi_1 | \phi_2 \rangle. \tag{7.26}$$

We will now assume that these functions are symmetry-adapted, i.e., ϕ_1^* and ϕ_2 belong to the irreducible representations R_1 and R_2, respectively. The product of the two functions will then transform as the product of these two representations, which formally is written as

$$R_1 \otimes R_2. \tag{7.27}$$

This product is called the direct product. One may now decompose this after the irreducible representations, i.e., one may set up a multiplication table like that in Table 7.5 for the group C_{3v} whose character table was shown in Table 7.1, and that was used in discussing the NH_3 molecule.

Table 7.5   Table of direct products for the C_{3v} group. The first row and the first column give the irreducible representations that are multiplied, and the entries give the results as decomposed into different irreducible representations. For $E \otimes E$ more irreducible representations are obtained as represented by the direct-sum symbols

	A_1	A_2	E
A_1	A_1	A_2	E
A_2	A_2	A_1	E
E	E	E	$A_1 \oplus A_2 \oplus E$

For any group there will be exactly one irreducible representation (which will be one-dimensional) for which all characters equal 1. This corresponds to functions that are invariant with respect to the symmetry operations. In the example of Tables 7.1 and 7.5 this is the A_1 representation. Only when integrating functions belonging to this over all space can the result be different from zero. This means that in Table 7.5 only those products of two functions that give rise to this representation (plus, eventually, also other representations) can give non-vanishing matrix elements. In Table 7.5 we see that this is only the case when the two representations are identical. Since, for this group, ϕ^* and ϕ transform according to the same irreducible transformation, this means that the two functions have to correspond to the same irreducible transformation. This result is in fact general [see, e.g., Table 7.6 for the group describing the symmetry operations of polycarbonitrile; here, however, it has to be remembered that when ϕ transforms according to $k+$ or $k-$, then ϕ^* transforms according to $(-k)+$ or $(-k)-$], so that we obtain the very important result

$$\langle \phi_1 | \phi_2 \rangle = 0 \quad \text{if} \quad R_1 \neq R_2. \tag{7.28}$$

Finally, we consider matrix elements of the type

$$\langle \phi_1 | \hat{F} | \phi_2 \rangle, \tag{7.29}$$

where once again we assume that ϕ_1 and ϕ_2 transform according to the irreducible representations R_1 and R_2, respectively.

Just as for the function $F(\vec{r})$ the operator $\hat{F}$ may be decomposed according to the irreducible representations [cf. Eq. (7.11)],

$$\hat{F} = \sum_{i=1}^{N_{i.r.}} \sum_{k=1}^{n_i} C_{ik} \hat{G}_{ik}. \tag{7.30}$$

This means that the matrix element of Eq. (7.29) can be written as a sum of matrix elements, each corresponding to the product of three irreducible representations,

$$R_1 \otimes R_F \otimes R_2, \tag{7.31}$$

where R_F is one of the irreducible representations in Eq. (7.30) for which at least one of the coefficients C_{ik} is non-zero.

Also the direct product in Eq. (7.31) can be decomposed, and once again we find that the corresponding matrix elements can only then be non-vanishing when

Table 7.6 Table of direct products for the group of symmetry operations for polycarbonitrile

	k_1+	k_1-
k_2+	$(k_1 + k_2)+$	$(k_1 + k_2)-$
k_2-	$(k_1 + k_2)-$	$(k_1 + k_2)+$

this direct product (when decomposed) contains the trivial irreducible representation where all characters equal 1.

A very important case is that where the operator $\hat{F}$ equals the Hamilton operator. In that case it turns out that the sum in Eq. (7.30) reduces to one term, i.e., $\hat{H}$ transforms only as the trivial irreducible representation. Since the direct product of this representation and any other representation R always equals R we end up with the important result

$$\langle \phi_1 | \hat{H} | \phi_2 \rangle = 0 \quad \text{if} \quad R_1 \neq R_2. \tag{7.32}$$

Combined with Eq. (7.28) this means that Hamilton and overlap matrix elements are only non-vanishing between functions that belong to the same irreducible representation. This is an advanced formulation of rules such as those stating that π and σ electrons do not interact, etc. Instead of presenting further mathematical details about group theory we shall discuss a couple of examples., In Chapter 19 we shall return to the application of group theory when we consider infinite, periodic systems.

7.3   AN EXAMPLE REVISITED

We shall discuss a very simple example in order to illustrate the concepts of group theory and their applications to electronic-structure calculations. We consider a single particle in one dimension moving in the potential given by

$$V(x) = (x^2 - 1)^2. \tag{7.33}$$

where A is some constant. This potential is a symmetric double-well potential with two minima at $x = \pm 1$ and a barrier at $x = 0$, and was discussed in Section 5.6.

The Schödinger equation becomes

$$\left[-\frac{\hbar^2}{2m} \frac{d^2}{dx^2} + V(x) \right] \psi(x) = E\psi(x). \tag{7.34}$$

The system possesses inversion symmetry about $x = 0$, and it is clearly seen (by simple insertion) that replacing

$$x \rightarrow -x \tag{7.35}$$

in Eq. (7.34), the Schrödinger equation is unchanged, i.e., the Hamilton operator is invariant under the symmetry operation, or, alternatively stated, the Hamilton operator transforms according to the identity representation.

Applying the variational principle we will assume that ψ can be written as a linear combination of four basis functions

$$\psi(x) = \sum_{i=1}^{4} c_i \chi_i(x) \tag{7.36}$$

with

$$\chi_1(x) = e^{-(x-1)^2}$$

$$\chi_2(x) = e^{-(x+1)^2}$$

$$\chi_3(x) = (x-1)e^{-(x-1)^2}$$

$$\chi_4(x) = (x+1)e^{-(x+1)^2} \tag{7.37}$$

χ_1 and χ_3 are functions centred around the minimum at $x = 1$, whereas χ_2 and χ_4 are centred around that at $x = -1$.

In order to calculate the coefficients c_i we would need all the matrix elements $\langle \chi_i | \chi_j \rangle$ and $\langle \chi_i | \hat{H} | \chi_j \rangle$. In principle, none of these vanish, and the calculation of the coefficients will require solving the 4×4 generalized eigenvalue problem of Eq. (5.6.5).

However, it is useful to make use of the symmetry properties of the system. Using a strategy similar to that leading to Eqs. (7.11) and (7.12) we therefore first construct the symmetry-adapted functions

$$\chi_{g1} = \chi_1 + \chi_2$$

$$\chi_{g2} = \chi_3 - \chi_4$$

$$\chi_{u1} = \chi_1 - \chi_2$$

$$\chi_{u2} = \chi_3 + \chi_3. \tag{7.38}$$

The two functions χ_{g1} and χ_{g2} are even with respect to the inversion about $x = 0$, whereas the other two are antisymmetric. That is, the first two transform according to the same (g) representation and the other two transform according to the u representation of Table 7.2. This means that with this new basis set [of Eq. (7.38)], the overlap matrix elements $\langle \chi_{r_1 i_1} | \chi_{r_2 i_2} \rangle$ and $\langle \chi_{r_1 i_1} | \hat{H} | \chi_{r_2 i_2} \rangle$ vanish unless $r_1 = r_2$. Thus, the 4×4 generalized eigenvalue problem is reduced to two 2×2 problems, which certainly represents a simplification, i.e.,

$$\begin{pmatrix} \langle \chi_{g1} | \hat{H} | \chi_{g1} \rangle & \langle \chi_{g1} | \hat{H} | \chi_{g2} \rangle \\ \langle \chi_{g2} | \hat{H} | \chi_{g1} \rangle & \langle \chi_{g2} | \hat{H} | \chi_{g2} \rangle \end{pmatrix} \cdot \begin{pmatrix} c_{g1} \\ c_{g2} \end{pmatrix} = E \cdot \begin{pmatrix} \langle \chi_{g1} | \chi_{g1} \rangle & \langle \chi_{g1} | \chi_{g2} \rangle \\ \langle \chi_{g2} | \chi_{g1} \rangle & \langle \chi_{g2} | \chi_{g2} \rangle \end{pmatrix} \cdot \begin{pmatrix} c_{g1} \\ c_{g2} \end{pmatrix} \tag{7.39}$$

and

$$\begin{pmatrix} \langle \chi_{u1} | \hat{H} | \chi_{u1} \rangle & \langle \chi_{u1} | \hat{H} | \chi_{u2} \rangle \\ \langle \chi_{u2} | \hat{H} | \chi_{u1} \rangle & \langle \chi_{u2} | \hat{H} | \chi_{u2} \rangle \end{pmatrix} \cdot \begin{pmatrix} c_{u1} \\ c_{u2} \end{pmatrix} = E \cdot \begin{pmatrix} \langle \chi_{u1} | \chi_{u1} \rangle & \langle \chi_{u1} | \chi_{u2} \rangle \\ \langle \chi_{u2} | \chi_{u1} \rangle & \langle \chi_{u2} | \chi_{u2} \rangle \end{pmatrix} \cdot \begin{pmatrix} c_{u1} \\ c_{u2} \end{pmatrix}. \tag{7.40}$$

Let us subsequently assume that for some (unimportant) reasons we need to calculate the matrix elements between the basis functions and the operator

$$\hat{O} = x. \tag{7.41}$$

This operator is antisymmetric with respect to the inversion, and transforms according to the u representation. With the original basis set of Eq. (7.37) we would need to calculate all 4×4 matrix elements, but using the symmetry properties and the symmetry-adapted basis functions of Eq. (7.41), we observe that $\langle \chi_{r_1 i_1} | \hat{O} | \chi_{r_2 i_2} \rangle$ vanishes unless the direct product of the three representations (i.e., $r_1 \otimes u \otimes r_2$) contains the g representation which, for the present example, is the case only when r_1 and r_2 are different representations. We need accordingly only to calculate 8 matrix elements, which is a simplification.

7.4   A WORKED EXAMPLE

We consider the water molecule of Fig. 7.7 where we in addition show a coordinate system that will be useful.

We shall assume that an electron occupies an orbital that is described with the help of $1s$ functions on the hydrogen atoms and $1s$, $2s$, $2p_x$, $2p_y$, and $2p_z$ functions on the oxygen atom. Assuming that we have only one function of each type this gives the following seven functions:

$$\chi_1 = 1s(\text{H1})$$

$$\chi_2 = 1s(\text{H2})$$

$$\chi_3 = 1s(\text{O})$$

$$\chi_4 = 2s(\text{O})$$

$$\chi_5 = 2p_x(\text{O})$$

$$\chi_6 = 2p_y(\text{O})$$

$$\chi_7 = 2p_z(\text{O}). \tag{7.42}$$

The orbital of interest can be calculated by solving the 7×7 secular equation

$$\underline{\underline{H}} \cdot \underline{c} = E \cdot \underline{\underline{O}} \cdot \underline{c}, \tag{7.43}$$

where $\underline{\underline{H}}$ and $\underline{\underline{O}}$ contains the matrix elements for the seven functions above. Their precise form are not important for the discussion here.

It is, however, useful to make use of the symmetry properties of the system. The system has four symmetry operations, i.e., the identity $\hat{I}$, the rotation about

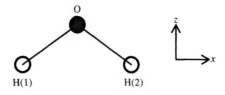

Figure 7.7   The water molecule H_2O. In order to clarify the discussion the two hydrogen atoms have been given different numbers and a coordinate system has been introduced

the H$-$O$-$H angle bisector $\hat{C}_2$, the reflection in the (x, z) plane $\hat{\sigma}_{xz}$, and that in the (y, z) plane $\hat{\sigma}_{yz}$. Any of those will mix the seven functions of Eq. (7.42), and we may describe this through

$$\tilde{\chi} = \underline{\underline{T}}_{\hat{O}} \cdot \underline{\chi}, \tag{7.44}$$

where the matrices $\underline{\underline{T}}_{\hat{O}}$ are

$$\underline{\underline{T}}_{\hat{i}} = \begin{pmatrix} 1 & 0 & 0 & 0 & 0 & 0 & 0 \\ 0 & 1 & 0 & 0 & 0 & 0 & 0 \\ 0 & 0 & 1 & 0 & 0 & 0 & 0 \\ 0 & 0 & 0 & 1 & 0 & 0 & 0 \\ 0 & 0 & 0 & 0 & 1 & 0 & 0 \\ 0 & 0 & 0 & 0 & 0 & 1 & 0 \\ 0 & 0 & 0 & 0 & 0 & 0 & 1 \end{pmatrix}$$

$$\underline{\underline{T}}_{\hat{C}_2} = \begin{pmatrix} 0 & 1 & 0 & 0 & 0 & 0 & 0 \\ 1 & 0 & 0 & 0 & 0 & 0 & 0 \\ 0 & 0 & 1 & 0 & 0 & 0 & 0 \\ 0 & 0 & 0 & 1 & 0 & 0 & 0 \\ 0 & 0 & 0 & 0 & -1 & 0 & 0 \\ 0 & 0 & 0 & 0 & 0 & -1 & 0 \\ 0 & 0 & 0 & 0 & 0 & 0 & 1 \end{pmatrix}$$

$$\underline{\underline{T}}_{\hat{\sigma}_{xz}} = \begin{pmatrix} 1 & 0 & 0 & 0 & 0 & 0 & 0 \\ 0 & 1 & 0 & 0 & 0 & 0 & 0 \\ 0 & 0 & 1 & 0 & 0 & 0 & 0 \\ 0 & 0 & 0 & 1 & 0 & 0 & 0 \\ 0 & 0 & 0 & 0 & 1 & 0 & 0 \\ 0 & 0 & 0 & 0 & 0 & -1 & 0 \\ 0 & 0 & 0 & 0 & 0 & 0 & 1 \end{pmatrix}$$

$$\underline{\underline{T}}_{\hat{\sigma}_{yz}} = \begin{pmatrix} 0 & 1 & 0 & 0 & 0 & 0 & 0 \\ 1 & 0 & 0 & 0 & 0 & 0 & 0 \\ 0 & 0 & 1 & 0 & 0 & 0 & 0 \\ 0 & 0 & 0 & 1 & 0 & 0 & 0 \\ 0 & 0 & 0 & 0 & -1 & 0 & 0 \\ 0 & 0 & 0 & 0 & 0 & 1 & 0 \\ 0 & 0 & 0 & 0 & 0 & 0 & 1 \end{pmatrix}. \tag{7.45}$$

Eq. (7.44) with Eq. (7.45) corresponds to Eq. (7.6).

Table 7.7 shows the character table for the group of the symmetry operations for the H_2O molecule. Subsequently we can construct symmetry projection operators as in Eq. (7.14) and applying these on the various basis functions leads to the following symmetry-adapted basis functions:

- four functions that belong to the A_1 representation: $\frac{1}{\sqrt{2}}(\chi_1 + \chi_2)$; χ_3; χ_4; χ_7;
- no function that belongs to the A_2 representation;

Table 7.7 The character table for the C_{2v} group, which describes the symmetry operations of the H_2O molecule of Fig. 7.7

Irreducible representation	$\hat{I}$	$\hat{C}_2$	$\hat{\sigma}_{xz}$	$\hat{\sigma}_{yz}$
A_1	1	1	1	1
A_2	1	1	-1	-1
B_1	1	-1	-1	1
B_2	1	-1	1	-1

- one function that belongs to the B_1 representation: χ_6;
- two functions that belong to the B_2 representation: $\frac{1}{\sqrt{2}}(\chi_1 - \chi_2)$; χ_5.

The Hamilton operator will transform according to the identity operation, and accordingly we will have Hamilton and overlap matrix elements only between functions that belong to the same representation. This means that the 7×7 secular equation can be reduced to one 4×4, one 2×2, and one 1×1 secular equation.

In Table 7.8 we show the direct products for the different irreducible representations of the C_{2v} group. This can be used when we now add an electrostatic field to the system. We describe this through the extra term

$$\hat{H}_1 = \vec{E} \cdot \vec{r} \qquad (7.46)$$

to the Hamilton operator. Here, $\vec{E}$ is the field vector. We shall distinguish between three cases.

For the first cases we have that $\vec{E}$ is parallel to the z axis. Then, $\hat{H}_1$ transforms according to the A_1 representation. Table 7.9 gives the irreducible representations of the direct products $R_1 \otimes A_1 \otimes R_2$.

Here we see that the A_1 representation occurs only along the diagonal, i.e., only for direct products between identical representations. Therefore, for the field along the z axis it is not able to change the symmetry properties of the system.

When $\vec{E}$ is parallel to the x axis, $\hat{H}_1$ transforms according to the B_2 representation. Table 7.10 gives the direct products in this case.

Table 7.8 Table of direct products for the C_{2v} group. The first row and the first column give the irreducible representations that are multiplied, and the entries give the results as decomposed into different irreducible representations

	A_1	A_2	B_1	B_2
A_1	A_1	A_2	B_1	B_2
A_2	A_2	A_1	B_2	B_1
B_1	B_1	B_2	A_1	A_2
B_2	B_2	B_1	A_2	A_1

Table 7.9 Table of direct products $R_1 \otimes A_1 \otimes R_2$ for the C_{2v} group. The first row and the first column give the irreducible representations that are multiplied, and the entries give the results as decomposed into different irreducible representations

	A_1	A_2	B_1	B_2
A_1	A_1	A_2	B_1	B_2
A_2	A_2	A_1	B_2	B_1
B_1	B_1	B_2	A_1	A_2
B_2	B_2	B_1	A_2	A_1

Table 7.10 Table of direct products $R_1 \otimes B_2 \otimes R_2$ for the C_{2v} group. The first row and the first column give the irreducible representations that are multiplied, and the entries give the results as decomposed into different irreducible representations

	A_1	A_2	B_1	B_2
A_1	B_2	B_1	A_2	A_1
A_2	B_1	B_2	A_1	A_2
B_1	A_2	A_1	B_2	B_1
B_2	A_1	A_2	B_1	B_2

Table 7.11 Table of direct products $R_1 \otimes B_1 \otimes R_2$ for the C_{2v} group. The first row and the first column give the irreducible representations that are multiplied, and the entries give the results as decomposed into different irreducible representations

	A_1	A_2	B_1	B_2
A_1	B_1	B_2	A_1	A_2
A_2	B_2	B_1	A_2	A_1
B_1	A_1	A_2	B_1	B_2
B_2	A_2	A_1	B_2	B_1

In this case, the A_1 representation occurs for products between different representations, meaning that these presentations will mix when the field is included. That is the A_1 and B_2 representations will mix, as will the B_1 and A_2 representations. Thus, the secular equations will in that case be a 6×6 one due to the A_1 and B_2 representations and a 1×1 one due to the A_2 and B_1 representations.

As Table 7.11 shows, also when the field is along the y direction (leading to that $\hat{H}_1$ transforms according to the B_1 representation), different representations will mix, i.e., the A_1 and B_1 ones will mix, as will the A_2 and B_2 ones.

Finally, for an arbitrary direction of the field, $\hat{H}_1$ will transform according to $A_1 \oplus B_1 \oplus B_2$, and the relevant table will be as a sum of those of Tables 7.9, 7.10, and 7.11 for example,

$$A_2 \otimes (A_1 \oplus B_1 \oplus B_2) \otimes B_1 = B_2 \oplus A_1 \oplus A_2. \tag{7.47}$$

Part II
BASIC METHODS

In this second part we shall present in detail the fundamentals behind the various electronic-structure methods that are currently applied. These are on the one side the so-called wavefunction-based methods (i.e., methods developed from the Hartree–Fock approximation) and on the other side the so-called density-based methods (i.e., methods developed from the density-functional theory of Hohenberg, Kohn, and Sham). The presentation is detailed since the methods form the basis for the subsequent parts.

8 The Schrödinger Equation and the Born–Oppenheimer Approximation

8.1 THE SCHRÖDINGER EQUATION

We have now all the mathematical tools that we need for seeking approximate solutions to the time-independent Schrödinger equation,

$$\hat{H}\Psi = E\Psi. \tag{8.1}$$

The variational principle allows us to approximate Ψ and gives some hints of how to evaluate the quality of a given approximate Ψ and how to improve a given approximation. But any practical implementation of this requires that we know the precise form of the Hamilton operator $\hat{H}$, which so far has not been presented in the general case (although we at various places have indirectly assumed that $\hat{H}$ was known!). This will be done in the present subsection.

We consider the non-relativistic case. This means, e.g., that spin dependences are largely neglected except that orbitals for fermions (like electrons) can be occupied by two particles — one with α spin and one with β spin. However, in some cases we will allow the position dependences of the orbitals for different spins to be different, i.e., allow for a spin polarization. Furthermore, spin–orbit interactions, etc., are not included. These will be discussed later.

Within the position-space representation (as we have used throughout this text) one obtains the non-relativistic Hamilton operator by writing down the *classical* total (kinetic plus potential) energy for the system of interest with the help of position and momentum variables. Subsequently, any momentum $\vec{p}$ for a particle of mass m is replaced by the operator $(\hbar/i)\vec{\nabla}$.

In the present case we have M nuclei. Each of those (labelled k) is assumed placed at the position $\vec{R}_k$ and to have the charge $Z_k e$. Furthermore, denoting its mass M_k and momentum $\vec{P}_k$, its kinetic energy becomes $(P_k^2/2M_k)$.

Equivalently, we have N electrons. They have the same charge, $-e$, and mass, m_e. The position of the ith electron is denoted by $\vec{r}_i$ and its momentum by $\vec{p}_i$, so that its kinetic energy is $(p_i^2/2m_e)$.

The total kinetic energy is accordingly

$$E_{\text{kin}} = \sum_{k=1}^{M} \frac{P_k^2}{2M_k} + \sum_{i=1}^{N} \frac{p_i^2}{2m_e}. \tag{8.2}$$

The classical potential energy of this system is simply the electrostatic energy due to the interactions between the charges. It is well known that for a set of charges q_n, $n = 1, \ldots, N_q$ placed at $\vec{s}_n$ this energy becomes the sum over all pairs

$$\sum_{n_1=1}^{N_q-1} \sum_{n_2=n_1+1}^{N_q} \frac{1}{4\pi\varepsilon_0} \frac{q_{n_1} q_{n_2}}{|\vec{s}_{n_1} - \vec{s}_{n_2}|} = \frac{1}{2} \sum_{n_1 \neq n_2=1}^{N_q} \frac{1}{4\pi\varepsilon_0} \frac{q_{n_1} q_{n_2}}{|\vec{s}_{n_1} - \vec{s}_{n_2}|}. \tag{8.3}$$

The two expressions are identical and differ only in the way we take care of how to add all contributions from the *different* pairs. In the second expression we therefore include the factor $\frac{1}{2}$ since the sum includes all pairs twice. Furthermore, ε_0 is the vacuum dielectric constant.

For the system of our interest we shall explicitly include two types of particles, nuclei and electrons. Then

$$E_{\text{pot}} = \frac{1}{2} \sum_{k_1 \neq k_2=1}^{M} \frac{1}{4\pi\varepsilon_0} \frac{Z_{k_1} Z_{k_2} e^2}{|\vec{R}_{k_1} - \vec{R}_{k_2}|} + \frac{1}{2} \sum_{i_1 \neq i_2=1}^{N} \frac{1}{4\pi\varepsilon_0} \frac{e^2}{|\vec{r}_{i_1} - \vec{r}_{i_2}|}$$
$$- \sum_{k=1}^{M} \sum_{i=1}^{N} \frac{1}{4\pi\varepsilon_0} \frac{Z_k e^2}{|\vec{R}_k - \vec{r}_i|}. \tag{8.4}$$

The first term on the right-hand side is the nucleus–nucleus interactions, the second one is the electron–electron interactions, and the third one is the nucleus–electron interactions.

In the total-energy expression

$$E_{\text{tot}} = E_{\text{kin}} + E_{\text{pot}} \tag{8.5}$$

we insert the gradient operators for the momenta and obtain then the Hamilton operator,

$$\hat{H} = -\sum_{k=1}^{M} \frac{\hbar^2}{2M_k} \nabla_{\vec{R}_k}^2 - \sum_{i=1}^{N} \frac{\hbar^2}{2m_e} \nabla_{\vec{r}_i}^2 + \frac{1}{2} \sum_{k_1 \neq k_2=1}^{M} \frac{1}{4\pi\varepsilon_0} \frac{Z_{k_1} Z_{n_2} e^2}{|\vec{R}_{k_1} - \vec{R}_{k_2}|}$$
$$+ \frac{1}{2} \sum_{i_1 \neq i_2=1}^{N} \frac{1}{4\pi\varepsilon_0} \frac{e^2}{|\vec{r}_{i_1} - \vec{r}_{i_2}|} - \sum_{k=1}^{M} \sum_{i=1}^{N} \frac{1}{4\pi\varepsilon_0} \frac{Z_k e^2}{|\vec{R}_k - \vec{r}_i|}. \tag{8.6}$$

Here, we have introduced a short-hand notation for the gradient-operators, i.e. (in Cartesian coordinates),

$$\vec{\nabla}_{\vec{r}_i} = \left(\frac{\partial}{\partial x_i}, \frac{\partial}{\partial y_i}, \frac{\partial}{\partial z_i} \right), \tag{8.7}$$

where

$$\vec{r}_i = (x_i, y_i, z_i) \tag{8.8}$$

is the position of the ith electron.

Similarly,

$$\vec{\nabla}_{\vec{R}_k} = \left(\frac{\partial}{\partial X_k}, \frac{\partial}{\partial Y_k}, \frac{\partial}{\partial Z_k} \right), \tag{8.9}$$

where

$$\vec{R}_k = (X_k, Y_k, Z_k) \tag{8.10}$$

is the position of the kth nucleus.

The total Hamilton operator of Eq. (8.6) consists of five terms: the kinetic-energy operator for the nucleus, the kinetic-energy operator for the electrons, the potential-energy operator for nucleus–nucleus interactions, the potential-energy operator for electron–electron interactions, and the potential-energy operator for nucleus–electron interactions. We write accordingly

$$\hat{H} = \hat{H}_{k,n} + \hat{H}_{k,e} + \hat{H}_{p,n-n} + \hat{H}_{p,e-e} + \hat{H}_{p,n-e} \tag{8.11}$$

with an obvious notation for the five terms.

8.2 THE BORN–OPPENHEIMER APPROXIMATION

The solution Ψ to the time-independent Schrödinger equation

$$\hat{H}\Psi = E\Psi \tag{8.12}$$

depends on the spin and position coordinates of all electrons, i.e., on

$$(\vec{r}_1, \sigma_1, \vec{r}_2, \sigma_2, \ldots, \vec{r}_N, \sigma_N) \equiv (\vec{x}_1, \vec{x}_2, \ldots, \vec{x}_N) \equiv \vec{x}, \tag{8.13}$$

and on the spin and position coordinates of all nuclei,

$$(\vec{R}_1, \Sigma_1, \vec{R}_2, \Sigma_2, \ldots, \vec{R}_M, \Sigma_M) \equiv (\vec{X}_1, \vec{X}_2, \ldots, \vec{X}_M) \equiv \vec{X}, \tag{8.14}$$

when introducing the short-hand notations from earlier.

Accordingly

$$\Psi = \Psi(\vec{X}, \vec{x}) \tag{8.15}$$

and the Schrödinger equation (8.12) takes the form

$$\hat{H}\Psi = (\hat{H}_{k,n} + \hat{H}_{k,e} + \hat{H}_{p,n-n} + \hat{H}_{p,e-e} + \hat{H}_{p,n-e})\Psi(\vec{X}, \vec{x}) = E \cdot \Psi(\vec{X}, \vec{x}). \tag{8.16}$$

We shall group the Hamilton operators into two parts so that Eq. (8.16) becomes

$$[(\hat{H}_{k,n} + \hat{H}_{p,n-n}) + (\hat{H}_{k,e} + \hat{H}_{p,e-e} + \hat{H}_{p,n-e})]\Psi(\vec{X}, \vec{x}) = E \cdot \Psi(\vec{X}, \vec{x}). \quad (8.17)$$

The first part depends solely on the nuclear coordinates whereas the second part also depends on the electronic ones.

We shall now introduce the Born–Oppenheimer approximation. The physical idea behind this is that the electrons move much faster than the nuclei, so that for a given set of positions of the nuclei the electrons adjust their positions 'immediately' to these before the nuclei move. One may apply two simple physical principles in order to justify this approximation. First, the uncertainty principle gives

$$\Delta x \cdot \Delta p = \Delta x \cdot \Delta(mv) = m\Delta x \cdot \Delta v \geq \hbar$$

$$\Delta X \cdot \Delta P = \Delta X \cdot \Delta(MV) = M\Delta X \cdot \Delta V \geq \hbar \quad (8.18)$$

where the first equation is for an electron and the second one for a nucleus. The masses of the nuclei are several thousand (1840 for the lightest nucleus, the proton) times larger than that of an electron. Hence, assuming that the inequalities in Eq. (8.18) can be approximately replaced by equalities, we obtain

$$\frac{\Delta X \Delta V}{\Delta x \Delta v} \simeq \frac{m}{M} \ll 1. \quad (8.19)$$

Furthermore, from classical statistical mechanics it is known that the average kinetic energy equals $\frac{3}{2}k_B T$, with k_B being the Boltzmann constant and T the temperature. That is,

$$\left\langle \tfrac{1}{2}MV^2 \right\rangle = \left\langle \tfrac{1}{2}mv^2 \right\rangle = \tfrac{3}{2}k_B T. \quad (8.20)$$

[Notice, however, that the electrons and nuclei are highly non-classical particles, so that Eq. (8.20) can be taken only as a crude estimate.] In order to get an order-of-magnitude estimate, we simply set

$$\Delta V = \langle V^2 \rangle^{1/2}$$

$$\Delta v = \langle v^2 \rangle^{1/2} \quad (8.21)$$

and obtain then, from Eqs. (8.19) and (8.20),

$$\frac{\Delta V}{\Delta v} \simeq \sqrt{\frac{m}{M}} \ll 1$$

$$\frac{\Delta X}{\Delta x} \simeq \sqrt{\frac{m}{M}} \ll 1. \quad (8.22)$$

That is, the nuclei move much more slowly than the electrons and are much more localized in space.

In order to transform these qualitative considerations over into the Schrödinger equation, we apply the factorization technique of Chapter 4. We write the solution to Eq. (8.17) as a product of two functions of which one is a wavefunction for only the nuclear coordinates and the other is a wavefunction that depends directly on the electronic coordinates and parametrically on the nuclear coordinates (this means that for different positions of the nuclei, the electronic wavefunction changes, but this will then also be the only dependence of this function on the nuclear coordinates). Hence, we write

$$\Psi(\vec{X}, \vec{x}) = \Psi_n(\vec{X}) \cdot \Psi_e(\vec{X}; \vec{x}). \tag{8.23}$$

Inserting this into Eq. (8.17) gives

$$\left[(\hat{H}_{k,n} + \hat{H}_{p,n-n}) + (\hat{H}_{k,e} + \hat{H}_{p,e-e} + \hat{H}_{p,n-e})\right] \Psi_n(\vec{X}) \cdot \Psi_e(\vec{X}; \vec{x})$$
$$= (\hat{H}_{k,n} + \hat{H}_{p,n-n})\Psi_n(\vec{X}) \cdot \Psi_e(\vec{X}; \vec{x})$$
$$+ (\hat{H}_{k,e} + \hat{H}_{p,e-e} + \hat{H}_{p,n-e})\Psi_n(\vec{X}) \cdot \Psi_e(\vec{X}; \vec{x})$$
$$\simeq \Psi_e(\vec{X}; \vec{x})(\hat{H}_{k,n} + \hat{H}_{p,n-n})\Psi_n(\vec{X})$$
$$+ \Psi_n(\vec{X})(\hat{H}_{k,e} + \hat{H}_{p,e-e} + \hat{H}_{p,n-e})\Psi_e(\vec{X}; \vec{x})$$
$$\equiv E \cdot \Psi_n(\vec{X}) \cdot \Psi_e(\vec{X}; \vec{x}). \tag{8.24}$$

In deriving this we have assumed that terms of the form

$$-\frac{\hbar^2}{2M_k} \nabla^2_{\vec{R}_k} \psi_e(\vec{X}; \vec{x}) \tag{8.25}$$

are small, i.e., that that part of the kinetic-energy of the nuclei that originates from the electronic part of the wavefunction can be neglected. The arguments for doing so are essentially those above: the nuclei move much more slowly than the electrons. There are, however, cases where this is not justified, and where special physical effects may occur as a consequence. These effects include structural distortions like the Jahn–Teller and Peierls' distortion as well as the occurrence of charge- or spin-density waves and of superconductivity.

But let us now assume that the approximations of Eq. (8.24) are justified. The approximation of Eq. (8.23) is an example of the factorization discussed in Chapter 4. As there, we shall proceed by dividing Eq. (8.24) by the function in Eq. (8.23), which gives

$$\frac{(\hat{H}_{k,n} + \hat{H}_{p,n-n})\Psi_n(\vec{X})}{\Psi_n(\vec{X})} + \frac{(\hat{H}_{k,e} + \hat{H}_{p,e-e} + \hat{H}_{p,n-e})\Psi_e(\vec{X}; \vec{x})}{\Psi_e(\vec{X}; \vec{x})} = E \tag{8.26}$$

or

$$\frac{(\hat{H}_{k,e} + \hat{H}_{p,e-e} + \hat{H}_{p,n-e})\Psi_e(\vec{X}; \vec{x})}{\Psi_e(\vec{X}; \vec{x})} = E - \frac{(\hat{H}_{k,n} + \hat{H}_{p,n-n})\Psi_n(\vec{X})}{\Psi_n(\vec{X})}. \tag{8.27}$$

Equivalent to the procedure of Chapter 4, we notice that the right-hand side does not depend on the electronic coordinates $\vec{x}$ (but may depend on the nuclear coordinates) and therefore that both sides must be independent of those, i.e.,

$$\frac{(\hat{H}_{k,e} + \hat{H}_{p,e-e} + \hat{H}_{p,n-e})\Psi_e(\vec{X};\vec{x})}{\Psi_e(\vec{X};\vec{x})} = E_e(\vec{X}). \tag{8.28}$$

This leads to the Schrödinger equation for the electrons,

$$(\hat{H}_{k,e} + \hat{H}_{p,e-e} + \hat{H}_{p,n-e})\Psi_e(\vec{X};\vec{x}) = E_e(\vec{X})\Psi_e(\vec{X};\vec{x}). \tag{8.29}$$

Explicitly written down it is,

$$\left[-\sum_{i=1}^{N} \frac{\hbar^2}{2m_e}\nabla_{\vec{r}_i}^2 + \frac{1}{2}\sum_{i_1 \neq i_2 = 1}^{N} \frac{1}{4\pi\varepsilon_0} \frac{e^2}{|\vec{r}_{i_1} - \vec{r}_{i_2}|} - \sum_{k=1}^{M}\sum_{i=1}^{N} \frac{1}{4\pi\varepsilon_0} \frac{Z_k e^2}{|\vec{R}_k - \vec{r}_i|} \right] \Psi_e(\vec{X};\vec{x})$$

$$= E_e(\vec{X}) \cdot \Psi_e(\vec{X};\vec{x}). \tag{8.30}$$

We see here that the dependence of Ψ_e on the nuclear coordinates is only through the electrostatic interactions between the electrons and the nuclei. This means that the nuclei generate an external potential in which the electrons move. For later purposes we add that this potential is

$$V(\vec{r}) = -\sum_{k=1}^{M} \frac{1}{4\pi\varepsilon_0} \frac{Z_k e^2}{|\vec{R}_k - \vec{r}|}. \tag{8.31}$$

Finally, the total energy E is found from Eq. (8.27),

$$E - \frac{(\hat{H}_{k,n} + \hat{H}_{p,n-n})\Psi_n(\vec{X})}{\Psi_n(\vec{X})} = E_e(\vec{X}). \tag{8.32}$$

By neglecting the kinetic energy of the nuclei completely, we end up with

$$\begin{aligned} E &= \frac{(\hat{H}_{k,n} + \hat{H}_{p,n-n})\Psi_n(\vec{X})}{\Psi_n(\vec{X})} + E_e(\vec{X}) \\ &= \frac{\hat{H}_{p,n-n}\Psi_n(\vec{X})}{\Psi_n(\vec{X})} + E_e(\vec{X}) \\ &= \hat{H}_{p,n-n} + E_e(\vec{X}) \\ &= \frac{1}{2}\sum_{k_1 \neq k_2 = 1}^{M} \frac{1}{4\pi\varepsilon_0} \frac{Z_{k_1} Z_{k_2} e^2}{|\vec{R}_{k_1} - \vec{R}_{k_2}|} + E_e(\vec{X}). \end{aligned} \tag{8.33}$$

Equations (8.29) and (8.23) comprise the Born–Oppenheimer approximation (Born and Oppenheimer, 1927). Its great advantage is that it allows for treating the nuclei as classical particles that give rise to a total-energy contribution [Eq. (8.33)] and an electrostatic field [Eq. (8.31)] in which the electrons move, but that their effects otherwise are ignored. Accordingly, we choose first (this will be discussed later) the positions of the nuclei. Once this is done we can solve the electronic Schrödinger equation (8.30) and calculate for these positions a total energy through Eq. (8.33). Subsequently, we may choose another set of nuclear positions, which will lead to new electronic wavefunctions and a new total energy (this is meant by the parametric dependence of the electronic wavefunction on the nuclear coordinates), and ultimately we may determine that structure (i.e., that set of nuclear positions) for which the total energy is lowest. This is then the theoretically optimized structure.

8.3   THE ADIABATIC APPROXIMATION

Within the Born–Oppenheimer approximation the nuclei were treated as fixed particles and all their quantum effects were neglected. The electronic wavefunction possesses, however, a parametric dependence on the nuclear coordinates. Using this dependence in calculating a part of the kinetic energy associated with the nuclear degrees of freedom [cf. Eq. (8.25)] results in an extra term to the total energy. When including this term the approach is called the adiabatic approximation (see, e.g., Hirschfelder and Meath, 1967), but only in few cases is there any important difference between the results for the Born–Oppenheimer approximation and those for the adiabatic approximation. Therefore, we shall not discuss the latter further here.

8.4   ATOMIC UNITS

The electronic Schrödinger equation derived in Section 8.2 looks very complicated:

$$\left[-\sum_{i=1}^{N} \frac{\hbar^2}{2m_e} \nabla_{\vec{r}_i}^2 + \frac{1}{2} \sum_{i_1 \neq i_2 = 1}^{N} \frac{1}{4\pi\varepsilon_0} \frac{e^2}{|\vec{r}_{i_1} - \vec{r}_{i_2}|} - \sum_{k=1}^{M} \sum_{i=1}^{N} \frac{1}{4\pi\varepsilon_0} \frac{Z_k e^2}{|\vec{R}_k - \vec{r}_i|} \right] \Psi_e(\vec{X}; \vec{x})$$

$$= E_e(\vec{X}) \cdot \Psi_e(\vec{X}; \vec{x}), \tag{8.34}$$

partly due to the occurrence of the fundamental constants $\hbar$, m_e, $4\pi\varepsilon_0$, and e. It has therefore become standard practice (Hartree, 1927; McWeeny, 1973) to set these equal to 1:

$$\hbar = 1$$

$$m_e = 1$$

$$|e| = 1$$

$$4\pi\varepsilon_0 = 1. \tag{8.35}$$

Then, energies are measured in hartrees (1 hartree = 27.21 eV) and lengths in bohr (1 bohr = 0.5292Å). The speed of light becomes $c = 1/\alpha = 137.036$, with α being the fine-structure constant.

There is nothing magic about this. Within classical mechanics we are used to SI units, i.e., lengths measured in metres, times in seconds, and masses in kilograms. We know then that energies are given in joules. We might however, also choose other basic units, e.g., lengths in units of 6.33 mm, a.s.o. These units may not be physically meaningful, but as long as we work consistently with them, there is nothing wrong about it. So also the atomic units of Eq. (8.35). Here, one may consider the fundamental length, time, mass, and charge units so chosen, that Eq. (8.35) is fulfilled.

For the sake of completeness we add that these atomic units are often called Hartree atomic units to distinguish them from the so-called Rydberg atomic units. In the latter, one sets $\hbar = 4\pi\varepsilon_e = 1$ as above, but instead $|e| = \sqrt{2}$ and $m_e = \frac{1}{2}$. Then, energies are given in rydberg (1 rydberg = 13.605eV) and lengths in bohr.

Inserting Eq. (8.35) into Eq. (8.34) leads to

$$\left[-\sum_{i=1}^{N} \frac{1}{2}\nabla_{\vec{r}_i}^2 + \frac{1}{2} \sum_{i_1 \neq i_2=1}^{N} \frac{1}{|\vec{r}_{i_1} - \vec{r}_{i_2}|} - \sum_{k=1}^{M}\sum_{i=1}^{N} \frac{Z_k}{|\vec{R}_k - \vec{r}_i|} \right] \Psi_e(\vec{X};\vec{x})$$

$$= E_e(\vec{X})\Psi_e(\vec{X};\vec{x}). \tag{8.36}$$

Already at this point, the equation *appears* simpler.

Next, we will omit explicitly writing the parametric dependence of the wave-function on the nuclear coordinates,

$$\Psi_e(\vec{X};\vec{x}) \rightarrow \Psi_e(\vec{x}), \tag{8.37}$$

although we will *remember* that the wavefunction will change when the nuclei are moved, and that the total energy is obtained from E_e by adding the electrostatic energy due to the nucleus–nucleus interactions. Then,

$$\left[-\sum_{i=1}^{N} \frac{1}{2}\nabla_{\vec{r}_i}^2 + \frac{1}{2} \sum_{i_1 \neq i_2=1}^{N} \frac{1}{|\vec{r}_{i_1} - \vec{r}_{i_2}|} - \sum_{k=1}^{M}\sum_{i=1}^{N} \frac{Z_k}{|\vec{R}_k - \vec{r}_i|} \right] \Psi_e(\vec{x}) = E_e \cdot \Psi_e(\vec{x}). \tag{8.38}$$

Next, we observe that

$$-\sum_{i=1}^{N} \frac{1}{2}\nabla_{\vec{r}_i}^2 - \sum_{k=1}^{M}\sum_{i=1}^{N} \frac{Z_k}{|\vec{R}_k - \vec{r}_i|} = \sum_{i=1}^{N} \left[-\frac{1}{2}\nabla_{\vec{r}_i}^2 - \sum_{k=1}^{M} \frac{Z_k}{|\vec{R}_k - \vec{r}_i|} \right]$$

$$= \sum_{i=1}^{N} \left[-\frac{1}{2}\nabla_{\vec{r}_i}^2 + V(\vec{r}_i) \right]$$

$$\equiv \sum_{i=1}^{N} \hat{h}_1(\vec{r}_i), \tag{8.39}$$

where we have introduced the electrostatic potential of the nuclei of Eq. (8.31),

$$V(\vec{r}) = -\sum_{k=1}^{M} \frac{Z_k}{|\vec{R}_k - \vec{r}|},\tag{8.40}$$

and where we in the last identity in Eq. (8.39) have used that the expression is a sum of identical single-particle operators $\hat{h}_1$ but acting on the different electrons,

$$\hat{h}_1(\vec{r}) = -\tfrac{1}{2}\nabla_{\vec{r}}^2 + V(\vec{r}) = -\tfrac{1}{2}\nabla^2 + V(\vec{r}).\tag{8.41}$$

Similarly,

$$\frac{1}{2}\sum_{i_1 \neq i_2=1}^{N} \frac{1}{|\vec{r}_{i_1} - \vec{r}_{i_2}|} \equiv \frac{1}{2}\sum_{i_1 \neq i_2=1}^{N} \hat{h}_2(\vec{r}_{i_1}, \vec{r}_{i_2})\tag{8.42}$$

with

$$\hat{h}_2(\vec{r}_i, \vec{r}_j) = \frac{1}{|\vec{r}_i - \vec{r}_j|}.\tag{8.43}$$

The electronic Schrödinger equation then takes the *apparently* much simpler form,

$$\left[\sum_{i=1}^{N} \hat{h}_1(\vec{r}_i) + \frac{1}{2}\sum_{i \neq j=1}^{N} \hat{h}_2(\vec{r}_i, \vec{r}_j)\right] \Psi_e(\vec{x}) = E_e \cdot \Psi_e(\vec{x}).\tag{8.44}$$

This is the equation we shall attempt to solve approximately in the following using the methods of the preceeding sections.

9 The Hartree, Hartree–Fock, and Hartree–Fock–Roothaan Methods

9.1 THE HARTREE APPROXIMATION

We shall now seek an approximate solution to the electronic Schrödinger equation

$$\hat{H}_e \Psi_e = E_e \Psi_e, \tag{9.1}$$

where

$$\hat{H}_e = \sum_{i=1}^{N} \hat{h}_1(\vec{r}_i) + \frac{1}{2} \sum_{i \neq j=1}^{N} \hat{h}_2(\vec{r}_i, \vec{r}_j). \tag{9.2}$$

The approximate solution,

$$\Psi_e \simeq \Phi, \tag{9.3}$$

is obtained by applying the variational method of Chapter 5, i.e. by varying either

$$\frac{\langle \Phi | \hat{H}_e | \Phi \rangle}{\langle \Phi | \Phi \rangle} \tag{9.4}$$

or

$$\langle \Phi | \hat{H}_e | \Phi \rangle - \lambda [\langle \Phi | \Phi \rangle - 1] \tag{9.5}$$

so that it becomes smallest possible. We accordingly seek a lowest estimate for the energy of the ground state.

As discussed in Chapter 5, the two approaches are equivalent. In Eq. (9.4) we simply vary the expectation value of the electronic Hamilton operator by considering *all* possible wavefunctions (of N electrons), that are not necessarily normalized, whereas in Eq. (9.5) we explicitly consider only those wavefunctions that are normalized (through the Lagrange multiplier λ, which takes care of the constraint that the wavefunction shall be normalized).

Whenever Φ is not exactly identical to Ψ_e, we cannot expect *all* expectation values

$$\frac{\langle \Phi | \hat{A} | \Phi \rangle}{\langle \Phi | \Phi \rangle}, \qquad (9.6)$$

with $\hat{A}$ being any N-electron operator, to be close to

$$\frac{\langle \Psi_e | \hat{A} | \Psi_e \rangle}{\langle \Psi_e | \Psi_e \rangle}. \qquad (9.7)$$

Therefore, one has to make a compromise: the approximate wavefunction Φ should be so exact that the calculated observables of interest are accurate enough. On the other hand, it should be so simple that the calculations do not become prohibitively involved. Where this compromise is set is often more a personal question and is also strongly dependent on the size and type of the system of interest, but there are nevertheless some general guidelines, which make the construction of the approximate wavefunction not completely arbitrary.

The Hartree approximation (Hartree, 1927, 1928, 1929) offers one way of constructing the approximate wavefunction. It can be interpreted as based on simple orbital energy diagrams like those of Fig. 9.1. We are used to thinking of N electrons as occupying N different orbitals that can each accommodate one electron. Thereby, we arrive at diagrams like those of Fig. 9.1. For the ground state (left part of Fig. 9.1) the electrons occupy the energetically lowest levels, whereas excited states (right part) are obtained by placing some of the electrons in energetically higher levels.

Denoting the wavefunctions for the individual orbitals by ϕ_i (note that, at the moment, we have not discussed where these come from) we may therefore suggest the following approximate wavefunction:

$$\Phi(\vec{x}_1, \vec{x}_2, \ldots, \vec{x}_N) = \phi_1(\vec{x}_1) \cdot \phi_2(\vec{x}_2) \ldots \phi_N(\vec{x}_N), \qquad (9.8)$$

where we have explicitly included the arguments of the different functions.

Applying the variational method in this case implies that we will consider variations in

$$\langle \Phi | \hat{H}_e | \Phi \rangle \qquad (9.9)$$

under the constraints that *all* the individual orbitals are orthonormal, i.e., that

$$\langle \phi_i | \phi_j \rangle = \delta_{i,j}. \qquad (9.10)$$

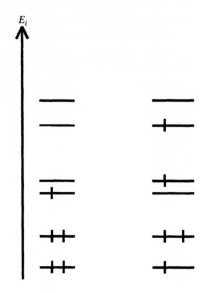

Figure 9.1 Orbitals, their occupations, and energies (the vertical scale) for some hypothetic systems

We have accordingly not one constraint as in Chapter 5, where we discussed the variational method, but $N(N-1)/2$ constraints. All these constraints can, however, easily be incorporated by considering the quantity

$$F = \langle \Phi | \hat{H}_e | \Phi \rangle - \sum_{i,j} \lambda_{ij} [\langle \phi_i | \phi_j \rangle - \delta_{i,j}]. \tag{9.11}$$

Compared to the simpler discussion in Chapter 5, where we introduced Lagrange multipliers, there is only a smaller difference. Namely, we will require that any of the derivatives of the quantity F of Eq. (9.11) with respect to any of the Lagrange multipliers λ_{ij} vanishes, instead of having only one such condition, as in Chapter 5.

The condition that F has a minimum means that when varying *any* of the orbitals of Eq. (9.8), i.e.

$$\phi_k(\vec{x}) \rightarrow \phi_k(\vec{x}) + \delta\phi_k(\vec{x}), \tag{9.12}$$

where $\delta\phi_k(\vec{x})$ is arbitrary but small, is

$$\delta F = 0. \tag{9.13}$$

The next step is now to insert Φ of Eq. (9.8) into Eq. (9.11) and study the condition (9.13) under the variations (9.12). This will lead us to the Hartree equations, which in principle determine the orbitals ϕ_i. However, the Hartree approximation (9.8) suffers from some severe problems, so that this approximation in the vast majority of cases is considered too crude. The Hartree method is therefore applied only in few special cases and we shall not consider it further here.

The main problem of the function Φ of Eq. (9.8) is that it does not take into account that the electrons are indistinguishable. It violates therefore one of the most fundamental principles of quantum theory. The fact that the electrons are indistinguishable means that Φ shall fulfill

$$\hat{P}_{ij}\Phi(\vec{x}_1, \vec{x}_2, \ldots, \vec{x}_{i-1}, \vec{x}_i, \vec{x}_{i+1}, \ldots, \vec{x}_{j-1}, \vec{x}_j, \vec{x}_{j+1}, \ldots, \vec{x}_N)$$

$$= \Phi(\vec{x}_1, \vec{x}_2, \ldots, \vec{x}_{i-1}, \vec{x}_j, \vec{x}_{i+1}, \ldots, \vec{x}_{j-1}, \vec{x}_i, \vec{x}_{j+1}, \ldots, \vec{x}_N)$$

$$\equiv -\Phi(\vec{x}_1, \vec{x}_2, \ldots, \vec{x}_{i-1}, \vec{x}_i, \vec{x}_{i+1}, \ldots, \vec{x}_{j-1}, \vec{x}_j, \vec{x}_{j+1}, \ldots, \vec{x}_N). \quad (9.14)$$

As indicated, the operator $\hat{P}_{ij}$ is a *permutation* operator that interchanges electron i and electron j. Thus, the wavefunction is to be antisymmetric, when interchanging *any* two electrons. And this condition is clearly not satisfied by the function of Eq. (9.8).

For the sake of completeness we add that the condition (9.14) is a consequence of the electrons being fermions. For bosons (e.g., vibrations or ^{4}He nuclei) the equivalent condition would be that the wavefunction was symmetric, i.e., the sign in the last identity in Eq. (9.14) should be a $+$. Put on a more formal mathematical level, $\hat{P}_{ij}$ and $\hat{H}$ are commuting, Hermitian operators. Therefore, we can find a set of functions that are simultaneously eigenfunctions to $\hat{H}$ and to $\hat{P}_{ij}$. The eigenvalues to $\hat{P}_{ij}$ are ± 1, and for fermions we will choose only those functions that correspond to the eigenvalue -1 of $\hat{P}_{ij}$.

9.2 THE HARTREE–FOCK METHOD

In order to satisfy the antisymmetry conditions, one improves the approximate function of Eq. (9.8) so that it satisfies this condition but otherwise retains most of the characteristics of the function of Eq. (9.8).

Let us start with a simple example to see how this works. We consider the simple function

$$f_1(x)f_2(y), \quad (9.15)$$

that depends on two variables x and y. Interchanging x and y gives

$$f_1(y)f_2(x), \quad (9.16)$$

which clearly differs from that of Eq. (9.15). We may, however, 'repair' the function of Eq. (9.15) so that it does become antisymmetric by subtracting that of Eq. (9.16), i.e., by considering

$$f_1(x)f_2(y) - f_1(y)f_2(x). \quad (9.17)$$

This function *is* antisymmetric when interchanging x and y. Furthermore, it resembles the original function (9.15) in some sense.

A slightly more complicated example is the function

$$f_1(x)f_2(y)f_3(z). \tag{9.18}$$

In order to make that antisymmetric with respect to the interchange of x and y we modify that to

$$f_1(x)f_2(y)f_3(z) - f_1(y)f_2(x)f_3(z). \tag{9.19}$$

This function is, however, not antisymmetric for the interchange of x and z. But when instead considering

$$f_1(x)f_2(y)f_3(z) - f_1(y)f_2(x)f_3(z) - f_1(z)f_2(y)f_3(x) + f_1(y)f_2(z)f_3(x), \tag{9.20}$$

we have constructed a function that is antisymmetric with respect to interchanging x and y or x and z. Finally, the antisymmetry with respect to interchanging y and z is obtained through the function

$$F(x, y, z) = f_1(x)f_2(y)f_3(z) - f_1(y)f_2(x)f_3(z) - f_1(z)f_2(y)f_3(x)$$
$$+ f_1(y)f_2(z)f_3(x) - f_1(x)f_2(z)f_3(y) + f_1(z)f_2(x)f_3(y). \tag{9.21}$$

Inspecting this function we realize that this is nothing but a determinant, i.e.,

$$F(x, y, z) = \begin{vmatrix} f_1(x) & f_2(x) & f_3(x) \\ f_1(y) & f_2(y) & f_3(y) \\ f_1(z) & f_2(z) & f_3(z) \end{vmatrix}. \tag{9.22}$$

Once this is realised, we can construct an antisymmetric function from the suggested form of Eq. (9.8):

$$\begin{vmatrix} \phi_1(\vec{x}_1) & \phi_2(\vec{x}_1) & \cdots & \phi_N(\vec{x}_1) \\ \phi_1(\vec{x}_2) & \phi_2(\vec{x}_2) & \cdots & \phi_N(\vec{x}_2) \\ \vdots & \vdots & \ddots & \vdots \\ \phi_1(\vec{x}_N) & \phi_2(\vec{x}_N) & \cdots & \phi_N(\vec{x}_N) \end{vmatrix}. \tag{9.23}$$

It turns out that it is useful to multiply this by $1/\sqrt{N!}$ (this will be shown below), so that we have the following approximate wavefunction:

$$\Phi(\vec{x}_1, \vec{x}_2, \ldots, \vec{x}_N) = \frac{1}{\sqrt{N!}} \begin{vmatrix} \phi_1(\vec{x}_1) & \phi_2(\vec{x}_1) & \cdots & \phi_N(\vec{x}_1) \\ \phi_1(\vec{x}_2) & \phi_2(\vec{x}_2) & \cdots & \phi_N(\vec{x}_2) \\ \vdots & \vdots & \ddots & \vdots \\ \phi_1(\vec{x}_N) & \phi_2(\vec{x}_N) & \cdots & \phi_N(\vec{x}_N) \end{vmatrix}$$
$$\equiv |\phi_1, \phi_2, \ldots, \phi_N|, \tag{9.24}$$

where in the last identity we have introduced a short-hand notation for the complete determinant.

The determinant of Eq. (9.24) is the so-called *Slater determinant*. We shall insert this into Eq. (9.11),

$$F = \langle \Phi | \hat{H}_e | \Phi \rangle - \sum_{i,j} \lambda_{ij} [\langle \phi_i | \phi_j \rangle - \delta_{i,j}] \tag{9.25}$$

and require that

$$\delta F = 0 \tag{9.26}$$

for *any* variation of *any* orbital

$$\phi_k(\vec{x}) \rightarrow \phi_k(\vec{x}) + \delta\phi_k(\vec{x}). \tag{9.27}$$

This is the principle behind the Hartree–Fock method and which results in the Hartree–Fock equations (Hartree, 1928; Fock, 1930). Consistent with this, the approximation (9.24) is called the Hartree–Fock approximation.

Like the Hartree approximation, the Hartree–Fock approximation also relies on the approximation of independent particles. Let us for instance consider the simple system of two electrons occupying the two orbitals of Fig. 9.2(a). When we now assume that one of the two electrons is in one of the two orbitals and furthermore assume that that electron is at a certain position [Fig. 9.2(b)], we will intuitively expect the other electron to react to that situation and seek to avoid the first electron. Therefore, we will expect that the other orbital rather looks like in Fig. 9.2(c) instead of as it is according to the independent-particle model of Fig. 9.2(a). It is thus clear, that the Hartree–Fock approximation is not perfect. Nevertheless, the Hartree–Fock approximation does contain a major part of the physics of most systems and is therefore very useful.

Before calculating the matrix elements of Eq. (9.25) it will be useful first to consider the Slater determinant in more detail. We shall thereby use some of the fundamental properties of determinants.

The Slater determinant of Eq. (9.24) contains in total $N!$ terms of which each is a product of N factors multiplied by either $+1$ or -1. The N factors include each of the functions ϕ_i exactly once and each of the arguments $\vec{x}_j$ also exactly once. Thus, each term has the form

$$(-1)^{P(i_1, i_2, \ldots, i_N)} \phi_{i_1}(\vec{x}_1) \phi_{i_2}(\vec{x}_2) \cdots \phi_{i_N}(\vec{x}_N). \tag{9.28}$$

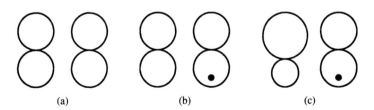

(a)   (b)   (c)

Figure 9.2 (a) Two different orbitals. The orbitals are spatially separated. (b) As (a) but with one of the two orbitals with the (assumued) position of one electron. (c) As (b) but showing schematically how the other orbital may change when including electronic interactions

$P(i_1, i_2, \ldots, i_N)$ is the number of permutations that are required in changing $1, 2, \ldots, N$ into $i_1, i_2, \ldots, i_N$. Below, we shall simply write $P(i)$ for simplicity. The Slater determinant of Eq. (9.24) then becomes

$$\Phi = \frac{1}{\sqrt{N!}} \sum_{i=1}^{N!} (-1)^{P(i)} \phi_{i_1}(\vec{x}_1) \phi_{i_2}(\vec{x}_2) \cdots \phi_{i_N}(\vec{x}_N), \tag{9.29}$$

where it is understood that the sum is over all different permutations of $1, 2, \ldots, N$, and where $i_1, i_2, \ldots, i_N$ depend on i.

We will need matrix elements of two types. First, we will need those for operators that are a sum of identical one-electron operators, i.e., for operators of the type

$$\hat{A} = \sum_{n=1}^{N} \hat{a}_1(n), \tag{9.30}$$

where $\hat{a}_1(n)$ acts only on the nth electron.

Inserting Φ of Eq. (9.29) we find

$$\langle \Phi | \hat{A} | \Phi \rangle = \frac{1}{N!} \left\langle \sum_{j=1}^{N!} (-1)^{P(j)} \phi_{j_1}(\vec{x}_1) \phi_{j_2}(\vec{x}_2) \cdots \phi_{j_N}(\vec{x}_N) \right|$$

$$\times \sum_{n=1}^{N} \hat{a}_1(n) \left| \sum_{i=1}^{N!} (-1)^{P(i)} \phi_{i_1}(\vec{x}_1) \phi_{i_2}(\vec{x}_2) \cdots \phi_{i_N}(\vec{x}_N) \right\rangle$$

$$= \frac{1}{N!} \sum_{n=1}^{N} \left\langle \sum_{j=1}^{N!} (-1)^{P(j)} \phi_{j_1}(\vec{x}_1) \phi_{j_2}(\vec{x}_2) \cdots \phi_{j_N}(\vec{x}_N) \right|$$

$$\times \hat{a}_1(n) \left| \sum_{i=1}^{N!} (-1)^{P(i)} \phi_{i_1}(\vec{x}_1) \phi_{i_2}(\vec{x}_2) \cdots \phi_{i_N}(\vec{x}_N) \right\rangle$$

$$= \frac{1}{N!} \sum_{n=1}^{N} \left\langle \sum_{j=1}^{N!} (-1)^{P(j)} \phi_{j_1}(\vec{x}_1) \phi_{j_2}(\vec{x}_2) \cdots \phi_{j_{n-1}}(\vec{x}_{n-1}) \right.$$

$$\times \phi_{j_n}(\vec{x}_n) \phi_{j_{n+1}}(\vec{x}_{n+1}) \cdots \phi_{j_N}(\vec{x}_N) \left| \hat{a}_1(n) \right| \sum_{i=1}^{N!} (-1)^{P(i)} \phi_{i_1}(\vec{x}_1)$$

$$\times \left. \phi_{i_2}(\vec{x}_2) \cdots \phi_{i_{n-1}}(\vec{x}_{n-1}) \phi_{i_n}(\vec{x}_n) \phi_{i_{n+1}}(\vec{x}_{n+1}) \cdots \phi_{i_N}(\vec{x}_N) \right\rangle$$

$$= \frac{1}{N!} \sum_{n=1}^{N} \sum_{i,j=1}^{N!} (-1)^{P(j)} (-1)^{P(i)} \langle \phi_{j_1} | \phi_{i_1} \rangle \cdot \langle \phi_{j_2} | \phi_{i_2} \rangle \cdots \langle \phi_{j_{n-1}} | \phi_{i_{n-1}} \rangle \cdot$$

$$\times \langle \phi_{j_n} | \hat{a}_1 | \phi_{i_n} \rangle \cdot \langle \phi_{j_{n+1}} | \phi_{i_{n+1}} \rangle \cdots \langle \phi_{j_N} | \phi_{i_N} \rangle. \tag{9.31}$$

Since the ϕ_k functions are required to be orthonormal,

$$\langle \phi_k | \phi_l \rangle = \delta_{k,l}, \tag{9.32}$$

Eq. (9.31) can be written as

$$\langle \Phi | \hat{A} | \Phi \rangle = \frac{1}{N!} \sum_{n=1}^{N} \sum_{i,j=1}^{N!} (-1)^{P(j)} (-1)^{P(i)} \delta_{j_1,i_1} \cdot \delta_{j_2,i_2} \cdots \delta_{j_{n-1},i_{n-1}}$$
$$\times \langle \phi_{j_n} | \hat{a}_1(n) | \phi_{i_n} \rangle \cdot \delta_{j_{n+1},i_{n+1}} \cdots \delta_{j_N,i_N}. \tag{9.33}$$

These N different terms for the N electrons are identical, and we may therefore replace the n summation by a factor N, i.e.,

$$\langle \Phi | \hat{A} | \Phi \rangle = \frac{1}{N!} N \sum_{i,j=1}^{N!} (-1)^{P(j)} (-1)^{P(i)} \delta_{j_1,i_1} \cdot \delta_{j_2,i_2} \cdots \delta_{j_{n-1},i_{n-1}}$$
$$\times \langle \phi_{j_n} | \hat{a}_1(n) | \phi_{i_n} \rangle \cdot \delta_{j_{n+1},i_{n+1}} \cdots \delta_{j_N,i_N}. \tag{9.34}$$

Furthermore, since each of the ϕ functions appears exactly once with a particular i index and once with a particular j index and since $i_1 = j_1, i_2 = j_2, \ldots, i_{n-1} = j_{n-1}, i_{n+1} = j_{n+1}, \ldots, i_N = j_N$, we must also have that $i_n = j_n$. This means also that $P(i) = P(j)$, which gives

$$\langle \Phi | \hat{A} | \Phi \rangle = \frac{1}{N!} N \sum_{i=1}^{N!} \langle \phi_{i_n} | \hat{a}_1(n) | \phi_{i_n} \rangle. \tag{9.35}$$

For a given i_n, all the other i_p (i.e., $i_1, i_2, \ldots, i_{n-1}, i_{n+1}, \ldots, i_N$) can be chosen freely, i.e., they give the same contribution to the matrix element in Eq. (9.35). There are $(N-1)!$ such terms, and we can thus replace the i summation by a single summation over only i_n if we simultaneously multiply by $(N-1)!$. This results in

$$\langle \Phi | \hat{A} | \Phi \rangle = \frac{1}{N!} N(N-1)! \sum_{i_n=1}^{N} \langle \phi_{i_n} | \hat{a}_1(n) | \phi_{i_n} \rangle = \sum_{i_n=1}^{N} \langle \phi_{i_n} | \hat{a}_1(n) | \phi_{i_n} \rangle. \tag{9.36}$$

Finally, renaming the summation index i_n as i we obtain

$$\langle \Phi | \hat{A} | \Phi \rangle = \sum_{i=1}^{N} \langle \phi_i | \hat{a}_1 | \phi_i \rangle, \tag{9.37}$$

i.e., the matrix element is a sum of contributions from the individual orbitals. This is actually also what we might have expected when remembering that the whole approach was based on inserting the individual electrons in different orbitals (whereby we have one electron per orbital) and subsequently repairing the wavefunction so that it satisfies the condition of being antisymmetric.

A special case is obtained when $\hat{A}$ is the identity (i.e., $a_1 = 1/N$). In that case we obtain

$$\langle \Phi | \Phi \rangle = 1, \tag{9.38}$$

which was the reason for including the prefactor $1/\sqrt{N!}$ with the determinant in Eq. (9.24).

The other type of matrix elements is very similar to that of Eq. (9.30), but contains instead a sum of identical two-electron operators,

$$\hat{A} = \frac{1}{2} \sum_{n \neq m = 1}^{N} \hat{a}_2(n, m). \tag{9.39}$$

As above we calculate the matrix element for this operator and obtain

$$
\begin{aligned}
\langle \Phi | \hat{A} | \Phi \rangle &= \frac{1}{N!} \left\langle \sum_{j=1}^{N!} (-1)^{P(j)} \phi_{j_1}(\vec{x}_1) \phi_{j_2}(\vec{x}_2) \cdots \phi_{j_N}(\vec{x}_N) \right| \\
&\quad \times \frac{1}{2} \sum_{n \neq m = 1}^{N} \hat{a}_2(n, m) \left| \sum_{i=1}^{N!} (-1)^{P(i)} \phi_{i_1}(\vec{x}_1) \phi_{i_2}(\vec{x}_2) \cdots \phi_{i_N}(\vec{x}_N) \right\rangle \\
&= \frac{1}{N!} \frac{1}{2} \sum_{n \neq m = 1}^{N} \left\langle \sum_{j=1}^{N!} (-1)^{P(j)} \phi_{j_1}(\vec{x}_1) \phi_{j_2}(\vec{x}_2) \cdots \phi_{j_{n-1}}(\vec{x}_{n-1}) \right. \\
&\quad \times \phi_{j_n}(\vec{x}_n) \phi_{j_{n+1}}(\vec{x}_{n+1}) \cdots \phi_{j_{m-1}}(\vec{x}_{m-1}) \phi_{j_m}(\vec{x}_m) \phi_{j_{m+1}}(\vec{x}_{m+1}) \cdots \phi_{j_N}(\vec{x}_N) \Big| \\
&\quad \times \hat{a}_2(n, m) \left| \sum_{i=1}^{N!} (-1)^{P(i)} \phi_{i_1}(\vec{x}_1) \phi_{i_2}(\vec{x}_2) \cdots \phi_{i_{n-1}}(\vec{x}_{n-1}) \right. \\
&\quad \times \phi_{i_n}(\vec{x}_n) \phi_{i_{n+1}}(\vec{x}_{n+1}) \cdots \phi_{i_{m-1}}(\vec{x}_{m-1}) \phi_{i_m}(\vec{x}_m) \phi_{i_{m+1}}(\vec{x}_{m+1}) \cdots \phi_{i_N}(\vec{x}_N) \Big\rangle \\
&= \frac{1}{N!} \frac{1}{2} \sum_{n \neq m = 1}^{N} \sum_{i,j=1}^{N!} (-1)^{P(j)} (-1)^{P(i)} \langle \phi_{j_1} | \phi_{i_1} \rangle \cdot \\
&\quad \times \langle \phi_{j_2} | \phi_{i_2} \rangle \cdots \langle \phi_{j_{n-1}} | \phi_{i_{n-1}} \rangle \cdot \langle \phi_{j_{m-1}} | \phi_{i_{m-1}} \rangle \cdot \langle \phi_{j_n}(\vec{x}_n) \phi_{j_m}(\vec{x}_m) | \hat{a}_2(n, m) | \\
&\quad \times \phi_{i_n}(\vec{x}_n) \phi_{i_m}(\vec{x}_m) \rangle \cdot \langle \phi_{j_{n+1}} | \phi_{i_{n+1}} \rangle \cdots \langle \phi_{j_{m+1}} | \phi_{i_{m+1}} \rangle \cdots \langle \phi_{j_N} | \phi_{i_N} \rangle \\
&= \frac{1}{N!} \frac{1}{2} \sum_{n \neq m = 1}^{N} \sum_{i,j=1}^{N!} (-1)^{P(j)} (-1)^{P(i)} \delta_{j_1, i_1} \cdot \delta_{j_2, i_2} \cdots \delta_{j_{n-1}, i_{n-1}} \cdot \delta_{j_{m-1}, i_{m-1}} \cdot \\
&\quad \times \langle \phi_{j_n}(\vec{x}_n) \phi_{j_m}(\vec{x}_m) | \hat{a}_2(n, m) | \phi_{i_n}(\vec{x}_n) \phi_{i_m}(\vec{x}_m) \rangle \cdot \\
&\quad \times \delta_{j_{n+1}, i_{n+1}} \cdot \delta_{j_{m+1}, i_{m+1}} \cdots \delta_{j_N, i_N}.
\end{aligned} \tag{9.40}
$$

Equivalent to above we observe that since all the i_k have to be equal to the corresponding j_k, except for two (no. n and m) and since all functions appear exactly once as bra and once as ket, we have only two possibilities for (j_n, j_m) and (i_n, i_m), i.e., either $(j_n, j_m) = (i_n, i_m)$ or $(j_n, j_m) = (i_m, i_n)$. In the first case, $P(i) = P(j)$ and in the second $P(i) = P(j) \pm 1$. This gives

$$\langle \Phi | \hat{A} | \Phi \rangle = \frac{1}{N!} \frac{1}{2} \sum_{n \neq m=1}^{N} \sum_{i=1}^{N!} \left[\langle \phi_{i_n}(\vec{x}_n) \phi_{i_m}(\vec{x}_m) | \hat{a}_2(n,m) | \phi_{i_n}(\vec{x}_n) \phi_{i_m}(\vec{x}_m) \rangle \right.$$
$$\left. - \langle \phi_{i_m}(\vec{x}_n) \phi_{i_n}(\vec{x}_m) | \hat{a}_2(n,m) | \phi_{i_n}(\vec{x}_n) \phi_{i_m}(\vec{x}_m) \rangle \right]. \quad (9.41)$$

Since all electron pairs (n,m) give the same contribution, we can replace the (n,m) summation by a factor over the number of such pairs, i.e., $N(N-1)$ (we do not divide by 2, since the sum include all pairs with $n \neq m$). Similarly, we can replace the i summation by one over only i_n and i_m if we simultaneously multiply by the number of terms that give the same (i_n, i_m), i.e., by $(N-2)!$. This gives

$$\langle \Phi | \hat{A} | \Phi \rangle = \frac{1}{N!} \frac{1}{2} N(N-1)(N-2)! \sum_{i_n \neq i_m=1}^{N}$$
$$\times \left[\langle \phi_{i_n}(\vec{x}_n) \phi_{i_m}(\vec{x}_m) | \hat{a}_2(n,m) | \phi_{i_n}(\vec{x}_n) \phi_{i_m}(\vec{x}_m) \rangle \right.$$
$$\left. - \langle \phi_{i_m}(\vec{x}_n) \phi_{i_n}(\vec{x}_m) | \hat{a}_2(n,m) | \phi_{i_n}(\vec{x}_n) \phi_{i_m}(\vec{x}_m) \rangle \right]. \quad (9.42)$$

Reorganizing the prefactors and renaming i_n and i_m as i and j yields

$$\langle \Phi | \hat{A} | \Phi \rangle = \frac{1}{2} \sum_{i \neq j=1}^{N} \left[\langle \phi_i \phi_j | \hat{a}_2 | \phi_i \phi_j \rangle - \langle \phi_j \phi_i | \hat{a}_2 | \phi_i \phi_j \rangle \right], \quad (9.43)$$

where we also have removed the explicit orbital dependence on the electron coordinates.

Finally, we notice that the extra terms $i = j$ in the summation will not contribute since the two bra–ket expressions in the square bracket will cancel each other in that case. Therefore, we can write Eq. (9.43) as

$$\langle \Phi | \hat{A} | \Phi \rangle = \frac{1}{2} \sum_{i,j=1}^{N} \left[\langle \phi_i \phi_j | \hat{a}_2 | \phi_i \phi_j \rangle - \langle \phi_j \phi_i | \hat{a}_2 | \phi_i \phi_j \rangle \right]. \quad (9.44)$$

Equivalently to the one-electron-operator case, this matrix element appears as a sum over all pairs of orbitals, instead of one over all pairs of electrons.

We are now ready to study the matrix elements of Eq. (9.15).

The Hamilton operator of Eq. (9.2) contains one- and two-electron operators,

$$\hat{H}_e = \sum_{i=1}^{N} \hat{h}_1(\vec{r}_1) + \frac{1}{2} \sum_{i \neq j=1}^{N} \hat{h}_2(\vec{r}_i, \vec{r}_j). \quad (9.45)$$

Therefore, the matrix element

$$\langle\Phi|\hat{H}_e|\Phi\rangle \tag{9.46}$$

of Eq. (9.25) can be calculated using Eqs. (9.37) and (9.44):

$$\langle\Phi|\hat{H}_e|\Phi\rangle = \sum_{i=1}^{N}\langle\phi_i|\hat{h}_1|\phi_i\rangle + \frac{1}{2}\sum_{i,j=1}^{N}\left[\langle\phi_i\phi_j|\hat{h}_2|\phi_i\phi_j\rangle - \langle\phi_j\phi_i|\hat{h}_2|\phi_i\phi_j\rangle\right]. \tag{9.47}$$

As discussed above we shall determine the orbitals ϕ_i so that F of Eq. (9.12) is stationary under small variations of the orbitals. This means that when varying *any* of them slightly,

$$\phi_k \to \phi_k + \delta\phi_k, \tag{9.48}$$

where $\delta\phi_k$ is arbitrary but small, F will not change when only considering first-order changes in F. That is,

$$\delta F = 0. \tag{9.49}$$

We need accordingly the changes in F when varying a specific orbitals, ϕ_k. We assume that ϕ_k changes according to Eq. (9.48) and find for the matrix element of Eq. (9.37)

$$
\begin{aligned}
\delta\langle\Phi|\hat{A}|\Phi\rangle &= \left[\langle\phi_1|\hat{a}_1|\phi_1\rangle + \langle\phi_2|\hat{a}_1|\phi_2\rangle + \cdots + \langle\phi_{k-1}|\hat{a}_1|\phi_{k-1}\rangle\right.\\
&\quad \left.+ \langle\phi_k + \delta\phi_k|\hat{a}_1|\phi_k + \delta\phi_k\rangle + \langle\phi_{k+1}|\hat{a}_1|\phi_{k+1}\rangle + \cdots + \langle\phi_N|\hat{a}_1|\phi_N\rangle\right]\\
&\quad - \left[\langle\phi_1|\hat{a}_1|\phi_1\rangle + \langle\phi_2|\hat{a}_1|\phi_2\rangle + \cdots + \langle\phi_{k-1}|\hat{a}_1|\phi_{k-1}\rangle + \langle\phi_k|\hat{a}_1|\phi_k\phi_k\rangle\right.\\
&\quad \left.+ \langle\phi_{k+1}|\hat{a}_1|\phi_{k+1}\rangle + \cdots + \langle\phi_N|\hat{a}_1|\phi_N\rangle\right]\\
&= \langle\phi_k + \delta\phi_k|\hat{a}_1|\phi_k + \delta\phi_k\rangle - \langle\phi_k|\hat{a}_1|\phi_k\rangle\\
&= \int\left[\phi_k(\vec{x}) + \delta\phi_k(\vec{x})\right]^*\hat{a}_1\left[\phi_k(\vec{x}) + \delta\phi_k(\vec{x})\right]d\vec{x} - \int\phi_k^*(\vec{x})\hat{a}_1\phi_k(\vec{x})d\vec{x}\\
&= \int\left[\phi_k^*(\vec{x})\hat{a}_1\delta\phi_k(\vec{x}) + \delta\phi_k^*(\vec{x})\hat{a}_1\phi_k(\vec{x}) + \delta\phi_k^*(\vec{x})\hat{a}_1\delta\phi_k(\vec{x})\right]d\vec{x}\\
&\simeq \int\left[\phi_k^*(\vec{x})\hat{a}_1\delta\phi_k(\vec{x}) + \delta\phi_k^*(\vec{x})\hat{a}_1\phi_k(\vec{x})\right]d\vec{x}\\
&= \langle\phi_k|\hat{a}_1|\delta\phi_k\rangle + \langle\delta\phi_k|\hat{a}_1|\phi_k\rangle, \tag{9.50}
\end{aligned}
$$

where we have included only the first-order terms in the change $\delta\phi_k$, i.e., ignored terms like $\delta\phi_k \cdot \delta\phi_k$. Furthermore, we have assumed — as is the case for the operators of our interest — that $\hat{a}_1$ is a linear operator, i.e., that

$$\hat{a}_1(c_1 \cdot f_1 + c_2 \cdot f_2) = c_1 \cdot \hat{a}_1 f_1 + c_2 \cdot \hat{a}_1 f_2, \tag{9.51}$$

where c_1 and c_2 are constants and f_1 and f_2 are functions on which $\hat{a}_1$ operates.

Finally, we will only consider Hermitian operators, so that Eq. (9.50) can be rewritten as

$$\delta \langle \Phi | \hat{A} | \Phi \rangle = \langle \delta \phi_k | \hat{a}_1 | \phi_k \rangle^* + \langle \delta \phi_k | \hat{a}_1 | \phi_k \rangle. \qquad (9.52)$$

Equivalently, we find for the two-electron-operator matrix elements of Eq. (9.44)

$$\begin{aligned}
\delta \langle \Phi | \hat{A} | \Phi \rangle &= \delta \left\{ \frac{1}{2} \sum_{i,j=1}^{N} \left[\langle \phi_i \phi_j | \hat{a}_2 | \phi_i \phi_j \rangle - \langle \phi_j \phi_i | \hat{a}_2 | \phi_i \phi_j \rangle \right] \right\} \\
&= \frac{1}{2} \sum_{i=1}^{N} \left[\langle \phi_i (\phi_k + \delta \phi_k) | \hat{a}_2 | \phi_i (\phi_k + \delta \phi_k) \rangle \right. \\
&\quad \left. - \langle (\phi_k + \delta \phi_k) \phi_i | \hat{a}_2 | \phi_i (\phi_k + \delta \phi_k) \rangle \right] \\
&\quad + \frac{1}{2} \sum_{j=1}^{N} \left[\langle (\phi_k + \delta \phi_k) \phi_j | \hat{a}_2 | (\phi_k + \delta \phi_k) \phi_j \rangle \right. \\
&\quad \left. - \langle \phi_j (\phi_k + \delta \phi_k) | \hat{a}_2 | (\phi_k + \delta \phi_k) \phi_j \rangle \right] \\
&\quad - \frac{1}{2} \sum_{i=1}^{N} \left[\langle \phi_i \phi_k | \hat{a}_2 | \phi_i \phi_k \rangle - \langle \phi_k \phi_i | \hat{a}_2 | \phi_i \phi_k \rangle \right] \\
&\quad - \frac{1}{2} \sum_{j=1}^{N} \left[\langle \phi_k \phi_j | \hat{a}_2 | \phi_k \phi_j \rangle - \langle \phi_j \phi_k | \hat{a}_2 | \phi_k \phi_j \rangle \right] \\
&\simeq \frac{1}{2} \sum_{i=1}^{N} \left[\langle \phi_i \delta \phi_k | \hat{a}_2 | \phi_i \phi_k \rangle + \langle \phi_i \phi_k | \hat{a}_2 | \phi_i \delta \phi_k \rangle \right. \\
&\quad \left. - \langle \delta \phi_k \phi_i | \hat{a}_2 | \phi_i \phi_k \rangle - \langle \phi_k \phi_i | \hat{a}_2 | \phi_i \delta \phi_k \rangle \right] \\
&\quad + \frac{1}{2} \sum_{j=1}^{N} \left[\langle \delta \phi_k \phi_j | \hat{a}_2 | \phi_k \phi_j \rangle + \langle \phi_k \phi_j | \hat{a}_2 | \delta \phi_k \phi_j \rangle \right. \\
&\quad \left. - \langle \phi_j \delta \phi_k | \hat{a}_2 | \phi_k \phi_j \rangle - \langle \phi_j \phi_k | \hat{a}_2 | \delta \phi_k \phi_j \rangle \right], \qquad (9.53)
\end{aligned}$$

where once again we have kept only first-order terms in the changes in the orbitals, and have assumed that $\hat{a}_2$ is a linear operator.

The two-electron operators of our interest obey

$$\langle f_1 f_2 | \hat{a}_2 | f_3 f_4 \rangle = \langle f_2 f_1 | \hat{a}_2 | f_4 f_3 \rangle, \qquad (9.54)$$

i.e., we may interchange pairs of single-electron functions *when* this is done simultaneously on the right- and left-hand side of the operator. Using this in

Eq. (9.53) gives

$$\delta\langle\Phi|\hat{A}|\Phi\rangle \simeq \sum_{i=1}^{N}\big[\langle\phi_i\delta\phi_k|\hat{a}_2|\phi_i\phi_k\rangle + \langle\phi_i\phi_k|\hat{a}_2|\phi_i\delta\phi_k\rangle$$
$$- \langle\delta\phi_k\phi_i|\hat{a}_2|\phi_i\phi_k\rangle - \langle\phi_k\phi_i|\hat{a}_2|\phi_i\delta\phi_k\rangle\big]. \tag{9.55}$$

Also here we only consider Hermitian operators, which then gives

$$\delta\langle\Phi|\hat{A}|\Phi\rangle \simeq \sum_{i=1}^{N}\big[\langle\phi_i\delta\phi_k|\hat{a}_2|\phi_i\phi_k\rangle + \langle\phi_i\delta\phi_k|\hat{a}_2|\phi_i\phi_k\rangle^*$$
$$- \langle\delta\phi_k\phi_i|\hat{a}_2|\phi_i\phi_k\rangle - \langle\delta\phi_k\phi_i|\hat{a}_2|\phi_i\phi_k\rangle^*\big]. \tag{9.56}$$

We shall now apply Eqs. (9.53) and (9.56) for F of Eq. (9.26),

$$F = \langle\Phi|\hat{H}_e|\Phi\rangle - \sum_{i,j}\lambda_{ij}\big[\langle\phi_i|\phi_j\rangle - \delta_{i,j}\big]. \tag{9.57}$$

Then

$$\delta F = \langle\delta\phi_k|\hat{h}_1|\phi_k\rangle^* + \langle\delta\phi_k|\hat{h}_1|\phi_k\rangle + \sum_{i=1}^{N}\big[\langle\phi_i\delta\phi_k|\hat{h}_2|\phi_i\phi_k\rangle$$
$$+ \langle\phi_i\delta\phi_k|\hat{h}_2|\phi_i\phi_k\rangle^* - \langle\delta\phi_k\phi_i|\hat{h}_2|\phi_i\phi_k\rangle$$
$$- \langle\delta\phi_k\phi_i|\hat{h}_2|\phi_i\phi_k\rangle^*\big] - \sum_{i=1}^{N}\big[\lambda_{ik}\langle\delta\phi_k|\phi_i\rangle^* + \lambda_{ki}\langle\delta\phi_k|\phi_i\rangle\big], \tag{9.58}$$

where the last term originates from the last term in Eq. (9.57) and where we have used that the $\delta_{i,j}$ are constants and that

$$\delta\sum_{i,j}\lambda_{ij}\big[\langle\phi_i|\phi_j\rangle - \delta_{i,j}\big] = \sum_{i}\big[\lambda_{ik}\langle\phi_i|\delta\phi_k\rangle + \lambda_{ki}\langle\delta\phi_k|\phi_i\rangle\big]$$
$$= \sum_{i}\big[\lambda_{ik}\langle\delta\phi_k|\phi_i\rangle^* + \lambda_{ki}\langle\delta\phi_k|\phi_i\rangle\big]. \tag{9.59}$$

The requirement of stationarity is that

$$\delta F = 0 \tag{9.60}$$

for *any* variation of *any* ϕ_k. This means that $\delta\phi_k$ as well as k are absolutely general. Explicitly written down, Eq. (9.60) resembles something like

$$\int f_1(x)\delta f(x)\,dx + \int f_2(x)\delta f^*(x)\,dx = 0 \tag{9.61}$$

for *any* $\delta f(x)$ and $\delta f^*(x)$. This can be fulfilled only if

$$f_1(x) \equiv f_2(x) \equiv 0 \tag{9.62}$$

for *all* x.

In order to apply this on Eq. (9.60), we shall rewrite δF in Eq. (9.58) in terms of integrals:

$$
\begin{aligned}
\delta F = & \int \delta\phi_k(\vec{x}_1)[\hat{h}_1\phi_k(\vec{x}_1)]^* \, d\vec{x}_1 + \int \delta\phi_k^*(\vec{x}_1)\hat{h}_1\phi_k(\vec{x}_1) \, d\vec{x}_1 \\
& + \sum_{i=1}^{N} \left\{ \int\int \phi_i^*(\vec{x}_2)\delta\phi_k^*(\vec{x}_1)[\hat{h}_2\phi_i(\vec{x}_2)\phi_k(\vec{x}_1)] \, d\vec{x}_1 \, d\vec{x}_2 \right. \\
& + \int\int \phi_i(\vec{x}_2)\delta\phi_k(\vec{x}_1)[\hat{h}_2\phi_i(\vec{x}_2)\phi_k(\vec{x}_1)]^* \, d\vec{x}_1 \, d\vec{x}_2 \\
& - \int\int \phi_i^*(\vec{x}_2)\delta\phi_k^*(\vec{x}_1)[\hat{h}_2\phi_i(\vec{x}_1)\phi_k(\vec{x}_2)] \, d\vec{x}_1 \, d\vec{x}_2 \\
& \left. - \int\int \delta\phi_k(\vec{x}_1)\phi_i(\vec{x}_2)[\hat{h}_2\phi_i(\vec{x}_1)\phi_k(\vec{x}_2)]^* \, d\vec{x}_1 \, d\vec{x}_2 \right\} \\
& - \sum_{i=1}^{N} [\lambda_{ik} \int \delta\phi_k(\vec{x}_1)\phi_i^*(\vec{x}_1) \, d\vec{x}_1 + \lambda_{ki} \int \delta\phi_k^*(\vec{x}_1)\phi_i(\vec{x}_1) \, d\vec{x}_1] \equiv 0.
\end{aligned}
\tag{9.63}
$$

This has to be fulfilled for *any* $\delta\phi_k(\vec{x}_1)$ and *any* $\delta\phi_k^*(\vec{x}_1)$, which only can be the case when each of the two equations

$$
\begin{aligned}
\hat{h}_1\phi_k(\vec{x}_1) + \sum_{i=1}^{N} & \left\{ \int \phi_i^*(\vec{x}_2)\hat{h}_2[\phi_i(\vec{x}_2)\phi_k(\vec{x}_1)] \, d\vec{x}_2 \right. \\
& \left. - \int \phi_i^*(\vec{x}_2)\hat{h}_2[\phi_i(\vec{x}_1)\phi_k(\vec{x}_2)] \, d\vec{x}_2 \right\} = \sum_{i=1}^{N} \lambda_{ki}\phi_i(\vec{x}_1)
\end{aligned}
\tag{9.64}
$$

and

$$
\begin{aligned}
\hat{h}_1\phi_k^*(\vec{x}_1) + \sum_{i=1}^{N} & \left\{ \int \phi_i(\vec{x}_2)\hat{h}_2[\phi_i^*(\vec{x}_2)\phi_k^*(\vec{x}_1)] \, d\vec{x}_2 \right. \\
& \left. - \int \phi_i(\vec{x}_2)\hat{h}_2[\phi_k^*(\vec{x}_2)\phi_i^*(\vec{x}_1)] \, d\vec{x}_2 \right\} = \sum_{i=1}^{N} \lambda_{ik}\phi_i^*(\vec{x}_1)
\end{aligned}
\tag{9.65}
$$

is fulfilled.

It is customary to introduce the orbital-dependent operators $\hat{J}_i$ by

$$
\begin{aligned}
\hat{J}_i \phi_k(\vec{x}_1) &= \int \phi_i^*(\vec{x}_2) \hat{h}_2 \phi_i(\vec{x}_2) \phi_k(\vec{x}_1) \, d\vec{x}_2 \\
&= \int \frac{|\phi_i(\vec{x}_2)|^2 \phi_k(\vec{x}_1)}{|\vec{r}_2 - \vec{r}_1|} \, d\vec{x}_2 \\
&= \int \frac{|\phi_i(\vec{x}_2)|^2}{|\vec{r}_2 - \vec{r}_1|} \, d\vec{x}_2 \phi_k(\vec{x}_1),
\end{aligned} \tag{9.66}
$$

where we have used the explicit expression for $\hat{h}_2$ from Eq. (8.43),

$$
\hat{h}_2(\vec{r}_i, \vec{r}_j) = \frac{1}{|\vec{r}_i - \vec{r}_j|}. \tag{9.67}
$$

$\hat{J}_i$ is accordingly seen to be the electrostatic potential of the orbital i in a given point $(\vec{r}_1)$.

Similarly, K_i is introduced by

$$
\begin{aligned}
\hat{K}_i \phi_k(\vec{x}_1) &= \int \phi_i^*(\vec{x}_2) \hat{h}_2 \phi_i(\vec{x}_1) \phi_k(\vec{x}_2) \, d\vec{x}_2 \\
&= \int \frac{\phi_i^*(\vec{x}_2) \phi_k(\vec{x}_2) \phi_i(\vec{x}_1)}{|\vec{r}_2 - \vec{r}_1|} \, d\vec{x}_2 \\
&= \int \frac{\phi_i^*(\vec{x}_2) \phi_k(\vec{x}_2)}{|\vec{r}_2 - \vec{r}_1|} \, d\vec{x}_2 \phi_i(\vec{x}_1) \\
&= \int \frac{\phi_i^*(\vec{x}_2) \hat{P}_{12}[\phi_i(\vec{x}_2) \phi_k(\vec{x}_1)]}{|\vec{r}_2 - \vec{r}_1|} \, d\vec{x}_2,
\end{aligned} \tag{9.68}
$$

where $\hat{P}_{12}$ is the permutation operator that interchanges the two following arguments (i.e., $\vec{x}_1$ and $\vec{x}_2$ of ϕ_k and ϕ_i).

The $\hat{J}_i$ and $\hat{K}_i$ operators are the so-called Coulomb and exchange operators, respectively. The Coulomb operators describe the classical electrostatic interactions between two charge distributions described by, e.g., $|\phi_i|^2$ and $|\phi_k|^2$, respectively, whereas the exchange operators have no classical analogue and their existence is a direct consequence of the quantum-mechanical requirement that the N-electron wavefunction is to be antisymmetric when interchanging any two electrons.

With the help of the Coulomb and exchange operators Eqs. (9.64) and (9.65) can be written as

$$
\left[\hat{h}_1 + \sum_{i=1}^{N} (\hat{J}_i - \hat{K}_i) \right] \phi_k = \sum_{i=1}^{N} \lambda_{ki} \phi_i \tag{9.69}
$$

and

$$\left[\hat{h}_1 + \sum_{i=1}^{N}(\hat{J}_i - \hat{K}_i)^*\right]\phi_k^* = \sum_{i=1}^{N}\lambda_{ik}\phi_i^*. \tag{9.70}$$

Taking the complex conjugate of Eq. (9.70), using that the complex conjugate of $\hat{h}_1$ is $\hat{h}_1$, and subtracting the resulting equation from Eq. (9.69) leads to

$$\sum_{i=1}^{N}(\lambda_{ki} - \lambda_{ik}^*)\phi_i = 0. \tag{9.71}$$

Since the orbitals are linearly independent (otherwise we would put more electrons into the same orbital and the orbitals would not be orthonormal), Eq. (9.71) can only be valid if

$$\lambda_{ki} = \lambda_{ik}^*, \tag{9.72}$$

i.e., the λs form an $N \times N$ Hermitian matrix, and Eqs. (9.69) and (9.70) are equivalent and complex conjugate to each other.

The operator

$$\hat{F} = \hat{h}_1 + \sum_{i=1}^{N}(\hat{J}_i - \hat{K}_i) \tag{9.73}$$

is the so-called Fock operator. It is Hermitian, which can easily be shown by using the fact that $\hat{h}_1$ is Hermitian and the properties of the $\hat{K}_i$ and $\hat{J}_i$ operators.

With the Fock operator Eq. (9.69) takes the simple form

$$\hat{F}\phi_k = \sum_{i=1}^{N}\lambda_{ki}\phi_i. \tag{9.74}$$

There are many different solutions to this equation, which leads to different sets of Lagrange multipliers λ_{ki}. In particular we can focus on that solution for which the Lagrange multipliers obey

$$\lambda_{ki} = \delta_{k,i}\varepsilon_k, \tag{9.75}$$

i.e., they are only non-zero for $k = i$ and in that case set equal to ε_k (this is just a new name for the Lagrange multiplier).

In this case the Hartree–Fock equations become

$$\hat{F}\phi_k = \varepsilon_k\phi_k, \tag{9.76}$$

which is a traditional eigenvalue problem. It resembles a Schrödinger equation, but we stress that the operator on the left-hand side is *not* the Hamilton operator but instead the Fock operator, that the equation is one for single particles and not for all N particles (the effects of all the other particles are included in $\hat{F}$), that the operator $\hat{F}$ is complicated in the sense that it depends on the solutions (through the $\hat{J}_i$ and $\hat{K}_i$ operators, that in turn depend on the orbital ϕ_i), and that we have not yet related the eigenvalue ε_k to any physical observable.

Equation (9.76) forms the standard Hartree–Fock equations. Compared with the more general form of Eq. (9.74), only the diagonal elements of the Lagrange multipliers are non-vanishing. The solutions to the more general equation (9.74) (we will denote these solutions $\tilde{\phi}_k$) are obtained from those of Eq. (9.76) through a unitary transformation,

$$\tilde{\phi}_l = \sum_{k=1}^{N} U_{lk}\phi_k \qquad (9.77)$$

where the matrix $\underline{\underline{U}}$ containing the coefficients U_{lk} is unitary,

$$\underline{\underline{U}}^\dagger \cdot \underline{\underline{U}} = \underline{\underline{U}} \cdot \underline{\underline{U}}^\dagger = \underline{\underline{1}}. \qquad (9.78)$$

Upon insertion it can easily be shown that the functions of Eq. (9.77) are solutions to Eq. (9.74) when the ϕ_k satisfy Eq. (9.76), but there are then off-diagonal Lagrange multipliers.

The functions of Eq. (9.77) represent a new set of orbitals compared to those of Eq. (9.74). However, any physical observable will be unchanged. This means that there is no direct physical reason for choosing one set instead of another. On the other hand, the solutions to Eq. (9.76) are particularly simple to interpret and therefore one focuses almost exclusively on those. But it is important to remember that other sets of orbitals that are of equal quality can be obtained through the transformation (9.77).

The Hartree–Fock equations (9.76) appear as being simple but when inserting the precise form of the different operators that enter them it becomes clear that they are complicated integro-differential equations that in most practical cases cannot be solved exactly. Therefore, simplified approaches are required. The famous one of Roothaan will be presented below in Section 9.4.

9.3 ORBITALS, TOTAL ENERGIES, AND KOOPMANS' THEOREM

The Fock operator in the Hartree–Fock equations,

$$\hat{F}\phi_k = \varepsilon_k\phi_k, \qquad (9.79)$$

is

$$\hat{F} = \hat{h}_1 + \sum_{i=1}^{N}(\hat{J}_i - \hat{K}_i) \qquad (9.80)$$

with

$$\hat{J}_i\phi_k(\vec{x}_1) = \int \frac{|\phi_i(\vec{x}_2)|^2}{|\vec{x}_2 - \vec{x}_1|} \, d\vec{x}_2 \phi_k(\vec{x}_1) \qquad (9.81)$$

and

$$\hat{K}_i\phi_k(\vec{x}_1) = \int \frac{\phi_i^*(\vec{x}_2)\hat{P}_{12}[\phi_i(\vec{x}_2)\phi_k(\vec{x}_1)]}{|\vec{x}_2 - \vec{x}_1|} \, d\vec{x}_2. \qquad (9.82)$$

The summation in Eq. (9.80) runs over the number of electrons N, which equals the number of occupied orbitals. We see now that the operators in Eqs. (9.81) and (9.82) depend on the precise form of these orbitals, which in turn are determined by solving the Hartree–Fock equation (9.79). This means that these equations can only be solved using a self-consistent procedure: starting with some (intelligent) initial guess of the orbitals, one generates the operators of Eqs. (9.81) and (9.82), which leads to new orbitals by solving the Hartree–Fock equations. These can be used in generating new operators, and so on. This procedure is repeated until the input and output orbitals agree to a given accuracy, at which point it is assumed that the Hartree–Fock calculations have been solved *self-consistently*.

However, at any point in this procedure (including at the end), solving the Hartree–Fock equations will in almost all cases lead to more than N orbitals, so that there is an arbitrariness in which of those are chosen to be occupied. When interested in the ground state one will choose those N orbitals to be occupied that belong to the lowest eigenvalues ε_k of Eq. (9.79). Why this is reasonable will be discussed below when we present the so-called Koopmans' theorem. But for excited states other orbitals will be occupied. Here, however, caution is required: the Hartree–Fock equations have been derived by applying the variational principle which, as presented in Chapter 5, is valid only for calculating the energetically lowest state of a given system, i.e., the ground state. The principle can be extended to be valid for calculating the energetically lowest state of a given irreducible representation of the group that describes the symmetry operations of the system (cf. Chapter 7).

Before discussing Koopmans' theorem we shall present various expressions for the total electronic energy,

$$E_{\mathrm{HF}} = \langle \Phi | H_e | \Phi \rangle, \tag{9.83}$$

where Φ is the Slater determinant of Eq. (9.24) and the subscript 'HF' denotes 'Hartree–Fock'.

This expression was given in Eq. (9.47):

$$\langle \Phi | \hat{H}_e | \Phi \rangle = \sum_{i=1}^{N} \langle \phi_i | \hat{h}_1 | \phi_i \rangle + \frac{1}{2} \sum_{i,j=1}^{N} \left[\langle \phi_i \phi_j | \hat{h}_2 | \phi_i \phi_j \rangle - \langle \phi_j \phi_i | \hat{h}_2 | \phi_i \phi_j \rangle \right]. \tag{9.84}$$

We may also use the Hartree–Fock equation (9.79) in calculating E_{HF}. We multiply the equation (9.79) for a given k with ϕ_k^* and integrate over all space. This gives

$$\langle \phi_k | \hat{F} | \phi_k \rangle = \varepsilon_k \langle \phi_k | \phi_k \rangle. \tag{9.85}$$

We sum over all the occupied orbitals (over k) and use the fact that the orbitals are normalized:

$$\sum_{k=1}^{N} \langle \phi_k | \hat{F} | \phi_k \rangle = \sum_{k=1}^{N} \varepsilon_k. \tag{9.86}$$

Subsequently, we insert the expression for $\hat{F}$ [Eq. (9.2)] together with

$$\hat{h}_2(\vec{r}_i, \vec{r}_j) = \frac{1}{|\vec{r}_i - \vec{r}_j|} = \hat{h}_2(\vec{r}_j, \vec{r}_i). \tag{9.87}$$

This gives

$$\sum_{k=1}^{N} \varepsilon_k = \sum_{k=1}^{N} \langle \phi_k | \hat{h}_1 | \phi_k \rangle + \sum_{k,l=1}^{N} \left[\langle \phi_k \phi_l | \hat{h}_2 | \phi_k \phi_l \rangle - \langle \phi_l \phi_k | \hat{h}_2 | \phi_k \phi_l \rangle \right]. \tag{9.88}$$

Comparing with Eq. (9.84), the sum of the eigenvalues in Eq. (9.88) is thus seen to be very similar to E_{HF} with, however, the important difference that the last sum in Eq. (9.88) (which is due to the electron–electron interactions) has been multiplied by two compared with E_{HF}. But it means that we have

$$E_{HF} = \sum_{k=1}^{N} \varepsilon_k - \frac{1}{2} \sum_{k,l=1}^{N} \left[\langle \phi_k \phi_l | \hat{h}_2 | \phi_k \phi_l \rangle - \langle \phi_l \phi_k | \hat{h}_2 | \phi_k \phi_l \rangle \right]. \tag{9.89}$$

The eigenvalues ε_k appear accordingly as being individual orbital energies whose sum gives one large contribution to the total electronic energy but which has to be modified by a double-counting term due to electron–electron interactions.

This is related to the results of Koopmans (1933). He considered a system of N electrons for which the Hartree–Fock equations were solved. As mentioned above, more than N orbitals will be found in the typical case. Koopmans compared the total electronic energy for this system with that obtained by either removing one of the electrons from a given orbital or adding one further electron to a given orbital. He found that

$$E_{HF}(N - 1) - E_{HF}(N) \simeq -\varepsilon_n \tag{9.90}$$

and

$$E_{HF}(N + 1) - E_{HF}(N) \simeq \varepsilon_m. \tag{9.91}$$

In Eq. (9.90) we have assumed that an electron of the nth orbital has been removed, and in Eq. (9.91) that an extra electron has been added to the mth orbital. It has also been assumed that the nth electron was occupied and the mth empty for the N-electron system. Finally, $E_{HF}(M)$ is the total electronic energy for the M-electron system.

Equations (9.90) and (9.91) are only valid to lowest order in the changes due to the changed number of electrons. Thus, changes of the orbitals due to the changed number of electrons are ignored (these changes are often called relaxation effects).

In order to prove Koopmans' theorem we proceed as follows. From Eq. (9.85) we find

$$\varepsilon_k = \langle \phi_k | \hat{F} | \phi_k \rangle = \langle \phi_k | \hat{h}_1 | \phi_k \rangle + \sum_{l=1}^{N} \left[\langle \phi_k \phi_l | \hat{h}_2 | \phi_k \phi_l \rangle - \langle \phi_l \phi_k | \hat{h}_2 | \phi_k \phi_l \rangle \right], \tag{9.92}$$

independently of whether the kth orbital is occupied or not (i.e., of whether $k \leq N$ or $k > N$).

Then, from Eq. (9.84),

$$E_{HF}(N-1) - E_{HF}(N)$$

$$= \sum_{\substack{i=1 \\ i \neq n}}^{N} \langle \phi_i | \hat{h}_1 | \phi_i \rangle + \frac{1}{2} \sum_{\substack{i,j=1 \\ i,j \neq n}}^{N} \left[\langle \phi_i \phi_j | \hat{h}_2 | \phi_i \phi_j \rangle - \langle \phi_j \phi_i | \hat{h}_2 | \phi_i \phi_j \rangle \right]$$

$$- \sum_{i=1}^{N} \langle \phi_i | \hat{h}_1 | \phi_i \rangle - \frac{1}{2} \sum_{i,j=1}^{N} \left[\langle \phi_i \phi_j | \hat{h}_2 | \phi_i \phi_j \rangle - \langle \phi_j \phi_i | \hat{h}_2 | \phi_i \phi_j \rangle \right]$$

$$= - \langle \phi_n | \hat{h}_1 | \phi_n \rangle - \frac{1}{2} \sum_{\substack{j=1 \\ j \neq n}}^{N} \left[\langle \phi_n \phi_j | \hat{h}_2 | \phi_n \phi_j \rangle - \langle \phi_j \phi_n | \hat{h}_2 | \phi_n \phi_j \rangle \right]$$

$$- \frac{1}{2} \sum_{\substack{i=1 \\ i \neq n}}^{N} \left[\langle \phi_i \phi_n | \hat{h}_2 | \phi_i \phi_n \rangle - \langle \phi_n \phi_i | \hat{h}_2 | \phi_i \phi_n \rangle \right]$$

$$= - \langle \phi_n | \hat{h}_1 | \phi_n \rangle - \sum_{\substack{j=1 \\ j \neq n}}^{N} \left[\langle \phi_n \phi_j | \hat{h}_2 | \phi_n \phi_j \rangle - \langle \phi_j \phi_n | \hat{h}_2 | \phi_n \phi_j \rangle \right]$$

$$= - \langle \phi_n | \hat{h}_1 | \phi_n \rangle - \sum_{j=1}^{N} \left[\langle \phi_n \phi_j | \hat{h}_2 | \phi_n \phi_j \rangle - \langle \phi_j \phi_n | \hat{h}_2 | \phi_n \phi_j \rangle \right]$$

$$= -\varepsilon_n, \tag{9.93}$$

i.e., Eq. (9.90) is obtained. The proof of Eq. (9.91) is very similar and will therefore not be presented here.

Koopmans' theorem, Eqs. (9.90) and (9.91), is only a first-order approximation. But it allows for interpreting the eigenvalues ε_k to the Hartree−Fock equations as orbital energies that are related to electronic transition energies, and it also provides a further argument for occupying the N orbitals of the N lowest ε_k in solving the Hartree−Fock calculations.

9.4 THE HARTREE−FOCK−ROOTHAAN METHOD

Solving the Hartree−Fock equations

$$\hat{F} \psi_k = \varepsilon_k \psi_k \tag{9.94}$$

requires finding not only the (real) constants ε_k but also each of the orbitals ϕ_k in every single point in position space as well as its spin-dependence. Obviously, it is an enormous amount of information that is sought and this, together with the complexity of the Hartree–Fock equations (see the discussion in the preceding subsection), makes solving them directly essentially impossible. In order to avoid these problems, Roothaan (1951) has proposed an alternative way of obtaining an approximate wavefunction Φ of the form of a Slater determinant,

$$\Phi(\vec{x}_1, \vec{x}_2, \ldots, \vec{x}_N) = \frac{1}{\sqrt{N!}} \begin{vmatrix} \phi_1(\vec{x}_1) & \phi_2(\vec{x}_1) & \ldots & \phi_N(\vec{x}_1) \\ \phi_1(\vec{x}_2) & \phi_2(\vec{x}_2) & \ldots & \phi_N(\vec{x}_2) \\ \vdots & \vdots & \ddots & \vdots \\ \phi_1(\vec{x}_N) & \phi_2(\vec{x}_N) & \ldots & \phi_N(\vec{x}_N) \end{vmatrix}$$

$$\equiv |\phi_1, \phi_2, \ldots, \phi_N|. \tag{9.95}$$

As before, Φ is obtained by considering F of Eq. (9.25),

$$F = \langle \Phi | \hat{H}_e | \Phi \rangle - \sum_{i,j} \lambda_{ij} [\langle \phi_i | \phi_j \rangle - \delta_{i,j}]. \tag{9.96}$$

Roothaan suggested, however, that instead of varying all orbitals in all points (i.e., varying infinitely many parameters), only a finite variation was considered. This was made possible by expanding the orbitals in a set of *fixed* basis functions, i.e.,

$$\psi_l(\vec{x}) = \sum_{p=1}^{N_b} \chi_p(\vec{x}) c_{pl}, \tag{9.97}$$

where the basis functions χ as well as their number N_b have been chosen in advance and only the expansion coefficients c_{pl} (whose number is finite) are varied.

We insert Eq. (9.97) into the expression for F,

$$F = \sum_{i=1}^{N} \langle \phi_i | \hat{h}_1 | \phi_i \rangle + \frac{1}{2} \sum_{i,j=1}^{N} [\langle \phi_i \phi_j | \hat{h}_2 | \phi_i \phi_j \rangle - \langle \phi_j \phi_i | \hat{h}_2 | \phi_i \phi_j \rangle]$$

$$- \sum_{i,j=1}^{N} \lambda_{ij} [\langle \phi_i | \phi_j \rangle - \delta_{i,j}] \tag{9.98}$$

and obtain

$$F = \sum_{i=1}^{N} \sum_{m,n=1}^{N_b} c_{mi}^* c_{ni} \langle \chi_m | \hat{h}_1 | \chi_n \rangle$$

$$+ \frac{1}{2} \sum_{i,j=1}^{N} \sum_{m,n,q,r=1}^{N_b} c_{mi}^* c_{ni} c_{qj}^* c_{rj} [\langle \chi_m \chi_q | \hat{h}_2 | \chi_n \chi_r \rangle - \langle \chi_q \chi_m | \hat{h}_2 | \chi_n \chi_r \rangle]$$

$$- \sum_{i,j=1}^{N} \lambda_{ij} \left[\sum_{m,n=1}^{N_b} c_{mi}^* c_{nj} \langle \chi_m | \chi_n \rangle - \delta_{i,j} \right]. \tag{9.99}$$

The requirement

$$\delta F = 0 \tag{9.100}$$

is then replaced by

$$\frac{\partial F}{\partial c_{pl}} = \frac{\partial F}{\partial c_{pl}^*} = 0, \tag{9.101}$$

where c_{pl} is the coefficient of the lth orbital to the pth basis function. These equations become

$$\sum_{m=1}^{N_b} c_{ml}^* \langle \chi_m | \hat{h}_1 | \chi_p \rangle + \sum_{i=1}^{N} \sum_{m,n,q=1}^{N_b} c_{ml}^* c_{ni}^* c_{qi} \left[\langle \chi_m \chi_n | \hat{h}_2 | \chi_p \chi_q \rangle \right.$$
$$\left. - \langle \chi_n \chi_m | \hat{h}_2 | \chi_p \chi_q \rangle \right] - \sum_{i=1}^{N} \lambda_{li} \sum_{m=1}^{N_b} c_{mi}^* \langle \chi_m | \chi_p \rangle = 0 \tag{9.102}$$

and

$$\sum_{m=1}^{N_b} c_{ml} \langle \chi_m | \hat{h}_1 | \chi_p \rangle + \sum_{i=1}^{N} \sum_{m,n,q=1}^{N_b} c_{ml} c_{ni} c_{qi}^* \left[\langle \chi_p \chi_q | \hat{h}_2 | \chi_m \chi_n \rangle \right.$$
$$\left. - \langle \chi_q \chi_p | \hat{h}_2 | \chi_m \chi_n \rangle \right] - \sum_{i=1}^{N} \lambda_{il} \sum_{m=1}^{N_b} c_{mi} \langle \chi_p | \chi_m \rangle = 0. \tag{9.103}$$

As in Section 9.2, where we derived the Hartree–Fock equations, one may prove that the Lagrange multipliers λ_{il} form a Hermitian matrix, and that, therefore, the two equations (9.102) and (9.103) are complex conjugate to each other. Furthermore, one can *choose* to determine that set of orbitals that correspond to

$$\lambda_{ij} = \varepsilon_i \delta_{i,j}, \tag{9.104}$$

and show that other orbitals are obtained through unitary transformations [Eq. (9.77)] of those thereby obtained. In total we may then rewrite Eq. (9.103) as

$$\sum_{m=1}^{N_b} \left\{ \langle \chi_p | \hat{h}_1 | \chi_m \rangle + \sum_{i=1}^{N} \sum_{n,q=1}^{N_b} c_{ni} c_{qi}^* \left[\langle \chi_p \chi_q | \hat{h}_2 | \chi_m \chi_n \rangle - \langle \chi_q \chi_p | \hat{h}_2 | \chi_m \chi_n \rangle \right] \right\} c_{ml}$$
$$= \varepsilon_l \sum_{m=1}^{N_b} \langle \chi_p | \chi_m \rangle c_{ml}. \tag{9.105}$$

This is the Hartree–Fock–Roothaan equations that have made Hartree–Fock calculations so widely used. As for the Hartree–Fock equations, they have to be solved self-consistently: From one set of solutions (i.e., the coefficients c_{ml}) one calculates the expression inside the large curly brackets on the left-hand side of Eq. (9.105) as well as the brackets on the right-hand side. Then, one obtains a generalized matrix eigenvalue equation,

$$\underline{\underline{F}} \cdot \underline{c}_l = \varepsilon_l \cdot \underline{\underline{O}} \cdot \underline{c}_l, \tag{9.106}$$

where $\underline{\underline{F}}$ contains the Fock matrix elements

$$F_{pm} = \langle \chi_p | \hat{h}_1 | \chi_m \rangle + \sum_{i=1}^{N} \sum_{n,q=1}^{N_b} c_{ni} c_{qi}^* \left[\langle \chi_p \chi_q | \hat{h}_2 | \chi_m \chi_n \rangle - \langle \chi_q \chi_p | \hat{h}_2 | \chi_m \chi_n \rangle \right] \tag{9.107}$$

and $\underline{\underline{O}}$ contains the overlap matrix elements

$$O_{pm} = \langle \chi_p | \chi_m \rangle. \tag{9.108}$$

Finally, $\underline{c}_l$ contains the (sought) coefficients of the lth orbital to the different basis functions.

Equation (9.106) can be solved using standard matrix routines, which yields a new set of coefficients, and so on until the procedure converges.

The great advantage of Roothaan's approach is that is replaces the problem of determining an infinite number of values (the values of *all* occupied orbitals in *all* position-space points) with that of determining only a finite number of parameter values (the expansion coefficients of the orbitals to the basis functions). This means that it is possible to improve any calculation in a systematic way: due to the variational principle increasing the set of basis functions will automatically lead to at least a better approximation for the total electronic energy. On the other hand, it also means that the quality of a given calculation depends heavily on the quality of the basis functions and, therefore, it is of ultimate importance to apply basis sets that are 'adequate' for the problem at hand. This is a far from trivial requirement, and much effort has been put into developing 'good' basis sets. Some typical choices of basis functions will be described in Chapter 10.

Finally, the accuracy of a Hartree–Fock–Roothaan calculation is never better than that indirectly assumed via the Hartree–Fock approximation. Thus, such a calculation can at most only offer an accurate solution to the Hartree–Fock equations, and the results are only reliable as long as the Hartree–Fock approximation is justified, i.e., that the electronic wavefunction can be approximated by a single Slater determinant containing single-electron orbitals. In Chapter 13 we shall discuss how to go beyond the Hartree–Fock approximations and when this may be necessary.

9.5 PHYSICAL PROPERTIES

Before turning to the problem of choosing appropriate basis sets we shall give some few examples of the kind of physical properties that can be calculated once the orbitals of the Slater determinant have been calculated.

The first quantity is the total energy itself. As discussed in Chapter 8, this contains two contributions when assuming that the Born–Oppenheimer approximation is valid [Eq. (8.33)]:

$$E(\vec{X}) = \frac{1}{2} \sum_{k_1 \neq k_2 = 1}^{M} \frac{Z_{k_1} Z_{k_2}}{|\vec{R}_{k_1} - \vec{R}_{k_2}|} + E_e(\vec{X}), \qquad (9.109)$$

where we explicitly have indicated that it depends on the positions of the nuclei, $\vec{R}_1, \vec{R}_2, \ldots, \vec{R}_M$. $\vec{X}$ is thereby a compact notation for all the nuclear degrees of freedom.

In our case, where we have assumed that the Hartree–Fock approximation is valid, E_e of Eq. (9.109) is the electronic total energy of Eqs. (9.83), (9.84), or (9.89),

$$E_e = E_{HF}. \qquad (9.110)$$

The simplest case one can think of (except for isolated atoms) is a diatomic molecule, AB. For this, E depends only on the bond length R, and is invariant with respect to arbitrary translations and rotations of the molecule (this is the well-known fact that the number of degrees for freedom for an M-atomic molecule equals $3M - 5$ for a linear molecule and otherwise $3M - 6$). A typical calculation will then give a curve like that of Fig. 9.3. The theoretically estimated equilibrium bond length R_e is then determined from

$$E(R_e) = \min. \qquad (9.111)$$

R_e is in most cases accurately given within the Hartree–Fock approximation although there are cases (e.g., systems containing transition-metal atoms) where severe errors may occur. These are cases where so-called correlation effects become important. Later (Chapter 11), we shall present methods that take these into account as well as a simple way of understanding and estimating when they become important.

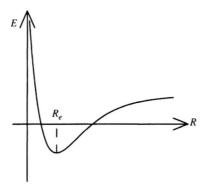

Figure 9.3 Schematic representation of how the total energy depends on the bond length for a diatomic molecule

One may also calculate the energy of the A–B bond, i.e., the dissociation energy D_e of the molecule,

$$D_e = E(A) + E(B) - E(AB) = E(\infty) - E(R_e). \qquad (9.112)$$

Within the Hartree–Fock approximation these are often less accurate than R_e.

Around the minimum in Fig. 9.3 we may expand the total energy in a Taylor series

$$E(R) \simeq E(R_e) + \frac{k}{2}(R - R_e)^2, \qquad (9.113)$$

where the series has been truncated after the second-order term, and where we have used that the first-order derivative vanishes at the minimum. Furthermore

$$k = \left. \frac{d^2E}{dR^2} \right|_{R=R_e}. \qquad (9.114)$$

This amounts to assuming that the two nuclei move in a harmonic potential and one can thereby determine the frequencies of these oscillations. From the standard expressions for the harmonic oscillator we obtain

$$\hbar\omega = \hbar\sqrt{\frac{k}{\mu}}, \qquad (9.115)$$

where the mass μ is the reduced mass of the two nuclei

$$\mu = \frac{M_A M_B}{M_A + M_B}. \qquad (9.116)$$

Some examples of calculated structures and vibrational frequencies are given in Tables 9.1 and 9.2. In particular in Table 9.2 we observe that the frequencies as obtained from the Hartree–Fock calculations are most often overestimated, which is a rather general result.

Table 9.1 Calculated structural properties of some small molecules. Lengths are given in ångströms and angles in degrees, and the results are from Daudel *et al.* (1983)

System	X–H bond length		H–X–H bond angle	
	Theory	Experiment	Theory	Experiment
H_2	0.730	0.742		
CH_3	1.072	1.079	120.0	120.0
CH_4	1.082	1.085	109.5	109.5
NH_2	1.015	1.024	108.6	103.4
NH_3	0.991	1.012	116.1	106.7
OH	0.967	0.971		
H_2O	0.948	0.957	111.5	104.5
HF	0.921	0.917		

Table 9.2 Calculated vibrational properties of some small molecules. The frequencies are given in cm^{-1}, and the results are from Daudel *et al.* (1983)

System	Theory	Experiment
H_2	4644	4405
CH_3	3321	3184
	3125	3002
	1470	1383
	776	580
CH_4	3372	3019
	3226	2917
	1718	1534
	1533	1306
NH_2	3676	3220
	3554	3173
	1651	1499
NH_3	3985	3444
	3781	3336
	1814	1627
	597	950
OH	3955	3735
H_2O	4143	3756
	3987	3657
	1678	1595
HF	4150	4138

Due to the zero-point motion (i.e., that for a harmonic oscillator the lowest energy is not 0 but $\frac{1}{2}\hbar\omega$), one can obtain a corrected total energy

$$E(R) \to E(R) + \tfrac{1}{2}\hbar\omega. \tag{9.117}$$

In the present case this is not important, but for larger systems where one wants to compare different structures that have more (different) types of vibrations and therefore more contributions to the zero-point motion [i.e., a term like that of Eq. (9.117) from each vibrational mode], these effects may be crucial in order to obtain a correct description of the relative stability of different structures.

We shall here, however, not study this further. Instead, we shall look at other quantities that are of interest when studying the properties of a given system. To these belong the orbital energies

$$\varepsilon_k \tag{9.118}$$

which, according to Koopmans' theorem can be related to ionization potentials and electron affinities. However, as discussed in Section 9.3, the agreement is not perfect since the theorem neglects effects related to the fact that the orbitals will change when the number of electrons is changed (i.e., so-called relaxation effects).

Another quantity of interest is the electron density in position space $\rho(\vec{r})$. This can be calculated as an expectation value

$$A = \langle \Phi | \hat{A} | \Phi \rangle. \tag{9.119}$$

For $\rho(\vec{r})$, $\hat{A}$ is a sum of single-particle operators,

$$\hat{A} = \sum_{i=1}^{N} \hat{a}_1(i), \tag{9.120}$$

where $\hat{a}_1(i)$ acts on only the ith electron. Within the Hartree–Fock approximation,

$$A = \sum_{i=1}^{N} \langle \phi_i | \hat{a}_1 | \phi_i \rangle. \tag{9.121}$$

For $\rho(\vec{r})$, $\hat{a}_1(i)$ is the operator for finding the ith electron at position $\vec{r}$, i.e.,

$$\hat{a}_1(i) = \delta(\vec{r} - \vec{r}_i) \tag{9.122}$$

in position-space representation. This gives, in Eq. (9.121),

$$\rho(\vec{r}) = \sum_{i=1}^{N} |\phi_i(\vec{r})|^2, \tag{9.123}$$

where $\phi_i(\vec{r})$ is the position-space dependence (i.e., the spin dependence has been integrated out) of ϕ_i.

In Fig. 9.4 we show some representative examples of calculated electron densities. We see, however, that these are often very structure-less and that they largely resemble the densities of superposed atoms. The effects of the chemical bonds are therefore hardly visible. In order to improve on that it has therefore become practice to compare the total density with that of the superposed atomic densities (i.e., the sum of the densities of the isolated atoms but placed at the positions where the nuclei are for the system of interest). This leads to the so-called difference density:

$$\Delta\rho(\vec{r}) = \rho(\vec{r}) - \sum_{m=1}^{M} \rho_{\text{atom } m}(\vec{r} - \vec{R}_m). \tag{9.124}$$

Two examples of this are shown in Fig. 9.5.

Another quantity of interest is the spin density. This may be defined as corresponding to

$$\hat{a}_1(i) = \delta(\vec{r} - \vec{r}_i)\hat{s}_{i,z}, \tag{9.125}$$

where $\hat{s}_{i,z}$ is the operator for the z component of the spin operator for the ith electron. The spin density is thus essentially the difference of the electron density

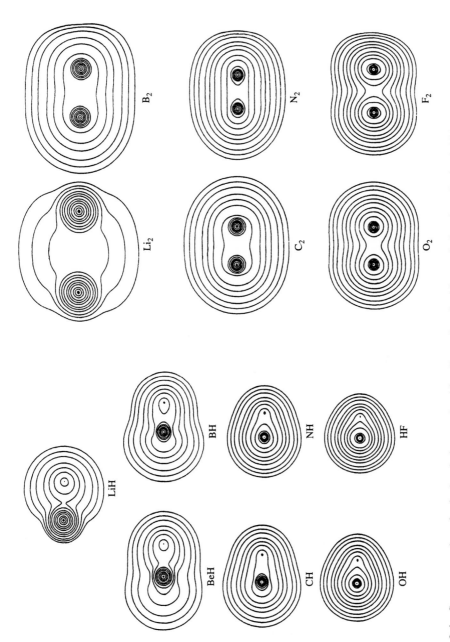

Figure 9.4 Some examples of calculated electron densities. Reproduced with permission from Bader (1970)

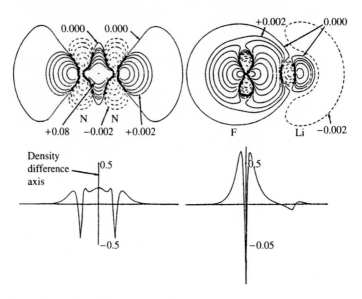

Figure 9.5 Calculated electron density differences for (left) N_2 and (right) FLi. Reproduced with permission from Bader (1970)

for the electrons with α spin minus that of the electrons with β spin:

$$\rho_s(\vec{r}) = \frac{1}{2} \sum_{i=\alpha-\text{spin}} |\phi_i(\vec{r})|^2 - \frac{1}{2} \sum_{i=\beta-\text{spin}} |\phi_i(\vec{r})|^2. \tag{9.126}$$

Examples of this are shown in Fig. 9.6.

The total dipole moment of a molecule is the sum of that of the nuclei plus that of the electrons. Considering, e.g., the y component, the latter is determined through

$$\hat{a}_1(i) = -y_i, \tag{9.127}$$

where y_i is the y coordinate of the ith electron (the elementary charge is, within the atomic units, set equal to 1). Since the contribution from the nuclei is

$$\sum_{m=1}^{M} Z_m Y_m, \tag{9.128}$$

(Z_m and Y_m is the atomic number and the y component of the position vector of the mth nucleus, respectively) the total y component of the dipole moment becomes

$$\mu_y = \sum_{m=1}^{M} Z_m Y_m - \sum_{i=1}^{N} \langle \phi_i | y | \phi_i \rangle. \tag{9.129}$$

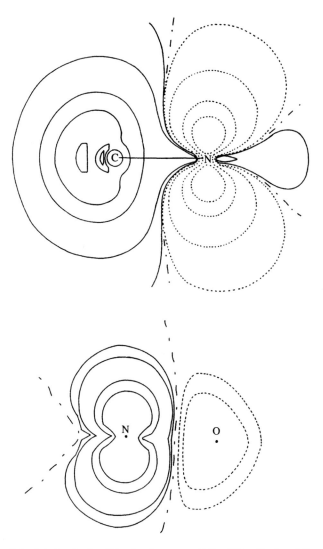

Figure 9.6 Calculated spin densities for (upper part) the CN radical and (lower part) the NO molecule. From Daudel *et al.* (1983). Reproduced by permission of John Wiley & Sons

The molecule consists of charged particles, i.e., nuclei and electrons. As such they produce an electrostatic field, whose potential easily can be calculated. The contribution from the nuclei is

$$\sum_{m=1}^{M} \frac{Z_m}{|\vec{R}_m - \vec{r}|},$$
(9.130)

where $\vec{r}$ is the point of interest. The contribution of the electrons can be determined through Eq. (9.121) with

$$\hat{a}_1(i) = \frac{-1}{|\vec{r}_i - \vec{r}|}. \tag{9.131}$$

This gives the following total electrostatic potential:

$$V(\vec{r}) = \sum_{m=1}^{M} \frac{Z_m}{|\vec{R}_m - \vec{r}|} - \sum_{i=1}^{N} \int \frac{|\phi_i(\vec{s})|^2}{|\vec{s} - \vec{r}|} \, d\vec{s}. \tag{9.132}$$

Figure 9.7 shows some few examples of this potential. It is seen that there are both regions of positive and regions of negative potential. As long as the system is neutral (which is the case for the systems of Fig. 9.7), the potential will fall off faster as r^{-1} far away from the system. The potential gives some ideas about where a charged particle (or a particle with a similar non-trivial potential) will attack; e.g., in a chemical reaction, a positive ion will feel attracted to the regions of negative potential, and vice versa.

Some further quantities of interest include the kinetic energy of the electrons for which

$$\hat{a}_1(i) = -\tfrac{1}{2}\nabla_{\vec{r}_i}^2, \tag{9.133}$$

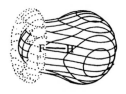

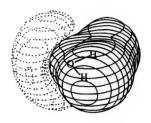

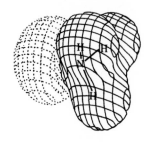

Figure 9.7   Electrostatic potentials for HF, H_2O, and NH_3. From Daudel *et al.* (1983). Reproduced by permission of John Wiley & Sons

and the potential energy related to the electron–nucleus interactions where

$$\hat{a}_1(i) = \sum_{m=1}^{M} \frac{-Z_m}{|\vec{R}_m - \vec{r}_i|}. \tag{9.134}$$

On the other hand, the potential energy related to the electron–electron interactions is determined by an operator that can be decomposed into identical two-electron operators,

$$\hat{A} = \frac{1}{2} \sum_{n \neq m=1}^{N} \hat{a}_2(n, m). \tag{9.135}$$

For this

$$\langle \Phi | \hat{A} | \Phi \rangle = \frac{1}{2} \sum_{i,j=1}^{N} \left[\langle \phi_i \phi_j | \hat{a}_2 | \phi_i \phi_j \rangle - \langle \phi_j \phi_i | \hat{a}_2 | \phi_i \phi_j \rangle \right], \tag{9.136}$$

as shown in Eq. (9.44).

For the electron–electron interactions we have

$$\hat{a}_2(i, j) = \frac{1}{|\vec{r}_i - \vec{r}_j|}. \tag{9.137}$$

9.6   RESTRICTED, UNRESTRICTED, EXTENDED, AND PROJECTED HARTREE–FOCK METHODS

As mentioned in the beginning of this chapter, the application of the variational principle means that one calculates not the exact but instead an approximate wavefunction for the N-electron system. Thereby, one has to make compromises according to how accurately various quantities are calculated and whether it can be accepted that physical principles are (may be only slightly) violated.

The total spin of the N-electron system is a quantity that can be used in evaluating the quality of a given approximate wavefunction. Alternatively, it can be used in putting extra restrictions on the approximate N-electron wavefunction. We shall here study this for the Hartree–Fock approximation.

We shall consider two operators, i.e., the square of the total spin,

$$\hat{S}^2 = \hat{S}_x^2 + \hat{S}_y^2 + \hat{S}_z^2, \tag{9.138}$$

where each of the components is a sum over all single-electron contributions,

$$\hat{S}_x = \sum_{i=1}^{N} \hat{s}_{i,x}$$

$$\hat{S}_y = \sum_{i=1}^{N} \hat{s}_{i,y}$$

$$\hat{S}_z = \sum_{i=1}^{N} \hat{s}_{i,z}, \tag{9.139}$$

as well as the z component just given.

It turns out that it is useful to consider the two so-called ladder operators

$$\hat{S}_+ = \hat{S}_x + i\hat{S}_y$$

$$\hat{S}_- = \hat{S}_x - i\hat{S}_y. \tag{9.140}$$

For an spin-eigenstate $\theta_{S,M}$ for which

$$\hat{S}^2 \theta_{S,M} = S(S+1)\theta_{S,M}$$

$$\hat{S}_z \theta_{S,M} = M\theta_{S,M} \tag{9.141}$$

we have

$$\hat{S}_+ \theta_{S,M} = [(S+M+1)(S-M)]^{1/2} \theta_{S,M+1}$$

$$\hat{S}_- \theta_{S,M} = [(S-M+1)(S+M)]^{1/2} \theta_{S,M-1}. \tag{9.142}$$

That is, the operators of Eq. (9.141) change the z component by ± 1 and keep the total spin.

With these operators we have

$$\hat{S}^2 = \hat{S}_+ \hat{S}_- - \hat{S}_z + \hat{S}_z^2 = \hat{S}_- \hat{S}_+ + \hat{S}_z + \hat{S}_z^2, \tag{9.143}$$

where we have used that $\hat{S}_+$ and $\hat{S}_-$ do not commute.

In the Slater determinant

$$\Phi(\vec{x}_1, \vec{x}_2, \ldots, \vec{x}_N) = \frac{1}{\sqrt{N!}} \begin{vmatrix} \phi_1(\vec{x}_1) & \phi_2(\vec{x}_1) & \ldots & \phi_N(\vec{x}_1) \\ \phi_1(\vec{x}_2) & \phi_2(\vec{x}_2) & \ldots & \phi_N(\vec{x}_2) \\ \vdots & \vdots & \ddots & \vdots \\ \phi_1(\vec{x}_N) & \phi_2(\vec{x}_N) & \ldots & \phi_N(\vec{x}_N) \end{vmatrix}$$

$$\equiv |\phi_1, \phi_2, \ldots, \phi_N| \tag{9.144}$$

we shall now explicitly include the spin dependence. We assume that we have n orbitals with α spin and m orbitals with β spin, i.e.,

$$\Phi(\vec{x}_1, \vec{x}_2, \ldots, \vec{x}_N) = |\phi_1^+ \alpha, \phi_2^+ \alpha, \ldots, \phi_n^+ \alpha, \phi_1^- \beta, \phi_2^- \beta, \ldots, \phi_n^- \beta, |, \tag{9.145}$$

where the functions ϕ_i^+ and ϕ_j^- depend *only* on position-space coordinates (i.e., no spin dependence), and where we allow for the orbitals of different spins to be different (hence, the superindices $+$ and $-$).

We must have

$$n + m = N. \tag{9.146}$$

But in addition, the *exact* N-electron wavefunction will be an eigenfunction to $\hat{S}^2$ and $\hat{S}_z$. It is, however, not obvious that this is the case also for the approximate wavefunction of Eq. (9.145). In order to study this we write first Φ as a sum over the $N!$ different terms, i.e., equivalently to Eq. (9.29),

$$\Phi = \frac{1}{\sqrt{N!}} \sum_{i=1}^{N!} (-1)^{P(i)} \phi_1^+(\vec{r}_{i_1})\alpha(i_1)\phi_2^+(\vec{r}_{i_2})\alpha(i_2)\cdots\phi_n^+(\vec{r}_{i_n})\alpha(i_n)$$

$$\times \phi_1^-(\vec{r}_{i_{n+1}})\beta(i_{n+1})\phi_2^-(\vec{r}_{i_{n+2}})\beta(i_{n+2})\cdots\phi_m^-(\vec{r}_{i_N})\beta(i_N), \tag{9.147}$$

where, in contrast to Eq. (9.29), we have used the fact that each term contains both all functions and all electron coordinates as arguments exactly once.

When letting the $\hat{S}_z$ operator of Eq. (9.139) act on a single term in the sum in Eq. (9.147), it returns this sum multiplied by $\frac{1}{2}$ for each α spin and by $-\frac{1}{2}$ for each β spin. This gives a total factor of $(n - m)/2$ for *any* term. Therefore,

$$\hat{S}_z\Phi = \frac{n - m}{2}\Phi, \tag{9.148}$$

i.e., the Slater determinant *is* an eigenfunction for the $\hat{S}_z$ operator independent of the precise form of the orbitals.

Φ is, however, not necessarily an eigenfunction for the $\hat{S}^2$ operator. We shall not study this in the general case, but only consider an extremely simple one, hoping that the problems become clear.

We consider the case that we have $n = m = 1$, i.e., two electrons. The Slater determinant is then

$$\Phi = \frac{1}{\sqrt{2}}\left[\phi_1^+(\vec{r}_1)\alpha(1)\phi_1^-(\vec{r}_2)\beta(2) - \phi_1^+(\vec{r}_2)\alpha(2)\phi_1^-(\vec{r}_1)\beta(1)\right]. \tag{9.149}$$

Since Φ is an eigenfunction to $\hat{S}_z$ (with the eigenvalue 0), it is it also for $\hat{S}_z^2$. But for $\hat{S}_+$ we find

$$\hat{S}_+\Phi = \left[\hat{s}_{1,+} + \hat{s}_{2,+}\right]\Phi$$

$$= \hat{s}_{1,+}\frac{1}{\sqrt{2}}\left[\phi_1^+(\vec{r}_1)\alpha(1)\phi_1^-(\vec{r}_2)\beta(2) - \phi_1^+(\vec{r}_2)\alpha(2)\phi_1^-(\vec{r}_1)\beta(1)\right]$$

$$+ \hat{s}_{2,+}\frac{1}{\sqrt{2}}\left[\phi_1^+(\vec{r}_1)\alpha(1)\phi_1^-(\vec{r}_2)\beta(2) - \phi_1^+(\vec{r}_2)\alpha(2)\phi_1^-(\vec{r}_1)\beta(1)\right]$$

$$= \frac{1}{\sqrt{2}}\left[-\phi_1^+(\vec{r}_2)\alpha(2)\phi_1^-(\vec{r}_1)\alpha(1) + \phi_1^+(\vec{r}_1)\alpha(1)\phi_1^-(\vec{r}_2)\alpha(2)\right] \tag{9.150}$$

where we have used Eq. (9.142) in giving

$$\hat{s}_+\alpha = 0$$
$$\hat{s}_+\beta = \alpha$$
$$\hat{s}_-\alpha = \beta$$
$$\hat{s}_-\beta = 0. \tag{9.151}$$

Subsequently, we apply $\hat{S}_-$ and obtain

$$\hat{S}_-\hat{S}_+\Phi = \hat{S}_- \frac{1}{\sqrt{2}}\left[-\phi_1^+(\vec{r}_2)\alpha(2)\phi_1^-(\vec{r}_1)\alpha(1) + \phi_1^+(\vec{r}_1)\alpha(1)\phi_1^-(\vec{r}_2)\alpha(2)\right]$$

$$= \left[\hat{s}_{1,-} + \hat{s}_{2,-}\right]\frac{1}{\sqrt{2}}\left[-\phi_1^+(\vec{r}_2)\alpha(2)\phi_1^-(\vec{r}_1)\alpha(1) + \phi_1^+(\vec{r}_1)\alpha(1)\phi_1^-(\vec{r}_2)\alpha(2)\right]$$

$$= \hat{s}_{1,-}\frac{1}{\sqrt{2}}\left[-\phi_1^+(\vec{r}_2)\alpha(2)\phi_1^-(\vec{r}_1)\alpha(1) + \phi_1^+(\vec{r}_1)\alpha(1)\phi_1^-(\vec{r}_2)\alpha(2)\right]$$

$$+ \hat{s}_{2,-}\frac{1}{\sqrt{2}}\left[-\phi_1^+(\vec{r}_2)\alpha(2)\phi_1^-(\vec{r}_1)\alpha(1) + \phi_1^+(\vec{r}_1)\alpha(1)\phi_1^-(\vec{r}_2)\alpha(2)\right]$$

$$= \frac{1}{\sqrt{2}}\left[-\phi_1^+(\vec{r}_2)\alpha(2)\phi_1^-(\vec{r}_1)\beta(1) + \phi_1^+(\vec{r}_1)\beta(1)\phi_1^-(\vec{r}_2)\alpha(2)\right]$$

$$+ \frac{1}{\sqrt{2}}\left[-\phi_1^+(\vec{r}_2)\beta(2)\phi_1^-(\vec{r}_1)\alpha(1) + \phi_1^+(\vec{r}_1)\alpha(1)\phi_1^-(\vec{r}_2)\beta(2)\right]$$

$$= \frac{1}{\sqrt{2}}\left[-\phi_1^+(\vec{r}_2)\alpha(2)\phi_1^-(\vec{r}_1)\beta(1) + \phi_1^+(\vec{r}_1)\beta(1)\phi_1^-(\vec{r}_2)\alpha(2)\right.$$

$$\left. - \phi_1^+(\vec{r}_2)\beta(2)\phi_1^-(\vec{r}_1)\alpha(1) + \phi_1^+(\vec{r}_1)\alpha(1)\phi_1^-(\vec{r}_2)\beta(2)\right]. \tag{9.152}$$

According to Eq. (9.143) this expression equals $\hat{S}^2\Phi$ (since $\hat{S}_z\Phi = 0$). Therefore, *only* if

$$\phi_1^+ = \phi_1^- \tag{9.153}$$

is Φ an eigenfunction to $\hat{S}^2$.

In the general case with

$$n = m \quad (= N/2) \tag{9.154}$$

one can equivalently show that Φ is an eigenfunction to $\hat{S}^2$ *if*

$$\phi_i^+ = \phi_i^-. \tag{9.155}$$

This is therefore often assumed. In that case the Hartree–Fock(–Roothaan) equations can be simplified by using the fact that the α and β functions are orthonormal and by performing all spin 'integrations' and summations. For example, the Hartree–Fock–Roothaan equation (9.105) become

$$\sum_{m=1}^{N_b} \left\{ \langle \chi_p | \hat{h}_1 | \chi_m \rangle + \sum_{i=1}^{N/2} \sum_{n,q=1}^{N_b} c_{ni} c_{qi}^* \left[2\langle \chi_p \chi_q | \hat{h}_2 | \chi_m \chi_n \rangle - \langle \chi_q \chi_p | \hat{h}_2 | \chi_m \chi_n \rangle \right] \right\} c_{ml}$$

$$= \varepsilon_l \sum_{m=1}^{N_b} \langle \chi_p | \chi_m \rangle c_{ml}, \tag{9.156}$$

where the N_b basis functions χ now are assumed to depend only on position coordinates. Compared with Eq. (9.105), the summations over the N orbitals have been replaced by ones over the $N/2$ different position-space functions.

The case (9.154) with the assumption (9.155) amounts to the so-called *restricted* Hartree–Fock (RHF) approximation. This is far the widest used Hartree–Fock method.

In all other cases one may either ignore the problem (in that case one talks about the *unrestricted* Hartree–Fock (UHF) method) or apply the technique of projection. The latter will now be discussed.

We know that *any* Hermitian, linear operator (e.g., $\hat{S}^2$) $\hat{A}$, defines a complete set of orthonormal functions f_i through

$$\hat{A} f_i = a_i f_i, \tag{9.157}$$

where

$$\langle f_i | f_j \rangle = \delta_{i,j}. \tag{9.158}$$

That the functions are complete means that *any* other function g can be expanded like

$$g = \sum_i c_i f_i, \tag{9.159}$$

with

$$c_i = \langle f_i | g \rangle. \tag{9.160}$$

We have

$$\hat{A} g = \hat{A} \left[\sum_i c_i f_i \right] = \sum_i c_i \hat{A} f_i = \sum_i c_i a_i f_i \neq \text{cst} \cdot g, \tag{9.161}$$

where we have used the fact that $\hat{A}$ is a linear operator. In our case $\hat{A}$ is the $\hat{S}^2$ operator, and g is the Slater determinant. Equation (9.161) is accordingly the statement that the Slater determinant in the general case is not an eigenfunction of the $\hat{S}^2$ operator.

Returning to the general case, we may project g onto a specific eigenfunction of $\hat{A}$. By this we mean formally multiplying $|g\rangle$ by $|f_k\rangle\langle f_k|$, i.e., we define the projection operators

$$\hat{P}_k = |f_k\rangle\langle f_k|. \tag{9.162}$$

It will be helpful to explicitly write down that the functions g and f_i depend on some variables, i.e.,

$$f_i = f_i(\vec{x})$$

$$g = g(\vec{x}). \tag{9.163}$$

Then

$$\hat{P}_k g(\vec{x}) = f_k(\vec{x}) \int f_k^*(\vec{y}) g(\vec{y})\, d\vec{y} = c_k f_k(\vec{x}), \tag{9.164}$$

by using Eq. (9.160).

Thus, $\hat{P}_k g$ returns a constant times an eigenfunction to $\hat{A}$. This is clearly also an eigenfunction to $\hat{A}$. Furthermore, it is not just any eigenfunction but it is the one that was used in defining the projection operator.

For the sake of completeness, we add that the sum of *all* the projection operators is the identity,

$$\left[\sum_k \hat{P}_k\right] g = \sum_k [\hat{P}_k g] = \sum_k c_k f_k = g = 1 \cdot g. \tag{9.165}$$

We can now apply this technique to the wavefunction Φ of Eq. (9.147). In order to illustrate the approach we shall consider the example of Eq. (9.149),

$$\Phi = \frac{1}{\sqrt{2}}\left[\phi_1^+(\vec{r}_1)\alpha(1)\phi_1^-(\vec{r}_1)\beta(2) - \phi_1^+(\vec{r}_1)\alpha(2)\phi_1^-(\vec{r}_1)\beta(1)\right]. \tag{9.166}$$

One can show that the spin function

$$\theta = \frac{1}{\sqrt{2}}\left[\alpha(1)\beta(2) + \alpha(2)\beta(1)\right] \tag{9.167}$$

is an eigenfunction of $\hat{S}^2$,

$$\hat{S}^2\theta = 2\theta = 1 \cdot (1+1)\theta \equiv S(S+1)\theta. \tag{9.168}$$

We therefore multiply Φ of Eq. (9.166) with θ^* and 'integrate' (i.e., sum over α and β spin functions for both electrons). This gives

$$\frac{1}{\sqrt{2}}\left[\phi_1^+(\vec{r}_1)\phi_1^-(\vec{r}_2) - \phi_1^+(\vec{r}_1)\phi_1^-(\vec{r}_2)\right]. \tag{9.169}$$

This function is subsequently multiplied with θ, resulting in

$$\frac{1}{\sqrt{2}}\left[\phi_1^+(\vec{r}_1)\phi_1^-(\vec{r}_2) - \phi_1^+(\vec{r}_1)\phi_1^-(\vec{r}_2)\right] \times \frac{1}{\sqrt{2}}\left[\alpha(1)\beta(2) + \alpha(2)\beta(1)\right]. \tag{9.170}$$

This function is no longer a single Slater determinant, but a sum of two,

$$\frac{1}{2}\left[\phi_1^+(\vec{r}_1)\alpha(1)\phi_1^-(\vec{r}_2)\beta(2) - \phi_1^+(\vec{r}_2)\alpha(2)\phi_1^-(\vec{r}_1)\beta(1)\right]$$

$$+ \frac{1}{2}\left[\phi_1^+(\vec{r}_1)\beta(1)\phi_1^-(\vec{r}_2)\alpha(2) - \phi_1^+(\vec{r}_2)\beta(2)\phi_1^-(\vec{r}_1)\alpha(1)\right], \qquad (9.171)$$

but is now an eigenfunction for $\hat{S}^2$. Due to the projection it will in the general case no longer be normalized, and one would have to renormalize the new function.

There are now two different ways of applying this technique. Either one can perform a standard unrestricted Hartree–Fock calculation (i.e., ignore the problem related to $\hat{S}^2$) and first after having completed the calculation (i.e., obtained the self-consistent solutions to the Hartree–Fock–Roothaan equations), perform the projection. This is the so-called *projected* Hartree–Fock method. Alternatively, one may at each step of the calculation perform the projection (i.e., doing the iterative process of solving the Hartree–Fock–Roothaan equations). This leads to the so-called *extended* Hartree–Fock method. But neither approach is optimal, and one should rather apply some of the methods that include correlation effects. These will be described in Chapter 13.

9.7 AN EXAMPLE

We shall here discuss an example that is constructed so that it is so simple that the calculations can be carried through by hand but thereby suffers from being unrealistic.

We consider a system with four electrons. These four electrons occupy pairwise two orbitals but have different spin dependences. Thus, the Slater determinant is

$$\Phi(\vec{x}_1, \vec{x}_2, \vec{x}_3, \vec{x}_4) = \frac{1}{\sqrt{4!}} \begin{vmatrix} \phi_1(\vec{r}_1)\alpha(1) & \phi_1(\vec{r}_1)\beta(1) & \phi_2(\vec{r}_1)\alpha(1) & \phi_2(\vec{r}_1)\beta(1) \\ \phi_1(\vec{r}_2)\alpha(2) & \phi_1(\vec{r}_2)\beta(2) & \phi_2(\vec{r}_2)\alpha(2) & \phi_2(\vec{r}_2)\beta(2) \\ \phi_1(\vec{r}_3)\alpha(3) & \phi_1(\vec{r}_3)\beta(3) & \phi_2(\vec{r}_3)\alpha(3) & \phi_2(\vec{r}_3)\beta(3) \\ \phi_1(\vec{r}_4)\alpha(4) & \phi_1(\vec{r}_4)\beta(4) & \phi_2(\vec{r}_4)\alpha(4) & \phi_2(\vec{r}_4)\beta(4) \end{vmatrix}$$

$$\equiv |\phi_1\alpha, \phi_1\beta, \phi_2\alpha, \phi_2\beta|. \qquad (9.172)$$

The orbitals $\phi_1\alpha$, $\phi_1\beta$, $\phi_2\alpha$, and $\phi_2\beta$ are expanded in the following four basis functions $\chi_1\alpha$, $\chi_1\beta$, $\chi_2\alpha$, and $\chi_2\beta$, whose precise form is unimportant for the present example.

Since the orbitals pairwise are assumed to have the same position-space dependence, but differing in the spin dependence, we can apply the restricted Hartree–Fock method. This means that Eq. (9.156) is applicable with $N_b = 2$. In detail we write

$$\phi_1(\vec{r}) = \chi_1(\vec{r}) \cdot c_{11} + \chi_2(\vec{r}) \cdot c_{21}$$

$$\phi_2(\vec{r}) = \chi_1(\vec{r}) \cdot c_{12} + \chi_2(\vec{r}) \cdot c_{22}. \qquad (9.173)$$

We denote

$$\langle \chi_p | \chi_q \rangle \equiv o_{pq}$$

$$\langle \chi_p | \hat{h}_1 | \chi_q \rangle \equiv h_{1,pq}$$

$$\langle \chi_p \chi_m | \hat{h}_2 | \chi_q \chi_n \rangle \equiv h_{2,pmqn}. \tag{9.174}$$

In a practical calculation these will be calculated once the precise form of the basis functions χ_1 and χ_2 has been defined. Here, however, we shall simply list values for those. These are given in Tables 9.3 and 9.4.

The matrix elements obey

$$o_{ij} = o_{ji}^*$$

$$h_{1,ij} = h_{1,ji}^*$$

$$h_{2,ijkl} = h_{2,jilk} = h_{2,klij}^* = h_{2,lkji}^*, \tag{9.175}$$

which not is limited to the present example.

We assume that we somehow have obtained the two orbitals in one iteration of the self-consistent calculation and that these are

$$\phi_1 = a_1 (\chi_1 \cdot 0.2 + \chi_2 \cdot 0.8)$$

$$\phi_2 = a_2 \left(\chi_1 \cdot 1.0 + \chi_2 \cdot \frac{4}{17} \right). \tag{9.176}$$

The two constants a_1 and a_2 are to be determined from the normalization,

$$1 \equiv a_1^2 \cdot [0.2 \cdot 0.2 \cdot o_{11} + 0.2 \cdot 0.8 \cdot o_{12} + 0.8 \cdot 0.2 \cdot o_{21} + 0.8 \cdot 0.8 \cdot o_{22}]$$

Table 9.3   Values for the matrix elements o_{pq} and $h_{1,pq}$ of Eq. (9.174)

p	q	o_{pq}	$h_{1,pq}$
1	1	1.0	−3.0
1	2	−0.5	−2.0
2	2	1.1875	−5.0

Table 9.4   Values for the matrix elements $h_{2,pmqn}$ of Eq. (9.174)

p	m	q	n	$h_{2,pmqn}$
1	1	1	1	0.5
1	1	1	2	0.3
1	1	2	2	0.2
1	2	2	1	0.2
1	2	1	2	0.4
1	2	2	2	0.3
2	2	2	2	0.5

$$= a_1^2 \cdot [0.2 \cdot 0.2 \cdot 1.0 + 0.2 \cdot 0.8 \cdot (-0.5) + 0.8 \cdot 0.2 \cdot (-0.5)$$
$$+ 0.8 \cdot 0.8 \cdot 1.1875]$$
$$= a_1^2 \cdot 0.64$$

$$1 \equiv a_2^2 \cdot \left[1 \cdot 1 \cdot o_{11} + 1 \cdot \frac{4}{17} \cdot o_{12} + \frac{4}{17} \cdot 1 \cdot o_{21} + \frac{4}{17} \cdot \frac{4}{17} \cdot o_{22} \right]$$

$$= a_2^2 \cdot \left[1 \cdot 1 \cdot 1.0 + 1 \cdot \frac{4}{17} \cdot (-0.5) + \frac{4}{17} \cdot 1 \cdot (-0.5) + \frac{4}{17} \cdot \frac{4}{17} \cdot 1.1875 \right]$$

$$= a_2^2 \cdot \frac{240}{289}, \tag{9.177}$$

giving (choosing the signs)

$$a_1 = \frac{5}{4}$$

$$a_2 = \frac{17}{4\sqrt{15}}. \tag{9.178}$$

The functions are orthogonal:

$$\langle \phi_1 | \phi_2 \rangle = a_1 a_2 \cdot \left[0.2 \cdot 1.0 \cdot o_{11} \right.$$

$$\left. + 0.2 \cdot \frac{4}{17} \cdot o_{12} + 0.8 \cdot 1.0 \cdot o_{21} + 0.8 \cdot 417 \cdot o_{22} \right]$$

$$= a_1 a_2 \cdot \left[0.2 \cdot 1.0 \cdot 1.0 + 0.2 \cdot \frac{4}{17} \cdot (-0.5) + 0.8 \cdot 1.0 \cdot (-0.5) \right.$$

$$\left. + 0.8 \cdot \frac{4}{17} \cdot 1.1875 \right] = a_1 \cdot a_2 \cdot 0 = 0. \tag{9.179}$$

Then the coefficients c_{ij} of Eq. (9.173) are

$$c_{11} = \frac{5}{4} \cdot 0.2 = \frac{1}{4}$$

$$c_{21} = \frac{5}{4} \cdot 0.8 = 1$$

$$c_{12} = \frac{17}{4\sqrt{15}} \cdot 1 = \frac{17}{4\sqrt{15}}$$

$$c_{22} = \frac{17}{4\sqrt{15}} \cdot \frac{4}{17} = \frac{1}{\sqrt{15}}. \tag{9.180}$$

In the iterative process of solving the Hartree–Fock–Roothaan equations, we insert these coefficients [together with the matrix elements of Eq. (9.174)] into

Eq. (9.156) and then obtain a new set of equations. In our case this is a 2×2 matrix equation,

$$\begin{pmatrix} -1.406667 & -1.193333 \\ -1.193333 & -3.386667 \end{pmatrix} \cdot \begin{pmatrix} c_1 \\ c_2 \end{pmatrix} = \varepsilon \cdot \begin{pmatrix} 1.0 & -0.5 \\ -0.5 & 1.1875 \end{pmatrix} \cdot \begin{pmatrix} c_1 \\ c_2 \end{pmatrix}.$$

$$(9.181)$$

Whereas the matrix on the right-hand side is simply the one containing the overlap matrix elements o_{ij} [cf. Eq. (9.156)], the one on the left-hand side is somewhat more complicated. As an example we calculate the $(1, 2)$ matrix element of that. From Eq. (9.156) we find

$$h_{1,12} + c_{11}c_{11}(2 \cdot h_{2,1121} - h_{2,1121}) + c_{11}c_{21}(2 \cdot h_{2,1221} - h_{2,2121})$$
$$+ c_{21}c_{11}(2 \cdot h_{2,1122} - h_{2,1122}) + c_{21}c_{21}(2 \cdot h_{2,1222} - h_{2,2122})$$
$$+ c_{12}c_{12}(2 \cdot h_{2,1121} - h_{2,1121}) + c_{22}c_{22}(2 \cdot h_{2,1221} - h_{2,2121})$$
$$+ c_{22}c_{12}(2 \cdot h_{2,1122} - h_{2,1122}) + c_{22}c_{22}(2 \cdot h_{2,1222} - h_{2,2122})$$
$$= h_{1,12} + (c_{11}c_{11} + c_{12}c_{12})(2 \cdot h_{2,1121} - h_{2,1121})$$
$$+ (c_{11}c_{21} + c_{12}c_{22})(2 \cdot h_{2,1221} - h_{2,2121})$$
$$+ (c_{21}c_{11} + c_{22}c_{12})(2 \cdot h_{2,1122} - h_{2,1122})$$
$$+ (c_{21}c_{21} + c_{22}c_{22})(2 \cdot h_{2,1222} - h_{2,2122})$$
$$= -2.0 + \left(\frac{1}{16} + \frac{289}{240} \right)(2 \cdot 0.3 - 0.3) + \left(\frac{1}{4} + \frac{17}{60} \right)(2 \cdot 0.2 - 0.4)$$
$$+ \left(\frac{1}{4} + \frac{17}{60} \right)(2 \cdot 0.2 - 0.2) + \left(1 + \frac{1}{15} \right)(2 \cdot 0.3 - 0.3)$$
$$= -1.193333.$$

$$(9.182)$$

Solving Eq. (9.181) will lead to two eigenvalues ε_1 and ε_2 and corresponding eigenvectors (c_{11}, c_{21}) and (c_{12}, c_{22}). The latter defines the eigenfunctions similar to Eq. (9.176) and once again they may have to be normalized as above. Then, the Hartree–Fock–Roothaan equations can be set up once more, and the whole procedure can be continued until the input eigenvectors agree to a given accuracy with the output eigenvectors.

10 Basis Sets

10.1 SLATER-TYPE ORBITALS

The Hartree–Fock approach is based on approximating the N-electron wavefunction Ψ_e with a single Slater determinant Φ, containing N single-electron orbitals ϕ_k. The Hartree–Fock equations should be solved in order to determine those as well as the single-particle energies ε_k. Roothaan suggested that in order to make the calculations possible for realistic systems, the orbitals should be expanded in a set of pre-defined basis functions χ_i,

$$\phi_k(\vec{x}) = \sum_{i=1}^{N_b} \chi_i(\vec{x}) c_{ik}. \tag{10.1}$$

So far we have not, however, discussed how these basis functions are chosen, although it is very important that they are chosen 'reasonably' if trustworthy results are to be obtained.

In choosing the basis functions one may be partly guided by physical or chemical intuition. In Fig. 10.1 we show a schematic representation of what we could expect as outcome of a calculation for a CO molecule. For the isolated C and O atoms we have $1s$, $2s$, and $2p$ electrons, and we will expect that the molecular orbitals are formed from these functions. Furthermore, due to the larger nuclear charge of O than of C, equivalent orbitals of the isolated C and O atoms will have lower energies for O than for C.

The atomic $1s$ orbitals are well localized in space and will hardly interact in the molecule. Therefore, the two lowest molecular orbitals will be essentially pure atomic $1s$ orbitals with, perhaps, a small interaction. The situation is different for the valence electrons. Here, by generating sp hybrids from the atomic $2s$ and $2p$ functions, we can obtain atomic orbitals that point from one atom either towards the other or away from it. The bonding combination of these that point towards each other will lead to a very strongly bonding orbital with

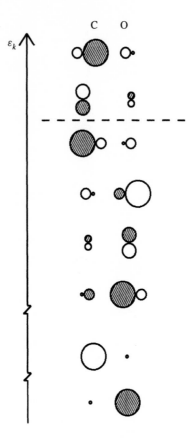

Figure 10.1   Schematic representation of the single-particle energies ε_k (on the vertical axis) as well as the corresponding orbitals for a CO molecule. We stress that the representation is only qualitative and even the relative ordering of the orbitals may be different for the real system. The horizontal dashed line separates occupied and unoccupied orbitals

the lowest energy above those of the 1s orbitals, whereas the corresponding anti-bonding combination will have the highest energy. In between those we have the bonding combination of the hybrids pointing away from the other atom, as well as the bonding combinations of the p orbitals perpendicular to the molecular axis (the latter will be fourfold degenerate when including spin). In addition we will also have — at somewhat higher energies — the corresponding antibonding combinations in the reverse order.

In total, we have seen that a picture based on constructing molecular orbitals from atomic functions seems to be adequate. Therefore, a very useful set of basis functions in Eq. (10.1) is one consisting of atomic orbitals. Thus, we may choose

$$\chi(\vec{r}) = R_{nl}(r)Y_{lm}(\theta, \phi), \tag{10.2}$$

where we have omitted the index on χ, assumed that the function χ belongs to an atom placed at the origin, and let Y_{lm} be a spherical harmonic function (i.e., the angular part of, e.g., a p_x, an s, or a $d_{3z^2-r^2}$ function). For an atom at another position, r, ϕ, and θ are the spherical coordinates with respect to that position.

The next question is how to choose the radial parts R_{nl} of χ in Eq. (10.2). A simple solution will be to use them as calculated, e.g., numerically for the isolated atoms, but it is not unlikely that they will change when passing to the molecule. On the other hand, for hydrogen-atom-like ions, R_{nl} is given as a (associated Laguerre) polynomial times an exponentially decaying function. This suggests using functions of the type

$$\chi(\vec{r}) = \chi_{\vec{R},\zeta,n,l,m}(\vec{r}) = \frac{(2\zeta)^{n+1/2}}{(2n!)^{1/2}} r^{n-1} e^{-\zeta r} Y_{lm}(\theta, \phi). \qquad (10.3)$$

Here, we may have different functions for different atoms (described by $\vec{R}$ and meaning, e.g., C and O for the CO molecule). Moreover, also for the same atom and (n, l) we may have more different functions (and thereby, due to the variational principle, improve the calculation) differing in the decay constants ζ. Finally, we may also include functions that for the isolated atoms are not occupied, e.g., p functions for hydrogen atoms and d functions for C and O atoms, etc.

The functions of Eq. (10.3) are the so-called Slater-type orbitals. As written there, they are normalized, but not orthogonal. Their largest problem is that it is prohibitively complicated to calculate the matrix elements entering the Fock matrix. Therefore, although they have the advantage of describing the eigenfunctions of systems that are physically related to the one of interest (i.e., of single-electron ions), their use has been somewhat limited.

10.2 GAUSSIAN-TYPE ORBITALS

One solution to the above (practical) problem is to replace the Slater-type orbitals of Eq. (10.3) with some functions that resemble the Slater-type orbitals but for which the matrix elements can be calculated relatively easily. The standard solution is to use Gaussian-type orbitals, also called Gaussians. These are defined as

$$\chi(\vec{r}) = \chi_{\vec{R},\alpha,n,l,m}(\vec{r}) = 2^{n+1} \frac{\alpha^{(2n+1)/4}}{[(2n-1)!!]^{1/2}(2\pi)^{1/4}} r^{n-1} e^{-\alpha r^2} Y_{lm}(\theta, \phi). \qquad (10.4)$$

Here the double factorial is

$$(2n-1)!! = (2n-1) \cdot (2n-3) \cdots 1. \qquad (10.5)$$

It is seen that the main difference to the Slater-type orbitals is that the argument of the exponential function contains an r^2 dependence instead of an r dependence. Due to this fact, *all* Gaussians have a flat tangent at the site of the nucleus.

This is clearly not the case for, e.g., the exact hydrogen $1s$ function. Since the potential diverges at the site of the nucleus, the region there is one where a large part of the total energy originates, so this so-called cusp problem (i.e., that the wavefunction has a wrong behaviour at the site of the nucleus) is a severe one. In order to circumvent it, one may use relatively many Gaussians, in particular ones that are very short-ranged (i.e., with large α).

In some cases the Y_{lm} functions are replaced by simple powers of x/r, y/r, and z/r. This means that there will be six d functions, x^2/r^2, y^2/r^2, z^2/r^2, xy/r^2, xz/r^2, and yz/r^2, instead of the usual five. But, it turns out that a certain linear combination of the six functions (the sum of the first three) forms a function of s and not d symmetry. Similarly, there will be ten and not seven f functions, but here one may construct three linear combinations that comprise three p-like functions.

Furthermore, more accurate results are often obtained when the basis set, as constructed from considerations of the orbitals of the isolated atoms, is augmented with so-called polarization functions, i.e., p functions on hydrogen atoms, d functions on atoms of the second and third row, and so on.

A standard notation for a Gaussian basis set would be something like (8,4,1) that could be applicable for, e.g., a carbon atom. Here, the three numbers mean that eight Gaussians of s symmetry, four of p symmetry, and 1 of d symmetry are used. The four of p symmetry differ in the α values and for each of these four, we have all three p functions (p_x, p_y, and p_z).

Often such a basis set is contracted. This means that one would take the Gaussians of Eq. (10.4) and form a smaller set of fixed linear combinations,

$$\tilde{\chi}_{\vec{R},k,n,l,m}(\vec{r}) = \sum_i u_{ki} \chi_{\vec{R},\alpha_i,n,l,m}(\vec{r}). \tag{10.6}$$

Here, the constants u_{ki} are fixed, and only the coefficients to the contracted functions $\tilde{\chi}$ are optimized within the Hartree–Fock–Roothaan calculations and k distinguishes different contracted functions. The corresponding notation is something like [4,2,1], where the numbers give the number of contracted functions [i.e., of linear combinations of the type (10.6)] of s, p, and d symmetry, respectively.

Often many different functions with the same n, l, and m but differing in the α are used. Then the α are often chosen according to

$$\alpha_p = a \cdot b^p, \tag{10.7}$$

where p labels the different α.

In Table 10.1 we show results of different calculations on an HF molecule with increased size of the basis set. It is first of all seen that the total energy does decrease with increasing basis-set size, as it should according to the variational principle. Simultaneously, the optimized bond-length appears to converge to some value, whereas the dipole moment shows a much slower convergence. This is perhaps not surprising since the total energy depends on an accurate

Table 10.1 Results of different calculations with increased size of the basis set for an HF molecule. E_e is the total energy (in hartrees) at the optimized value for the bond length (denoted r_e; in bohr), and μ_e is the dipole moment (in elementary-charge units times bohr). A primitive set of the type $6s3p/3s$ means that 6 Gaussians of s type and three of p type on fluorine and 3 of s type on hydrogen have been used. These were subsequently contracted to a set like, e.g., $2s1p/1s$, meaning two contracted Gaussians of s type on fluorine, 1 of p type on fluorine, and 1 of s type on hydrogen. The results are from Daudel *et al.* (1983)

Primitive set	Contracted set	E_e (a.u.)	R_e (a.u.)	μ_e (a.u.)
$6s3p/3s$	$2s1p/1s$	−98.572844	1.8055	0.49258
$12s6p/6s$	$2s1p/1s$	−99.501718	1.8028	0.51000
$8s4p/4s$	$3s2p/2s$	−99.887286	1.7410	0.89971
$10s4p/4s$	$3s2p/2s$	−99.983425	1.7386	0.90487
$9s5p/4s$	$3s2p/3s$	−100.018895	1.7467	0.95544
$9s5p/4s$	$3s2p/2s$	−100.020169	1.7475	0.96334
$9s5p/5s$	$3s2p/3s$	−100.020665	1.7376	0.96256
$9s5p/5s$	$4s3p/2s$	−100.022946	1.7390	0.93645
$11s6p/5s$	$4s2p/3s$	−100.026364	1.7422	0.91244
$9s5p/4s2p$	$3s2p/2s1p$	−100.034266	1.7257	0.87851
$10s6p/5s$	$5s3p/3s$	−100.036872	1.7380	0.93757
$10s6p/5s$	$5s4p/3s$	−100.037008	1.7371	0.93656
$9s5p/4s2p$	$4s3p/2s1p$	−100.040470	1.7046	0.83604
$11s6p/5s2p$	$4s2p/3s1p$	−100.044050	1.7168	0.84243
$10s6p/5s2p$	$5s4p/3s1p$	−100.044751	1.7206	0.81251
$9s5p2d/4s2p$	$3s2p1d/2s1p$	−100.049112	1.7053	0.74383
$9s5p2d/4s2p$	$4s3p1d/2s1p$	−100.049799	1.7046	0.74154
$11s6p2d/5s2p$	$4s2p1d/3s1p$	−100.057755	1.7036	0.69515
$10s6p1d/5s2p$	$5s4p1d/3s1p$	−100.059724	1.7078	0.74436
$10s6p2d/5s2p$	$5s3p1d/3s1p$	−100.062343	1.7027	0.74871

description of the orbitals close to the nuclei, whereas the dipole moment has its largest contributions from the regions further away from the nuclei. The example illustrates thus that the variational method first of all gives accurate total energies, whereas other properties may be far from converged when the total energy seems to be converged.

One may partly return to using Slater-type orbitals by expanding those in terms of Gaussians, whereby all the integrals can be evaluated. This leads to basis sets like, e.g., STO-3G, which means that each Slater-type orbital is expanded in three Gaussians. A basis set of the type STO 4-31G is one that could be used for a second-row atom. The notation means that the core electrons (the $1s$ electrons) are described by expanding the Slater-type orbitals in four Gaussians, whereas the valence s and p (i.e., $2s$ and $2p$) orbitals are expanded in three and one Gaussians, respectively. A larger basis set can be obtained by adding polarization functions. For instance, the notation 4-31G** indicates that d functions have been added to the basis set for second- and third-row atoms (the first *), whereas the second * shows that p polarization functions have been added to the basis set for hydrogen atoms.

10.3   PLANE WAVES

The Slater-type orbitals are closely related to the true eigenfunctions for single-electron ions, and as such have been devised using chemical or physical intuition. Also the Gaussians are well localized to specific atoms, although the potential for which they are eigenfunctions (a harmonic potential) is less relevant in electronic-structure calculations for molecules and solids.

A completely different approach is used when using plane waves

$$\chi_{\vec{k}}(\vec{r}) = e^{i\vec{k}\cdot\vec{r}} \tag{10.8}$$

as basis functions. These are completely delocalized and can therefore not be ascribed individual atoms. They are the correct eigenfunctions for free electrons and as such they can be considered good basis functions for studies on metallic crystalline materials.

Nevertheless, they are also used in studies on molecular systems, although mainly with density-functional (see later) methods. They have two advantages: it is very easy to calculate all kinds of matrix elements, where the so-called fast-Fourier-transform techniques are of great help. Second, the size of the basis set can be increased systematically in a very simple way. Their main disadvantages are that the basis sets very easily become very large, and that it is less trivial to interpret the results in terms of chemical bonds, etc.

10.4   NUMERICAL BASIS FUNCTIONS

A few methods use basis functions that are represented completely numerically and obtained for the isolated atoms as well as some relevant ions. Thus, for CO one may construct numerical basis functions by considering the isolated C and O atoms as well as C^+ and O^- ions. All basis functions are then represented in the form

$$\chi(\vec{r}) = R_{nl}(r)Y_{lm}(\theta, \phi), \tag{10.9}$$

where only the radial part $R_{nl}(r)$ is given numerically. It is obvious that these methods require that all integrals are calculated numerically and, consequently, that methods are devised with which this can be done efficiently. On the other hand, the methods offer a transparent way of interpreting the results. As for the plane waves, these basis sets are mainly used in density-functional methods.

10.5   AUGMENTED WAVES

A compromise between the basis sets of the last two subsections is offered by the augmented waves. The philosophy behind these is to focus on the potential felt by a valence electron. Close to the nuclei this potential is dominated by that of the nucleus and the core electrons and is therefore largely spherically symmetric.

Further away from all nuclei (in the so-called interstitial region) the potential varies much more slowly. One may therefore consider a simplified potential

$$V(\vec{r}) = \begin{cases} V_s(|\vec{r} - \vec{R}|) & \text{for } |\vec{r} - \vec{R}| \le s_{\vec{R}} \\ V_0 & \text{in the interstitial region.} \end{cases} \qquad (10.10)$$

Here $V_s(|\vec{r} - \vec{R}|)$ is the spherically symmetric part of the potential around the nucleus at $\vec{R}$, $s_{\vec{R}}$ is the radius of a sphere where the spherically symmetric part of the potential is used, and V_0 is a constant.

The single-particle Schrödinger equation for this potential

$$\left[-\tfrac{1}{2}\nabla^2 + V(\vec{r})\right] \chi(\vec{r}) = \varepsilon \cdot \chi(\vec{r}) \qquad (10.11)$$

can be solved numerically inside the spheres and analytically in the interstitial region. This leads then to (basis) functions that are given as either plane waves

$$e^{i\vec{k}\cdot\vec{r}} \qquad (10.12)$$

or as spherical waves

$$h_l^{(1)}(\kappa r)Y_{lm}(\theta, \phi) \qquad (10.13)$$

(for a spherical wave centred at the origin) in the interstitial region. These functions are subsequently augmented inside the spheres with numerically given functions. In Eq. (10.13), $h_l^{(1)}$ is a spherical Hankel function. For the special case $\kappa \to 0$, the Hankel function changes into r^{-l-1}.

If the plane waves of Eq. (10.12) are used, one arrives at augmented plane-wave methods (APW), and with the spherical waves one arrives at augmented spherical-wave (ASW) or muffin-tin-orbital (LMTO) methods.

Their advantages are the transparency in interpreting the results and that only small-sized basis sets are necessary for obtaining accurate results. The main disadvantage is the separation of space into spheres and interstitial region, whereby the calculation of matrix elements becomes non-trivial. Finally, we add that these basis sets are mainly applied with density-functional methods, where the single-particle equation (10.11) has a simple meaning, and that the potential of Eq. (10.10) is used solely in constructing the basis functions — not necessarily in solving the Schrödinger equation.

10.6 SYMMETRY

Solving the Hartree–Fock–Roothaan equations

$$\sum_{m=1}^{N_b} \left\{ \langle \chi_p|\hat{h}_1|\chi_m\rangle + \sum_{i=1}^{N}\sum_{n,q=1}^{N_b} c_{ni}c_{qi}^* \left[\langle\chi_p\chi_q|\hat{h}_2|\chi_m\chi_n\rangle - \langle\chi_q\chi_p|\hat{h}_2|\chi_m\chi_n\rangle\right] \right\} c_{ml}$$

$$= \varepsilon_l \sum_{m=1}^{N_b} \langle\chi_p|\chi_m\rangle c_{ml} \qquad (10.14)$$

requires that a number of one-electron integrals,

$$\langle \chi_p | \hat{h}_1 | \chi_m \rangle \tag{10.15}$$

and

$$\langle \chi_p | \chi_m \rangle, \tag{10.16}$$

as well as a number of two-electron integrals,

$$\langle \chi_p \chi_q | \hat{h}_2 | \chi_m \chi_n \rangle \tag{10.17}$$

are calculated.
Here,

$$\hat{h}_1(\vec{r}) = -\frac{1}{2}\nabla^2 - \sum_{k=1}^{M} \frac{Z_k}{|\vec{R}_k - \vec{r}|} \tag{10.18}$$

and

$$\hat{h}_2(\vec{r}_i, \vec{r}_j) = \frac{1}{|\vec{r}_i - \vec{r}_j|}. \tag{10.19}$$

In order to reduce the number of those that are non-vanishing it is of advantage to use symmetry-adapted basis functions as described in Chapter 7. This means that, from basis functions that are centred at different atoms, (but are equivalent due to symmetry), we construct linear combinations so that these transform according to given irreducible representations of the group describing the symmetry of the system. This procedure does not work, however, for plane waves (since these are not centred at any site). Since both $\hat{h}_1$ and $\hat{h}_2$ transform according to the 'trivial' irreducible representation, we obtain that the matrix elements of Eqs. (10.15) and (10.16) only then can be non-zero if χ_p and χ_m transform according to the same irreducible representation. For those of Eq. (10.17) we may state the condition in the form that the direct product of the irreducible representations of χ_p and χ_q and the direct product of the irreducible representations of χ_m and χ_n contain at least one common irreducible representation.

As one example we mention that these results are very simple to apply for systems with one-dimensional translational or screw-axis symmetry (cf. Chapter 7). Here, the continuous variable k is used in characterizing the irreducible representations, and we find thus that the matrix elements of Eqs. (10.15) and (10.16) can be non-zero only if $k_m - k_p = 0$ (modulo 2π), and those of Eqs. (10.17) if $k_m + k_n - k_p - k_q = 0$ (modulo 2π). Here, k_r is k for the symmetry-adapted basis function χ_r.

10.7   BASIS SET SUPERPOSITION ERROR

A special situation may occur when studying the interaction between different systems using approximate wavefunctions obtained with smaller basis sets. Let us consider just a single system and for this a single orbital that is described approximately through a finite set of basis functions

$$\psi_i(\vec{x}) = \sum_{j=1}^{N_b} \chi_j(\vec{x}) c_{ji}. \tag{10.20}$$

Whenever this basis set is extended through further functions, the quality of the calculation is improved and the total electronic energy is lowered (this is the variational principle). Thus, when one more system is added for which we also describe the electronic properties through a finite basis set, the mere fact that any of the orbitals of each of the two parts can obtain an improved description through incorporating more basis functions (i.e., those that are centred on the other part) means that the total energy of the two systems together may be lower than the sum of the total energies of the two systems individually. And when the two parts are brought together this will mean that the interaction energies may become overestimated. This is the so-called 'basis set superposition error'.

One way to improve this is to consider each of the two isolated systems separately with basis sets that are extended with those of the of the other part but without including the electrons or nuclei of that part. This is the counterpoise method (Boys and Bernardi, 1970).

11 Semiempirical Methods

11.1 THE HÜCKEL METHOD

Within the Hartree–Fock approximation we solve the electronic Schrödinger equation

$$\hat{H}_e \Psi_e = E_e \Psi_e, \tag{11.1}$$

with

$$\hat{H}_e = \sum_{i=1}^{N} \hat{h}_1(\vec{r}_1) + \frac{1}{2} \sum_{i \neq j=1}^{N} \hat{h}_2(\vec{r}_i, \vec{r}_j), \tag{11.2}$$

approximately by assuming that the N-electron wavefunction can be written as a single Slater determinant,

$$\Psi_e \simeq \Phi. \tag{11.3}$$

As demonstrated in Section 9.2, this leads to the Hartree–Fock equations,

$$\hat{F}\phi_k = \varepsilon_k \phi_k, \tag{11.4}$$

where the Fock operator is

$$\hat{F} = \hat{h}_1 + \sum_{i=1}^{N} (\hat{J}_i - \hat{K}_i). \tag{11.5}$$

The Coulomb and exchange operators $\hat{J}_i$ and $\hat{K}_i$ depend on the solutions to Eq. (11.4),

$$\hat{J}_i = \int \frac{|\phi_i(\vec{x}_2)|^2}{|\vec{x}_2 - \vec{x}|} \, d\vec{x}_2, \tag{11.6}$$

and

$$\hat{K}_i = \int \frac{\phi_i^*(\vec{x}_2)\hat{P}_{12}\phi_i(\vec{x}_2)}{|\vec{x}_2 - \vec{x}|} \, d\vec{x}_2. \tag{11.7}$$

Here, $\hat{P}_{12}$ is the permutation operator that interchanges the arguments of the two next functions.

Solving the Hartree–Fock equations for a given set of nuclear positions,

$$\vec{R} = (\vec{R}_1, \vec{R}_2, \ldots, \vec{R}_M) \tag{11.8}$$

leads to orbitals ϕ_i, orbital energies ε_i, and the total electronic energy E_e for precisely this structure. In some cases one is interested in only a qualitative picture of the bonding properties of the system of interest and a very exact solution of the Hartree–Fock equations is not needed. In that case semiempirical methods can be useful.

Let us consider the naphthalene molecule of Fig. 11.1. This molecule is simple in the sense that it contains only carbon and hydrogen atoms. Some first ideas of the electronic structure can be obtained as follows.

Each carbon atom possesses six electrons of which two occupy the $1s$ core orbitals. These orbitals are so localized in space (or, equivalently, have such low energies) that they do not participate in the chemical bonds. Of the other four, three occupy sp^2 hybrids that lie in the plane of the molecule and the last occupies a p orbital that is perpendicular to the plane of the molecule. We may actually use group theory at this point for the reflection symmetry operation in the plane of the molecule. This symmetry allows us to separate the orbitals into those that are symmetric and those that are antisymmetric with respect to this reflection. For naphthalene the sp^2 hybrids are symmetric whereas the last p orbitals are antisymmetric. Per tradition, the former are classified as σ orbitals and the latter as π orbitals.

The single $1s$ orbital of each hydrogen atom is also of σ symmetry. Group theory tells us now that there will be no Hamilton or overlap matrix elements between functions of σ and π symmetry. Therefore, we can study each set of functions separately.

The carbon sp^2 hybrids and the hydrogen $1s$ orbitals will form strong covalent bonds between the atoms. These will accordingly have deep orbital energies. On the other hand, the bonds formed by the π orbitals will be much weaker and their energies will lie much higher. Thus, these bonds are the weakest ones and when disturbing the system in any way (through experiments, solution, chemical reactions, structural modifications, etc.) the π orbitals will be those that change

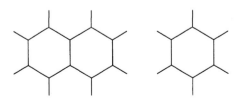

Figure 11.1   The naphthalene (left) and benzene (right) molecule

first. Therefore, for many purposes it is sufficient to study only the π electrons. This is the basic idea of the Hückel model (Hückel, 1931, 1932).

In order to transform these considerations into a working scheme we consider instead of the Hartree–Fock equations (11.1) the Hartree–Fock–Roothaan equations,

$$\underline{\underline{F}} \cdot \underline{c}_l = \varepsilon_l \cdot \underline{\underline{O}} \cdot \underline{c}_l, \tag{11.9}$$

where $\underline{\underline{F}}$ contains the Fock matrix elements

$$F_{pm} = \langle \chi_p | \hat{h}_1 | \chi_m \rangle + \sum_{i=1}^{N} \sum_{n,q=1}^{N_b} c_{ni} c_{qi}^* \left[\langle \chi_p \chi_q | \hat{h}_2 | \chi_m \chi_n \rangle - \langle \chi_q \chi_p | \hat{h}_2 | \chi_m \chi_n \rangle \right] \tag{11.10}$$

and $\underline{\underline{O}}$ contains the overlap matrix elements

$$O_{pm} = \langle \chi_p | \chi_m \rangle. \tag{11.11}$$

These equations have been obtained by expanding the sought single-particle eigenfunctions to the Hartree–Fock equations in a set of basis functions,

$$\phi_l = \sum_{p=1}^{N_b} \chi_p c_{pl}. \tag{11.12}$$

We now realize that the Fock matrix elements of Eq. (9.107) are

$$\int \chi_p^*(\vec{x}) \left[\hat{h}_1 + \int \sum_{i=1}^{N} \sum_{n,q=1}^{N_b} c_{ni}^* c_{qi} \frac{\chi_q^*(\vec{x}_1)\chi_n(\vec{x}_1) - \chi_q^*(\vec{x}_1)\hat{P}_{12}\chi_n(\vec{x}_1)}{|\vec{x}_1 - \vec{x}|} \, d\vec{x}_1 \right] \chi_m(\vec{x}) \, d\vec{x}. \tag{11.13}$$

In our case, i.e., naphthalene of Fig. 11.1, we will focus on the π electrons, so that the basis functions χ_p of Eq. (11.12) are π orbitals centred on the different carbon atoms (i.e., one π orbital per carbon atom). Then the matrix elements of Eq. (11.13) are the matrix elements between the π orbitals of the different orbitals. As Eq. (11.13) shows, these depend on the *total* electronic distribution. However, they may be approximated. We may, e.g., assume that these matrix elements often vanish and, furthermore, that when they are non-vanishing they depend only on the two functions χ_p and χ_m, but not on all the other electrons or orbitals. This corresponds to assuming that the π electrons move around in some kind of effective potential that is independent of the precise structure of the molecule, that is that the π electrons of naphthalene and those of benzene (cf. Fig. 11.1) 'behave in the same way'.

The formal approximations are first of all

$$\langle \chi_p | \chi_m \rangle = \delta_{p,m}, \tag{11.14}$$

i.e., the π functions are assumed to be orthonormal.

Furthermore

$$\langle \chi_p | \hat{F} | \chi_m \rangle = \begin{cases} \alpha & m = p \\ \beta & \chi_p \text{ and } \chi_m \text{ neighbours} \\ 0 & \text{otherwise.} \end{cases} \qquad (11.15)$$

Here, $\hat{F}$ is the Fock operator, i.e., the operator inside the square bracket in Eq. (11.13). In Eq. (11.15) 'neighbours' means that the two functions χ_p and χ_m belong to two carbon atoms that are neighbours.

The approximations (11.14) and (11.15) are the basic approximations of the Hückel method. They correspond to assuming that all interactions are of very short range, and that the matrix elements depend only on the topology of the molecule, i.e., the π functions of one molecule (e.g., naphthalene) are very similar to those of another (e.g., benzene).

α is the matrix elements for a function with itself. Changing α will lead only to an overall rigid shift of the energy scale for the single-particle energies ε_i. One may, accordingly, set it equal to 0. α is called the on-site energy or Coulomb integral.

β, on the other hand, defines the size of the interactions. Changing it will lead to an overall expansion or contraction of the energy scale, and as for α one may therefore fix β at a single value, 1. Then, all energies are measured in units of β. β is called the (nearest-neighbour) hopping integral, transfer integral, or resonance integral.

So far we have assumed that the system of our interest consists of three-fold-coordinated carbon atoms as well as hydrogen atoms. Replacing, e.g., one or more CH units by nitrogen atoms (e.g., change benzene C_6H_6 into pyridine C_5H_5N) will lead to some modifications but we can still apply the basic ideas of the Hückel method. The only changes will be that α will take one value when the atom of the two functions is a carbon atom and another value when the atom is a nitrogen atom. Similarly, β will take one value when the two nearest neighbours both are carbon atoms, another value when they both are nitrogen atoms (this case is not of relevance in this example), and a third value when one is a carbon atom and the other a nitrogen atom. Similar approximations are made for other substituents, and also when replacing some of the H atoms with other sidegroups (e.g., CH_3 groups) the changes can be incorporated into the α and β parameters.

The Hückel approach offers a simple way of obtaining some qualitative insight into the properties of a given molecule. It is based on many crude approximations, and can therefore not be used in giving quantitative information, but it is nevertheless a powerful, simple tool for many first-hand applications. It is mainly developed for treating π electrons of organic molecules. It can, however, be used also in other context. We need to separate the system of our interest into units (carbon atoms, for the Hückel approach), where each unit has one electron of interest (the π electrons in the Hückel approach). We shall return to this in Chapter 12.

The Hückel approach corresponds to replacing the true many-electron Hamilton or Fock operator by a simpler one,

$$\hat{H}_{\text{Hückel}} = \sum_{i=1}^{N} \hat{h}_{\text{eff}}(i).$$

(11.16)

Then the N-electron wavefunction Ψ obeying

$$\hat{H}_{\text{Hückel}} \Psi = E\Psi$$

(11.17)

is given exactly as a single Slater-determinant,

$$\Psi(\vec{x}_1, \vec{x}_2, \ldots, \vec{x}_N) = |\phi_1, \phi_2, \ldots, \phi_N|,$$

(11.18)

and the orbitals ϕ_i are determined through the single-particle equations

$$\hat{h}_{\text{eff}} \phi_i = \varepsilon_i \phi_i.$$

(11.19)

These can be solved by expanding ϕ_i in the basis of π functions of the different atoms and using the matrix elements above.

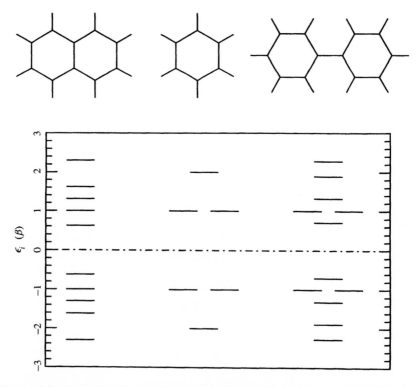

Figure 11.2 Examples of the calculated single-particle energies as obtained with the Hückel method for benzene (middle part), naphthalene (left part), and biphenyl (right part). The vertical axes give the orbital energies (in units of β)

Solving these equations leads to a set of orbital energies ε_i as well as the eigenfunctions. However, since the precise form of the π basis functions is not specified, the eigenfunctions are only given as coefficients. In total the calculations will give information like that of Fig. 11.2.

11.2   THE EXTENDED HÜCKEL METHOD

The original Hückel method was developed for describing the π electrons of organic molecules. The basic ideas may, however, also be used for other systems. Thus, also the extended Hückel method (Hoffmann, 1963, 1964) amounts to considering an operator of the form of Eq. (11.16),

$$\hat{H}_{\mathrm{eH}} = \sum_{i=1}^{N} \hat{h}_{\mathrm{eff}}(i), \tag{11.20}$$

instead of the true N-electron Hamilton or one-electron Fock operator. The main difference to the simplest Hückel method is that the extended Hückel method is not restricted to organic molecules or to systems with only one electron or orbital per atom.

Accordingly, within the extended Hückel method we calculate *all* orbitals (except for the core orbitals, maybe) by solving the single-particle equations,

$$\hat{h}_{\mathrm{eff}}\phi_i = \varepsilon_i\phi_i. \tag{11.21}$$

The solutions ϕ_i are expanded in a set of atomic basis functions:

$$\phi_i = \sum_{j=1}^{N_b} \chi_j c_{ji}. \tag{11.22}$$

For a water molecule, H_2O, this set would consist of one $1s$ function per hydrogen atom, and one $1s$, one $2s$, and three $2p$ functions for the oxygen atom (neglecting spin). We stress that their precise form will not be specified.

As in Section 9.4, the coefficients c_{ji} and the eigenvalues ε_i are calculated by considering

$$F = \langle\Phi|\hat{H}_{\mathrm{eH}}|\Phi\rangle - \sum_{i,j}\lambda_{ij}\left[\langle\phi_i|\phi_j\rangle - \delta_{i,j}\right]. \tag{11.23}$$

We can apply the methods of Chapter 9 in obtaining

$$F = \sum_{i=1}^{N}\langle\phi_i|\hat{h}_{\mathrm{eff}}|\phi_i\rangle - \sum_{i,j=1}^{N}\lambda_{ij}\left[\langle\phi_i|\phi_j\rangle - \delta_{i,j}\right], \tag{11.24}$$

where the fact that $\hat{H}_{\mathrm{eH}}$ contains only single-particle operators and no two-electron operators leads to significant simplifications compared with Eq. (9.98).

We proceed as in Section 9.4 and obtain the analogues of the Hartree–Fock–Roothaan equations (9.106),

$$\sum_{n=1}^{N_b} \langle \chi_p | \hat{h}_{\text{eff}} | \chi_n \rangle c_{nl} = \varepsilon_l \sum_{n=1}^{N_b} \langle \chi_p | \chi_n \rangle c_{nl}. \tag{11.25}$$

These equations can be solved once the matrix elements $\langle \chi_p | \hat{h}_{\text{eff}} | \chi_n \rangle$ and $\langle \chi_p | \chi_n \rangle$ are known. For these we will use approximations equivalent to those of the simple Hückel model,

$$\langle \chi_p | \chi_n \rangle = \delta_{p,n} \tag{11.26}$$

and

$$\langle \chi_p | \hat{h}_{\text{eff}} | \chi_p \rangle = -\tfrac{1}{2}(I_p + A_p) \tag{11.27}$$

for the on-site matrix elements. In Eq. (11.27), I_p is the ionization potential of the isolated atom, i.e., the energy required for removing an electron from the orbital χ_p. Equivalently, A_p is the electron affinity, i.e., the energy for adding an electron to the orbital χ_p.

For the off-diagonal matrix elements we choose

$$\langle \chi_p | \hat{h}_{\text{eff}} | \chi_n \rangle = \tfrac{1}{2} K \left[\langle \chi_p | h_{\text{eff}} | \chi_p \rangle + \langle \chi_n | h_{\text{eff}} | \chi_n \rangle \right] O(\chi_p, \chi_n), \tag{11.28}$$

where $O(\chi_p, \chi_n)$ is the overlap between the two functions. It is not identical to that of Eq. (11.28) but has to be calculated from some atomic orbitals. $O(\chi_p, \chi_n)$ will decay as a function of the distance between the two atoms where the two functions are centred. Moreover, it will depend on the relative orientation of the two functions, as illustrated in Fig. 11.3. In this figure we show the example of χ_p and χ_n being a p function and an s function. The p function is supposed to make the angle θ with the line joining the two atoms. In *this* case $O(\chi_p, \chi_n)$ and hence also the matrix element of Eq. (11.28) depends on $\cos\theta$. The famous Slater–Koster tables (Slater and Koster, 1954) give the general angular dependences of such matrix elements for arbitrary $Y_{l,m}$ functions at different sites with specific angular dependences.

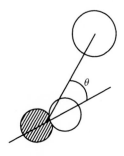

Figure 11.3 A p function on one atom and an s function on another. The overlap between them will depend on the angle θ

The parameter K in Eq. (11.28) is a constant. Empirically it has been found that

$$K = 1.75 \qquad (11.29)$$

is a reasonable choice.

Once the matrix elements have been specified, we can calculate the orbitals ϕ_i and their energies ε_i. Compared to the simple Hückel method the 'only' difference is that the extended Hückel method allows for treating any compound and not only the π electrons of organic molecules. In contrast to the Hartree–Fock–Roothaan method, the simplest Hückel and the extended Hückel methods do not require that the single-particle equations are solved self-consistently: a single diagonalization of the matrix eigenvalue equation (11.25) gives the sought information.

11.3   THE PPP METHOD

The Hückel method was devised to treat π electrons of organic molecules. In order to simplify the calculations, an effective Hamilton operator was constructed, which was the sum of N one-electron operators. Thus, the problems related to the two-electron operators were avoided. The Pariser–Parr–Pople (ppp) (Pople, 1953; Pariser and Parr, 1953) method is an extension of the Hückel method for the π electrons of organic molecules, but including two-electron operators.

We consider only the π electrons and assume that the π orbitals can be written as a linear combination of atomic π orbitals,

$$\phi_l = \sum_{p=1}^{N_b} \chi_p c_{pl}. \qquad (11.30)$$

The Hartree–Fock–Roothaan equations are written as

$$\sum_{m=1}^{N_b} \left\{ \langle \chi_p | \hat{h}_{\text{eff},1} | \chi_m \rangle + \sum_{i=1}^{N_\pi} \sum_{n,q=1}^{N_b} c_{ni} c_{qi}^* \left[\langle \chi_p \chi_q | \hat{h}_{\text{eff},2} | \chi_m \chi_n \rangle \right. \right.$$

$$\left. \left. - \langle \chi_q \chi_p | \hat{h}_{\text{eff},2} | \chi_m \chi_n \rangle \right] \right\} c_{ml}$$

$$= \varepsilon_l \sum_{m=1}^{N_b} \langle \chi_p | \chi_m \rangle c_{ml}. \qquad (11.31)$$

Compared with the original equations (9.106), we have introduced two changes. Thus, we consider *only* the π electrons, so that the total number of electrons N has been replaced by that of the π electrons N_π. This means also that the effects of all the other electrons (i.e., of the σ electrons) have to be included somewhere else. We therefore introduce also some effective single-electron Hamilton operators $\hat{h}_{\text{eff},1}$ and two-electron Hamilton operators $\hat{h}_{\text{eff},2}$ so that the correct electronic structure of the π electrons is obtained. We shall not specify these operators, but just state that these are constructed 'somehow'.

Equation (11.31) can be solved if the matrix elements $\langle \chi_p | \hat{h}_{\text{eff},1} | \chi_n \rangle$ and $\langle \chi_n \chi_q | \hat{h}_{\text{eff},2} | \chi_m \chi_p \rangle$ are known. And as in the Hückel and extended Hückel method these are replaced by parameters. An often applied major approximation is that the matrix elements are assumed to vanish unless the functions belong to nearest neighbours. Furthermore, they are considered constants independent of the system and its precise structure. There are, however, different ways of parametrizing the remaining ones, and we shall not go into further details here.

11.4   THE ZDO AND INDO METHODS

The extended Hückel method was obtained by generalizing the simple Hückel method for π electrons to general systems. Similarly, one may consider the zero-differential-overlap (ZDO) methods as generalizations of the PPP method. For these methods, the starting point is the Hartree–Fock–Roothaan equations, whereas the Hückel-based methods consider an effective single-particle approach (cf. Section 11.2). This means that in the Hartree–Fock–Roothaan equations,

$$
\sum_{m=1}^{N_b} \left\{ \langle \chi_p | \hat{h}_1 | \chi_m \rangle + \sum_{i=1}^{N} \sum_{n,q=1}^{N_b} c_{ni} c_{qi}^* \left[\langle \chi_p \chi_q | \hat{h}_2 | \chi_m \chi_n \rangle \right. \right.
$$

$$
\left. \left. - \langle \chi_q \chi_p | \hat{h}_2 | \chi_m \chi_n \rangle \right] \right\} c_{ml}
$$

$$
= \varepsilon_l \sum_{m=1}^{N_b} \langle \chi_p | \chi_m \rangle c_{ml}, \tag{11.32}
$$

the one- and two-electron matrix elements are not calculated but replaced by parameters. The basis functions χ_p are 'realistic' atomic functions, e.g., hydrogen $1s$ functions and oxygen $1s$, $2s$, and $2p$ functions for water, although their precise form is not specified.

The different methods differ in the way the parameters are specified. In the ZDO methods all except for very few matrix elements are set equal to 0, i.e.,

$$
\langle \chi_p | \chi_n \rangle = \delta_{p,n}
$$

$$
\langle \chi_p | \hat{h}_1 | \chi_n \rangle = 0 \quad \text{for} \quad p \neq n
$$

$$
\langle \chi_n \chi_q | \hat{h}_2 | \chi_m \chi_p \rangle = 0 \quad \text{for} \quad n \neq m \quad \text{or} \quad q \neq p. \tag{11.33}
$$

This corresponds to assuming that everything that is not very large is set equal to 0 ('zero differential overlap').

The remaining parameters are obtained by requiring that the calculations reproduce certain (experimentally or theoretically obtained) quantities for a given set of molecules. This can be, e.g., structural properties, ionization potentials, dipole moments, or others. This means also that most often a given method is optimized for describing a single or just some few properties for an often limited class of systems, whereas results of studies of other properties or types of systems are less reliable. The most well-known of these methods are the CNDO methods (complete neglect of differential overlap) due to Pople *et al.* (1965).

Here, e.g., the CNDO/2 method of Pople and Segal (1966) was constructed to yield accurate structural properties. In Tables 11.1 and 11.2 we show some examples of calculated bond lengths and bond angles obtained with this method. On the other hand, the CNDO/1 method of Pople and Segal (1965) was constructed to give accurate ionization potentials. In Table 11.3 we compare the CNDO/1 and CNDO/2 methods for this property.

As a further example, Table 11.4 gives some heats of atomization from the CNDO/2 method, i.e., the energy required for separating the molecule into isolated atoms.

Table 11.1 Equilibrium bond lengths (in Å) from the CNDO/2 method (calculated) and experiment. From Segal (1967)

Molecule	Bond	Experiment	Calculation
HF	H$-$F	0.917	1.004
N$_2$	N$-$N	1.094	1.140
CO	C$-$O	1.128	1.190
O$_2$	O$-$O	1.207	1.132
C$_2$H$_2$	C$-$H	1.058	1.093
C$_2$H$_2$	C$-$C	1.208	1.198
C$_2$H$_4$	C$-$H	1.086	1.110
C$_2$H$_4$	C$-$C	1.339	1.320
C$_2$H$_6$	C$-$H	1.091	1.117
C$_2$H$_6$	C$-$C	1.536	1.476
HCN	C$-$H	1.065	1.093
HCN	C$-$N	1.156	1.180
H$_2$CO	C$-$H	1.102	1.116
H$_2$CO	C$-$O	1.210	1.251

Table 11.2 Equilibrium bond angles (in deg.) from the CNDO/2 method (calculated) and experiment. From Pople and Segal (1966)

Molecule	Bond angle	Experiment	Calculated
H$_2$O	H$-$O$-$H	104.45	107.1
CO$_2$	O$-$C$-$O	180	180
NO$_2{}^+$	O$-$N$-$O	180	180
O$_3$	O$-$O$-$O	116.8	114.0
NH$_3$	H$-$N$-$H	106.6	106.7
BF$_3$	F$-$B$-$F	120	120
NO$_3{}^-$	O$-$N$-$O	120	120
NF$_3$	F$-$N$-$F	102.5	104.0

Table 11.3 Dipole moments (in debyes) from the CNDO/1 and CNDO/2 methods and experiment. From Klopman and Evans (1977)

Molecule	CNDO/1	CNDO/2	Experiment
Water	1.79	2.08	1.82
Ammonia	1.99	2.08	1.47
Propene		0.36	0.364
Toluene		0.22	0.31
Fluoromethane		1.66	1.855
Fluorobenzene		1.66	1.66
Methanol		1.94	1.69
Phenol		1.73	1.55
Acetaldehyde		2.53	2.68
Acetone		2.90	2.90
Benzaldehyde		2.50	2.72

Table 11.4 Heats of atomization (in kcal/mole) from the CNDO/2 method and experiment. From Wiberg (1968)

Molecule	Calculated	Experiment
Ethane	713.60	712.20
Propane	1006.91	1006.07
Ethylene	563.14	563.44
Acetylene	416.33	406.50
Allene	710.44	706.18
Benzene	1373.14	1370.90

Table 11.5 Singlet-singlet excitation energies (in eV) from the CNDO/S method and experiment. From Del Bene and Jaffé (1969)

Molecule	$n \rightarrow \pi$ calc.	$n \rightarrow \pi$ exp.	$\pi \rightarrow \pi$ calc.	$\pi \rightarrow \pi$ exp.
Formaldehyde	3.4	3.5	9.3	—
Formic acid	4.2	4.8	8.2	8.0
Allene	4.9	5.9	6.6	7.2
Ketene	3.5	3.2	5.6	5.8
Diazomethane	2.2	2.6	4.4	4.6
Pyrrole	—	—	5.0	5.7
			5.4	6.5
			7.0	7.1
			7.0	—
Furan			5.2	5.9
			5.8	6.5
			7.3	7.4
			7.3	—

Table 11.6 Heats of formation (in kcal/mole) from experiment, AM1 calculations, MNDO calculations, and parameter-free Hartree–Fock calculations with either 3-21G or 6-31G basis sets. From Dewar *et al.* (1985)

Molecule	Exp.	AM1	MNDO	3-21G	6-31G
Ethane	−20.04	−17.4	−19.7	−19.8	−18.1
Cyclopropane	12.7	17.8	11.2	4.3	10.3
Hydrogen peroxide	−32.5	−35.3	−38.2	−13.9	−29.4

Table 11.7 Ionization potential (in electron-volts) from experiment, AM1 calculations, and MNDO calculations. From Dewar *et al.* (1985)

Molecule	Exp.	AM1	MNDO
Ethane	12.10	11.77	12.70
Benzene	9.24	9.65	9.39

Table 11.8 Dipole moment (in debyes) from experiment, AM1 calculations, and MNDO calculations. From Dewar *et al.* (1985)

Molecule	Exp.	AM1	MNDO
Toluene	0.36	0.27	0.06
CO	0.11	0.06	0.19
Phenole	1.45	1.24	1.67

Finally, the CNDO/S method of Del Bene and Jaffé (1968a,b) was constructed in order to reproduce accurate spectroscopic information. In Table 11.5 we show some excitation energies obtained with this method.

By including more parameters, i.e., by being less strict than in Eq. (11.33), more accurate studies can be done at the costs of increasing complexity and less transparency of the methods. Thereby one arrives at, e.g., the intermediate neglect of differential overlap (INDO) method of Pople and coworkers (1965), the partial neglect of differential overlap (PNDO) method of Dewar and Klopman (1967), or the modified intermediate neglect of differential overlap (MINDO) method of Baird and Dewar (1969). We shall not discuss these further. We shall only mention the Austin model 1 (AM1) method of Dewar *et al.* (1985) since this is one of the currently mostly applied method. It has been extended and improved over the years since its introduction in 1985, so that it is presently an accurate and efficient method for obtaining structural information for very many different types of systems; not only organic molecules or molecules of light atoms. In Tables 11.6, 11.7, and 11.8 we have collected some few calculated properties of this method together with those of other methods.

12 Creation and Annihilation Operators

12.1 PROJECTION OPERATORS

In the present section we shall briefly introduce the so-called second quantization. This is not directly related to electronic-structure calculations but provides a useful notation that we shall apply in the following section. Moreover, it is often met in the literature and will therefore be briefly introduced here.

The simplest Hückel model is developed for studying the electronic orbitals and their energies of the π electrons in organic molecules. The method is based on defining an effective Hamilton operator that reproduces a defined set of properties correctly. This operator,

$$\hat{H}_{\text{Hückel}} = \sum_{i=1}^{N} \hat{h}_{\text{eff}}(i), \tag{12.1}$$

contains only single-particle operators, $\hat{h}_{\text{eff}}$, and it is assumed that the exact solutions to the single-particle equations,

$$\hat{h}_{\text{eff}} \phi_i = \varepsilon_i \phi_i \tag{12.2}$$

can be written *exactly* as a *finite* linear combination of the π functions on the different atoms,

$$\phi_i = \sum_{j=1}^{M} \chi_j c_{ji}. \tag{12.3}$$

Here, χ_j is the π function of the jth atom, and we have in total M atoms.

We stress that we assume that the expressions (12.3) are exact. This means that we assume that the orbitals of interest contain no other types of terms, or, alternatively, that the functions $\{\chi_j\}$ constitute a complete set for the problem of interest.

In Section 9.6 we discussed projection operators that were constructed from the eigenfunctions of a given operator. Here, we shall do it slightly differently by not assuming that the functions are eigenfunctions for a given operator. Instead, we assume that we have a complete set of functions, i.e., the set $\{\chi_k\}$. We assume that the functions are orthonormal. Since this set is complete for the system and properties of interest, *any* quantity can be expanded as

$$g(\vec{r}) = \sum_{j=1}^{M} \chi_j(\vec{r}) d_j, \tag{12.4}$$

or in a bra–ket notation,

$$|g\rangle = \sum_{j=1}^{M} d_j |\chi_j\rangle. \tag{12.5}$$

By multiplying by $\langle \chi_l |$ and using the orthonormality of the functions, we obtain

$$\langle \chi_l | g \rangle = \sum_{k=1}^{M} d_j \langle \chi_l | \chi_k \rangle = d_l. \tag{12.6}$$

Accordingly,

$$|g\rangle = \sum_{k=1}^{N} d_k |\chi_k\rangle = \sum_{k=1}^{M} |\chi_k\rangle\langle\chi_k|g\rangle = \left[\sum_{k=1}^{M} |\chi_k\rangle\langle\chi_k|\right] |g\rangle = 1 \cdot |g\rangle, \tag{12.7}$$

giving that the sum of the projection operators,

$$\hat{P}_k = |\chi_k\rangle\langle\chi_k|, \tag{12.8}$$

is the identity

$$1 = \sum_{k=1}^{M} \hat{P}_k. \tag{12.9}$$

We shall now use the projection operators of Eq. (12.8) in obtaining an expression for the Hückel Hamilton operator $\hat{h}_{\text{eff}}$. Thereby the identity (12.9) will be applied. The expression that will be derived is that of the so-called second quantized form.

12.2 THE HÜCKEL METHOD

The Hückel method amounts to solving the single-particle equations,

$$\hat{h}_{\text{eff}} |\phi_i\rangle = \varepsilon_i |\phi_i\rangle. \tag{12.10}$$

We rewrite the left-hand side and use the projection operators above together with the identity (12.8),

$$
\begin{aligned}
\hat{h}_{\text{eff}}|\phi_i\rangle &= 1 \cdot \hat{h}_{\text{eff}} \cdot 1 \cdot |\phi_i\rangle \\
&= \sum_{k=1}^{M} \hat{P}_k \cdot \hat{h}_{\text{eff}} \cdot \sum_{l=1}^{M} \hat{P}_l \cdot |\phi_i\rangle \\
&= \sum_{k=1}^{M} |\chi_k\rangle\langle\chi_k| \cdot \hat{h}_{\text{eff}} \cdot \sum_{l=1}^{M} |\chi_l\rangle\langle\chi_l| \cdot |\phi_i\rangle \\
&= \sum_{k=1}^{M} |\chi_k\rangle\langle\chi_k|\hat{h}_{\text{eff}} \sum_{l=1}^{M} |\chi_l\rangle\langle\chi_l|\phi_i\rangle \\
&= \sum_{k,l=1}^{M} |\chi_k\rangle\langle\chi_k|\hat{h}_{\text{eff}}|\chi_l\rangle\langle\chi_l|\phi_i\rangle \\
&= \sum_{k,l=1}^{M} |\chi_k\rangle h_{kl} \langle\chi_l|\phi_i\rangle,
\end{aligned}
\tag{12.11}
$$

where the matrix elements

$$
h_{kl} = \langle\chi_k|\hat{h}_{\text{eff}}|\chi_l\rangle \tag{12.12}
$$

are those between the basis functions (i.e., equal to α or β for this specific model). We can rewrite Eq. (12.11) as

$$
\begin{aligned}
\hat{h}_{\text{eff}}|\phi_i\rangle &= [\hat{h}_{\text{eff}}]|\phi_i\rangle \\
&= \left[\sum_{k,l=1}^{M} |\chi_k\rangle h_{kl} \langle\chi_l| \right] |\phi_i\rangle \\
&= \left[\sum_{k,l=1}^{M} h_{kl}|\chi_k\rangle\langle\chi_l| \right] |\phi_i\rangle.
\end{aligned}
\tag{12.13}
$$

We have not made any assumptions about the function ϕ_i, so Eq. (12.13) is generally valid. This means that

$$
\hat{h}_{\text{eff}} = \sum_{k,l=1}^{M} h_{kl}|\chi_k\rangle\langle\chi_l|. \tag{12.14}
$$

We introduce the operators,

$$
\begin{aligned}
\hat{a}_k^\dagger &= |\chi_k\rangle \\
\hat{a}_k &= \langle\chi_k|.
\end{aligned}
\tag{12.15}
$$

Then,

$$\hat{h}_{\text{eff}} = \sum_{k,l=1}^{M} h_{kl}\hat{a}_k^\dagger\hat{a}_l. \tag{12.16}$$

The operators $\hat{a}_k^\dagger$ and $\hat{a}_k$ of Eq. (12.15) are the so-called *creation* and *annihilation* operators. For our purposes they will occur only in pairs as in Eq. (12.16), so let us demonstrate what such a pair does. We consider a general function g that can be expanded according to Eq. (12.15),

$$|g\rangle = \sum_{j=1}^{M} d_j|\chi_j\rangle. \tag{12.17}$$

We consider

$$\hat{a}_k^\dagger\hat{a}_l|g\rangle = \hat{a}_k^\dagger\hat{a}_l \sum_{j=1}^{M} d_j|\chi_j\rangle$$

$$= |\chi_k\rangle\langle\chi_l| \sum_{j=1}^{M} d_j|\chi_j\rangle$$

$$= \sum_{j=1}^{M} d_j|\chi_k\rangle\langle\chi_l|\chi_j\rangle$$

$$= d_l|\chi_k\rangle. \tag{12.18}$$

That is, 'the electrons that belonged to atom l are transferred to atom k', or, alternatively, we annihilate the electrons on the lth atom and create them again on the kth atom.

Equation (12.16) is the Hückel Hamilton operator in what is called second quantization form. It is seen to contain a linear combination of pairs of creation and annihilation operators for the individual atomic orbitals multiplied by the matrix element for the two atomic orbitals. The basic assumption is that the atomic orbitals form a *complete* set of *orthonormal* functions for the properties of interest. It is not restricted to the simple Hückel model, as long as these two assumptions are fulfilled. Thus, for the extended Hückel model the single-particle Hamilton operator takes exactly the same form as in Eq. (12.16), except that M is to be replaced by the number of (orthonormal) basis functions,

$$\hat{h}_{\text{eff}} = \sum_{k,l=1}^{N_b} h_{kl}\hat{a}_k^\dagger\hat{a}_l. \tag{12.19}$$

Also the PPP and the other semi-empirical methods of the preceding section can be defined with the help of annihilation and creation operators. Since these, however, contain two-electron matrix elements, their form will be more complicated, i.e., containing terms of the form

$$h_{klmn}\hat{a}_k^{\dagger}\hat{a}_l^{\dagger}\hat{a}_m\hat{a}_n. \tag{12.20}$$

We shall not discuss these further here. They are important for various phenomenological studies and they have certain advantages over the more 'traditional' formulation of Chapter 11, but we consider this beyond the scope of the present manuscript.

12.3 ELECTRONIC EXCITATIONS AND CONFIGURATIONS

There is, however, one place where the annihilation and creation operators will be useful, and that occurs when studying electronic configurations as extracted from calculations within the Hartree–Fock approximation. These will be important when we discuss correlation effects in the following section.

A Hartree–Fock(–Roothaan) calculation will yield a set of orbitals as in Fig. 12.1. At first, a set of orbitals are obtained without their occupancies (i.e., the left part of Fig. 12.1) but, as discussed in Chapter 9, one will choose the orbitals as shown in the middle part of Fig. 12.1 to be occupied when interested in the ground state. One may, however, also consider an excited state obtained by transferring an electron from an energetically lower (occupied) orbital to an energetically higher (vacant) orbital, whereby diagrams like that of the right part of Fig. 12.1 are obtained.

This last process may be thought of as consisting of two; i.e., an annihilation of the electron in one of the energetically lower orbitals, and a subsequent creation of the electron in one of the energetically higher orbitals, as shown in Fig. 12.2.

We stress that this process is not a *real* one, but rather introduced as a convenient mathematical model.

We will assume that we have ordered the orbitals according to increasing orbital energies,

$$\varepsilon_1 \le \varepsilon_2 \cdots \le \varepsilon_N \cdots \le \varepsilon_{N_b}, \tag{12.21}$$

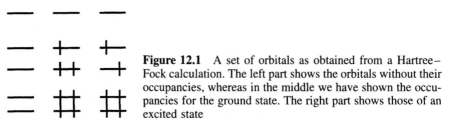

Figure 12.1 A set of orbitals as obtained from a Hartree–Fock calculation. The left part shows the orbitals without their occupancies, whereas in the middle we have shown the occupancies for the ground state. The right part shows those of an excited state

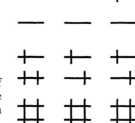

Figure 12.2 The process of generating the right diagram of Fig. 12.1 from the middle one in that figure. The middle one here corresponds to the situation where one electron has been annihilated

where N_b is the number of basis functions in the calculation, which also gives the number of orbitals that can be calculated.

The orbitals are denoted ϕ_i. Then, the configuration of the left side of Fig. 12.2 is given by the Slater determinant

$$\Phi_0 = |\phi_1, \phi_2, \ldots, \phi_N|. \tag{12.22}$$

Assuming that the configuration of the middle part of Fig. 12.2 corresponds to that of the left part except that the electron of the nth orbital has been removed ($n \leq N$), we can write this as

$$\begin{aligned}
\hat{a}_n \Phi_0 &= \hat{a}_n |\phi_1, \phi_2, \ldots, \phi_N| \\
&= \hat{a}_n |\phi_1, \phi_2, \ldots, \phi_{n-1}, \phi_n, \phi_{n+1}, \ldots, \phi_N| \\
&= |\phi_1, \phi_2, \ldots, \phi_{n-1}, \phi_{n+1}, \ldots, \phi_N|.
\end{aligned} \tag{12.23}$$

Here, the operator $\hat{a}_n$ removes the electron of the nth orbital.

Similarly, we will assume that the excited electron of the right configuration of Fig. 12.1 occupies the νth orbital ($\nu > N$). This configuration shall then be written as

$$\begin{aligned}
|\phi_1, \phi_2, \ldots, &\phi_{n-1}, \phi_{n+1}, \ldots, \phi_N, \phi_\nu| \\
&= \hat{a}_\nu^\dagger |\phi_1, \phi_2, \ldots, \phi_{n-1}, \phi_{n+1}, \ldots, \phi_N| \\
&= \hat{a}_\nu^\dagger \hat{a}_n |\phi_1, \phi_2, \ldots, \phi_{n-1}, \phi_n, \phi_{n+1}, \ldots, \phi_N| \\
&= \hat{a}_\nu^\dagger \hat{a}_n |\phi_1, \phi_2, \ldots, \phi_N| \\
&= \hat{a}_\nu^\dagger \hat{a}_n \Phi_0 \\
&\equiv \Phi_n^\nu,
\end{aligned} \tag{12.24}$$

where $\hat{a}_\nu^\dagger$ creates an electron in the νth orbital. The last identity in Eq. (12.24) introduces a simple notation for the resulting Slater determinant when starting from that of Eq. (12.22): The lower index implies that the orbital ϕ_n has been emptied, whereas the upper index implies that the orbital ϕ_ν has been occupied instead.

Hereby we have introduced creation and annihilation operators as a convenient tool for describing excited configurations in terms of the ground-state configuration. As in Section 12.2, these operators occur always in pairs. In the present manuscript we shall not use these operators for anything but as a mathematical tool — they are *not* supposed to describe physical processes!

Finally, we can introduce the general notation

$$\Phi_{k,l,m,\ldots}^{\kappa,\lambda,\mu,\ldots} = \hat{a}_{\kappa}^{\dagger}\hat{a}_{k}\hat{a}_{\lambda}^{\dagger}\hat{a}_{l}\hat{a}_{\mu}^{\dagger}\hat{a}_{m}\ldots\Phi_{0} \qquad (12.25)$$

for excited configurations where more than one electron has been excited.

13 Correlation Effects

13.1 MORE CONFIGURATIONS

In Chapter 9 we derived the Hartree–Fock equations that were obtained when approximating the N-Electron eigenfunction Ψ to the electronic Schrödinger equation

$$\hat{H}_e \Psi = E_e \Psi \tag{13.1}$$

through a single Slater determinant

$$\Psi \simeq \Phi = |\phi_1, \phi_2, \ldots, \phi_N|. \tag{13.2}$$

In Section 9.6 we saw that, although Φ might provide an accurate estimate for the total electronic energy E_e of Eq. (13.1), other properties might be less well described. In particular, within the unrestricted Hartree–Fock approximation (where the position-space dependence of spin-up and spin-down orbitals was allowed to differ) Φ was in the general case not an eigenfunction of the total $\hat{S}^2$ operator. Using the technique of projection it was possible to 'repair' that deficiency, but the price could be, as we saw through an example, that the resulting wavefunction no longer possessed the form (13.2) but was a sum of more Slater determinants (in our example, of two Slater determinants).

When writing the N-electron wavefunction as a linear combination of more Slater determinants, we go beyond the Hartree–Fock approximation — the Hartree–Fock approximation is a special case where the coefficients to the different Slater determinants all are 0 except for one that equals 1. Assuming that the Hartree–Fock equations can be solved exactly (this corresponds to having a complete set of basis functions within the Hartree–Fock–Roothaan approach, and one talks about reaching the *Hartree–Fock limit*), the differences between the single-determinant wavefunction and the many-determinant wavefunction are called correlation effects. In some cases these differences are small; in others not. In this section we shall discuss two aspects of this: how to include these and how to estimate when they might be important.

Before turning to the first of these two aspects we consider one more (very simple) example, i.e., two hydrogen atoms with the interatomic distance D. For D very large, the two hydrogen atoms are essentially non-interacting, whereas for $D \simeq 0.7$ Å, we have an H_2 molecule. For *any* D we have two electrons that within the Hartree–Fock approximation occupy two orbitals that differ only in the spin dependence, i.e.

$$\Phi = |\phi_1\alpha, \phi_1\beta|$$

$$= \frac{1}{\sqrt{2}}\left[\phi_1(1)\alpha(1)\phi_1(2)\beta(2) - \phi_1(2)\alpha(2)\phi_1(1)\beta(1)\right]$$

$$= \frac{1}{\sqrt{2}}\phi_1(1)\phi_1(2)\left[\alpha(1)\beta(2) - \alpha(2)\beta(1)\right]. \tag{13.3}$$

The molecular orbital ϕ_1 will here be supposed to be written as the bonding combination of two atom-centred s functions, one on each atom; cf. Fig. 13.1. That is,

$$\phi_1 = (2 + 2S)^{-1/2}(\chi_1 + \chi_2), \tag{13.4}$$

which is normalized,

$$1 = \langle\phi_1|\phi_1\rangle$$

$$= \frac{1}{2 + 2S}\left[\langle\chi_1|\chi_1\rangle + \langle\chi_2|\chi_2\rangle + \langle\chi_1|\chi_2\rangle + \langle\chi_2|\chi_1\rangle\right]$$

$$= \frac{1}{2 + 2S}[1 + 1 + S + S] \tag{13.5}$$

with

$$S = \langle\chi_1|\chi_2\rangle. \tag{13.6}$$

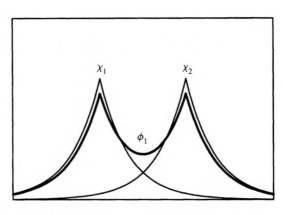

Figure 13.1 The molecular orbital ϕ_1 for the H_2 system as constructed from the two atomic orbitals χ_1 and χ_2

Inserting Eq. (13.4) into Eq. (13.3) gives

$$
\begin{aligned}
\Phi &= \frac{1}{2 + 2S} [\chi_1(1) + \chi_2(1)] [\chi_1(2) + \chi_2(2)] [\alpha(1)\beta(2) - \alpha(2)\beta(1)] \\
&= \frac{1}{2 + 2S} [\chi_1(1)\chi_1(2) + \chi_2(1)\chi_2(2) + \chi_2(1)\chi_1(2) + \chi_1(1)\chi_2(2)] \\
&\quad \times [\alpha(1)\beta(2) - \alpha(2)\beta(1)] \\
&= \frac{1}{2 + 2S} \Big\{ [\chi_1(1)\chi_1(2) + \chi_2(1)\chi_2(2)] + [\chi_2(1)\chi_1(2) + \chi_1(1)\chi_2(2)] \Big\} \\
&\quad \times [\alpha(1)\beta(2) - \alpha(2)\beta(1)]. \tag{13.7}
\end{aligned}
$$

Inside the curly bracket we have two terms, each inside a square bracket. The first of those describes distributions of the electrons for which both electrons are on the same atom, i.e., either on atom 1 or on atom 2, whereas the second term describes distributions for which the two electrons are on different atoms. Since the first term effectively corresponds to the H^+-H^- and H^--H^+ arrangements this term is called an ionic term. The other one, on the other hand, corresponds to $H-H$ arrangements and is accordingly called a covalent term.

For small D (i.e., as for the molecule), this description might be acceptable, but for large D one would intuitively expect that the system consists of two neutral atoms. Accordingly, the fact that the ionic and covalent terms are given the same weight in Eq. (13.7) does not seem to be appropriate.

One may improve on this by considering a more general wavefunction, i.e.,

$$
C \Big\{ c[\chi_1(1)\chi_1(2) + \chi_2(1)\chi_2(2)] + [\chi_2(1)\chi_1(2) + \chi_1(1)\chi_2(2)] \Big\} \\
\times [\alpha(1)\beta(2) - \alpha(2)\beta(1)], \tag{13.8}
$$

where the parameter c depends on D and approaches 0 for $D \to \infty$. Moreover, C is a normalization constant, which for the present discussion is unimportant.

We may now realize that equivalent to the bonding orbital ϕ_1 of Eq. (13.4) we may also construct the corresponding antibonding one,

$$
\phi_2 = (2 - 2S)^{-1/2}(\chi_1 - \chi_2). \tag{13.9}
$$

A Slater determinant constructed from this is

$$
\begin{aligned}
\Phi' &= \frac{1}{2 - 2S} [\chi_1(1) - \chi_2(1)] [\chi_1(2) - \chi_2(2)] [\alpha(1)\beta(2) - \alpha(2)\beta(1)] \\
&= \frac{1}{2 - 2S} [\chi_1(1)\chi_1(2) + \chi_2(1)\chi_2(2) - \chi_2(1)\chi_1(2) - \chi_1(1)\chi_2(2)] \\
&\quad \times [\alpha(1)\beta(2) - \alpha(2)\beta(1)]
\end{aligned}
$$

$$= \frac{1}{2 - 2S} \Big\{ \big[\chi_1(1)\chi_1(2) + \chi_2(1)\chi_2(2) \big] - \big[\chi_2(1)\chi_1(2) + \chi_1(1)\chi_2(2) \big] \Big\}$$
$$\times \big[\alpha(1)\beta(2) - \alpha(2)\beta(1) \big]. \tag{13.10}$$

Comparing with Eq. (13.7) it can be seen that the wavefunction of Eq. (13.8) can be written as

$$c_1 \Phi + c_2 \Phi' \tag{13.11}$$

with c_1 and c_2 related to c and C through

$$C \cdot c = \frac{c_1}{2 + 2S} + \frac{c_2}{2 - 2S}$$
$$C = \frac{c_1}{2 + 2S} - \frac{c_2}{2 - 2S}. \tag{13.12}$$

This wavefunction is thus one that consists of more than one Slater determinant, i.e., one containing correlation effects.

The two Slater determinants of Eqs. (13.7) and (13.10) correspond to the two configurations of Fig. 13.2. For very large D, the single-particle energies of ϕ_1 and ϕ_2 approach each other (the energies of the bonding and the antibonding orbitals do not differ), whereas for small D they have markedly different energies. Their behaviour is sketched in Fig. 13.3.

In Eq. (13.8) we expect c to be markedly different from 1 only for large D. This means that $|c_2|$ of Eq. (13.11) is markedly different from 0 only for large D. Comparing Figs. 13.2 and 13.3 it may now be suggested that excited configurations (i.e., configurations that differ from that of the ground state where the N energetically lowest orbitals are occupied) become important when the energy difference between occupied and unoccupied orbitals for the ground-state configuration is small. That this in fact is the case will be shown below by using the so-called Møller–Plesset (MP) perturbation theory. But before doing so we shall study the general case of configuration interaction (CI).

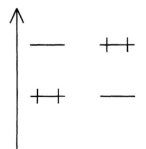

Figure 13.2 The two configurations of Φ and Φ' of Eqs. (13.7) and (13.10), respectively. The vertical axis gives the single particle energies of ϕ_1 and ϕ_2

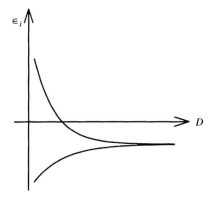

Figure 13.3   The single-particle energies of ϕ_1 and ϕ_2 as a function of D

13.2   CONFIGURATION INTERACTION (CI)

Let us start with the Hartree–Fock approximation and let us assume that we have solved the Hartree–Fock–Roothaan equations for some system. This means that we have obtained a Slater determinant of the form

$$\Phi_0 = |\phi_1, \phi_2, \ldots, \phi_N|, \tag{13.13}$$

where the single-electron orbitals ϕ_i are expanded in some basis set,

$$\phi_i(\vec{x}) = \sum_{j=1}^{N_b} \chi_j(\vec{x}) c_{ji} \tag{13.14}$$

($\vec{x}$ being, as usual, the combined position and spin coordinate).

Solving the Hartree–Fock–Roothaan equations self-consistently will give us $N_b \geq N$ orthonormal single-electron orbitals, as shown in the left part of Fig. 13.4. For the ground state (the middle part of Fig. 13.4) we occupy the N energetically lowest of those, which results in the wavefunction of Eq. (13.13). We may, however, also use energetically higher orbitals in constructing Slater determinants for excited states. With the notation of Section 12.3 [e.g., Eq. (12.25)] these are

$$\Phi_{k,l,m,n,\ldots}^{\kappa,\lambda,\mu,\nu,\ldots} = \hat{a}_\kappa^\dagger \hat{a}_k \hat{a}_\lambda^\dagger \hat{a}_l \hat{a}_\mu^\dagger \hat{a}_m \hat{a}_\nu^\dagger \hat{a}_n \cdots \Phi_0. \tag{13.15}$$

A more general wavefunction is thus

$$\Phi = \sum_{k,l,m,n,\ldots} \sum_{\kappa,\lambda,\mu,\nu,\ldots} c_{k,l,m,n,\ldots}^{\kappa,\lambda,\mu,\nu,\ldots} \Phi_{k,l,m,n,\ldots}^{\kappa,\lambda,\mu,\nu,\ldots}, \tag{13.16}$$

i.e., a wavefunction that contains all the possible configurations that can be constructed from the orbitals of the Hartree–Fock–Roothaan equations. It should

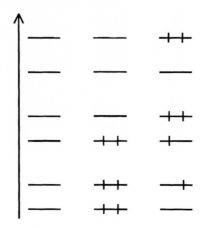

Figure 13.4 A set of orbitals as obtained from a Hartree–Fock–Roothaan calculation. The left part shows the orbitals without their occupancies, whereas in the middle we have shown the occupancies for the ground state. The right part shows those of an excited state

be added that the excited configurations of Eq. (13.15) are obtained from the single-electron orbitals that are calculated using the Fock operator for the *ground* state. Thereby *all* single-electron orbitals — both those that for the ground state are occupied and those that are empty — become orthonormal which represents an important simplification in the calculations as we shall see. That this is so follows from the fact that they are eigenfunctions to the same (i.e., Fock) operator.

Instead of using the complicated notation of Eq. (13.16) we shall write

$$\Phi = \sum_{I=0}^{N_{CI}} C_I \Phi_I, \qquad (13.17)$$

where we have used the compact index I for the different configurations, and where

$$N_{CI} + 1 = \binom{N_b}{N} \qquad (13.18)$$

is the number of configurations that can be constructed from the orbitals that are obtained from the Hartree–Fock–Roothaan equations.

When setting all $C_I = 0$ except for $C_0 = 1$, we have the Hartree–Fock approximation.

In the general case we apply the variational method on the quantity

$$K = \langle \Phi | \hat{H}_e | \Phi \rangle - \lambda [\langle \Phi | \Phi \rangle - 1] \qquad (13.19)$$

as discussed in Section 5.5. Thereby, *only* the parameters C_I are varied, but not the c_{ji} of Eq. (13.14) used in expanding the single-electron orbitals.

Analogously to Eq. (5.50) we find

$$\frac{\partial}{\partial C_K^*} \left[\left\langle \sum_I C_I \Phi_I \left| \hat{H}_e \right| \sum_J C_J \Phi_J \right\rangle - \lambda \left(\left\langle \sum_I C_I \Phi_I \middle| \sum_J C_J \Phi_J \right\rangle - 1 \right) \right]$$

$$= \frac{\partial}{\partial C_K^*} \left[\left\langle \sum_I C_I \Phi_I \left| \hat{H}_e \right| \sum_J C_J \Phi_J \right\rangle - \lambda \left\langle \sum_I C_I \Phi_I \middle| \sum_J C_J \Phi_J \right\rangle \right]$$

$$= \frac{\partial}{\partial C_K^*} \sum_{I,J} C_I^* C_J \left[\langle \Phi_I | \hat{H}_e | \Phi_J \rangle - \lambda \langle \Phi_I | \Phi_J \rangle \right]$$

$$= \sum_J C_J \left[\langle \Phi_K | \hat{H}_e | \Phi_J \rangle - \lambda \langle \Phi_K | \Phi_J \rangle \right] \equiv 0, \qquad (13.20)$$

giving the matrix eigenvalue problem

$$\underline{\underline{H}} \cdot \underline{C} = \lambda \cdot \underline{\underline{O}} \cdot \underline{C}, \qquad (13.21)$$

where

$$\underline{\underline{H}}_{KJ} = \langle \Phi_K | \hat{H}_e | \Phi_J \rangle$$

$$\underline{\underline{O}}_{KJ} = \langle \Phi_K | \Phi_J \rangle. \qquad (13.22)$$

Furthermore, as shown in Section 5.5, the eigenvalue λ is the sought energy,

$$\lambda = \frac{\langle \Phi | \hat{H}_e | \Phi \rangle}{\langle \Phi | \Phi \rangle}. \qquad (13.23)$$

Let us now study the matrix elements of Eq. (13.22).

The Hamilton operator is written as

$$\hat{H}_e = \sum_{i=1}^{N} \hat{h}_1(\vec{r}_i) + \frac{1}{2} \sum_{i \neq j=1}^{N} \hat{h}_2(\vec{r}_i, \vec{r}_j), \qquad (13.24)$$

where furthermore

$$\hat{h}_2(\vec{r}_i, \vec{r}_j) = \hat{h}_2(\vec{r}_j, \vec{r}_i) = \frac{1}{|\vec{r}_i - \vec{r}_j|}. \qquad (13.25)$$

Analogously to Section 9.2 we first consider the general case of an operator that is a sum of identical single-electron operators,

$$\hat{A} = \sum_{n=1}^{N} \hat{a}_1(n). \qquad (13.26)$$

For this we find [cf. Eqs. (9.30)–(9.34)]

$$
\langle \Phi_J | \hat{A} | \Phi_I \rangle = \frac{1}{N!} \left\langle \sum_{j=1}^{N!} (-1)^{P(j)} \phi_{j_1}(\vec{x}_1) \phi_{j_2}(\vec{x}_2) \cdots \phi_{j_N}(\vec{x}_N) \right|
$$

$$
\times \sum_{n=1}^{N} \hat{a}_1(n) \left| \sum_{i=1}^{N!} (-1)^{P(i)} \phi_{i_1}(\vec{x}_1) \phi_{i_2}(\vec{x}_2) \cdots \phi_{i_N}(\vec{x}_N) \right\rangle
$$

$$
= \frac{1}{N!} \sum_{n=1}^{N} \left\langle \sum_{j=1}^{N!} (-1)^{P(j)} \phi_{j_1}(\vec{x}_1) \phi_{j_2}(\vec{x}_2) \cdots \phi_{j_N}(\vec{x}_N) \right|
$$

$$
\times \hat{a}_1(n) \left| \sum_{i=1}^{N!} (-1)^{P(i)} \phi_{i_1}(\vec{x}_1) \phi_{i_2}(\vec{x}_2) \cdots \phi_{i_N}(\vec{x}_N) \right\rangle
$$

$$
= \frac{1}{N!} \sum_{n=1}^{N} \left\langle \sum_{j=1}^{N!} (-1)^{P(j)} \phi_{j_1}(\vec{x}_1) \phi_{j_2}(\vec{x}_2) \cdots \phi_{j_{n_1}}(\vec{x}_{n_1}) \phi_{j_n}(\vec{x}_n) \right.
$$

$$
\times \phi_{j_{n+1}}(\vec{x}_{n+1}) \cdots \phi_{j_N}(\vec{x}_N) \Big|
$$

$$
\times \hat{a}_1(n) \left| \sum_{i=1}^{N!} (-1)^{P(i)} \phi_{i_1}(\vec{x}_1) \phi_{i_2}(\vec{x}_2) \cdots \phi_{i_{n_1}}(\vec{x}_{n_1}) \phi_{i_n}(\vec{x}_n) \right.
$$

$$
\times \phi_{i_{n+1}}(\vec{x}_{n+1}) \cdots \phi_{i_N}(\vec{x}_N) \Big\rangle
$$

$$
= \frac{1}{N!} \sum_{n=1}^{N} \sum_{i,j=1}^{N!} (-1)^{P(j)} (-1)^{P(i)} \langle \phi_{j_1} | \phi_{i_1} \rangle
$$

$$
\times \langle \phi_{j_2} | \phi_{i_2} \rangle \cdots \langle \phi_{j_{n-1}} | \phi_{i_{n_1}} \rangle \cdot \langle \phi_{j_n} | \hat{a}_1 | \phi_{i_n} \rangle \cdot \langle \phi_{j_{n+1}} | \phi_{i_{n+1}} \rangle \cdots \langle \phi_{j_N} | \phi_{i_N} \rangle
$$

$$
= \frac{1}{N!} \sum_{n=1}^{N} \sum_{i,j=1}^{N!} (-1)^{P(j)} (-1)^{P(i)} \delta_{j_1,i_1} \delta_{j_2,i_2} \cdots \delta_{j_{n-1},i_{n-1}} \langle \phi_{j_n} | \hat{a}_1 | \phi_{i_n} \rangle
$$

$$
\times \delta_{j_{n+1},i_{n+1}} \cdots \delta_{j_N,i_N}
$$

$$
= \frac{1}{N!} N \sum_{i,j=1}^{N!} (-1)^{P(j)} (-1)^{P(i)} \delta_{j_1,i_1} \delta_{j_2,i_2} \cdots \delta_{j_{n-1},i_{n-1}} \langle \phi_{j_n} | \hat{a}_1 | \phi_{i_n} \rangle
$$

$$
\times \delta_{j_{n+1},i_{n+1}} \cdots \delta_{j_N,i_N}. \tag{13.27}
$$

This equation shows first of all (by setting $\hat{a}_1 = \frac{1}{N}$), that different configurations are orthonormal,

$$
\langle \Phi_I | \Phi_j \rangle = \delta_{I,J}. \tag{13.28}
$$

Moreover, for any single-electron operator there will be non-vanishing matrix elements only between configurations that differ at most in one orbital. Writing the configurations with the use of the creation and annihilation operators as in Eq. (13.16), this means that there will be non-vanishing matrix elements only between configurations that can be obtained from each other by applying at most one pair of creation and annihilation operators.

Exactly the same techniques can be applied for operators of the form

$$\hat{A} = \frac{1}{2} \sum_{n \neq m=1}^{N} \hat{a}_2(n, m). \tag{13.29}$$

Then,

$$\langle \Phi_J | \hat{A} | \Phi_I \rangle = \frac{1}{N!} \frac{N(N-1)}{2} \sum_{i,j=1}^{N!} (-1)^{P(j)} (-1)^{P(i)} \delta_{j_1,i_1} \delta_{j_2,i_2} \cdots \delta_{j_{n-1},i_{n-1}} \delta_{j_{m-1},i_{m-1}}$$

$$\times \langle \phi_{j_n} \phi_{j_m} | \hat{a}_2 | \phi_{i_n} \phi_{i_m} \rangle \delta_{j_{n+1},i_{n+1}} \delta_{j_{m+1},i_{m+1}} \cdots \delta_{j_N,i_N}, \tag{13.30}$$

i.e., these are non-vanishing only for configurations differing in at most two orbitals.

We shall now study a special case. First of all, the Hamilton operator is

$$\hat{H}_e = \sum_{i=1}^{N} \hat{h}_1(i) + \frac{1}{2} \sum_{i \neq j=1}^{N} \hat{h}_2(i, j), \tag{13.31}$$

whereas the orbitals ϕ_i are solutions to the Hartree–Fock equations,

$$\hat{F} \phi_i = \varepsilon_i \phi_i \tag{13.32}$$

with

$$\hat{F} = \hat{h}_1 + \sum_{j=1}^{N} (\hat{J}_j - \hat{K}_j). \tag{13.33}$$

In Eq. (13.33), the j summation runs over all those orbitals that for the ground state, Φ_0, are occupied.

We may substitute Eq. (13.33) in Eq. (13.32) and rearrange, giving

$$\hat{h}_1(i) \phi_i = \varepsilon_i \phi_i - \sum_{j=1}^{N} (\hat{J}_j - \hat{K}_j) \phi_i. \tag{13.34}$$

Here,

$$\hat{J}_j \phi_i(\vec{x}_1) = \int \frac{|\phi_j(\vec{x}_2)|^2}{|\vec{r}_2 - \vec{r}_1|} \, d\vec{x}_2 \phi_i(\vec{x}_1)$$

$$\hat{K}_j \phi_i(\vec{x}_1) = \int \frac{\phi_j^*(\vec{x}_2) \phi_i(\vec{x}_2)}{|\vec{r}_2 - \vec{r}_1|} \, d\vec{x}_2 \phi_j(\vec{x}_1). \tag{13.35}$$

These formulas can be used in studying a special case of the Hamilton matrix elements

$$\langle \Phi_J | \hat{H}_e | \Phi_I \rangle, \tag{13.36}$$

i.e., that where one of the configurations is the ground-state configuration, and the other is a single excited one,

$$\Phi_I = \Phi_0$$
$$\Phi_J = \Phi_n^v \tag{13.37}$$

(i.e., one electron has been excited from the nth orbital to the vth one).

We then have from Eqs. (13.27) and (13.30),

$$\langle \Phi_J | \hat{H}_e | \Phi_I \rangle = \frac{1}{N!} N \sum_{i,j=1}^{N!} (-1)^{P(j)} (-1)^{P(i)} \delta_{j_1,i_1} \delta_{j_2,i_2} \cdots \delta_{j_{n-1},i_{n-1}} \langle \phi_{j_n} | \hat{h}_1 | \phi_{i_n} \rangle$$

$$\times \delta_{j_{n+1},i_{n+1}} \cdots \delta_{j_N,i_N} + \frac{1}{N!} \frac{N(N-1)}{2} \sum_{i,j=1}^{N!} (-1)^{P(j)} (-1)^{P(i)}$$

$$\times \delta_{j_1,i_1} \delta_{j_2,i_2} \cdots \delta_{j_{n-1},i_{n-1}} \delta_{j_{m-1},i_{m-1}} \langle \phi_{j_n} \phi_{j_m} | \hat{h}_2 | \phi_{i_n} \phi_{i_m} \rangle$$

$$\times \delta_{j_{n+1},i_{n+1}} \cdots \delta_{j_{m-1},i_{m-1}} \delta_{j_{m+1},i_{m+1}} \cdots \delta_{j_N,i_N}. \tag{13.38}$$

Φ_I and Φ_J differ only by one orbital, so the terms of the second contribution on the right-hand side above are then non-vanishing only when $j_1 = i_1$, $j_2 = i_2$, ..., $j_{n-1} = i_{n-1}$, $j_{n+1} = i_{n+1}$, ..., $j_N = i_N$. By using that under these conditions, all indices i_p, except for i_n and i_m, can be chosen in $(N-2)!$ different ways, we obtain

$$\frac{1}{N!} \frac{N(N-1)}{2} \sum_{i,j=1}^{N!} (-1)^{P(j)} (-1)^{P(i)} \delta_{j_1,i_1} \delta_{j_2,i_2} \cdots \delta_{j_{n-1},i_{n-1}} \delta_{j_{m-1},i_{m-1}}$$

$$\times \langle \phi_{j_n} \phi_{j_m} | \hat{h}_2 | \phi_{i_n} \phi_{i_m} \rangle \delta_{j_{n+1},i_{n+1}} \cdots \delta_{j_{m-1},i_{m-1}} \delta_{j_{m+1},i_{m+1}} \cdots \delta_{j_N,i_N}$$

$$= \frac{1}{2} \sum_{i=1}^{N} [\langle \phi_v \phi_i | \hat{h}_2 | \phi_n \phi_i \rangle - \langle \phi_i \phi_v | \hat{h}_2 | \phi_n \phi_i \rangle + \langle \phi_i \phi_v | \hat{h}_2 | \phi_i \phi_n \rangle - \langle \phi_v \phi_i | \hat{h}_2 | \phi_i \phi_n \rangle]$$

$$= \sum_{i=1}^{N} [\langle \phi_v \phi_i | \hat{h}_2 | \phi_n \phi_i \rangle - \langle \phi_i \phi_v | \hat{h}_2 | \phi_n \phi_i \rangle], \tag{13.39}$$

where we have used Eq. (13.25) so that the two expressions only differing in the simultaneous interchange of both functions of the bra and of the two functions of the ket are identical.

We can now use the definitions of the Coulomb and exchange operators of Eq. (13.35) in rewriting Eq. (13.39) as (this works since Φ_0 is the ground state that has been used in defining the Fock operator, that in turn was used in determining the single-particle electrons).

$$\sum_{i=1}^{N}\left[\langle\phi_v\phi_i|\hat{h}_2|\phi_n\phi_i\rangle - \langle\phi_i\phi_v|\hat{h}_2|\phi_n\phi_i\rangle\right] = \sum_{i=1}^{N}\langle\phi_v|\hat{J}_i - \hat{K}_i|\phi_n\rangle$$

$$= \left\langle\phi_v\left|\sum_{i=1}^{N}(\hat{J}_i - \hat{K}_i)\right|\phi_n\right\rangle$$

$$= \langle\phi_v|\hat{F} - \hat{h}_1|\phi_n\rangle, \tag{13.40}$$

where we have used Eq. (13.33).

The first expression on the right-hand side of Eq. (13.38) is similarly

$$\langle\phi_v|\hat{h}_1|\phi_n\rangle, \tag{13.41}$$

where we have once again used that the two functions of Eq. (13.37) only differ by one orbital.

In total,

$$\langle\Phi_J|\hat{H}_e|\Phi_I\rangle = \langle\phi_v|\hat{F} - \hat{h}_1|\phi_n\rangle + \langle\phi_v|\hat{h}_1|\phi_n\rangle$$

$$= \langle\phi_v|\hat{F}|\phi_n\rangle$$

$$= \langle\phi_v|\varepsilon_n|\phi_n\rangle$$

$$= \varepsilon_n\langle\phi_v|\phi_n\rangle$$

$$= 0. \tag{13.42}$$

This means, that both overlap [due to Eq. (13.28)] and Hamilton matrix elements vanish between the ground-state configuration and any single-excited one. The consequences of this will be discussed further in Section 13.4.

Returning to the general description of the CI method we add only that, having determined all the relevant matrix elements of Eq. (13.21) (i.e., not only those discussed in detail here), we may obtain the wavefunction (13.17) as well as its corresponding energy. The total approach is the so-called *configuration interaction* (CI) method. Notice that in contrast to the Hartree–Fock calculations, the CI calculations do not require any self-consistency and Eq. (13.21) is not to be solved iteratively.

13.3 MULTIPLE-CONFIGURATION METHOD (MC-SCF)

Within the CI method we vary the coefficients C_I to the various configurations,

$$\Phi = \sum_I C_I\Phi_I. \tag{13.43}$$

On the other hand, the expansion coefficients c_{ji} of the individual orbitals to the basis functions,

$$\phi_i = \sum_j \chi_j c_{ji}, \qquad (13.44)$$

are kept as obtained within the Hartree–Fock approximation. One may, however, choose to vary these, too. This increases the computational efforts significantly, but does, due to the variational principle, allow for an improved wavefunction.

When this method is applied, the so-called *multiple-configuration* (MC-SCF) method is obtained. Compared with the CI method, the calculation of the various expansion coefficients is significantly more complicated, and, as for the Hartree–Fock–Roothaan method, one has to obtain these using an iterative approach, i.e., the solution has to be self-consistent (this gives the label 'SCF', which means self-consistent field).

13.4   SIZE CONSISTENCY; CAS-SCF

Let us try to study how complicated a full CI calculation can become. We consider for simplicity the water molecule, H_2O. It has $N = 10$ electrons and only $M = 3$ nuclei, and is as such a very small molecule. We could imagine using four Gaussians per hydrogen atom for the $1s$ orbitals, similarly, four Gaussians for the $1s$ orbital of oxygen, three for the oxygen $2s$ orbital, and three for each of the oxygen $2p$ orbitals. This gives in total $2 \times 4 + 4 + 3 + 3 \times 3 = 24$ functions, or, including spin, 48 basis functions. Solving the Hartree–Fock–Roothaan equations leads to 48 orbitals of which the 10 energetically lowest are occupied for the ground state.

From the 48 orbitals and 10 electrons we can, however, construct in total

$$\binom{48}{10} = 6\,540\,715\,896 \qquad (13.45)$$

configurations. This number is so large that it is absolutely impossible to carry a full CI calculation through, where *all* coefficients

$$\Phi = \sum_{k,l,m,n,\ldots} \sum_{\kappa,\lambda,\mu,\nu,\ldots} c_{k,l,m,n,\ldots}^{\kappa,\lambda,\mu,\nu,\ldots} \Phi_{k,l,m,n,\ldots}^{\kappa,\lambda,\mu,\nu,\ldots} \qquad (13.46)$$

are varied. Instead one would choose to consider only those configurations where a smaller number of electrons are excited or in other ways attempting to reduce the number (e.g., allowing only configurations with doubly occupied orbitals will reduce the number of configurations to 42 504).

By only considering single excitations, the expansion would become

$$\Phi = c_0 \Phi_0 + \sum_k \sum_\kappa c_k^\kappa \Phi_k^\kappa. \qquad (13.47)$$

However, as shown in Section 13.2, the Hamilton and overlap matrix elements between *any* of the single-excited configuration and the ground-state configuration vanish, so the wavefunction (13.47) does not bring any improvement over the Hartree–Fock wavefunction

$$\Phi = \Phi_0. \tag{13.48}$$

This is so, since the secular equation becomes

$$
\begin{pmatrix}
E_0 & 0 & 0 & \cdots & 0 \\
0 & E_{11} & E_{12} & \cdots & E_{1P} \\
0 & E_{21} & E_{22} & \cdots & E_{2P} \\
\vdots & \vdots & \vdots & \ddots & \vdots \\
0 & E_{P1} & E_{P2} & \cdots & E_{PP}
\end{pmatrix}
\cdot
\begin{pmatrix}
c_0 \\ c_1 \\ c_2 \\ \vdots \\ c_P
\end{pmatrix}
= E \cdot
\begin{pmatrix}
1 & 0 & 0 & \cdots & 0 \\
0 & 1 & 0 & \cdots & 0 \\
0 & 0 & 1 & \cdots & 0 \\
\vdots & \vdots & \vdots & \ddots & \vdots \\
0 & 0 & 0 & \cdots & 1
\end{pmatrix}
\cdot
\begin{pmatrix}
c_0 \\ c_1 \\ c_2 \\ \vdots \\ c_P
\end{pmatrix},
$$

$$\tag{13.49}$$

where $P = N \cdot (N_b - N)$ is the number of single-excited configurations.

Only when also including double excitations (or higher) can the total energy be lowered. In that case also the single-excited configurations may obtain non-vanishing coefficients.

An important problem related to restricting the number of excitations is that of size consistency. In order to understand this problem in detail we study the system of Fig. 13.5. This consists of two identical parts that are supposed to be non-interacting. We assume also that we include only single and double excitations (although our arguments are not restricted to this case). When treating each part of the system of Fig. 13.5 separately, the configurations shown are included in the calculation.

On the other hand, when we treat the full system, we will expect the total energy to be twice what it would be when considering each part separately (due to the lack of interactions between the two parts). However, the configuration of

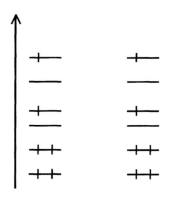

Figure 13.5 Two identical non-interacting systems, which both are in double-excited configurations

Fig. 13.5 corresponds then to a quadruple excitation and is as such not included in the calculation. Therefore, the system will be treated differently in the two cases.

It would not help to allow for quadruple excitations for the doubled system since this would also include contributions where four electrons of the one part and none of the other part were excited. And these configurations are clearly not included in the calculations for the separate parts.

When including *all* excitations, these *size-inconsistency* problems disappear. However, as shown above, the calculations become prohibitively involved. On the other hand, the complete-active-space self-consistent-field (CAS-SCF) method (Roos *et al.*, 1980) offers a non-trivial way of circumventing this problem by carefully choosing the configurations that are included in the CI expression. This method, however, will not be discussed further.

13.5   THE COUPLED-CLUSTER METHOD

Another way of circumventing the size-inconsistency problem is to apply the so-called coupled-cluster method (Čiček, 1966). We rewrite the expansion of Eq. (13.16) as

$$\Phi = \sum_{k,l,m,n,\dots} \sum_{\kappa,\lambda,\mu,\nu,\dots} c_{k,l,m,n,\dots}^{\kappa,\lambda,\mu,\nu,\dots} \Phi_{k,l,m,n,\dots}^{\kappa,\lambda,\mu,\nu,\dots}$$

$$= c_0\Phi_0 + \sum_k \sum_\kappa c_k^\kappa \Phi_k^\kappa + \sum_{k,l}^{\kappa,\lambda} c_{k,l}^{\kappa,\lambda} \Phi_{k,l}^{\kappa,\lambda} + \cdots$$

$$= \mathcal{N} \left[\Phi_0 + \sum_k \sum_\kappa d_k^\kappa \hat{a}_\kappa^\dagger \hat{a}_k \Phi_0 + \sum_{k,l}^{\kappa,\lambda} d_{k,l}^{\kappa,\lambda} \hat{a}_\kappa^\dagger \hat{a}_k \hat{a}_\lambda^\dagger \hat{a}_l \Phi_0 + \cdots \right]$$

$$= \mathcal{N} [1 + \hat{T}_1 + \hat{T}_2 + \cdots] \phi_0 \tag{13.50}$$

where $\mathcal{N}$ is a normalization constant, and

$$\hat{T}_1 = \sum_k \sum_\kappa d_k^\kappa \hat{a}_\kappa^\dagger \hat{a}_k$$

$$\hat{T}_2 = \sum_{k,l}^{\kappa,\lambda} d_{k,l}^{\kappa,\lambda} \hat{a}_\kappa^\dagger \hat{a}_k \hat{a}_\lambda^\dagger \hat{a}_l$$

$$\cdots \tag{13.51}$$

are operators involving the creation and annihilation operators of Section 12.5. The parameters that will be determined are the d, and $\hat{T}_n$ is an operator that involves n-electron excitations.

One may now truncate the sum in Eq. (13.50) after a smaller number of terms (e.g., including only single- and double-excitations). This corresponds to setting

$$\Phi = \mathcal{N}[1 + \hat{T}_1 + \hat{T}_2 + \cdots + \hat{T}_P]\phi_0. \tag{13.52}$$

Thereby we are back to the problem of size-inconsistency. For the sake of simplicity we define

$$\hat{T} = 1 + \hat{T}_1 + \hat{T}_2 + \cdots + \hat{T}_P. \tag{13.53}$$

Instead of the wavefunction of Eq. (13.52) we consider

$$\Phi = \mathcal{N}' e^{\hat{T}} \Phi_0. \tag{13.54}$$

Here, per definition,

$$e^{\hat{T}} = 1 + \hat{T} + \tfrac{1}{2}\hat{T}\hat{T} + \cdots. \tag{13.55}$$

One may show that through this choice, the size-inconsistency problem has disappeared. Compared with the general form above, the form (13.54) restricts the configurations that are included and, furthermore, establish relations between the various parameters for excitations involving fewer and more electrons. Otherwise, the calculations proceed just as in the CI case, and we shall therefore not discuss this approach further.

13.6 MØLLER–PLESSET PERTURBATION THEORY

Often the Hartree–Fock approximation provides an accurate description of the system and the effects of the inclusion of correlations as, e.g., with the CI or MC-SCF methods, are of secondary importance. Accordingly, the correlation effects may be considered a smaller perturbation and as such treated using the perturbation theory of Chapter 6. This is the approach of Møller and Plesset (1934) for the inclusion of correlation effects.

For the sake of simplification we shall here consider the ground state but mention that the method in principle can be applied for any state, i.e., also for an excited state. Our starting point is the Hartree–Fock equations

$$\hat{F}\phi_l = \varepsilon_l \phi_l. \tag{13.56}$$

Here,

$$\hat{F} = \hat{h}_1 + \sum_{i=1}^{N}(\hat{J}_i - \hat{K}_i) \tag{13.57}$$

is a single-electron operator, and the sum in Eq. (13.57) runs over all occupied orbitals, i.e., over the energetically lowest ones. $\hat{J}_i$ and $\hat{K}_i$ depend on the

ith orbital,

$$\hat{J}_i\phi_k(\vec{x}_1) = \int \frac{|\phi_i(\vec{x}_2)|^2}{|\vec{r}_1 - \vec{r}_2|} \, d\vec{x}_2 \phi_k(\vec{x}_1)$$

$$\hat{K}_i\phi_k(\vec{x}_1) = \int \frac{\phi_i^*(\vec{x}_2)\phi_k(\vec{x}_2)}{|\vec{r}_1 - \vec{r}_2|} \, d\vec{x}_2 \phi_i(\vec{x}_1). \tag{13.58}$$

Solving Eq. (13.56) gives, however, not only the N occupied orbitals, but — in principle — a complete set of orbitals, since $\hat{F}$ is a Hermitian operator. Thus, the set $\{\phi_l\}$ constitutes a complete set of single-electron wavefunctions.

The operator $\hat{F}$ is a single-electron operator, which we may formally write as

$$\hat{F} = \hat{F}(\text{electron } i) \equiv \hat{F}(i). \tag{13.59}$$

We define now first the N-electron operator

$$\hat{G}' = \sum_{i=1}^{N} \hat{F}(i). \tag{13.60}$$

Also this is a Hermitian operator, and the N-electron Slater determinants

$$|\phi_{i_1}, \phi_{i_2}, \ldots, \phi_{i_N}| \tag{13.61}$$

define a complete set of eigenfunctions with

$$\hat{G}'|\phi_{i_1}, \phi_{i_2}, \ldots, \phi_{i_N}| = (\varepsilon_{i_1} + \varepsilon_{i_2} + \cdots + \varepsilon_{i_N})|\phi_{i_1}, \phi_{i_2}, \ldots, \phi_{i_N}|. \tag{13.62}$$

This is easily shown using Eq. (13.56).

In particular, for the ground state we have

$$\hat{G}'|\phi_1, \phi_2, \ldots, \phi_N| = \left[\sum_{i=1}^{N} \varepsilon_i\right] |\phi_1, \phi_2, \ldots, \phi_N|. \tag{13.63}$$

On the other hand, from Eq. (9.89) we know that

$$\sum_{i=1}^{N} \varepsilon_i = E_{\text{HF}} + \frac{1}{2} \sum_{i,j=1}^{N} \left[\langle \phi_i\phi_j|\hat{h}_2|\phi_i\phi_j\rangle - \langle \phi_j\phi_i|\hat{h}_2|\phi_i\phi_j\rangle\right]$$

$$= E_{\text{HF}} + E', \tag{13.64}$$

where

$$E' = \frac{1}{2} \sum_{i,j=1}^{N} \left[\langle \phi_i\phi_j|\hat{h}_2|\phi_i\phi_j\rangle - \langle \phi_j\phi_i|\hat{h}_2|\phi_i\phi_j\rangle\right] \tag{13.65}$$

is some number.

Since the total electronic energy from the Hartree–Fock approximation shall be our starting point in the perturbation calculation, we see from Eq. (13.64) that it is convenient to consider

$$\hat{G} = \hat{G}' - E'. \tag{13.66}$$

This has the same eigenfunctions as $\hat{G}'$ but the eigenvalues have been shifted by $-E'$.

The precise form of $\hat{G}$ is

$$\hat{G} = \sum_{i=1}^{N} \hat{F}(i) - E'$$

$$= \sum_{i=1}^{N} \hat{h}_1(i) + \sum_{i,j=1}^{N} [\hat{J}_j(i) - \hat{K}_j(i)]$$

$$- \frac{1}{2} \sum_{i,j=1}^{N} \left[\langle \phi_i \phi_j | \hat{h}_2 | \phi_i \phi_j \rangle - \langle \phi_j \phi_i | \hat{h}_2 | \phi_i \phi_j \rangle \right]. \tag{13.67}$$

This will be compared with the true N-electron Hamilton operator,

$$\hat{H}_e = \sum_{i=1}^{N} \hat{h}_1(i) + \frac{1}{2} \sum_{i \neq j=1}^{N} \hat{h}_2(i, j). \tag{13.68}$$

In order to apply perturbation theory we write

$$\hat{H}_e = \hat{G} + \Delta\hat{H} \tag{13.69}$$

with

$$\Delta\hat{H} = \frac{1}{2} \sum_{i \neq j=1}^{N} \hat{h}_2(i, j) - \sum_{i,j=1}^{N} [\hat{J}_j(i) - \hat{K}_j(i)]$$

$$+ \frac{1}{2} \sum_{i,j=1}^{N} \left[\langle \phi_i \phi_j | \hat{h}_2 | \phi_i \phi_j \rangle - \langle \phi_j \phi_i | \hat{h}_2 | \phi_i \phi_j \rangle \right]. \tag{13.70}$$

First-order perturbation theory gives that the ground-state energy changes by

$$\langle \Phi_0 | \Delta\hat{H} | \Phi_0 \rangle. \tag{13.71}$$

By using the precise form of the operators in Eq. (13.70) one may now show that this term vanishes. The proof of this is very similar to the one we applied in Section 13.2 in showing that the Hamilton matrix elements between the ground state and any single-excited configuration vanish. It will therefore not be presented here.

First the second-order term is non-vanishing. This term is

$$\sum_{i,j} \sum_{\alpha,\beta} \frac{\langle \Phi_0 | \Delta \hat{H} | \Phi_{i,j}^{\alpha,\beta} \rangle \langle \Phi_{i,j}^{\alpha,\beta} | \Delta \hat{H} | \Phi_0 \rangle}{E_0 - E_{ij}^{\alpha\beta}}, \tag{13.72}$$

where we have used the fact that only those configurations $\Phi_{i,j}^{\alpha,\beta}$, where exactly two electrons have been excited from the ground state Φ_0 have non-vanishing matrix elements in the denominator. To prove this, we use the fact that $\Delta \hat{H}$ consists of two-electron operators and that, according to the results of Section 13.2, there will be non-vanishing elements only when the two determinants differ by at most two orbitals. And the same arguments as in Section 13.2 can also be used in proving that there is no contribution from single-excited configurations.

Furthermore,

$$E_0 = \sum_{i=1}^{N} \varepsilon_i - E' = E_{HF} \tag{13.73}$$

and

$$E_{ij}^{\alpha\beta} = E_{HF} + \varepsilon_\alpha + \varepsilon_\beta - \varepsilon_i - \varepsilon_j \tag{13.74}$$

are the eigenvalues of $\hat{G}$.

The most important aspect is now that the denominator in Eq. (13.72) equals

$$-\varepsilon_\alpha - \varepsilon_\beta + \varepsilon_i + \varepsilon_j \tag{13.75}$$

due to Eq. (13.74). The two first energies are energies of orbitals that for the ground state are vacant, whereas the two last ones are those of occupied orbitals. This means that the denominator is small (and, hence, the correlation effects are large) when there is a small energy difference between occupied and unoccupied orbitals. This explains partly the results for the $H + H$ system in Section 13.1, where we found that correlation was important for large interatomic distances, where the energies of the occupied and unoccupied orbitals approached each other.

The result tells us also that for, e.g., compounds containing transition-metal atoms, where there are many (empty and occupied) d orbitals close to each other, we will expect that correlation effects are important, or, alternatively, that the Hartree–Fock approximation (i.e., the single-Slater-determinant approximation) is not a very good one.

Finally, we add that higher-order terms to the perturbation series can be calculated. These will become increasingly complex, but already second-order perturbation theory provides a very important improvement over the pure Hartree–Fock approximation, and in very many cases results with this method are highly accurate. Usually, calculations with second-order Møller–Plesset perturbation theory are given the label MP2, and equivalent for higher-order calculations (i.e., MP3, MP4, . . .).

14 Where are the Electrons and Atoms?

14.1 REDUCED DENSITY MATRICES

With the configuration expansion of the N-electron wavefunction

$$\Phi = \sum_I C_I \Phi_I \tag{14.1}$$

we have obtained a complicated description of the system of our interest and the wavefunction (14.1) becomes complicated to interpret. In Eq. (14.1), each Φ_I is a Slater determinant,

$$\Phi_I = |\phi_{I,1}, \phi_{I,2}, \ldots, \phi_{I,N}|, \tag{14.2}$$

and the single-electron orbitals $\phi_{I,i}$ are in turn given as linear combinations of some predefined basis functions,

$$\phi_{I,i}(\vec{r}) = \sum_j \chi_j(\vec{r}) c_{I,i,j}, \tag{14.3}$$

where, in contrast to earlier, we have included an extra configuration index (I) in order to distinguish between the different terms in Eq. (14.1).

The electron density is calculated as the expectation value

$$\langle \Phi | \hat{A} | \Phi \rangle \tag{14.4}$$

where

$$\hat{A} = \sum_{n=1}^{N} \hat{a}_1(n) \tag{14.5}$$

which in the present case is

$$\hat{a}_1(n) = \delta(\vec{r} - \vec{r}_n). \tag{14.6}$$

Using Eq. (13.27) the expectation value in Eq. (14.4) becomes

$$
\rho(\vec{r}) = \sum_{I,J} C_J^* C_I \frac{1}{(N-1)!} \sum_{i,j=1}^{N!} (-1)^{P(j)}(-1)^{P(i)}
$$
$$
\times \langle \phi_{I,i_1} | \phi_{J,j_1} \rangle \langle \phi_{I,i_2} | \phi_{J,j_2} \rangle \cdots \langle \phi_{I,i_{n-1}} | \phi_{J,j_{n-1}} \rangle \langle \phi_{j_n} | \hat{a}_1 | \phi_{i_n} \rangle
$$
$$
\times \langle \phi_{I,i_{n+1}} | \phi_{J,j_{n+1}} \rangle \cdots \langle \phi_{I,i_N} | \phi_{J,j_R} \rangle
$$
$$
= \sum_{I,J} C_J^* C_I \frac{1}{(N-1)!} \sum_{i,j=1}^{N!} (-1)^{P(j)}(-1)^{P(i)}
$$
$$
\times \langle \phi_{I,i_1} | \phi_{J,j_1} \rangle \langle \phi_{I,i_2} | \phi_{J,j_2} \rangle \cdots \langle \phi_{I,i_{n-1}} | \phi_{J,j_{n-1}} \rangle \phi_{j_n}^*(\vec{r}) \phi_{i_n}(\vec{r})
$$
$$
\times \langle \phi_{I,i_{n+1}} | \phi_{J,j_{n+1}} \rangle \cdots \langle \phi_{I,i_N} | \phi_{J,j_R} \rangle. \tag{14.7}
$$

In deriving this equation (cf. Section 13.2) we have used that the N electrons give identical contributions. Therefore, Eq. (14.8) could also be derived from the spin- and position-space density (remember, $\vec{x}$ is a combined index that includes both spin- and position-space coordinates),

$$
\rho(\vec{x}_1) = N \int \int \int \cdots \int \Phi^*(\vec{x}_1, \vec{x}_2, \vec{x}_3, \ldots, \vec{x}_N)
$$
$$
\times \Phi(\vec{x}_1, \vec{x}_2, \vec{x}_3, \ldots, \vec{x}_N) \, d\vec{x}_2 \, d\vec{x}_3 \cdots d\vec{x}_N. \tag{14.8}
$$

Here, we have integrated over all but the first electron, and then used the fact that the electrons are equivalent by subsequently multiplying by N.

The position-space density is subsequently obtained from $\rho(\vec{x}_1)$ by 'integrating' (i.e., summing) over the spin variable.

$\rho(\vec{x}_1)$ is a special case (the diagonal elements) of the so-called (first order) reduced density matrix (Löwdin, 1955),

$$
\rho(\vec{x}_1) = \rho^{(1)}(\vec{x}_1, \vec{x}_1), \tag{14.9}
$$

with

$$
\rho^{(1)}(\vec{x}_a, \vec{x}_b) = N \int \int \int \cdots \int \Phi^*(\vec{x}_a, \vec{x}_2, \vec{x}_3, \ldots, \vec{x}_N)
$$
$$
\times \Phi(\vec{x}_b, \vec{x}_2, \vec{x}_3, \ldots, \vec{x}_N) \, d\vec{x}_2 \, d\vec{x}_3 \cdots d\vec{x}_N, \tag{14.10}
$$

i.e., the first argument of Φ is no longer the same in the two factors of the integrand.

Notice that the name 'matrix' indicates that the function has two arguments (i.e., $\vec{x}_a$ and $\vec{x}_b$), although in contrast to more 'conventional' matrices, these are not discrete but continuous so that the 'matrix' is an $\infty \times \infty$ matrix.

In terms of $\rho^{(1)}(\vec{x}_1, \vec{x}_2)$, the expectation value (14.4) with $\hat{A}$ given by Eq. (14.5) but otherwise general becomes

$$
\langle \Phi | \hat{A} | \phi \rangle = \int \left[\hat{a}_1(\vec{x}_b) \rho^{(1)}(\vec{x}_a, \vec{x}_b) \right] \big|_{\vec{x}_a = \vec{x}_b} \, d\vec{x}_b, \tag{14.11}
$$

i.e., first $\hat{a}_1$ operates on the $\vec{x}_b$-dependence of the reduced density matrix, then $\vec{x}_a$ is set equal to $\vec{x}_b$, and, finally, one integrates over $\vec{x}_b$.

14.2 NATURAL ORBITALS

Equivalently to Eq. (14.7) we may write Eq. (14.10) as

$$\rho^{(1)}(\vec{x}_a, \vec{x}_b) = N \sum_{I,J} C_J^* C_I \int \int \int \cdots \int \Phi_J^*(\vec{x}_a, \vec{x}_2, \vec{x}_3, \ldots, \vec{x}_N)$$

$$\times \Phi_I(\vec{x}_b, \vec{x}_2, \vec{x}_3, \ldots, \vec{x}_N) \, d\vec{x}_2 \, d\vec{x}_3 \cdots d\vec{x}_N. \tag{14.12}$$

This expression is a double summation over products of two Slater determinants. The Slater determinants may contain one or more orbitals in common, so that the expression in Eq. (14.12) may actually also be considered a linear combination of products of two single-electron orbitals [cf. Eq. (14.7)]. That is, we may write

$$\rho^{(1)}(\vec{x}_a, \vec{x}_b) = \sum_{i,j} n_{ij} \phi_i^*(\vec{x}_a) \phi_j(\vec{x}_b), \tag{14.13}$$

where n_{ij} depends on the coefficients C_I and on whether the orbital ϕ_I appears in the configuration Φ_I.

The important point is that n_{ij} is independent of $(\vec{x}_a, \vec{x}_b)$. We may therefore consider n_{ij} as an element of a matrix whose size is as large as we have different orbitals. In a Hartree–Fock–Roothaan calculation this number equals that of the basis functions, N_b.

It is useful to diagonalizing this matrix, i.e., to write

$$\underline{\underline{n}} = \underline{\underline{U}}^\dagger \cdot \underline{\underline{\tilde{n}}} \cdot \underline{\underline{U}}, \tag{14.14}$$

where

$$\tilde{n}_{ij} = \tilde{n}_{ii} \delta_{i,j}, \tag{14.15}$$

and where $\underline{\underline{U}}$ is a unitary matrix,

$$\sum_k U_{ik}^* U_{jk} = \sum_k U_{ki}^* U_{kj} = \delta_{i,j}. \tag{14.16}$$

We insert Eqs. (14.14), (14.15), and (14.16) into Eq. (14.13) and obtain

$$\rho^{(1)}(\vec{x}_a, \vec{x}_b) = \sum_{i,j} n_{ij} \phi_I^*(\vec{x}_a) \phi_j(\vec{x}_b)$$

$$= \sum_{i,j} \sum_{k,l} U_{ki}^* \tilde{n}_{kl} U_{lj} \phi_i^*(\vec{x}_a) \phi_j(\vec{x}_b)$$

$$= \sum_{i,j} \sum_k U_{ki}^* \tilde{n}_{kk} U_{kj} \phi_i^*(\vec{x}_a) \phi_j(\vec{x}_b)$$

$$= \sum_k \sum_{i,j} U^*_{ki} \tilde{n}_{kk} U_{kj} \phi^*_i(\vec{x}_a) \phi_j(\vec{x}_b)$$

$$= \sum_k \tilde{n}_{kk} \left[\sum_i U_{ki} \phi_i(\vec{x}_a) \right]^* \left[\sum_j U_{kj} \phi_j(\vec{x}_b) \right]$$

$$= \sum_k \tilde{n}_k \tilde{\phi}^*_k(\vec{x}_a) \tilde{\phi}_k(\vec{x}_b), \tag{14.17}$$

where we have used that only the diagonal terms of $\tilde{n}_{kl}$ are non-vanishing and therefore set $\tilde{n}_k = \tilde{n}_{kk}$. Furthermore, we have defined the new set of orbitals,

$$\tilde{\phi}_k(\vec{x}) = \sum_i U_{ki} \phi_l(\vec{x}). \tag{14.18}$$

Since $\underline{\underline{U}}$ is unitary, these new orbitals are orthogonal, because the original ones are so.

The important point is that the expression (14.17) does not contain any cross-products. Thus, the electron density becomes a simple superposition of orbital contributions,

$$\rho(\vec{r}) = \sum_k \tilde{n}_k |\tilde{\phi}_k(\vec{r})|^2. \tag{14.19}$$

Here, $\tilde{n}_k$ is called the occupation number (or occupancy) of the kth orbital.

The orbitals $\tilde{\phi}_k$ are called the *natural orbitals* (Löwdin, 1955). Here, the label 'natural' comes from the ease with which Eqs. (14.18) and (14.19) can be interpreted.

Since electrons are Fermions, no orbital can accommodate more than one electron. Therefore,

$$0 \le \tilde{n}_k \le 1. \tag{14.20}$$

In the Hartree–Fock approximation, the natural orbitals are those directly calculated by solving the Hartree–Fock(–Roothaan) equations, and the occupancies $\tilde{n}_k$ are all either 0 or 1. Moreover, the orbitals are then solutions to the Hartree–Fock(–Roothaan) equations.

Including more configurations in the expansion of Φ, the occupancies become in general different from 0 and 1. Moreover, in the general case, they are not solutions to some single-particle eigenfunction equation. In all cases, the sum of all $\tilde{n}_k$ equals the total number of electrons.

14.3   MULLIKEN POPULATIONS

The separation of the total electron density into a superposition of those of individual natural orbitals is a first step in the direction of something that intuitively can be interpreted. It is, however, not 'physical' in the sense that the description in terms of natural orbitals offers a more exact description or that it is more closely related to experimental observables — it is rather a convenient tool that makes the interpretation of the results simpler, although the importance of this will not be understated.

This is also the case for the Mulliken populations (Mulliken, 1953). The basic philosophy behind these is to separate the electron density into atomic contributions. Already at this point it may be obvious that it is not clear how an electron density that is spread out all over space can be split into different atomic components, and, therefore, the Mulliken population analysis is somewhat arbitrary. It does, therefore, not offer exact charges of the individual atoms and an exact decomposition of the electron density, but rather it provides trends. In particular, comparisons between different but related compounds may be useful in giving useful information on bonding that can otherwise be obtained only with difficulty.

We may use the decomposition into natural orbitals as a starting point,

$$\rho(\vec{r}) = \sum_k \tilde{n}_k |\tilde{\phi}_k(\vec{r})|^2. \qquad (14.21)$$

Here,

$$\tilde{\phi}_k(\vec{x}) = \sum_i U_{ki}\phi_i(\vec{x}). \qquad (14.22)$$

The important (and crucial) point is that the orbitals ϕ_l are calculated within (e.g.) the Hartree–Fock–Roothaan approach and as such written as a linear combination of some basis functions,

$$\phi_i(\vec{x}) = \sum_j \chi_j(\vec{x})c_{ji}. \qquad (14.23)$$

Combining Eqs. (14.22) and (14.23) gives

$$\tilde{\phi}_k = \sum_i U_{ki} \sum_j \chi_j c_{ji}$$

$$= \sum_j \left[\sum_i U_{ki} c_{ji}\right] \chi_j$$

$$\equiv \sum_j \tilde{c}_{jk} \chi_j. \qquad (14.24)$$

We shall now specify the basis functions χ_j in some more detail. These are supposed to be atom-centred and only in that case does the Mulliken analysis make any sense. For a water molecule, H_2O, this means that some of the basis functions are centred on one of the hydrogen atoms, some of them on the other hydrogen atom, and some of them on the oxygen atom. On each atom we may have more different basis functions, e.g., four $1s$ functions, three $2s$ functions, and three of each type of $2p$ functions on the oxygen atom. We write accordingly

$$\chi_j(\vec{x}) \equiv \chi_{p,l,m,\alpha}(\vec{x}), \qquad (14.25)$$

where p specifies the atom, (l, m) the angular dependence, and α everything else (e.g., the decay constants for Gaussians as well as the spin dependence).

Any of the natural orbitals is normalized, giving

$$1 = \langle \tilde{\phi}_k | \tilde{\phi}_k \rangle$$

$$= \left\langle \sum_{p_1, l_1, m_1, \alpha_1} \tilde{c}_{p_1, l_1, m_1, \alpha_1, k} \chi_{p_1, l_1, m_1, \alpha_1} \middle| \sum_{p_2, l_2, m_2, \alpha_2} \tilde{c}_{p_2, l_2, m_2, \alpha_2, k} \chi_{p_2, l_2, m_2, \alpha_2} \right\rangle$$

$$= \sum_{p_1, p_2} \left\{ \sum_{l_1, l_2} \sum_{m_1, m_2} \sum_{\alpha_1, \alpha_2} \tilde{c}^*_{p_1, l_1, m_1, \alpha_1, k} \tilde{c}_{p_2, l_2, m_2, \alpha_2, k} \langle \chi_{p_1, l_1, m_1, \alpha_1} | \chi_{p_2, l_2, m_2, \alpha_2} \rangle \right\}$$

$$= \sum_p n_{p,k} + \sum_{p_1 \neq p_2} n'_{p_1, p_2, k}, \tag{14.26}$$

where

$$n_{p,k} = \sum_{l_1, l_2} \sum_{m_1, m_2} \sum_{\alpha_1, \alpha_2} \tilde{c}^*_{p, l_1, m_1, \alpha_1, k} \tilde{c}_{p, l_2, m_2, \alpha_2, k} \langle \chi_{p, l_1, m_1, \alpha_1} | \chi_{p, l_2, m_2, \alpha_2} \rangle \tag{14.27}$$

is the so-called net population of atom p from the kth orbital (notice that this includes only the basis functions from the same, pth, atom). Furthermore,

$$n'_{p_1, p_2, k} = \sum_{l_1, l_2} \sum_{m_1, m_2} \sum_{\alpha_1, \alpha_2} \tilde{c}^*_{p_1, l_1, m_1, \alpha_1, k} \tilde{c}_{p_2, l_2, m_2, \alpha_2, k} \langle \chi_{p_1, l_1, m_1, \alpha_1} | \chi_{p_2, l_2, m_2, \alpha_2} \rangle \tag{14.28}$$

defines the so-called overlap population

$$n_{p_1, p_2, k} = n'_{p_1, p_2, k} + n'_{p_2, p_1, k} \tag{14.29}$$

between atom p_1 and atom p_2 for the kth orbital.

The idea behind this approach is to use the basis functions in defining the populations. Thus, the populations become strongly dependent on the choice of the basis functions.

Above we have only separated the total electron density of a given (natural) orbital into atomic net and overlap components. One may decompose the density further by, e.g., also splitting it into angular components. How this is done should be obvious from the equations above. For instance from Eq. (14.28)

$$n'_{p_1, p_2, k} = \sum_{l_1, l_2} \sum_{m_1, m_2} \sum_{\alpha_1, \alpha_2} \tilde{c}^*_{p_1, l_1, m_1, \alpha_1, k} \tilde{c}_{p_2, l_2, m_2, \alpha_2, k} \langle \chi_{p_1, l_1, m_1, \alpha_1} | \chi_{p_2, l_2, m_2, \alpha_2} \rangle$$

$$= \sum_{l_1, l_2} \left\{ \sum_{m_1, m_2} \sum_{\alpha_1, \alpha_2} \tilde{c}^*_{p_1, l_1, m_1, \alpha_1, k} \tilde{c}_{p_2, l_2, m_2, \alpha_2, k} \langle \chi_{p_1, l_1, m_1, \alpha_1} | \chi_{p_2, l_2, m_2, \alpha_2} \rangle \right\}$$

$$\equiv \sum_l n'_{p_1, p_2, l, k} + \sum_{l_1 \neq l_2} n'_{p_1, p_2, l_1, l_2, k}. \tag{14.30}$$

The decomposition into atomic net and overlap components corresponds to writing,

$$1 = \sum_p n_{p,k} + \sum_{p_1 \neq p_2} n'_{p_1,p_2,k}$$

$$= \sum_p n_{p,k} + \sum_{p_1 > p_2} n_{p_1,p_2,k}. \qquad (14.31)$$

This can be rewritten as

$$1 = \sum_p n_{p,k} + \sum_{p_1 \neq p} n'_{p_1,p,k}$$

$$= \sum_p \left[n_{p,k} + \sum_{p_1 \neq p} n'_{p_1,p,k} \right]$$

$$\equiv \sum_p N_{p,k} \qquad (14.32)$$

with

$$N_{p,k} = n_{p,k} + \sum_{p_1 \neq p} n'_{p_1,p,k} = n_{p,k} + \frac{1}{2} \sum_{p_1 \neq p} n_{p_1,p,k}, \qquad (14.33)$$

being the so-called gross population on atom p from the kth orbital. It is seen that this corresponds to (arbitrarily) split each overlap population into two identical parts that each is ascribed to one of the two atoms.

Let us consider a simple example. We consider a two-atomic molecule and one orbital that is written as a sum of two basis function, one on each atom. That is,

$$\phi = c_A \chi_A + c_B \chi_B. \qquad (14.34)$$

We assume that c_A and c_B are real and that

$$\langle \chi_A | \chi_A \rangle = 1$$

$$\langle \chi_B | \chi_B \rangle = 1$$

$$\langle \chi_A | \chi_B \rangle = S \qquad (14.35)$$

with S being real, too.

From

$$1 = \langle \phi | \phi \rangle = c_A^2 + c_B^2 + 2c_A c_B S \qquad (14.36)$$

we find

$$n_A = c_A^2$$

$$n_B = c_B^2$$

$$n_{A,B} = 2c_A c_B S \qquad (14.37)$$

as the net and overlap populations, respectively.

In addition we find [from Eq. (14.36)],

$$c_B = -c_A S \pm [1 - c_A^2 + c_A^2 S^2]^{1/2}. \qquad (14.38)$$

The gross population on atom A is then

$$N_A = n_A + \tfrac{1}{2} n_{A,B} = c_A^2 + c_A c_B S$$
$$= c_A^2 - c_A^2 S^2 \pm c_A S[1 - c_A^2 + c_A^2 S^2]^{1/2}. \qquad (14.39)$$

When $|c_A| \ll |c_B|$ and the last term on the right-hand side of Eq. (14.39) is negative, it may now happen that $N_A < 0$, i.e., we have a negative (!) number of electrons on a specific atom. This is, e.g., the case for $c_A = 0.1$ and $S = 0.8$ which gives $c_B = -1.078$, and subsequently $n_A = 0.01$, $n_B = 1.163$, $n_{A,B} = -0.173$, $N_B = 1.077$, and $N_A = -0.077$. This is an extreme case, but it illustrates the problems that *may* arise when studying the Mulliken populations.

One way of improving this is to split the overlap populations according to the net populations. That is, by using

$$n_{p_1,p_2,k} = \frac{n_{p_1,k}}{n_{p_1,k} + n_{p_2,k}} n_{p_1,p_2,k} + \frac{n_{p_2,k}}{n_{p_1,k} + n_{p_2,k}} n_{p_1,p_2,k} \qquad (14.40)$$

where the first term is ascribed to the p_1th atom, and the second to the p_2th atom.

For the example above we would then obtain 0.008 electrons on the A atom and 0.992 electrons on the B atom, which is certainly an improvement. On the other hand, it should be stressed that all these modifications are nothing but just modifications, and that the numbers are *not* to be interpreted as *exact* numbers of electrons on the individual atoms.

14.4 LÖWDIN POPULATIONS

The straightforward decomposition of the total electron density of a single (natural) orbital into net and overlap populations was obtained as follows:

$$1 = \sum_p n_{p,k} + \sum_{p_1 \neq p_2} n_{p_1,p_2,l}, \qquad (14.41)$$

where

$$n_{p,k} = \sum_{l_1,l_2} \sum_{m_1,m_2} \sum_{\alpha_1,\alpha_2} \tilde{c}^*_{p,l_1,m_1,\alpha_1,k} \tilde{c}_{p,l_2,m_2,\alpha_2,k} \langle \chi_{p,l_1,m_1,\alpha_1} | \chi_{p,l_2,m_2,\alpha_2} \rangle \qquad (14.42)$$

and

$$n'_{p_1,p_2,k} = \sum_{l_1,l_2} \sum_{m_1,m_2} \sum_{\alpha_1,\alpha_2} \tilde{c}^*_{p_1,l_1,m_1,\alpha_1,k} \tilde{c}_{p_2,l_2,m_2,\alpha_2,k} \langle \chi_{p_1,l_1,m_1,\alpha_1} | \chi_{p_2,l_2,m_2,\alpha_2} \rangle. \qquad (14.43)$$

The first expression defines the net populations and the second one the overlap populations. Combining these, one could obtain the gross populations. This last step could, however, be avoided if the basis functions were orthonormal. Then, the overlap populations would vanish and the net and gross populations would become identical.

This is the idea behind the Löwdin populations. One defines a new set of basis functions,

$$\chi'_j = \sum_i \chi_i V_{ji} \tag{14.44}$$

from the original ones. As seen in this equation, the new basis functions χ' are linear combinations of *all* the original ones χ, and it is therefore no longer obvious how one should ascribe a single basis function to a specific atom. However, it has been shown that through the so-called Löwdin's symmetric orthonormalization (Löwdin, 1950), the new basis functions resemble the original ones the most. Therefore, one constructs the new basis functions according to this prescription, and then simply inserts these new ones instead of the original ones.

Equation (14.44) can be written in matrix form,

$$\underline{\underline{\chi}}' = \underline{\underline{V}} \cdot \underline{\underline{\chi}}. \tag{14.45}$$

In order to determine $\underline{\underline{V}}$ as given by Löwdin's symmetric orthonormalization we proceed as follows. We define the overlap matrix of the original basis functions, $\underline{\underline{O}}$, with

$$O_{ij} = \langle \chi_i | \chi_j \rangle. \tag{14.46}$$

This matrix is diagonalized,

$$\underline{\underline{O}} = \underline{\underline{U}}^\dagger \cdot \underline{\underline{\Lambda}} \cdot \underline{\underline{U}}, \tag{14.47}$$

where $\underline{\underline{\Lambda}}$ is a diagonal matrix,

$$\Lambda_{ij} = \lambda_i \delta_{i,j} \tag{14.48}$$

with the λ being the eigenvalues of $\underline{\underline{O}}$. Furthermore, $\underline{\underline{U}}$ is a unitary matrix.

Subsequently, we define

$$\underline{\underline{V}} = \underline{\underline{U}}^\dagger \cdot \underline{\underline{\Lambda}}^{-1/2} \cdot \underline{\underline{U}}, \tag{14.49}$$

where $\underline{\underline{\Lambda}}^{-1/2}$ is a diagonal matrix with the elements

$$\left(\underline{\underline{\Lambda}}^{-1/2} \right)_{ij} = \lambda_i^{-1/2} \delta_{i,j}. \tag{14.50}$$

Thereby, we have defined the matrix $\underline{\underline{V}}$ and we can calculate the Löwdin populations. We have to remember that the expansion coefficients c_{ji} in the expansion

$$\phi_i = \sum_j \chi_j c_{ji} \tag{14.51}$$

have to be modified, when replacing χ_j by χ'_j. That is, from

$$\chi_j = \sum_k \chi'_k \left(\underline{\underline{V}}^{-1}\right)_{jk}, \tag{14.52}$$

we get

$$\begin{aligned}\phi_i &= \sum_k \left[\sum_j c_{ji}\left(\underline{\underline{V}}^{-1}\right)_{jk}\right] \chi'_k \\ &= \sum_k c'_{ki}\chi'_k\end{aligned} \tag{14.53}$$

by introducing the new expansion coefficients.

Since the set of basis functions χ'_l is orthonormal, there is no overlap populations within the Löwdin procedure, but only net (= gross) populations, which is clearly an advantage. On the other hand, the transformation to the new basis set may introduce some unwanted artifacts.

There are other ways of decomposing the total electron density into atomic fragments. All of them have some advantages and some disadvantages. On the other hand, none of them provides 'exact' atomic charges that can be measured uniquely in an experiment. Rather, they are tools for interpreting outcomes of electronic-structure calculations. Here, we shall not discuss them further but instead approach the experimental studies a little.

14.5 DYSON ORBITALS

Within the Hartree–Fock approximation we obtain a set of orbitals by solving the Hartree–Fock equations. The Hartree–Fock equations were derived by minimizing the expectation value of the electronic energy

$$\langle \Phi | \hat{H}_e | \Phi \rangle \tag{14.54}$$

where Φ is a single Slater determinant,

$$\Phi = |\phi_1, \phi_2, \ldots, \phi_N|, \tag{14.55}$$

and where we in addition required that the orbitals were orthonormal,

$$\langle \phi_i | \phi_j \rangle = \delta_{i,j}. \tag{14.56}$$

The constraints (14.56) were included via Lagrange multipliers λ_{ij} [one multiplier for each constraint of Eq. (14.56)], and we showed that when these were *chosen* so that

$$\lambda_{ij} = \varepsilon_i \delta_{i,j} \tag{14.57}$$

it was particularly simple to interpret their values as (according to Koopmans' theorem) ionization potential and electron affinities for occupied and empty orbitals, respectively. One may accordingly consider these orbitals particularly relevant.

Going beyond the Hartree–Fock approximation, i.e., including more Slater determinants in a CI expansion, the natural orbitals appeared as those that most easily could be interpreted. However, although they, from a mathematical point of view as well as by considering them a convenient tool in interpreting the theoretical results, are useful, it is not clear whether they have any relevance in describing experimental results. To this end the Dyson orbitals appear to be more suited.

We consider a system of N electrons. Its wavefunction in the ground state is assumed to be

$$\Psi_0^{(N)}(\vec{x}_1, \vec{x}_2, \vec{x}_3, \ldots, \vec{x}_N), \qquad (14.58)$$

where the lower index 0 specifies the ground state, and the upper index marks that we have N electrons. We shall not bother about the precise form of $\Psi_0^{(N)}$.

In an experiment we may remove one electron, e.g., electron no. 1. This ionization process may simultaneously leave the remaining $(N-1)$-electron system in an excited state, which we will write as

$$\Psi_I^{(N-1)}(\vec{x}_2, \vec{x}_3, \ldots, \vec{x}_N), \qquad (14.59)$$

where I characterizes the excited state.

That part of $\Psi_0^{(N)}$ that is *not* contained in $\Psi_I^{(N-1)}$ may now be considered the orbital for the single electron that has been removed from the system, as it was *before* it was removed. Mathematically formulated this is

$$\Psi_I(\vec{x}_1) = \sqrt{N} \int \int \cdots \int \Psi_I^{(N-1)*}(\vec{x}_2, \vec{x}_3, \ldots, \vec{x}_N)$$

$$\times \Psi_0^{(N)}(\vec{x}_1, \vec{x}_2, \vec{x}_3, \ldots, \vec{x}_N) \, d\vec{x}_2 \, d\vec{x}_3 \cdots d\vec{x}_N. \qquad (14.60)$$

This is the definition of a Dyson orbital (see, e.g., Duffy *et al.*, 1994). In contrast to the previous definitions of orbitals, this depends both on the ground state of the N-electron system, and on the (maybe excited) state of the $(N-1)$-electron system. Finally, the factor $\sqrt{N}$ is included in order to obtain an improved normalization (see below).

One can calculate the Dyson orbitals by calculating the wavefunctions for both systems and subsequently performing the integration of Eq. (14.60). This requires thus the calculations for *two* systems, and is as such more complicated. Alternatively, one may calculate it by starting out with the N-electron system, and calculate directly (via a so-called Green's function technique) the changes of this due to the removal of one electron. We shall not discuss this here further for two reasons: it is complicated, and the calculation of Dyson orbitals is not common.

Green's function will be introduced below in another context (when treating various types of distortions for crystalline materials), and the interested reader *may* be able to imagine using the techniques derived there for the calculation of Dyson orbitals. Instead, we shall consider the Dyson orbitals for some simple systems in order to develop a feeling for their properties.

The first system of interest will be one with $N = 2$. The ground state is supposed written as a single Slater determinant:

$$\Psi_0^{(N)} = |\phi_1, \phi_2| = \frac{1}{\sqrt{2}} \left[\phi_1(\vec{x}_1)\phi_2(\vec{x}_2) - \phi_1(\vec{x}_2)\phi_2(\vec{x}_1) \right] . \tag{14.61}$$

The (excited) $(N - 1)$-electron system has only one electron, and we have

$$\Psi_I^{(N-1)} = \tilde{\phi}(\vec{x}_2). \tag{14.62}$$

The Dyson orbital is now defined as

$$\Psi_I(\vec{x}_1) = \sqrt{2}\frac{1}{\sqrt{2}} \int \tilde{\phi}^*(\vec{x}_2) \left[\phi_1(\vec{x}_1)\phi_2(\vec{x}_2) - \phi_1(\vec{x}_2)\phi_2(\vec{x}_1) \right] d\vec{x}_2$$

$$= \langle \tilde{\phi}|\phi_2\rangle\phi_1(\vec{x}_1) - \langle \tilde{\phi}|\phi_1\rangle\phi_2(\vec{x}_1), \tag{14.63}$$

i.e., a linear combination of the two original orbitals.

We can now imagine various scenarios. For example, *if* the ionization process leads to nothing but a simple removal of one electron from one orbital, whereas the other orbital is completely unchanged, then the Dyson orbital equals the orbital from which the electron is removed (e.g., when the electron is removed from the ϕ_1 orbital, then $\tilde{\phi} = \phi_2$, and $\Psi_I = \phi_1$). This is, however, an unrealistic situation, since the remaining electron will relax, i.e., change the shape of its orbital due to the changed potential (due to the one electron less). In some cases it may also be in an excited orbital. For example, if ϕ_1 and ϕ_2 are s orbitals only differing by the spin dependence, whereas $\tilde{\phi}$ is a p function, then the Dyson orbital is identically zero! Thereby we obtain a further property of the Dyson orbitals: they are not necessarily normalized.

As a further example we consider an $N = 3$ electron system, where both $\Psi_0^{(N)}$ and $\Psi_I^{(N-1)}$ are supposed given as Slater determinants:

$$\Psi_0^{(N)} = \frac{1}{\sqrt{3!}} \begin{vmatrix} \phi_1(\vec{x}_1) & \phi_2(\vec{x}_1) & \phi_3(\vec{x}_1) \\ \phi_1(\vec{x}_2) & \phi_2(\vec{x}_2) & \phi_3(\vec{x}_2) \\ \phi_1(\vec{x}_3) & \phi_2(\vec{x}_3) & \phi_3(\vec{x}_3) \end{vmatrix} \tag{14.64}$$

and

$$\Psi_I^{(N-1)} = \frac{1}{\sqrt{2!}} \begin{vmatrix} \tilde{\phi}_2(\vec{x}_2) & \tilde{\phi}_3(\vec{x}_2) \\ \tilde{\phi}_2(\vec{x}_3) & \tilde{\phi}_3(\vec{x}_3) \end{vmatrix} . \tag{14.65}$$

This gives

$$
\begin{aligned}
\psi_I(\vec{x}_1) = \tfrac{1}{2}\Big\{ &\big[\langle\tilde{\phi}_2|\phi_2\rangle\langle\tilde{\phi}_3|\phi_3\rangle - \langle\tilde{\phi}_2|\phi_3\rangle\langle\tilde{\phi}_3|\phi_2\rangle - \langle\tilde{\phi}_3|\phi_2\rangle\langle\tilde{\phi}_2|\phi_3\rangle \\
&+ \langle\tilde{\phi}_3|\phi_3\rangle\langle\tilde{\phi}_2|\phi_2\rangle \big]\phi_1(\vec{x}_1) + \big[\langle\tilde{\phi}_2|\phi_3\rangle\langle\tilde{\phi}_3|\phi_1\rangle - \langle\tilde{\phi}_2|\phi_1\rangle\langle\tilde{\phi}_3|\phi_3\rangle \\
&- \langle\tilde{\phi}_3|\phi_3\rangle\langle\tilde{\phi}_2|\phi_1\rangle + \langle\tilde{\phi}_3|\phi_1\rangle\langle\tilde{\phi}_2|\phi_3\rangle \big]\phi_2(\vec{x}_1) + \big[\langle\tilde{\phi}_2|\phi_1\rangle\langle\tilde{\phi}_3|\phi_2\rangle \\
&- \langle\tilde{\phi}_2|\phi_2\rangle\langle\tilde{\phi}_3|\phi_1\rangle - \langle\tilde{\phi}_3|\phi_1\rangle\langle\tilde{\phi}_2|\phi_2\rangle + \langle\tilde{\phi}_3|\phi_2\rangle\langle\tilde{\phi}_2|\phi_1\rangle \big]\phi_3(\vec{x}_1) \Big\} \\
= &\big[\langle\tilde{\phi}_2|\phi_2\rangle\langle\tilde{\phi}_3|\phi_3\rangle - \langle\tilde{\phi}_2|\phi_3\rangle\langle\tilde{\phi}_3|\phi_2\rangle \big]\phi_1(\vec{x}_1) \\
&+ \big[\langle\tilde{\phi}_2|\phi_3\rangle\langle\tilde{\phi}_3|\phi_1\rangle - \langle\tilde{\phi}_2|\phi_1\rangle\langle\tilde{\phi}_3|\phi_3\rangle \big]\phi_2(\vec{x}_1) \\
&+ \big[\langle\tilde{\phi}_2|\phi_1\rangle\langle\tilde{\phi}_3|\phi_2\rangle - \langle\tilde{\phi}_2|\phi_2\rangle\langle\tilde{\phi}_3|\phi_1\rangle \big]\phi_3(\vec{x}_1),
\end{aligned}
\tag{14.66}
$$

i.e., a linear combination of the initial single-particle orbitals, whose precise form depends sensitively on the final state.

We may arrive at a different interpretation of the Dyson orbital through the following considerations. We consider the $(N-1)$-electron system after the ionization. It is supposed to be in the excited state described by

$$
\Psi_I^{(N-1)}(\vec{x}_2, \vec{x}_3, \ldots, \vec{x}_N).
\tag{14.67}
$$

The electron that has been emitted occupies another orbital ϕ_0 that is 'far away' from the $(N-1)$-electron system. One may suggest that the total N-electron wavefunction is

$$
\Psi_I^{(N-1)}(\vec{x}_2, \vec{x}_3, \ldots, \vec{x}_N)\phi_0(\vec{x}_1).
\tag{14.68}
$$

This is, however, not antisymmetric, so instead the wavefunction becomes

$$
\begin{aligned}
\frac{1}{\sqrt{N}} \Big[&\Psi_I^{(N-1)}(\vec{x}_2, \vec{x}_3, \ldots, \vec{x}_N)\phi_0(\vec{x}_1) \\
&- \Psi_I^{(N-1)}(\vec{x}_1, \vec{x}_3, \ldots, \vec{x}_N)\phi_0(\vec{x}_2) \\
&- \Psi_I^{(N-1)}(\vec{x}_2, \vec{x}_1, \ldots, \vec{x}_N)\phi_0(\vec{x}_3) \\
&- \cdots - \Psi_I^{(N-1)}(\vec{x}_2, \vec{x}_3, \ldots, \vec{x}_1)\phi_0(\vec{x}_N) \Big].
\end{aligned}
\tag{14.69}
$$

Before the ionization the N-electron system occupied the ground-state, i.e.,

$$
\Psi_0^{(N)}(\vec{x}_1, \vec{x}_2, \vec{x}_3, \ldots, \vec{x}_N).
\tag{14.70}
$$

On the other hand, this is just one wavefunction out of a complete set of N-electron wavefunctions $\{\Psi_K^{(N)}\}$ ($K = 0$ for the ground state, whereas other states have $K \neq 0$). The wavefunction of Eq. (14.69) may therefore be expanded in this set, and in particular we may ask ourselves about the coefficient of this expansion to the function of Eq. (14.70). This coefficient is

$$
\begin{aligned}
C_0 = \frac{1}{\sqrt{N}} \int \int \cdots \int [\Psi_I^{(N-1)}(\vec{x}_2, \vec{x}_3, \ldots, \vec{x}_N)\phi_0(\vec{x}_1) \\
- \Psi_I^{(N-1)}(\vec{x}_1, \vec{x}_3, \ldots, \vec{x}_N)\phi_0(\vec{x}_2) \\
- \Psi_I^{(N-1)}(\vec{x}_2, \vec{x}_1, \ldots, \vec{x}_N)\phi_0(\vec{x}_3) \\
- \cdots - \Psi_I^{(N-1)}(\vec{x}_2, \vec{x}_3, \ldots, \vec{x}_1)\phi_0(\vec{x}_N)]^* \\
\times \Psi_0^{(N)}(\vec{x}_1, \vec{x}_2, \vec{x}_3, \ldots, \vec{x}_N)\, d\vec{x}_1\, d\vec{x}_2 \cdots d\vec{x}_N.
\end{aligned}
\tag{14.71}
$$

By using the fact that the electrons are indistinguishable, we find

$$
\begin{aligned}
C_0 &= \frac{N}{\sqrt{N}} \int \int \cdots \int [\Psi_I^{(N-1)}(\vec{x}_2, \vec{x}_3, \ldots, \vec{x}_N)\phi_0(\vec{x}_1)]^* \\
&\quad \times \Psi_0^{(N)}(\vec{x}_1, \vec{x}_2, \vec{x}_3, \ldots, \vec{x}_N)\, d\vec{x}_1\, d\vec{x}_2 \cdots d\vec{x}_N \\
&= \sqrt{N} \int \phi_0^*(\vec{x}_1) \left[\int \cdots \int \Psi_I^{(N-1)*}(\vec{x}_2, \vec{x}_3, \ldots, \vec{x}_N)\Psi_0^{(N)} \right. \\
&\quad \left. \times (\vec{x}_1, \vec{x}_2, \vec{x}_3, \ldots, \vec{x}_N)\, d\vec{x}_2 \cdots d\vec{x}_N \right] d\vec{x}_1 \\
&= \langle \phi_0 | \psi_I \rangle.
\end{aligned}
\tag{14.72}
$$

Thus, the Dyson orbital is relevant when studying the final system after the ionization, as well as that before the ionization.

14.6 ATOMS IN MOLECULES

Both the Mulliken populations and (although to a lesser extent) the Löwdin populations depend on the basis sets used in carrying the calculations through. Alternatively, one may study the total electron density itself (independently of the basis functions that have been used in constructing it) and attempt to separate this into atomic components. This is the basic idea behind Bader's atoms-in-molecules approach (Bader and Nguyen-Dang, 1981; Bader, 1990), which should be much less sensitive to the precise definitions of the basis functions.

If we consider the total electron density of a diatomic molecule, it will have two sharp maxima at the positions of the nuclei. On the line joining the two nuclei, there will be a minimum somewhere, and intuitively it would appear as reasonable to split the total electron density into two components at the point of this minimum, so that each part is ascribed to one of the two atoms.

More general, from the total electron density $\rho(\vec{r})$, one determines the set of points satisfying

$$
\vec{\nabla}\rho(\vec{r}) = \vec{0}.
\tag{14.73}
$$

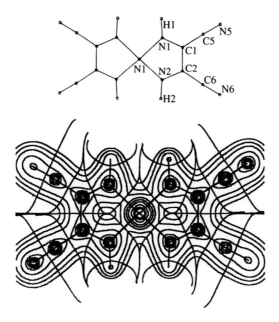

Figure 14.1 An example of how the total electron density of a given molecule can be split into different atomic parts within the concept of atoms in molecules as well as the construction of the bonds. The upper part shows the molecule of interest, $Ni(C_4N_4H_2)_2$. Reprinted with permission from Hwang and Wang (1998). Copyright 1998 American Chemical Society

These (so-called critical) points define a set of (open or closed) surfaces in the three-dimensional position space. Furthermore, at a given point $\vec{r}$, the vector $\vec{\nabla}\rho(\vec{r})$ will be perpendicular to the surface of constant electron density ρ.

Starting at a given point $\vec{r}$ and following the so-called gradient path $\vec{\nabla}\rho$ one arrives at one of the attractors, which is either one of the nuclei, the point infinity, or a point lying somewhere between two nuclei. In the last case one talks about a bond critical point. A curve consisting of critical points that passes through a pair of nuclei and a bond critical point is defined as a bond. Finally, there are curves (or surfaces) containing the bond critical points but ending at infinity. These surfaces divide space into different parts, which can each be ascribed to a single atom. In Fig. 14.1 we show an example of such a separation of space into different atomic parts together with the contour curves of the electron density and the bonds.

14.7 ELECTRON-LOCALIZATION FUNCTION (ELF)

Recently, the electron-localization function (ELF; Becke and Edgecombe, 1990) has been pursued as a useful tool in visualizing orbitals. We shall assume that the total N-particle wavefunction can be written as a single Slater determinant, i.e.,

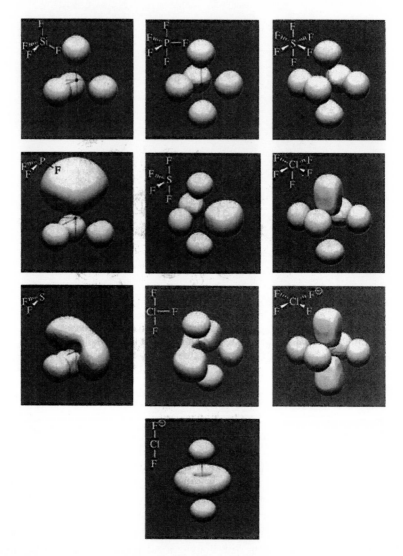

Figure 14.2 A set of examples of the electron-localization function (ELF) for some small molecules. Reproduced by permission of VCH from Fässler and Savin (1997)

that either the Hartree–Fock approximation is assumed valid or that a density-functional method (see Chapter 15) is applied.

ELF is a function in position space defined as

$$\text{ELF}(\vec{r}) = \left[1 + \left(\frac{D(\vec{r})}{D_h(\vec{r})} \right)^2 \right]^{-1}. \tag{14.74}$$

Here, $D(\vec{r})$ can be interpreted as a local kinetic energy density,

$$D(\vec{r}) = \frac{1}{2} \sum_{l=1}^{N} |\vec{\nabla}\phi_i(\vec{r})|^2 - \frac{1}{8} \frac{|\vec{\nabla}\rho(\vec{r})|^2}{\rho(\vec{r})}, \qquad (14.75)$$

whereas $D_h(\vec{r})$ is the kinetic energy density of a homogeneous electron gas with the density as that of the point of interest,

$$D_h(\vec{r}) = \frac{3}{10} \left[3\pi^2 \rho(\vec{r}) \right]^{5/3}. \qquad (14.76)$$

It can be shown that $D(\vec{r})/D_h(\vec{r})$ is a measure for the probability of finding an electron with the same spin as one placed at $\vec{r}$ within a given volume about the point of interest $\vec{r}$. This ratio is accordingly large at points $\vec{r}$ where more electrons with the same spin may be found, i.e., in regions between electron pairs. Therefore, ELF is small in points occupied by electron pairs, which means that regions of large ELF separate electron pairs. Accordingly, depicting the ELF will essentially lead to illustrations of the electron pairs, i.e., of the chemical bonds and of the lone pairs, as well as of closed shells.

ELF is a function in the three-dimensional position space. Through the definition of Eq. (14.74) it lies between 0 and 1. In order to draw it, it has become practice to draw the surfaces defined by ELF equal to some constant, typically about 0.8. In Fig. 14.2 we show a set of examples of the ELF for some simple molecules.

15 Density Functional Theory

15.1 THOMAS–FERMI AND $X\alpha$ METHODS

So far we have attempted to solve the electronic Schrödinger equation

$$\hat{H}_e \Psi_e = E_e \Psi_e \qquad (15.1)$$

more or less accurately. The (approximate) solution Ψ_e is an N-electron wavefunction and depends accordingly on $3N$ position-space coordinates and N spin coordinates. For just medium-sized systems this function is therefore extremely complex: For a water molecule it is a function of 30 position-space coordinates and 10 spin coordinates, for a benzene molecule is depends on 126 position-space coordinates and 42 spin coordinates, whereas for a crystal it depends on of the order of 10^{24} coordinates.

Having obtained the wavefunction it is, in principle, possible to calculate any experimental observable although, due to practical limitations, many of the calculated properties are less accurate than may be desirable.

A basic problem is that the wavefunction is very much more complex than is necessary when calculating experimental observables. Thus, most operators for the experimental observables depend on the coordinates of only one or two electrons, i.e., of the $3N$ position-space and N spin coordinates, at most 6 position-space and 2 spin coordinates are required. That is, the operators are of the form

$$\hat{A} = \sum_{n=1}^{N} \hat{a}_1(n) \qquad (15.2)$$

or

$$\hat{A} = \frac{1}{2} \sum_{n \neq m=1}^{N} \hat{a}_2(n, m). \qquad (15.3)$$

When calculating the expectation values

$$\langle \Psi_e | \hat{A} | \Psi_e \rangle \tag{15.4}$$

most of the complexity of Ψ_e is redundant.

A particularly simple case occurs when $\hat{A}$ is written as in Eq. (15.2) and when, furthermore, $\hat{a}_1$ depends only on the position-space coordinates,

$$\hat{a}_1(n) = a_1(\vec{r}_n). \tag{15.5}$$

This is actually the case for very many physically and chemically relevant observables.

Since

$$a_1(\vec{r}_n) = \int a_1(\vec{s})\delta(\vec{r}_n - \vec{s})\,d\vec{s}, \tag{15.6}$$

we find (cf. Section 14.1)

$$\langle \Psi_e | \hat{A} | \Psi_e \rangle = \int \int \cdots \int \Psi_e^*(\vec{x}_1, \vec{x}_2, \ldots, \vec{x}_N) \Psi_e(\vec{x}_1, \vec{x}_2, \ldots, \vec{x}_N)$$

$$\times \sum_{n=1}^{N} a_1(\vec{r}_n)\,d\vec{x}_1\,d\vec{x}_2 \cdots d\vec{x}_N$$

$$= \int \left[\int \int \cdots \int \Psi_e^*(\vec{x}_1, \vec{x}_2, \ldots, \vec{x}_N) \Psi_e(\vec{x}_1, \vec{x}_2, \ldots, \vec{x}_N) \right.$$

$$\left. \times \sum_{n=1}^{N} a_1(\vec{s})\delta(\vec{r}_n - \vec{s})\,d\vec{r}_1\,d\vec{r}_1 \cdots d\vec{r}_N \right] d\vec{s}$$

$$= \int \left[\int \int \cdots \int \Psi_e^*(\vec{x}_1, \vec{x}_2, \ldots, \vec{x}_N) \Psi_e(\vec{x}_1, \vec{x}_2, \ldots, \vec{x}_N) \right.$$

$$\left. \times \sum_{n=1}^{N} \delta(\vec{r}_n - \vec{s})\,d\vec{r}_1\,d\vec{r}_1 \cdots d\vec{r}_N \right] a_1(\vec{s})\,d\vec{s}$$

$$= \int \rho(\vec{s})a_1(\vec{s})\,d\vec{s}, \tag{15.7}$$

i.e., only the position-space density is required.

One may now suggest that, somehow, one may avoid determining the complete N-electron wavefunction but instead can determine only the position-space density and from that obtain all information that is of interest. This means that instead of solving the Schrödinger Eq. (15.1) for the wavefunction one would have to solve another equation that determines directly the electron density $\rho(\vec{r})$.

The proposal that this might be possible goes back to the beginning of the modern quantum theory. Thomas and Fermi suggested (Thomas, 1926, Fermi,

1927, Gombás, 1949) that for 'larger' systems (i.e., for N not too small), the number of electrons was so large that the system could be treated using statistical arguments. Then, the electron density is the number of electrons per small volume element, and by assuming that this number is large a statistical treatment is justified. Furthermore, one can then derive an approximate expression for the total energy of such a gas of electrons that move in a given external field (which in our case is the electrostatic field generated by the nuclei). This total-energy expression then becomes one depending solely on the electron density $\rho(\vec{r})$. In its original form it is

$$E_{\text{TF}}[\rho(\vec{r})] = C_F \int \rho^{5/3}(\vec{r})\,d\vec{r} + \int V_{\text{ext}}(\vec{r})\rho(\vec{r})\,d\vec{r} + \frac{1}{2}\int\int \frac{\rho(\vec{r}_1)\rho(\vec{r}_2)}{|\vec{r}_1 - \vec{r}_2|}\,d\vec{r}_1\,d\vec{r}_2.$$

(15.8)

Here, C_F is a constant,

$$C_F = \tfrac{3}{10}(3\pi^2)^{2/3},$$

(15.9)

and $V_{\text{ext}}(\vec{r})$ is the external potential generated by the nuclei,

$$V_{\text{ext}}(\vec{r}) = \sum_{k=1}^{M} \frac{-Z_k}{|\vec{R}_k - \vec{r}|}.$$

(15.10)

The first term in Eq. (15.8) originates from the kinetic energy, and the last one from the electrostatic interactions between the electrons.

Subsequently, we may minimize E_{TF} under the constraint that the correct total number of electrons is given:

$$\int \rho(\vec{r})\,d\vec{r} = N.$$

(15.11)

By using a Lagrange multiplier (μ) for the constraint, we thus require

$$\delta\left\{E_{\text{tot}} - \mu\left[\int \rho(\vec{r})\,d\vec{r} - N\right]\right\} = 0,$$

(15.12)

i.e., that the expression within the curly brackets has its minimum.

One may now attempt to solve the equation (15.12) for a given system, but for those of interest here it turns out that the results are very inaccurate. After all, the whole formalism is constructed as being approximate, and only as long as the approximations are valid can one expect reasonable results. First of all, the assumption that there is a large number of electrons per small volume element is rarely justified.

Later developments have led to improved results, but one should stress that the whole approach is constructed as an approximate description and therefore also the results are at most approximate.

The approach of Thomas and Fermi is a pure density-based one, where the problem of calculating the N-electron wavefunction is replaced by that of calculating the electron density in three-dimensional position space. It is not constructed as being an exact alternative to solving the Schrödinger equation, but rather as an approximation.

Also the $X\alpha$ approach of Slater (1951) and Gáspár (1954) is constructed as an approximation, but in this case to the problem of solving the Hartree–Fock equations. These equations are

$$\hat{F}\phi_i = \varepsilon_i\phi_i \tag{15.13}$$

where

$$\hat{F} = \hat{h}_1 + \sum_j (\hat{J}_j - \hat{K}_j). \tag{15.14}$$

Now, since we assume that the Hartree–Fock approximation is valid,

$$\left[\sum_j \hat{J}_j\right]\phi_i(\vec{r}_1) = \left[\sum_j \int \frac{|\phi_j(\vec{r}_2)|^2}{|\vec{r}_2 - \vec{r}_1|}\, d\vec{r}_2\right]\phi_i(\vec{r}_1)$$

$$= \left[\int \frac{\sum_j |\phi_j(\vec{r}_2)|^2}{|\vec{r}_2 - \vec{r}_1|}\, d\vec{r}_2\right]\phi_i(\vec{r}_1)$$

$$= \left[\int \frac{\rho(\vec{r}_2)}{|\vec{r}_2 - \vec{r}_1|}\, d\vec{r}_2\right]\phi_i(\vec{r}_1)$$

$$= V_C(\vec{r}_1)\phi_i(\vec{r}_1), \tag{15.15}$$

where V_C is the classical electrostatic (Coulomb) potential from the electron density $\rho(\vec{r})$,

$$V_C(\vec{r}_1) = \int \frac{\rho(\vec{r}_2)}{|\vec{r}_2 - \vec{r}_1|}\, d\vec{r}_2. \tag{15.16}$$

This potential satisfies Poisson's equation,

$$\nabla^2 V_C(\vec{r}) = -4\pi\rho(\vec{r}). \tag{15.17}$$

On the other hand,

$$\left[\sum_j \hat{K}_j\right]\phi_i(\vec{r}_1) = \sum_j \int \frac{\phi_j^*(\vec{r}_2)\phi_i(\vec{r}_1)}{|\vec{r}_2 - \vec{r}_1|}\, d\vec{r}_2\phi_j(\vec{r}_1) \tag{15.18}$$

cannot be written as a product of ϕ_i times some potential that depends solely on the electron density. However, Slater suggested *approximating* the expression in Eq. (15.18) as

$$\left[\sum_j \hat{K}_j\right]\phi_i(\vec{r}_1) \simeq V_x[\rho(\vec{r}_1)]\phi_i(\vec{r}_1), \tag{15.19}$$

where V_x is some functional of ρ. Slater argued that

$$V_x(\vec{r}_1) = -\frac{3}{2}\alpha \left[\frac{3}{\pi}\rho(\vec{r}_1) \right]^{1/3} \tag{15.20}$$

was a good approximation with

$$\alpha = 1. \tag{15.21}$$

Gáspár showed shortly later that a more correct approximation was obtained with

$$\alpha = \tfrac{2}{3}. \tag{15.22}$$

Later, other values between these two have been suggested.

As with the Thomas–Fermi theory, the $X\alpha$ method is constructed as an approximation to solving the Schrödinger equation and not as an exact theory.

15.2   THE HOHENBERG–KOHN THEOREMS

With the two approaches above the electron density was given a more central role, representing the fact that ultimately the electron density and not the complete wavefunction is the interesting observable. Moreover, the equations that resulted were easier to solve than the Schrödinger or Hartree–Fock equations. However, in both cases it was not meant as being more than an approximation to the 'true' wavefunction-based approaches. This changed with the appearance of the theorems of Hohenberg and Kohn (1964), where it was shown that it is *possible* to calculate *any* ground-state property through knowledge of only the electron density. This means that we do *not* need the full wavefunction but only the electron density. Therefore, first calculating the full wavefunction and subsequently the electron density seems to be an unnecessarily complicated procedure and it should be possible to calculate the electron density directly. However, the theorems of Hohenberg and Kohn show only that it is possible to calculate any ground-state property, but not how.

We shall now reproduce the two fairly simple theorems of Hohenberg and Kohn. We consider a system of N electrons that move in some external potential. For the system of interest the external potential is that of the nuclei, as given by Eq. (15.10), but also for other cases (e.g., when the system is exposed to an external electrostatic or gravitational field), we can write it as a sum of one-electron potentials,

$$\sum_{i=1}^{N} V_{\text{ext}}(\vec{r}_i). \tag{15.23}$$

We do not know this potential in advance. We assume that we have measured some electron density in an experiment, but we do not know what system it belongs to. However, we will assume that it is the density of the ground state.

We know N since

$$N = \int \rho(\vec{r}) \, d\vec{r}. \tag{15.24}$$

We can then specify the kinetic-energy part of the total N-electron Hamilton operator. Since we have N electrons, it must be

$$\sum_{i=1}^{N} -\frac{1}{2} \nabla^2_{\vec{r}_i}. \tag{15.25}$$

In addition, that part of the Hamilton operator that originates from the electron–electron interactions is

$$\sum_{i>j=1}^{N} \frac{1}{|\vec{r}_i - \vec{r}_j|} \equiv V(\vec{r}_1, \vec{r}_2, \ldots, \vec{r}_N). \tag{15.26}$$

The total Hamilton operator is thus

$$\hat{H} = \sum_{i=1}^{N} -\frac{1}{2} \nabla^2_{\vec{r}_i} + \sum_{i=1}^{N} V_{\text{ext}}(\vec{r}_i) + V(\vec{r}_1, \vec{r}_2, \ldots, \vec{r}_N). \tag{15.27}$$

We do not know V_{ext}, but we know the electron density $\rho(\vec{r})$. Hohenberg and Kohn proved that for a given $\rho(\vec{r})$, that is the density of the ground state for some system, we cannot have two different external potentials V_{ext}, i.e., that part of the Hamilton operator is also uniquely specified. Their proof proceeds by assuming that the opposite is true and then showing that this leads to a self-contradiction. We assume hence that we have two different Hamilton operators,

$$\hat{H}_1 = \sum_{i=1}^{N} -\frac{1}{2} \nabla^2_{\vec{r}_i} + \sum_{i=1}^{N} V_{\text{ext},1}(\vec{r}_i) + V(\vec{r}_1, \vec{r}_2, \ldots, \vec{r}_N) \tag{15.28}$$

and

$$\hat{H}_2 = \sum_{i=1}^{N} -\frac{1}{2} \nabla^2_{\vec{r}_i} + \sum_{i=1}^{N} V_{\text{ext},2}(\vec{r}_i) + V(\vec{r}_1, \vec{r}_2, \ldots, \vec{r}_N). \tag{15.29}$$

The external potentials $V_{\text{ext},1}$ and $V_{\text{ext},2}$ are assumed differing by more than an additive constant. Then, we have two *different* wavefunctions for the ground states,

$$\hat{H}_1 \Psi_1 = E_1 \Psi_1$$
$$\hat{H}_2 \Psi_2 = E_2 \Psi_2 \tag{15.30}$$

where

$$\Psi_1 \neq \Psi_2. \tag{15.31}$$

Notice that we will not have to specify these; we just need to know that they exist. However, both Ψ_1 and Ψ_2 give the *same* electron density (this is the basic assumption),

$$\rho(\vec{r}) = \sum_{i=1}^{N} \int \int \cdots \int \Psi_1^*(\vec{r}_1, \vec{r}_2, \ldots, \vec{r}_N) \delta(\vec{r}_i - \vec{r}) \Psi_1(\vec{r}_1, \vec{r}_2, \ldots, \vec{r}_N)$$

$$\times \, d\vec{r}_1 \, d\vec{r}_2 \cdots d\vec{r}_N$$

$$= \sum_{i=1}^{N} \int \int \cdots \int \Psi_2^*(\vec{r}_1, \vec{r}_2, \ldots, \vec{r}_N) \delta(\vec{r}_i - \vec{r}) \Psi_2(\vec{r}_1, \vec{r}_2, \ldots, \vec{r}_N)$$

$$\times \, d\vec{r}_1 \, d\vec{r}_2 \cdots d\vec{r}_N. \tag{15.32}$$

The variational principle tells us that

$$\langle \Psi | \hat{H}_1 | \Psi \rangle > \langle \Psi_1 | \hat{H}_1 | \Psi_1 \rangle = E_1, \tag{15.33}$$

where Ψ is any N-electron wavefunction *different* from Ψ_1. We now *choose*

$$\Psi = \Psi_2 \tag{15.34}$$

and then obtain

$$
\begin{aligned}
E_1 &< \langle \Psi_2 | \hat{H}_1 | \Psi_2 \rangle = \langle \Psi_2 | \hat{H}_1 - \hat{H}_2 + \hat{H}_2 | \Psi_2 \rangle \\
&= \langle \Psi_2 | \hat{H}_1 - \hat{H}_2 | \Psi_2 \rangle + \langle \Psi_2 | \hat{H}_2 | \Psi_2 \rangle \\
&= \langle \Psi_2 | \sum_{i=1}^{N} V_{\text{ext},1}(\vec{r}_i) - \sum_{i=1}^{N} V_{\text{ext},2}(\vec{r}_i) | \Psi_2 \rangle + E_2 \\
&= \int \rho(\vec{r}) [V_{\text{ext},1}(\vec{r}) - V_{\text{ext},2}(\vec{r})] \, d\vec{r} + E_2.
\end{aligned}
\tag{15.35}
$$

We have here used that the difference between $\hat{H}_1$ and $\hat{H}_2$ is only the difference in the external potentials, and that Ψ_1 and Ψ_2 produce the same electron density.

We repeat the above procedure for the case where Ψ_1 and Ψ_2 have been interchanged,

$$\langle \Psi | \hat{H}_2 | \Psi \rangle > \langle \Psi_2 | \hat{H}_2 | \Psi_2 \rangle = E_2, \tag{15.36}$$

where Ψ is any N-electron wavefunction *different* from Ψ_2. We choose

$$\Psi = \Psi_1 \tag{15.37}$$

and obtain then

$$
\begin{aligned}
E_2 &< \langle \Psi_1 | \hat{H}_2 | \Psi_1 \rangle = \langle \Psi_1 | \hat{H}_2 - \hat{H}_1 + \hat{H}_1 | \Psi_1 \rangle \\
&= \langle \Psi_1 | \hat{H}_2 - \hat{H}_1 | \Psi_1 \rangle + \langle \Psi_1 | \hat{H}_1 | \Psi_1 \rangle
\end{aligned}
$$

$$= \langle \Psi_1 | \sum_{i=1}^{N} V_{ext,2}(\vec{r}_i) - \sum_{i=1}^{N} V_{ext,1}(\vec{r}_i) | \Psi_1 \rangle + E_1$$

$$= \int \rho(\vec{r})[V_{ext,2}(\vec{r}) - V_{ext,1}(\vec{r})] \, d\vec{r} + E_1. \tag{15.38}$$

Eq. (15.35) gives

$$E_1 - E_2 < \int \rho(\vec{r})[V_{ext,1}(\vec{r}) - V_{ext,2}(\vec{r})] \, d\vec{r} \tag{15.39}$$

or (multiplying by -1),

$$E_2 - E_1 > \int \rho(\vec{r})[V_{ext,2}(\vec{r}) - V_{ext,1}(\vec{r})] \, d\vec{r}. \tag{15.40}$$

On the other hand, Eq. (15.38) gives

$$E_2 - E_1 < \int \rho(\vec{r})[V_{ext,2}(\vec{r}) - V_{ext,1}(\vec{r})] \, d\vec{r}. \tag{15.41}$$

Eq. (15.40) and Eq. (15.41) cannot both be true. Therefore, we have a contradiction, and something in our assumptions must be wrong. That is, we can not have two *different* external potentials that produce the same electron density. This means that the electron density $\rho(\vec{r})$ defines *all* terms in the Hamilton operator, and therefore we can *in principle* determine the complete N-electron wavefunction for the ground state by only knowing the electron density. And once this is known any ground-state property can be calculated.

This is the first Hohenberg–Kohn theorem. It states that once you know the ground-state electron density in position space *any* ground-state property is uniquely defined. That is, any ground-state property is a functional of the electron density in position space. The theorem is just an existence theorem: it tells that there exists some functional relating the electron density with any ground-state property, but not what this functional looks like. And so far, no *exact* functionals have been derived for most properties, although many good approximations exist.

The content of the first Hohenberg–Kohn theorem can be illustrated through the following example. We consider the N_2 and CO molecules. They have the same numbers of electrons and nuclei, but whereas the former has a symmetric electron density the latter has not. Thereby, we can distinguish between the two. However, adding an external electrostatic potential along the bond for the N_2 molecule, also the electron density of this becomes polarized, and it is no longer obvious how to distinguish between the two molecules. But according to the first Hohenberg–Kohn theorem, it is possible to distinguish between them uniquely.

Let us now *assume* that we know the functional of ρ that gives us the total electronic energy E_e for the ground state. We know also that this density can — in principle — be calculated from some N-electron wavefunction, i.e., from the ground-state wavefunction for the system of our interest, Ψ_0. Thus,

$$E_e = \langle \Psi_0 | \hat{H} | \Psi_0 \rangle. \tag{15.42}$$

Since Ψ_0 is the ground state, E_e must be the smallest possible value of

$$\langle\Psi|\hat{H}|\Psi\rangle, \tag{15.43}$$

i.e.,

$$E_e = \min_\Psi \langle\Psi|\hat{H}|\Psi\rangle, \tag{15.44}$$

where we have specified that we vary Ψ.

Furthermore, we know that Ψ leads to the correct electron density ρ. We will therefore write explicitly

$$E_e = \min_{\Psi\to\rho} \langle\Psi|\hat{H}|\Psi\rangle. \tag{15.45}$$

Thereby we have *formally* written E_e as a functional of ρ,

$$E_e = E_e[\rho] = \min_{\Psi\to\rho} \langle\Psi|\hat{H}|\Psi\rangle. \tag{15.46}$$

What happens if we take that functional but insert a wrong density ρ'? Since $\rho \neq \rho'$, the two densities cannot be constructed from the same wavefunction. Therefore, that wavefunction Ψ' that leads to the minimum of

$$E_e[\rho'] = \min_{\Psi'\to\rho'} \langle\Psi'|\hat{H}|\Psi'\rangle. \tag{15.47}$$

is not that of the ground state of the system. Therefore, any expectation value $\langle\Psi'|\hat{H}|\Psi'\rangle$ of Eq. (15.47) is larger than that of Eq. (15.46), i.e., we have a variational principle for the density functionals,

$$E_e[\rho'] \geq E_e[\rho]. \tag{15.48}$$

This is the second Hohenberg–Kohn theorem.

It states that once the functional that relates the electron density in position space with the total electronic energy is known, one may calculate it approximately by inserting approximate densities ρ'. Furthermore, just as for the variational method for wavefunctions, one may improve any actual calculation by minimizing $E_e[\rho']$.

15.3 FUNCTIONAL DERIVATIVES

Before continuing it will be useful to introduce the concept of functional derivatives. Compared with the well-known derivatives where a given function is differentiated with respect to one or more variables, we here consider a functional of some function and differentiate this with respect to the function.

Let us first consider a function of several parameters,

$$y = y(x_1, x_n, \ldots, x_N). \tag{15.49}$$

We will assume that each of the parameters x_i is a function of some further parameter, t, i.e.,

$$x_i = x_i(t). \qquad (15.50)$$

Then,

$$\frac{dy}{dt} = \sum_{i=1}^{N} \frac{\partial y}{\partial x_i} \frac{dx_i}{dt}. \qquad (15.51)$$

Here, y is a *function*, i.e., it depends on a set of parameter.

We may modify this example slightly and let the set of x_i become a continuous set, i.e., y becomes a *functional* of the function x. In principle, this corresponds to letting $N \rightarrow \infty$. We will still let x be some function of t and then obtain, as a generalization of Eq. (15.51),

$$\frac{dy}{dt} = \int \frac{\delta y}{\delta x} \frac{dx}{dt} \, dx, \qquad (15.52)$$

where

$$\frac{\delta y}{\delta x} \qquad (15.53)$$

is the functional derivative of y with respect to x.

In the general case, the functional derivative of the functional F that depends on the function f is defined as

$$\frac{\delta}{\delta f(x)} F[f(x)] = \lim_{\delta f(x) \rightarrow 0} \frac{F[f(x) + \delta f(x)] - F[f(x)]}{\delta f(x)}, \qquad (15.54)$$

i.e., very similar to a 'traditional' derivative. Moreover, we see also that the functional derivative could have been applied in Section 9.1, when we derived the Hartree–Fock equations. Actually, the quantities we calculated there were nothing other than functional derivatives, although we did not call them such. Finally, we study the special case that $y = x$ in Eq. (15.52). Then,

$$\frac{dx}{dt} = \int \frac{\delta x}{\delta x'} \frac{dx'}{dt} \, dx'. \qquad (15.55)$$

This identity can be valid in the general case only if

$$\frac{\delta x}{\delta x'} = \delta(x - x'). \qquad (15.56)$$

15.4   THE KOHN–SHAM METHOD

The Hohenberg–Kohn theorems provide a formalistic proof for the correctness of the approach of Thomas and Fermi, but do not provide any practical scheme for calculating ground-state properties from the electron density. This was, on the other hand, provided by the approach of Kohn and Sham (1965).

The second theorem of Hohenberg and Kohn states that $E_e[\rho']$ has a minimum for the correct ground-state density ρ. This means that

$$\delta E_e[\rho] \equiv E_e[\rho + \delta\rho] - E_e[\rho] = 0, \qquad (15.57)$$

where we only allow those changes $\delta\rho$ that do not change the total number of electrons; i.e., we require that

$$\int \rho(\vec{r}) \, d\vec{r} = N. \qquad (15.58)$$

Equations (15.57) and (15.58) may be combined by using a Lagrange multiplier (μ) to

$$\delta \left\{ E_e[\rho(\vec{r})] - \mu \left[\int \rho(\vec{r}) \, d\vec{r} - N \right] \right\} = 0. \qquad (15.59)$$

This has to be true for *any* variation in ρ. We can therefore use the concept of functional derivatives in rewriting Eq. (15.59) as

$$\frac{\delta}{\delta\rho(\vec{r})} \left\{ E_e[\rho(\vec{r})] - \mu \left[\int \rho(\vec{r}) \, d\vec{r} - N \right] \right\} = 0. \qquad (15.60)$$

Unfortunately, we do not know E_e as a functional of $\rho(\vec{r})$. It contains, however, as one term, the kinetic energy T. The kinetic energy is an observable (e.g., through momentum-space spectroscopy), and as such it *must* be a functional of the electron density in position space,

$$T = T[\rho(\vec{r})]. \qquad (15.61)$$

Another term is that due to the external potential,

$$\int V_{\text{ext}}(\vec{r})\rho(\vec{r}) \, d\vec{r}, \qquad (15.62)$$

which is written as a functional of $\rho(\vec{r})$.

A third term is the classical Coulomb interaction energy,

$$\frac{1}{2} \int \int \frac{\rho(\vec{r}_1)\rho(\vec{r}_2)}{|\vec{r}_1 - \vec{r}_2|} \, d\vec{r}_1 \, d\vec{r}_2 = \frac{1}{2} \int V_{\text{C}}(\vec{r})\rho(\vec{r}) \, d\vec{r}, \qquad (15.63)$$

where we have used the definition of the Coulomb potential of Eqs. (15.15) and (15.16), and where the factor $\frac{1}{2}$ is included in order to avoid double-counting.

But as we saw for the Hartree–Fock method, there may be further terms due to exchange effects. And when going beyond the Hartree–Fock approximation, also correlation effects lead to extra (correlation) contributions. We shall simply call these E'_{xc} ('xc' for exchange correlation), and since both E_e and all the other terms are functionals of ρ, so must E'_{xc} be.

In total we therefore have

$$E_e = T[\rho(\vec{r})] + \int V_{\text{ext}}(\vec{r})\rho(\vec{r})\,d\vec{r} + \int V_C(\vec{r})\rho(\vec{r})\,d\vec{r} + E'_{\text{xc}}[\rho(\vec{r})]. \qquad (15.64)$$

Then,

$$\frac{\delta E_e}{\delta\rho(\vec{r})} = \frac{\delta T}{\delta\rho(\vec{r})} + \frac{\delta}{\delta\rho(\vec{r})}\left[\int V_{\text{ext}}(\vec{r}')\rho(\vec{r}')\,d\vec{r}'\right.$$
$$\left. + \frac{1}{2}\int\int\frac{\rho(\vec{r}_1)\rho(\vec{r}_2)}{|\vec{r}_1 - \vec{r}_2|}\,d\vec{r}_1\,d\vec{r}_2\right] + \frac{\delta E'_{\text{xc}}}{\delta\rho(\vec{r})}. \qquad (15.65)$$

In order to calculate the functional derivative of the argument of the square brackets, we change

$$\rho(\vec{r}) \rightarrow \rho(\vec{r}) + \delta\rho(\vec{r}), \qquad (15.66)$$

calculate the first-order changes i.e., neglecting second- and higher-order changes in $\delta\rho(\vec{r})$, and obtain then from Eq. (15.59)

$$\frac{\delta T}{\delta\rho} + V_{\text{ext}}(\vec{r}) + V_C(\vec{r}) + \frac{\delta E'_{\text{xc}}}{\delta\rho} = \mu \qquad (15.67)$$

with V_C given by Eq. (15.16).

The trick of Kohn and Sham is now as follows. They considered a *fictitious* system of *non-interacting* particles. They assumed that this system has the *same* density as the real system and the *same* energy as the real system. In order to ensure that they have the same density and energy as the real system, these particles are assumed moving in some external potential $V_{\text{eff}}(\vec{r})$. For this model system we may also apply the procedure above. However, since the particles are non-interacting, their total-energy expression is considerably simpler,

$$E_e = T_0[\rho(\vec{r})] + \int V_{\text{eff}}(\vec{r})\rho(\vec{r})\,d\vec{r}. \qquad (15.68)$$

Notice that the kinetic energy is not identical to that of Eq. (15.64), since these particles are different. Analogous to Eq. (15.67) we arrive at

$$\frac{\delta T_0}{\delta\rho} + V_{\text{eff}}(\vec{r}) = \mu. \qquad (15.69)$$

Comparing Eqs. (15.67) and (15.69) we arrive at

$$V_{\text{eff}}(\vec{r}) = \frac{\delta T}{\delta\rho} - \frac{\delta T_0}{\delta\rho} + V_{\text{ext}}(\vec{r}) + V_C(\vec{r}) + \frac{\delta E'_{\text{xc}}}{\delta\rho}. \qquad (15.70)$$

The trick is now that for the model system the Hamilton operator is particularly simple,

$$\hat{H} = \sum_{i=1}^{N}\left[-\frac{1}{2}\nabla^2_{\vec{r}_i} + V_{\text{eff}}(\vec{r}_i)\right] \equiv \sum_{i=1}^{N}\hat{h}_{\text{eff}}(i). \qquad (15.71)$$

Accordingly, there are only single-particle operators. The solution to the Schrödinger equation for this system

$$\hat{H}\Psi = E_e\Psi \tag{15.72}$$

can therefore be written exactly as a single Slater determinant,

$$\Psi = |\phi_1, \phi_2, \ldots, \phi_N|, \tag{15.73}$$

where

$$\hat{h}_{\text{eff}}\phi_i = \varepsilon_i\phi_i \tag{15.74}$$

is the single-particle equation that determines the single-particle orbitals. This equation is easily derived as in Section 9.2 for the Hartree–Fock method, but with the important simplification that we have no two-particle operators.

Furthermore,

$$\rho(\vec{r}) = \sum_{i=1}^{N} |\phi_i(\vec{r})|^2, \tag{15.75}$$

where the summation runs over the N orbitals with the lowest eigenvalues ε_i.

The main problem is that we do not know the exact form of the effective potential V_{eff}, i.e., we do not know the terms

$$\frac{\delta T}{\delta\rho} - \frac{\delta T_0}{\delta\rho} + \frac{\delta E'_{\text{xc}}}{\delta\rho} \equiv \frac{\delta E_{\text{xc}}}{\delta\rho} \equiv V_{\text{xc}}(\vec{r}) \tag{15.76}$$

of Eq. (15.70), i.e., the so-called exchange-correlation energy and potential. However, there exist approximations to these that are good approximations in very many cases. On the other hand, by introducing an approximation in the Schrödinger equation for the model particles, we do not know whether improved calculations (e.g., using larger basis sets) also lead to improved results (compared with, e.g., experiment or other calculations).

Two further aspects shall be mentioned at this point. First, the success of the Kohn–Sham approach is based on the assumption that it is possible to construct the model system of non-interacting particles moving in an effective external potential. Thus, it is indirectly assumed that for *any* ground-state density there exists such an effective potential, i.e., that any ρ is what is called V-representative. There exists (specifically constructed) examples where this is not the case, but in most (all?) practical applications, this represents no problem.

Second, the Lagrange multiplier μ above is the chemical potential for the electrons,

$$\mu = \frac{\partial E_e}{\partial N}. \tag{15.77}$$

This follows from Eq. (15.59),

$$\mu = \frac{\delta E_e[\rho(\vec{r})]}{\delta\rho(\vec{r})}, \tag{15.78}$$

i.e., μ gives the changes in the total (electronic) energy upon changing the number of electrons. Notice that, although the right-hand side of Eq. (15.78) appears as a function of $\vec{r}$, the left-hand side is a constant. This must therefore also be the case for the right-hand side.

15.5 EXTENSIONS; SPIN AND SYMMETRY

The Hohenberg–Kohn theorems as we have derived them apply to the ground state of a given system that is supposed to be spin-unpolarized. This means that when separating the total electron density into one component for spin-up orbitals and one for spin-down orbitals,

$$\rho(\vec{r}) = \rho_\alpha(\vec{r}) + \rho_\beta(\vec{r}), \tag{15.79}$$

we have assumed that

$$\rho_\alpha(\vec{r}) = \rho_\beta(\vec{r}). \tag{15.80}$$

However, when this is not the case, i.e., the spin-polarization

$$m(\vec{r}) = \rho_\alpha(\vec{r}) - \rho_\beta(\vec{r}) \tag{15.81}$$

is different from 0, then the Hohenberg–Kohn theorems have to be extended. The arguments will not be carried through here. The result is, however, that E_e (and any other ground-state property) becomes a functional not only of ρ but also of m, i.e.,

$$E_e = E_e[\rho(\vec{r}), m(\vec{r})]. \tag{15.82}$$

Another extension of the theorems applies to different symmetries. In Chapter 7 we applied group theory in studying symmetry properties of our systems. We showed that states (single-particle orbitals or complete N-electron wavefunctions) belonging to different irreducible representations of the group for the symmetry operations of the system of interest do not interact. This means that we can consider each irreducible representation separately. For each of those, the variational principle of Chapter 5 will apply (since any state of a given irreducible representation will only contain contributions from exactly that irreducible representation). We can use that in generalizing the Hohenberg–Kohn theorems that then apply for the energetically lowest state of *each* irreducible representation. This means that if we have the representations $R_1, R_2, \ldots$ for a given system, then we can apply density-functional theory in studying the energetically lowest state for the R_1 representation, that for the R_2 representation and so on, but not energetically higher ones of any of the representations.

15.6 LOCAL AND NON-LOCAL APPROXIMATIONS

The Coulomb potential is given by

$$V_C(\vec{r}) = \int \frac{\rho(\vec{s})}{|\vec{r} - \vec{s}|} \, d\vec{s}, \tag{15.83}$$

i.e., at the point $\vec{r}$ it depends on the electron density ρ at *every* point of space.

Also the exchange-correlation potential at a given point

$$V_{\mathrm{xc}}(\vec{r}) = \frac{\delta E_{\mathrm{xc}}[\rho]}{\delta \rho(\vec{r})} \tag{15.84}$$

may depend on the electron density at every point of space, although it is not known whether this is the case.

A very simple case is that of the so-called homogeneous electron gas. This system, which may be only approximately realized in some metals, is one for which the electron density is constant in space. Then, the dependence of the exchange-correlation potential in a given point on the spatial variations of ρ disappears. Furthermore, it is possible to study this system accurately using methods that will not be described here. This means that we can calculate V_{xc} (which here is independent of $\vec{r}$) for this system as a function of the constant density. This leads to a curve like that of Fig. 15.1.

For the spin-polarized system similar calculations may be carried through for a homogeneous electron gas that has a spatially constant spin polarization. Thereby, the curve of Fig. 15.1 changes into a two-dimensional surface with the density defining one coordinate and the spin-polarization the other.

It is custom to introduce the so-called electron-gas parameter r_s, also for systems where the electron density not is a constant. It describes the radius of a sphere that contains exactly one electron, i.e.

$$\frac{4\pi}{3} r_s^3 = \frac{1}{\rho}, \tag{15.85}$$

and, when the density is position-dependent, so will r_s be.

Then, the results like the curve of Fig. 15.1 can be fitted with a function of r_s. There exists various forms, of which the $X\alpha$ form of Section 15.1 is one. This corresponds to neglecting correlation effects and only including those of exchange. Another form is due to von Barth and Hedin (1972) and is

$$V_{\mathrm{xc}} = \left[-\frac{1}{\pi} \left(\frac{9\pi}{4} \right)^{1/3} \frac{1}{r_s} - \gamma c^F \cdot F\left(\frac{r_s}{75} \right) + \gamma c^P \cdot F\left(\frac{r_s}{30} \right) \right] (2x)^{1/3}$$

$$- c^P \ln\left(1 + \frac{30}{r_s} \right) + \gamma \left[c^F \cdot F\left(\frac{r_s}{75} \right) - c^P \cdot F\left(\frac{r_s}{30} \right) \right]$$

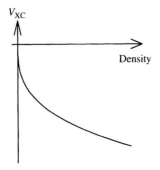

V_{XC}

Density

Figure 15.1   Schematic view of V_{xc} as a function of the constant electron density for the homogeneous electron gas

$$+ \left[-c^F \cdot \ln \left(1 + \frac{75}{r_s} \right) + c^P \cdot \ln \left(1 + \frac{30}{r_s} \right) \right.$$

$$\left. + \frac{4}{3} c^F \cdot F \left(\frac{r_s}{75} \right) - \frac{4}{3} c^P \cdot F \left(\frac{r_s}{30} \right) \right] f(x) \tag{15.86}$$

with x being related to the spin-polarization,

$$x = \frac{1}{2} + \frac{m(\vec{r})}{2\rho(\vec{r})}, \tag{15.87}$$

r_s is given by Eq. (15.87), the various constants given through (in atomic units, a.u.)

$$c^F = 0.0127$$

$$c^P = 0.0252$$

$$\gamma = \frac{4}{3(2^{1/3} - 1)}, \tag{15.88}$$

and the two functions f and F being

$$f(x) = \frac{1}{1 - 2^{-1/3}} \left(x^{4/3} + (1 - x)^{4/3} - 2^{-1/3} \right)$$

$$F(z) = (1 + z^3) \ln \left(1 + \frac{1}{z} \right) + \frac{z}{2} - z^2 - \frac{1}{3}. \tag{15.89}$$

A more recent approximation is due to Vosko, *et al.* (1980), but their shapes as functions of r_s and x differ only marginally.

When passing to a 'real' system (e.g., a molecule or a solid), the electron density is no longer constant. However, one may now apply a so-called local-density or local-spin-density approximation. Then, in the point $\vec{r}$ we use the potential V_{xc} as it would be for the homogeneous electron gas with the density (and, eventually, spin polarization) equal to that of the point $\vec{r}$. This means that spatial variations in ρ (and m) are completely ignored. This is the so-called local-(spin-)density approximation (LDA, LSDA, LSD), that has been very successful for describing many properties of many systems (in particular solids), but fails in describing many others.

With the local-density approximation spatial variations in ρ are ignored. When comparing with, e.g., the expression for the Coulomb potential, Eq. (15.83), the local-density approximation appears as a very crude approximation. One can, however, use a number of arguments in showing that it is better than it may at a first view appear to be. A main reason is that for any quantities like, e.g., the total energy the precise description of the exchange-correlation effects in every single point is of only secondary importance, since one first of all calculates expectation values, i.e., integrals. And then it turns out that the average behaviour of the exchange-correlation effects are well described with a local-density approximation. We consider, however, further details of these arguments beyond the scope of the present representation and will therefore abandon those here. The arguments explain also partly the success that the density-functional methods have had over the last several years.

Nevertheless, as a first step in introducing dependences on spatial variations, one may let V_{xc} depend also on $|\vec{\nabla}\rho|$, $\nabla^2\rho$, ... Thereby one arrives at the so-called non-local-density approximations or generalized gradient approximations (GGA). Some accurate functionals of this type were developed during the second half of the 1980s (e.g., by Becke, 1988 and by Perdew, 1991 and Perdew and Wang, 1986), and with these a number of failures of the local-density approximation have been removed.

The local-density or non-local-density approximations are parametrized approximations. Nevertheless, the parameters are not adjusted in order to describe certain physical observables of the system of interest or of related systems. Thus, in that sense the currently applied density-functional methods are parameter-free, although approximate. Their advantage is that they include correlation effects, and that, particularly for larger systems or systems containing heavier atoms, they do not become prohibitively involved to apply.

Per tradition, density-based methods have been applied mainly in physics, whereas wavefunction-based methods have been applied mainly in chemistry. Both approaches have advantages and disadvantages, and it is not possible to choose one uniquely as generally superior to the other.

15.7   FITTING

The Hartree–Fock approximation amounts to solve the Hartree–Fock equations

$$\left[\hat{h}_1 + \sum_{j=1}^{N} \left(\hat{J}_j - \hat{K}_j\right)\right] \phi_i = \varepsilon_i \phi_i. \tag{15.90}$$

Here,

$$\sum_{j=1}^{N} \hat{J}_j \phi_i(\vec{r}) = \int \frac{\rho(\vec{s})}{|\vec{r} - \vec{s}|} \, d\vec{s}\,\phi_i(\vec{r}) = V_C(\vec{r})\phi_i(\vec{r}), \tag{15.91}$$

where we have introduced the Coulomb potential that obeys Poisson's equation

$$\nabla^2 V_C(\vec{r}) = -4\pi\rho(\vec{r}). \tag{15.92}$$

Similarly,

$$\sum_{j=1}^{N} \hat{K}_j \phi_i(\vec{r}) = \sum_{j=1}^{N} \int \frac{\phi_j^*(\vec{s})\phi_i(\vec{s})}{|\vec{r} - \vec{s}|} \, d\vec{s}\,\phi_j(\vec{r}). \tag{15.93}$$

These equations are solved by expanding the solutions ϕ_i in some basis functions,

$$\phi_i = \sum_{k} \chi_k c_{ki}, \tag{15.94}$$

whereby solving the Hartree–Fock(–Roothaan) equations can be cast into that of solving a matrix-eigenvalue problem. In setting up the matrices we need matrix elements of the types

$$\langle \chi_i | \chi_j \rangle$$

$$\langle \chi_i | \hat{h}_1 | \chi_j \rangle$$

$$\langle \chi_i \chi_j | \hat{h}_2 | \chi_k \chi_l \rangle. \tag{15.95}$$

These matrix elements can in principle be calculated for a given basis set, although in some cases (e.g., for Slater-type orbitals) it is complicated.

Within the density-functional method we have to solve the Kohn–Sham equations

$$\left[\hat{h}_1 + V_C(\vec{r}) + V_{xc}(\vec{r})\right] \phi_i(\vec{r}) = \varepsilon_i \phi_i(\vec{r}) \tag{15.96}$$

which are seen to be very similar to the Hartree–Fock equations.

As above, the solutions are expanded in a basis set [Eq. (15.94)] and the task of solving the Kohn–Sham equations becomes that of solving a matrix-eigenvalue problem. In this case we need the following types of matrix elements:

$$\langle \chi_i | \chi_j \rangle$$

$$\langle \chi_i | \hat{h}_1 | \chi_j \rangle$$

$$\langle \chi_i | V_C | \chi_j \rangle$$

$$\langle \chi_i | V_{xc} | \chi_j \rangle. \tag{15.97}$$

The first three types of those are equivalent to those of the Hartree–Fock approximation, but the last one differs. This is, furthermore, extremely difficult to calculate analytically. Already within the $X\alpha$ or local-density approximation, V_{xc} is a highly nonlinear function of $\rho(\vec{r})$ with

$$\rho(\vec{r}) = \sum_{i=1}^{N} |\phi_i(\vec{r})|^2 = \sum_{i=1}^{N} \sum_{j,k} c_{ji}^* c_{ki} \chi_j^*(\vec{r}) \chi_k(\vec{r}). \tag{15.98}$$

Within the $X\alpha$ approximation, V_{xc} is proportional to $\rho^{1/3}$, which for most types of basis functions can *not* be expressed in a simple analytical form. Consider as the simplest possible case an electron density from one single orbital with this orbital being given as a linear combination of two Gaussians at different sites,

$$\rho(\vec{r}) = |\phi(\vec{r})|^2, \tag{15.99}$$

with

$$\phi(\vec{r}) = c_1 e^{-\alpha_1 |\vec{r} - \vec{R}_1|^2} + c_2 e^{-\alpha_2 |\vec{r} - \vec{R}_2|^2}. \tag{15.100}$$

For c_1 and c_2 real, we find

$$[\rho(\vec{r})]^{1/3} = \left[c_1 e^{-\alpha_1 |\vec{r} - \vec{R}_1|^2} + c_2 e^{-\alpha_2 |\vec{r} - \vec{R}_2|^2}\right]^{2/3} = ? \tag{15.101}$$

In order to solve this problem it has become practice to fit V_{xc} with some approximate form:

$$V_{xc} \simeq \tilde{V}_{xc} = \sum_{n=1}^{N_{xc}} d_n^{xc} f_n^{xc} \qquad (15.102)$$

where f_n^{xc} are some chosen so-called auxiliary basis functions for which the exchange-correlation matrix elements of Eq. (15.97) can be solved and for which the fit (15.102) is reasonably effective and accurate. Of course, the introduction of this auxiliary basis set may introduce new sources of errors in a calculation and one has therefore to ensure that these errors are negligible.

Once having introduced the strategy of fitting, it may also be applied in calculating the Coulomb potential V_C. This is done by fitting the electron density,

$$\rho \simeq \tilde{\rho} = \sum_{n=1}^{N_{\rho}} d_n^{\rho} f_n^{\rho}, \qquad (15.103)$$

where the functions f_n^{ρ} are chosen so that Poisson's equation for the approximate electron density,

$$\nabla^2 \tilde{V}_C = -4\pi\tilde{\rho} \qquad (15.104)$$

can be solved analytically. Thereby the Coulomb potential gets the form,

$$V_C \simeq \tilde{V}_C = \sum_{n=1}^{N_C} d_n^C f_n^C. \qquad (15.105)$$

As above, also here the success depends sensitively on the quality of the fit.

15.8   THE QUASI-PARTICLES

With the approach of Kohn and Sham one solves the single-particle Kohn–Sham equations

$$\left[\hat{h}_1 + V_C(\vec{r}) + V_{xc}(\vec{r})\right]\phi_i(\vec{r}) = \varepsilon_i\phi_i(\vec{r}). \qquad (15.106)$$

These were obtained as the equations for a non-existing model system of non-interacting particles. In the strict sense, the only connection to the real system is that the two systems have the same density,

$$\rho(\vec{r}) = \sum_{i=1}^{N} |\phi_i(\vec{r})|^2, \qquad (15.107)$$

and energy,

$$E_e = \sum_{i=1}^{N} \langle\psi_i|\hat{h}_1|\psi_i\rangle + \frac{1}{2}\int V_C(\vec{r})\rho(\vec{r})\,d\vec{r} + E_{xc}[\rho(\vec{r})]. \qquad (15.108)$$

For the latter we have

$$\sum_{i=1}^{N} \langle \psi_i | \hat{h}_1 | \psi_i \rangle = \sum_{i=1}^{N} \langle \psi_i | -\frac{1}{2} \nabla^2 | \psi_i \rangle - \sum_{i=1}^{N} \langle \psi_i | \sum_{k=1}^{M} \frac{Z_k}{|\vec{r} - \vec{R}_k|} | \psi_i \rangle$$

$$= \sum_{i=1}^{N} \langle \psi_i | -\frac{1}{2} \nabla^2 | \psi_i \rangle - \sum_{k=1}^{M} \int \frac{Z_k}{|\vec{r} - \vec{R}_k|} \rho(\vec{r}) \, d\vec{r}. \quad (15.109)$$

Furthermore, it is practice to write

$$E_{xc}[\rho(\vec{r})] = \int \varepsilon_{xc}[\rho(\vec{r})] \rho(\vec{r}) \, d\vec{r}, \quad (15.110)$$

whereby

$$V_{xc}(\vec{r}) \equiv \frac{\delta E[\rho(\vec{r})]}{\delta \rho(\vec{r})} = \rho(\vec{r}) \cdot \frac{\delta \varepsilon[\rho(\vec{r})]}{\delta \rho(\vec{r})} + \varepsilon_{xc}[\rho(\vec{r})]. \quad (15.111)$$

Within the local-density approximation, ε_{xc} is a *function* (thus, not a functional) of the electron density in the point of interest. Passing to the non-local approximations, ε_{xc} depends also on $|\vec{\nabla}\rho|$, $\nabla^2 \rho$, ..., in the point of interest.

Finally, we add that for spin-polarized systems, E_{xc} and V_{xc} contain also a dependence on the local spin polarization.

The Kohn–Sham equations are very similar to the Hartree–Fock equations,

$$\left[\hat{h}_1 + V_C - \sum_{j=1}^{N} \hat{K}_j \right] \phi_i = \varepsilon_i \phi_i, \quad (15.112)$$

except for the difference in the treatment of exchange and correlation effects. In particular, the $X\alpha$ method, which is a density-functional method, is constructed as an approximation to the Hartree–Fock method. It is therefore not surprising that the solutions to the Kohn–Sham equations are often given an interpretation similar to that applied to the solutions to the Hartree–Fock equations.

This means first of all that the single-particle energies ε_i from the Kohn–Sham equations are interpreted as being ionization energies and electron affinities for occupied and empty orbitals, respectively. One may actually prove that this interpretation holds true but only for one single energy, i.e., the energy of the energetically highest occupied orbital is — assuming that we have the exact density functional — the first ionization potential (Almbladh and von Barth, 1985). Interpreting all the other single-particle energies as related to excitation energies is an approximation that, however, has proved to be good in very many — but *not* all — cases.

On the other hand, Janak (1978) has shown that *any* single-particle eigenvalue to the Kohn–Sham equations obeys

$$\varepsilon_i = \frac{\partial E_e}{\partial n_i}, \quad (15.113)$$

i.e., it can be considered the partial, orbital-specific electron affinity or ionization potential in contrast to the integral ones as is the case for Koopmans' theorem [according to which $\varepsilon_i = E_e(N) - E_e(N - 1)$ or $\varepsilon_i = E_e(N + 1) - E_e(N)$ for occupied or occupied orbitals, respectively].

One may then compare the single-particle energies of Hartree–Fock and those of density-functional calculations. Then one finds that for the occupied orbitals they often agree well with those of the Hartree–Fock calculations but they most often span a slightly (about 20%) larger energy interval than those of density-functional calculations. Comparing with experiment, it turns out that the density-functional results actually often (at least for large systems) compare better than the Hartree–Fock results. On the other hand, the Kohn–Sham eigenvalues of the occupied orbitals are most often rigidly shifted upwards in energy compared both with experiment and with Hartree–Fock values. For the unoccupied orbitals one finds that the energy gap separating occupied and unoccupied orbitals in Hartree–Fock calculations is significantly overestimated and in density-functional calculations significantly underestimated. This is the so-called band-gap problem that we shall return to later.

Equivalent to this procedure it is also practice to consider the densities of the individual orbitals as those of electronic orbitals. But as for the Hartree–Fock method, the concept of orbitals is not uniquely defined (remember, any unitary transformation among the occupied orbitals leads to unchanged physical observables). Therefore, this interpretation can only be used in obtaining insight into the chemical bonds, etc., and not as giving absolute numbers for experimental observables.

15.9   PHYSICAL PROPERTIES

Per construction, the density-functional methods are first of all methods for calculating total energies and total electron densities. Therefore, we shall first see how well they perform in calculating total energies.

In Table 15.1 we show the calculated binding energy for first-row dimers for the experimental bond lengths. These energies are accordingly the total energy of the isolated atoms minus that of the molecule with the experimental bond length. The table shows a well-known fact, that local-(spin-)density calculations overbind, i.e., the binding energy is too large. Experience has shown that the amount of overbinding to a good approximation is a constant for a given bond, e.g., it takes one value for a C–C double bond, another for a C–C single bond, a third for a C–H bond, etc. On the other hand, Hartree–Fock calculations underestimate the binding energies.

As discussed in Chapter 13, systems based on transition-metal atoms may be difficult to treat by Hartree–Fock methods since the transition-metal atoms have a large number of empty and occupied orbitals close to each other and, correspondingly, the Hartree–Fock approximation is a poor one. In Tables 15.2 and 15.3 we therefore compare experimental, density-functional, Hartree–Fock,

Table 15.1 Experimental and calculated binding energies (in eV) for experimental ground states of the first-row dimers. Hartree–Fock (HF) calculations for Be_2 give a purely repulsive energy curve. The results are from Jones and Gunnarsson (1989)

Molecule	Exp.	LSD	$X\alpha$	HF
H_2	4.75	4.91	3.59	3.64
Li_2	1.07	1.01	0.21	0.17
Be_2	0.10	0.50	0.43	
B_2	3.09	3.93	3.79	0.89
C_2	6.32	7.19	6.00	0.79
N_2	9.91	11.34	9.09	5.20
O_2	5.22	7.54	7.01	1.28
F_2	1.66	3.32	3.04	-1.37

Table 15.2 Experimental and calculated bond length (R_e), binding energy (D_e), and vibrational frequency (ω_e) for the $^1\Sigma_g^+$ state of Cu_2. The density-functional calculations are all based on a local-spin-density approximation, and those marked $X\alpha$ use that approximation. HF are calculations with the Hartree–Fock approximation, whereas HF + corr. include correlation effects. From Jones and Gunnarsson (1989)

Method	R_e (a.u.)	ω_e (cm^{-1})	D_e (eV)
Exp.	4.195	265	1.97
LSD	4.10–4.30	248–330	2.30–2.65
$X\alpha$	4.12–4.20	286–290	2.10–2.16
HF	4.58–4.61	198	0.51–0.56
HF+corr.	4.23–4.62	200–242	0.15–2.07

Table 15.3 As Table 15.2, but for the $^1\Sigma_g^+$ state of Cr_2. From Jones and Gunnarsson (1989)

Method	R_e (a.u.)	ω_e (cm^{-1})	D_e (eV)
Exp.	3.17	470	1.56±0.3, 1.44±0.02
LSD	3.17–3.21	441–470	1.80–2.80
$X\alpha$	5.10–5.20	92–110	0.4–1.0
HF	<1.5–2.95	7.50	
HF+corr.	3.04–6.14	70–396	0.1–1.86

and *ab initio* calculations extended with correlation effects on Cu_2 and Cr_2. We list the bond length (i.e., that for which the total energy has its minimum), the binding energy (defined as above, but for the optimized bond length), and the vibrational frequency (simply related to the curvature of the total-energy curve around the minimum). In particular for Cr_2 the Hartree–Fock approximation is bad. This is confirmed by density-functional calculations with the $X\alpha$ approximation, which essentially corresponds to an approximate Hartree–Fock method without correlation effects.

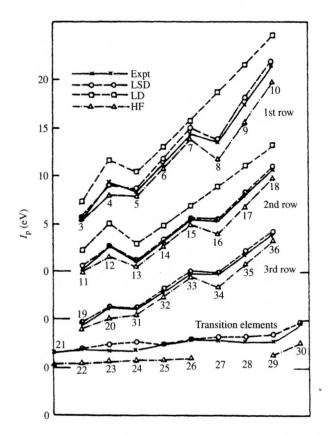

Figure 15.2 First ionization energy of atoms in the local-density (LD), local-spin-density (LSD), and Hartree–Fock (HF) approximation compared with experiment. The numbers show the atomic numbers. For reasons of clarity, the zero of energy is shifted by 5, 10, and 15 eV for the second row, the third row, and the transition-metal row, respectively. The LD results for the first and second rows are increased by an additional 2 eV. Reproduced with permission from Jones and Gunnarsson (1989)

Finally, we show in Fig. 15.2 the first ionization potential from various types of calculations and from experiment.

15.10 SELF-INTERACTION

In deriving the Hartree–Fock equations,

$$\hat{F}\phi_i = \varepsilon_i\phi_i \tag{15.114}$$

with

$$\hat{F} = \hat{h}_1 + \sum_j (\hat{J}_j - \hat{K}_j) \tag{15.115}$$

we used that

$$\hat{J}_i\phi_i = \hat{K}_i\phi_i. \tag{15.116}$$

That this is so follows from the definitions of the operators,

$$\hat{J}_j\phi_i(\vec{r}) = \int \frac{|\phi_j(\vec{s})|^2}{|\vec{r} - \vec{s}|} \, d\vec{s}\phi_i(\vec{r})$$

$$\hat{K}_j\phi_i(\vec{r}) = \int \frac{\phi_j^*(\vec{s})\phi_i(\vec{s})}{|\vec{r} - \vec{s}|} \, d\vec{s}\phi_j(\vec{r}). \tag{15.117}$$

Without this identity the equations would have been

$$\left[\hat{h}_1 + \sum_{j\neq i}(\hat{J}_j - \hat{K}_j)\right]\phi_i = \varepsilon_i\phi_i, \tag{15.118}$$

i.e., we would have different operators for different orbitals.

Within the approximate density-functional methods, the exchange operator $\hat{K}_j$ is replaced by an approximate one. Thereby, the identity (15.116) is no longer exactly satisfied. By including the $i = j$ term in Eq. (15.118) we allow the electron to interact with itself, in the Coulomb part and also in the exchange part. But the two terms cancel. For the approximate functionals they may not cancel, and this may thus lead to a significant error (we include erroneously a part of the interaction of an electron with itself, and since the electron always 'is close to itself', the error can become important). In order to improve that it has been proposed to remove this so-called self-interaction (Cowan, 1967; Lindgren, 1971; Bryant and Mahan, 1978; Perdew and Zunger, 1981). Within the local-density approximation (and we shall not go beyond that here), this leads to the following extra term in the total electronic energy

$$E_{\text{SIC}} = \sum_{i=1}^{N} \left[-\frac{1}{2} \int \int \frac{|\phi_i(\vec{r}_1)|^2|\phi_i(\vec{r}_2)|^2}{|\vec{r}_1 - \vec{r}_2|} \, d\vec{r}_1 \, d\vec{r}_2 + \frac{3}{4}\left(\frac{6}{\pi}\right)^{1/3} \int |\phi_i(\vec{r})|^{8/3} \, d\vec{r} \right], \tag{15.119}$$

where the first term is the correct interaction of the orbital with itself and the last term is the local-density approximation to that term. 'SIC' means *self-interaction corrected*.

This contribution to the total electronic energy has to be added to the expression of Eq. (15.108), and the Kohn–Sham equations take then the form

$$\left[\hat{h}_1 + V_C(\vec{r}) + V_{\text{xc}}(\vec{r}) + V_{\text{SIC},i}(\vec{r})\right]\phi_i(\vec{r}) = \sum_j \varepsilon_{ij}\phi_j(\vec{r}), \tag{15.120}$$

where the extra potential is

$$V_{\text{SIC},i}(\vec{r}) = \frac{\delta E_{\text{SIC}}}{\delta\rho_i(\vec{r})} = -\int \frac{|\phi_i(\vec{s})|^2}{|\vec{r} - \vec{s}|} \, d\vec{s} + \left[\frac{6}{\pi}|\phi_i(\vec{r})|^2\right]^{1/3} \tag{15.121}$$

with

$$\rho_i(\vec{r}) = |\phi_i(\vec{r})|^2. \tag{15.122}$$

The main problem with this approach is that the operator of the left-hand side of Eq. (15.120) becomes orbital-dependent (i.e., different orbitals satisfy different equations), whereby it is not guaranteed that the orbitals are orthonormal and one has to have the more complex right-hand side that explicitly guarantees this orthonormality (i.e., not only the terms with ε_{ij} for $i = j$). This means also that the problem of solving these equations is not a standard eigenvalue problem.

There exists some few studies based on this approach. For crystalline materials the self-interaction vanishes identically for completely delocalized orbitals, but the correction may become important for certain cases where transitions to localized electrons take place (e.g., a Mott transition) and for systems containing well-localized $3d$ or $4f$ valence electrons.

15.11   HYBRID METHODS

Until the middle of the 1980s almost exclusively all density-functional studies were carried through using one of the various local-(spin-)density approximations for the exchange and correlation effects. Most often, the results depended only little on the precise form of the approximation, and the results were in many cases accurate, although there were cases where the density-functional calculations failed.

During the second half of the 1980s different non-local or gradient-corrected approximations were developed. These gave results that — for the systems where the local approximations worked — were as accurate as those, and were in addition able to describe the properties of some of the systems where the local approximations failed. These were first of all systems containing weakly interacting parts, e.g., hydrogen-bonded systems as well as all types of binding energies.

Since the beginning of the 1990s a new approach has been taken (Becke, 1993) that leads to even more accurate energies and structures. This approach combines the Hartree–Fock and the density-functional treatments of exchange effects, whereas correlation effects are treated within the density-functional scheme. It is based on the so-called *adiabatic connection approach* (Harris, 1984), which will be described below.

The Hamilton operator for the electrons of the system of our interest is

$$\hat{H} = \sum_{i=1}^{N} \hat{h}_1(i) + \frac{1}{2} \sum_{i \neq j=1}^{N} \hat{h}_2(i, j). \tag{15.123}$$

We rewrite $\hat{H}$ in terms of the kinetic-energy operator,

$$\hat{T} = -\frac{1}{2} \sum_{i=1}^{N} \nabla_i^2, \tag{15.124}$$

the external potential (which in many cases is that of the nuclei),

$$\hat{V}_{\text{ext}} = \sum_{i=1}^{N} V_{\text{ext}}(i), \tag{15.125}$$

that part of the electrostatic electron–electron interactions that can be described using the classical Coulomb potential,

$$\hat{V}_{\text{C}} = \sum_{i=1}^{N} V_{\text{C}}, \tag{15.126}$$

and the remaining (exchange-correlation) part

$$\hat{V}_{\text{ee}}. \tag{15.127}$$

This gives the total Hamilton operator,

$$\hat{H} = \hat{T}^0 + \hat{V}_{\text{ext}} + \hat{V}_{\text{C}} + \hat{V}_{\text{ee}}, \tag{15.128}$$

where T^0 is that (major) part of the kinetic energy that equals the kinetic energy of the non-interacting particles.

On the other hand, within the Kohn–Sham approach the total Hamilton operator for the (non-interacting) model particles is

$$\hat{H} = \sum_{i=1}^{N} \hat{h}_{\text{eff}}(i). \tag{15.129}$$

It can be rewritten as an expression like that of Eq. (15.128):

$$\hat{H} = \hat{T}^0 + \hat{V}_{\text{ext}} + \hat{V}_{\text{C}} + \hat{V}_{\text{xc}}, \tag{15.130}$$

where the main difference is that we here have a local one-particle potential

$$\hat{V}_{\text{xc}} = \sum_{i=1}^{N} V_{\text{xc}}(i), \tag{15.131}$$

instead of the $\hat{V}_{\text{ee}}$ operator.

The last term in Eq. (15.128) *cannot* be written as a sum of one-particle operators, but is a true many-body operator. It contains not only that part of the electrostatic electron–electron interactions that is not contained in $\hat{V}_{\text{C}}$, but also the difference in the kinetic energy of the system of interacting electrons and that of the non-interacting particles.

We stress that the system whose Hamilton operator is that of Eq. (15.130) is one of non-interacting particles, whereas that of Eq. (15.128) is one of fully interacting particles. The two systems are, however, constructed such that they have the same density.

We will now construct a whole set of systems, characterized by the parameter λ,

$$0 \leq \lambda \leq 1. \tag{15.132}$$

All systems shall have the same density, and they are described by the class of Hamilton operators

$$\hat{H}_\lambda = \hat{T}^0 + \lambda \hat{V}_{ee} + \hat{V}_\lambda + \hat{V}_C + \hat{V}_{ext}. \tag{15.133}$$

For $\lambda = 1$ we have the system of fully interacting particles, and for $\lambda = 0$ that of non-interacting particles. Furthermore, $\hat{V}_\lambda$ is supposed to be a sum of single-particle operators,

$$\hat{V}_\lambda = \sum_{i=1}^{N} \hat{V}_\lambda(i). \tag{15.134}$$

Moreover, it is supposed to be a multiplicative operator, i.e.,

$$\hat{V}_\lambda \phi(\vec{r}) = V_\lambda(\vec{r}) \cdot \phi(\vec{r}), \tag{15.135}$$

and is thus a generalization to interacting particles of the exchange-correlation potential for the non-interacting particles.

In order to guarantee that the same density is obtained independent of λ, $\hat{V}_\lambda$ is so adjusted that this is the case.

Finally, for any λ we assume that we can solve the Schrödinger equation for the ground state,

$$\hat{H}_\lambda \Psi_\lambda = E_\lambda \Psi_\lambda. \tag{15.136}$$

All these assumptions are nothing but assumptions. It is not sure that there does not exist physically relevant systems where they are not satisfied. Thus, it has not been proved that the whole class of potentials V_λ satisfying Eq. (15.135) exists. Moreover, it may also happen that for certain values of λ the ground state is found for other densities than that for $\lambda = 0$ (i.e., that state crossings occur). These cases shall, however, not be considered here and we shall simply proceed assuming that all our assumptions are satisfied.

We shall now make use of the Hellmann–Feynman theorem. This will be described in more detail in a later section, but here we shall just use the result of it, i.e., that

$$\frac{\partial}{\partial \lambda} \langle \Psi_\lambda | \hat{H}_\lambda | \Psi_\lambda \rangle = \left\langle \Psi_\lambda \left| \frac{\partial \hat{H}_\lambda}{\partial \lambda} \right| \Psi_\lambda \right\rangle. \tag{15.137}$$

Then, we can use a formula like

$$f(b) = f(a) + \int_a^b \frac{df}{dx} \, dx \tag{15.138}$$

in obtaining

$$\langle\Psi_1|\hat{H}_1|\Psi_1\rangle = \langle\Psi_0|\hat{H}_0|\Psi_0\rangle + \int_0^1 \frac{\partial}{\partial\lambda}\langle\Psi_\lambda|\hat{H}_\lambda|\Psi_\lambda\rangle\,d\lambda$$

$$= \langle\Psi_0|\hat{H}_0|\Psi_0\rangle + \int_0^1 \langle\Psi_\lambda|\frac{\partial\hat{H}_\lambda}{\partial\lambda}|\Psi_\lambda\rangle\,d\lambda$$

$$= \langle\Psi_0|\hat{H}_0|\Psi_0\rangle + \int_0^1 \langle\Psi_\lambda|\hat{H}_{ee} + \frac{\partial\hat{V}_\lambda}{\partial\lambda}|\Psi_\lambda\rangle\,d\lambda$$

$$= \langle\Psi_0|\hat{H}_0|\Psi_0\rangle + \int_0^1 \langle\Psi_\lambda|\hat{H}_{ee}|\Psi_\lambda\rangle\,d\lambda$$

$$+ \int_0^1 \langle\Psi_\lambda|\frac{\partial\hat{V}_\lambda}{\partial\lambda}|\Psi_\lambda\rangle\,d\lambda. \tag{15.139}$$

Here,

$$\int_0^1 \langle\Psi_\lambda|\frac{\partial\hat{V}_\lambda}{\partial\lambda}|\Psi_\lambda\rangle\,d\lambda = \int_0^1 \int \rho(\vec{r})\frac{\partial V_\lambda(\vec{r})}{\partial\lambda}\,d\vec{r}\,d\lambda. \tag{15.140}$$

Since $\rho(\vec{r})$ per construction is independent of λ (we assumed that independent of λ we obtain the same density), the expression of Eq. (15.140) equals

$$\int \rho(\vec{r})\int_0^1 \frac{\partial V_\lambda(\vec{r})}{\partial\lambda}\,d\lambda\,d\vec{r} = \int \rho(\vec{r})V_1(\vec{r})\,d\vec{r} - \int \rho(\vec{r})V_0(\vec{r})\,d\vec{r}. \tag{15.141}$$

We have accordingly

$$\langle\Psi_1|\hat{H}_1|\Psi_1\rangle = \langle\Psi_0|\hat{H}_0|\Psi_0\rangle + \int_0^1 \langle\Psi_\lambda|\hat{V}_{ee}|\Psi_\lambda\rangle\,d\lambda$$

$$+ \int V_1(\vec{r})\rho(\vec{r})\,d\vec{r} - \int V_0(\vec{r})\rho(\vec{r})\,d\vec{r}. \tag{15.142}$$

We now insert the different terms of the Hamilton operators $\hat{H}_1$ and $\hat{H}_0$ [i.e., Eqs. (15.128) and (15.130), respectively], and by requiring that V_C and V_{ext} are the same for both systems, we obtain

$$\langle\Psi_1|\hat{T}^0|\Psi_1\rangle + \langle\Psi_1|\hat{V}_{ee}|\Psi_1\rangle + \int \rho(\vec{r})V_{ext}(\vec{r})\,d\vec{r} + \int \rho(\vec{r})V_C(\vec{r})\,d\vec{r}$$

$$+ \int \rho(\vec{r})V_1(\vec{r})\,d\vec{r}$$

$$= \langle\Psi_0|\hat{T}^0|\Psi_0\rangle + \int \rho(\vec{r})V_{ext}(\vec{r})\,d\vec{r} + \int \rho(\vec{r})V_C(\vec{r})\,d\vec{r} + \int \rho(\vec{r})V_0(\vec{r})\,d\vec{r}$$

$$+ \int_0^1 \langle\Psi_\lambda|\hat{V}_{ee}|\Psi_\lambda\rangle\,d\lambda + \int \rho(\vec{r})V_1(\vec{r})\,d\vec{r} - \int \rho(\vec{r})V_0(\vec{r})\,d\vec{r}, \tag{15.143}$$

or

$$\langle \Psi_1 | \hat{V}_{ee} | \Psi_1 \rangle = \int_0^1 \langle \Psi_\lambda | \hat{V}_{ee} | \Psi_\lambda \rangle \, d\lambda. \tag{15.144}$$

This is the adiabatic connection formula. The left-hand side is the expectation value for the true wavefunction of the interacting electrons for the operator that describes all exchange and correlation effects (i.e., all those effects that per definition are not included in the Coulomb potential, in the external potential, or in that part of the kinetic energy that is as for the non-interacting particles). On the other hand, the right-hand side is an integral over a whole class of matrix elements for varying strength of electron–electron interactions.

At first sight, Eq. (15.144) appears as bringing nothing but increased complexity. However, the idea behind the hybrid methods is that one may try to approximate the integral of the right-hand side. First of all, one separates $\hat{V}_{ee}$ into two contributions, one from exchange effects, and one from correlation effects:

$$\hat{V}_{ee} = \hat{V}_x + \hat{V}_c, \tag{15.145}$$

and each part is treated separately.

For the correlation effects one uses the 'standard' density-functional approach (i.e., a local or a non-local approximate potential), but for the exchange effects one approximates the integral on the right-hand side of Eq. (15.144). The approximation is very crude and corresponds to considering only the end points, i.e., the integral is approximated according to

$$\int_0^1 f(x) \, dx \simeq a \cdot f(1) - b \cdot f(0), \tag{15.146}$$

where a and b are chosen so that the results are accurate in some sense.

Applying this for the exchange interactions we need to specify the exchange energy for the system of interacting electrons as well as that of the non-interacting particles. The first is given by the expression from the Hartree–Fock theory, i.e.,

$$E_{x;HF} = \frac{1}{2} \sum_{i,j=1}^{N} \int \int \frac{\phi_i^*(\vec{r}_1) \phi_j^*(\vec{r}_2) \phi_i(\vec{r}_2) \phi_j(\vec{r}_1)}{|\vec{r}_1 - \vec{r}_2|} \, d\vec{r}_1 \, d\vec{r}_2. \tag{15.147}$$

On the other hand, for that of the non-interacting particles we will use a local-spin-density approximation, i.e.

$$E_{x;LSD} = E_{x;LSD}[\rho(\vec{r})], \tag{15.148}$$

which is, as indicated, a functional of the electron density.

In order to include even more flexibility in the functional we will consider some of the more recent non-local-density approximations both for exchange and correlation effects. These can be written as the local-density approximation plus something else that depends on ρ, $|\vec{\nabla}\rho|$, $\nabla^2 \rho$, ..., and can, furthermore, be separated into one contribution from exchange effects and one from correlation effects. In total we thus have

$$E_{xc} = a_1 E_{x;HF} + a_2 E_{x;LSD} + E_{c;LSD} + a_x \Delta E_{x;NLSD} + a_c \Delta E_{c;NLSD}, \quad (15.149)$$

where $E_{c;LSD}$ is the local-spin-density approximation for correlation effects, and $\Delta E_{x;NLSD}$ and $\Delta E_{c;NLSD}$ is the non-local corrections to the exchange and correlation energies, respectively. The reason for introducing the coefficients a_x and a_c is that through the term $E_{x;HF}$ we have already included some non-local effects.

By finally requiring that we include all exchange effects, i.e.,

$$a_1 + a_2 = 1, \quad (15.150)$$

we can write E_{xc} as

$$E_{xc} = E_{x;LSD} + a_0 \left[E_{x;HF} - E_{x;LSD} \right] + E_{c;LSD} + a_x \Delta E_{x;NLSD} + a_c \Delta E_{c;NLSD} \quad (15.151)$$

with $a_1 = a_0$ and $a_2 = 1 - a_0$.

Table 15.4 Atomization energies in kcal/mol as obtained by Becke (1993) using his B3-approach for a set of smaller molecules

Molecule	Exp.	Hybrid	Molecule	Exp.	Hybrid
H_2	103.5	101.6	LiH	56.0	52.9
CH	79.9	79.9	$CH_2(^3B_1)$	179.6	184.1
$CH_2(^1A_1)$	170.6	168.2	CH_3	289.2	292.6
CH_4	392.5	393.5	NH	79.0	81.3
NH_2	170.0	173.1	NH_3	276.7	276.8
OH	101.3	101.9	H_2O	219.3	217.0
HF	135.2	133.3	Li_2	24.0	17.9
LiF	137.6	131.7	C_2H_2	388.9	389.0
C_2H_4	531.9	534.3	C_2H_6	666.3	668.7
CN	176.6	176.7	HCN	301.8	302.4
CO	256.2	253.4	HCO	270.3	273.7
H_2CO	357.2	357.9	CH_3OH	480.8	480.8
N_2	225.1	230.0	N_2H_4	405.4	407.2
NO	150.1	151.5	O_2	118.0	123.1
H_2O_2	252.3	249.8	F_2	36.9	35.6
CO_2	381.9	385.1	$SiH_2(^1A_1)$	144.4	142.8
$SiH_2(^3B_1)$	123.4	126.4	SiH_3	214.0	213.3
SiH_4	302.8	300.0	PH_2	144.7	146.8
PH_3	227.4	225.6	H_2S	173.2	172.7
HCl	102.2	102.0	Na_2	16.6	13.2
Si_2	74.0	76.3	P_2	116.1	112.2
S_2	100.7	105.8	Cl_2	57.2	58.6
NaCl	97.5	92.6	SiO	190.5	184.2
CS	169.5	166.9	SO	123.5	126.5
ClO	63.3	66.6	ClF	60.3	60.7
Si_2H_6	500.1	496.7	CH_3Cl	371.0	373.2
CH_3SH	445.1	446.2	HOCl	156.3	156.2
SO_2	254.0	251.4	BeH	46.9	54.5

This expression is due to Becke (1993), and it includes three constants, a_0, a_x, and a_c. Therefore, calculations with this approximation are given the acronym B3xxx, where 'xxx' denotes which non-local-density approximation is used and 'B3' denotes the 3-parameter approximation of Becke. Most often the GGA of Lee, Yang, and Parr (1988) is applied, resulting in the B3LYP functional.

In his original work, Becke found

$$a_0 = 0.20$$

$$a_x = 0.72$$

$$a_c = 0.81. \tag{15.152}$$

These parameters were determined in order to reproduce energies of a given set of smaller molecules as accurately as possible. His results are reproduced in Tables 15.4–6. The atomization energies are the energies required to dissociate the molecules into isolated neutral atoms, the ionization potentials are those required to remove one electron from the highest occupied orbital, and the proton affinities are the energies required to add a proton to the system.

Finally, Table 15.7 gives a set of examples of calculated structural properties with different functionals.

Table 15.5 First ionization potentials in eV as obtained by Becke (1993) using his B3-approach for a set of smaller molecules

Molecule	Exp.	Hybrid	Molecule	Exp.	Hybrid
H	13.60	13.71	He	24.59	24.71
Li	5.39	5.56	Be	9.32	9.02
B	8.30	8.71	C	11.26	11.58
N	14.54	14.78	O	13.61	13.95
F	17.42	17.58	Ne	21.56	21.60
Na	5.14	5.27	Mg	7.65	7.57
Al	5.98	6.12	Si	8.15	8.25
P	10.49	10.57	S	10.36	10.48
Cl	12.97	13.04	Ar	15.76	15.80
CH_4	12.62	12.47	NH_3	10.18	10.12
OH	13.01	13.09	H_2O	12.62	12.54
HF	16.04	15.99	SiH_4	11.00	10.85
PH	10.15	10.31	PH_2	9.82	10.03
PH_3	9.87	9.81	SH	10.37	10.43
$SH_2(^2B_1)$	10.47	10.42	$SH_2(^2A_1)$	12.78	12.64
HCl	12.75	12.74	C_2H_2	11.40	11.23
C_2H_4	10.51	10.36	CO	14.01	14.05
$N_2(^2\Sigma_g)$	15.58	15.77	$N_2(^2\Pi_u)$	16.70	16.65
O_2	12.07	12.46	P_2	10.53	10.41
S_2	9.36	9.58	Cl_2	11.50	11.35
ClF	12.66	12.55	CS	11.33	11.34

Table 15.6 Proton affinities in kcal/mol as obtained by Becke (1993) using his B3-approach for a set of smaller molecules

Molecule	Exp.	Hybrid
H_2	100.8	100.9
C_2H_2	152.3	157.0
NH_3	202.5	204.4
H_2O	165.1	165.7
SiH_4	154.0	153.9
PH_3	187.1	186.1
H_2S	168.8	168.9
HCl	133.6	134.6

Table 15.7 Bond lengths (in Å) for some small molecules as calculated with non-local density functionals (BP86 and BRP86) and with hybrid methods (B3P86 and BR3P86) together with experimental values (Exp.). The results are from Neumann and Handy (1995)

Molecule	BP86	BRP86	B3P86	BR3P86	Exp.
H_2	0.747	0.741	0.744	0.741	0.741
LiH	1.606	1.608	1.597	1.598	1.596
BeH	1.356	1.353	1.350	1.349	1.343
CH	1.137	1.131	1.126	1.123	1.120
NH	1.051	1.046	1.041	1.038	1.036
OH	0.984	0.981	0.974	0.971	0.970
HF	0.931	0.929	0.921	0.919	0.917
Li_2^+	2.747	2.756	2.737	2.747	2.673
LiF	1.580	1.582	1.575	1.576	1.564
CN	1.174	1.170	1.162	1.158	1.172
CO	1.135	1.130	1.125	1.123	1.128
N_2	1.103	1.101	1.092	1.089	1.098
NO	1.160	1.158	1.145	1.143	1.151
O_2	1.224	1.223	1.202	1.199	1.208
F_2	1.422	1.426	1.392	1.389	1.412
HCl	1.290	1.287	1.281	1.279	1.275
Si_2	2.291	2.305	2.263	2.269	2.246
Na_2	3.079	3.054	3.071	3.050	3.079
P_2	1.908	1.915	1.888	1.890	1.893
S_2	1.925	1.936	1.904	1.907	1.889
Cl_2	2.034	2.047	2.012	2.015	1.988
NaCl	2.396	2.406	2.382	2.385	2.361
SiO	1.528	1.528	1.511	1.509	1.510
CS	1.550	1.554	1.534	1.535	1.535
SO	1.512	1.514	1.490	1.489	1.481
ClO	1.601	1.602	1.582	1.578	1.570
ClF	1.667	1.672	1.540	1.639	1.628

16 Some Simplifications and Technical Details

16.1 FROZEN-CORE APPROXIMATION

Within the Hartree–Fock approximation one solves the Hartree–Fock equations

$$\hat{F}\phi_i = \varepsilon_i \phi_i \tag{16.1}$$

with

$$\hat{F} = \hat{h}_1 + \sum_j (\hat{J}_j - \hat{K}_j) \tag{16.2}$$

and

$$\hat{J}_j \phi_i(\vec{r}) = \int \frac{|\phi_j(\vec{s})|^2}{|\vec{r} - \vec{s}|} \, d\vec{s} \phi_i(\vec{r})$$

$$\hat{K}_j \phi_i(\vec{r}) = \int \frac{\phi_j^*(\vec{s})\phi_i(\vec{s})}{|\vec{r} - \vec{s}|} \, d\vec{s} \phi_j(\vec{r}). \tag{16.3}$$

Also when going beyond the Hartree–Fock approximation, i.e., when including correlation effects, some set of one-particle wavefunctions is to be determined. To a varying degree of approximation these can be interpreted as describing electronic orbitals.

These points apply also to the Kohn–Sham equations,

$$\left[\hat{h}_1 + V_C(\vec{r}) + V_{xc}(\vec{r})\right]\phi_i(\vec{r}) = \varepsilon_i \phi_i(\vec{r}) \tag{16.4}$$

where the potentials V_C and V_{xc} depend on the total electron density

$$\rho(\vec{r}) = \sum_{i=1}^{N} |\phi_i(\vec{r})|^2. \tag{16.5}$$

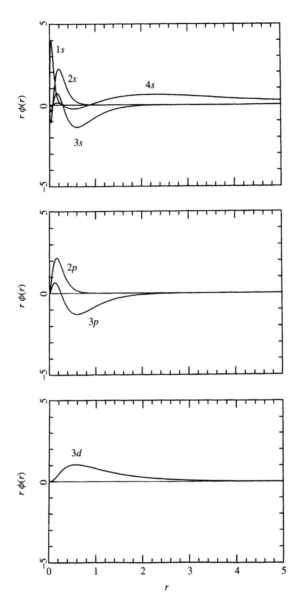

Figure 16.1 s, p, and d functions for an isolated Zn atom. All quantities are given in a.u.

In both cases some of the orbitals will hardly change when passing from the isolated atoms to the compound of interest. Consider, e.g., the orbitals of an isolated Zn atom, Fig. 16.1. For crystalline Zn the interatomic distance is 5.05 a.u. Therefore, the $1s$, $2s$, and $3s$ functions of one Zn atom will hardly overlap with any wavefunctions of a neighbouring Zn atom and, accordingly, be very similar for the isolated atom and for an atom in the crystal. The same holds true for the $2p$ and $3p$ functions. Only the $4s$ and $3d$ functions are so extended that they will overlap with functions of the neighbouring atoms. Another way of putting it is to say that only the single-particle energies of the $4s$ and $3d$ functions lie in the valence region, which, roughly, means above 30–40 eV (i.e., above 1.5 a.u.) below the vacuum level (the energy zero defined as the energy infinitely far away from the system of interest).

Since the core orbitals are almost unchanged when passing from the isolated atoms to the crystal, it is a good approximation to assume that for a given system they are identical to those of the isolated atoms. Thereby, only the wavefunctions of the valence electrons will be optimized in solving the Hartree–Fock, CI, Kohn–Sham, or ... equations. They will still contribute to the matrix elements, but, e.g., for the Hartree–Fock approximation, in applying the variational principle

$$\delta \left\{ \langle \Phi | \hat{H}_e | \Phi \rangle - \sum_{i,j} \lambda_{ij} [\langle \phi_i | \phi_j \rangle - \delta_{i,j}] \right\} = 0, \qquad (16.6)$$

we will only consider variations of the valence orbitals and not of the core orbitals. Similar simplifications apply to the other methods. Effectively, the core electrons will act as providing an external potential in which the 'interesting' valence electrons move.

For the density-functional methods a further simplification arises. Here, the total-electronic-energy expression is a functional of the total electron density and thus, in principle, independent of the orbitals. Therefore, we do not need the explicit expressions for the wavefunctions of the core electrons, but only their total density. Then,

$$E_e[\rho(\vec{r})] = E_e[\rho_c(\vec{r}) + \rho_v(\vec{r})], \qquad (16.7)$$

where ρ_c and ρ_v is the core- and valence-electron density, respectively. ρ_c is kept as for the isolated atoms, and only ρ_v is calculated. This results in a significant simplification so that there is hardly any difference in the computational demands in treating carbon and lead.

In all cases, the approximation of treating the wavefunctions or densities of the core electrons as being identical to their isolated-atom quantities is known as a *frozen-core* approximation.

16.2   PSEUDOPOTENTIALS

In Chapter 3 we discussed the eigenvalue problem in general. One of the results was that the eigenfunctions are orthogonal. This holds, of course, also for the single-particle eigenfunctions to the Hartree–Fock or Kohn–Sham equations, i.e., we have

$$\langle \phi_i | \phi_j \rangle = \delta_{i,j}. \qquad (16.8)$$

For the atomic wavefunctions this orthogonality constraint is automatically satisfied for wavefunctions with different angular dependences, and for those of the
same angular dependence the constraint becomes satisfied by introducing nodes in
the radial dependence of the functions. This can, e.g., be seen above in Fig. 16.1,
where the 1s function is everywhere positive, the 2s function has one radial node,
the 3s function has two radial nodes, and so on. One can therefore think of the
occurrence of the nodes as being caused by the condition that the wavefunctions
are orthogonal.

On the other hand, Fig. 16.1 shows also that those parts of the atomic wavefunctions that are interesting from a chemical-bond point of view lie outside the
regions of the radial nodes. Since it is difficult to expand the rapidly oscillating
wavefunction with all its radial nodes in a given basis set (e.g., Gaussians, plane
waves, etc.), it would be desirable to get rid of the oscillations. For instance, for
a Mo or Nb 5s function, cf. Fig. 16.2, it would be desirable to replace the true
atomic wavefunction (the dashed curve) with a smoother one (the full curve) that
agrees with the true one in the region of the chemical bond.

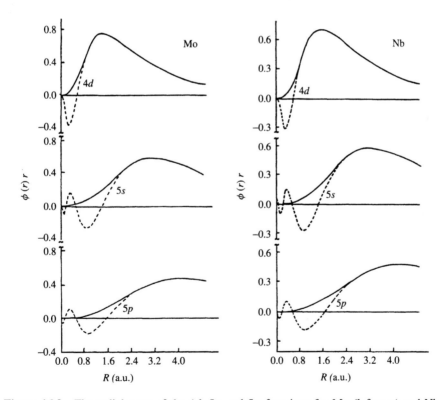

Figure 16.2   The radial parts of the 4d, 5s, and 5p functions for Mo (left part) and Nb
(right part). The dashed curves show the true functions, and the full curves show the
smoothed ones. Reproduced with permission from Fu and Ho (1983). Copyright 1983 by
the American Physical Society

Let us resume. We want to solve the single-particle equation

$$\hat{h}\psi = \varepsilon\psi, \qquad (16.9)$$

where $\hat{h}$ is 'some' single-particle Hamilton-like operator, for instance the Fock operator in Hartree–Fock theory or the Kohn–Sham operator in density-functional theory.

ψ has a number of radial nodes because it has to be orthogonal to the core orbitals. This means that we can write ψ as a linear combination of something that is smooth plus a linear combination of the core wavefunctions,

$$\psi = \phi + \sum_c b_c \psi_c, \qquad (16.10)$$

where the sum runs only over the core orbitals, b_c is some (yet) unknown constants, and ϕ_c is the core wavefunctions.

The constants b_c can be determined by using that the core wavefunctions are orthonormal, and that ψ is orthogonal to the core wavefunctions. We multiply Eq. (16.10) by any of the core wavefunctions, $\psi_{c_0}^*$, and integrate. This gives

$$0 \equiv \langle \psi_{c_0} | \psi \rangle = \langle \psi_{c_0} | \phi \rangle + \sum_c b_c \langle \psi_{c_0} | \psi_c \rangle$$

$$= \langle \psi_{c_0} | \phi \rangle + b_{c_0}, \qquad (16.11)$$

or

$$b_{c_0} = -\langle \psi_{c_0} | \phi \rangle. \qquad (16.12)$$

An alternative way of deriving this is to start out with the smooth function ϕ. This function is, however, to be made orthogonal to all core wavefunctions. Applying a Schmidt orthogonalization leads then to

$$\phi \rightarrow \psi = \phi - \sum_c \langle \psi_c | \phi \rangle \psi_c, \qquad (16.13)$$

i.e., Eqs. (16.10)–(16.12) once more.

We now insert Eq. (16.10) into the left-hand side of the single-particle equation (16.9) and obtain

$$\hat{h}\psi(\vec{r}) = \hat{h}\left[\phi(\vec{r}) + \sum_c b_c \psi_c(\vec{r})\right]$$

$$= \hat{h}\phi(\vec{r}) + \sum_c b_c \hat{h}\psi_c(\vec{r})$$

$$= \hat{h}\phi(\vec{r}) + \sum_c b_c \varepsilon_c \psi_c(\vec{r})$$

$$= \hat{h}\phi(\vec{r}) - \sum_c \int \psi_c^*(\vec{r}_1)\phi(\vec{r}_1)\,d\vec{r}_1 \varepsilon_c \psi_c(\vec{r}). \qquad (16.14)$$

The right-hand side is, equivalently,

$$\varepsilon\psi(\vec{r}) = \varepsilon \left[\phi(\vec{r}) + \sum_c b_c \psi_c(\vec{r}) \right]$$

$$= \varepsilon\phi(\vec{r}) + \sum_c b_c \varepsilon \psi_c(\vec{r})$$

$$= \varepsilon\phi(\vec{r}) - \sum_c \int \psi_c^*(\vec{r}_1)\phi(\vec{r}_1)\,d\vec{r}_1\,\varepsilon\psi_c(\vec{r}). \tag{16.15}$$

The expressions of Eqs. (16.14) and (16.15) are to be identical, giving

$$\hat{h}\phi(\vec{r}) + \sum_c (\varepsilon - \varepsilon_c) \int \psi_c^*(\vec{r}_1)\phi(\vec{r}_1)\,d\vec{r}_1\,\psi_c(\vec{r}) = \varepsilon\phi(\vec{r}). \tag{16.16}$$

The left-hand side contains the single-particle operator $\hat{h}$ plus a non-local operator all acting on $\phi(\vec{r})$. $\hat{h}$ can in turn be written as a kinetic-energy operator plus 'something more'. In total, Eq. (16.16) can therefore be written as

$$\left[-\tfrac{1}{2}\nabla^2 + \hat{V}_{ps}(\vec{r}) \right]\phi(\vec{r}) = \varepsilon\phi(\vec{r}). \tag{16.17}$$

This is exactly a single-particle equation for the smooth part of the wavefunction. The price to be paid is that $\hat{V}_{ps}$ is complicated. It contains both that part of $\hat{h}$ that is not the kinetic-energy part (i.e., in Hartree–Fock theory this is the external potential from the nuclei, the Coulomb potential and the exchange operators, and in density-functional theory it is the external potential from the nuclei, the Coulomb potential and the exchange-correlation potential), and the non-local operator

$$\sum_c (\varepsilon - \varepsilon_c) \int \psi_c^*(\vec{r}_1)\hat{P}_{12}\phi(\vec{r}_1)\,d\vec{r}_1, \tag{16.18}$$

where $\hat{P}_{12}$ is the permutation operator.

However, it is often a good approximation to write

$$\hat{V}_{ps}(\vec{r}) \simeq V_{ps}(\vec{r}), \tag{16.19}$$

i.e., the operator is replaced by a simple multiplicative potential. The development of accurate approximations of this form has been an ongoing research activity for the last decades, and in its present state the approximation is accurate. Nevertheless, there is still a price to be paid, and that is that the potential will depend on the type of function that it shall act on. $V_{ps}(\vec{r})$ of Eq. (16.19) is the so-called pseudopotential (Phillips and Kleinman, 1959). In Fig. 16.3 we show the potentials for the Mo and Nb atoms.

Finally, one may also introduce non-local pseudopotential as a generalization of Eq. (16.18). Then,

$$\hat{V}_{ps}\phi(\vec{r}) = \int V_{ps}(\vec{r}, \vec{r}\,')\phi(\vec{r}\,')\,d\vec{r}\,'. \tag{16.20}$$

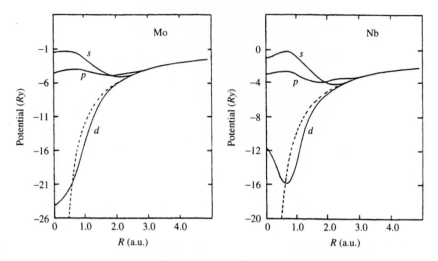

Figure 16.3 The pseudopotentials for a Mo (left part) and Nb atom (right part). The dashed curves show the Coulomb potential from a point charge containing the nucleus and the core electrons, whereas the full curves show the potentials for s, p, and d valence functions of Fig. 16.2. Reproduced with permission from Fu and Ho (1983). Copyright 1983 by the American Physical Society

The great advantage connected with the introduction of pseudopotentials is that the sought wavefunctions become smooth. Therefore, first of all when expanding these in a basis set of plane waves, the number of basis functions needs not to be as large as when the wavefunctions contain all the radial nodes. Accordingly, pseudopotentials are very often combined with plane waves and this combination has developed into a powerful technique for electronic-structure calculations both for molecules and for solids.

16.3 (LINEARIZED)-AUGMENTED-WAVE METHODS: LMTO AND LAPW

By introducing pseudopotentials one arrives at wavefunctions that are smooth and therefore can be expanded more easily in, e.g., plane waves. This is computationally a great simplification but although the number of plane waves that are required for accurate results can be reduced, it still is so large that a direct interpretation of the results in terms of chemical bonds, etc., is non-trivial.

The augmented-wave methods are based on the opposite strategy: By using basis functions that are adapted to the potential their number can be kept small and the results become physically and chemically transparent.

The starting point is the so-called muffin-tin potential. We consider a crystalline compound for which the atoms are fairly closely packed. A two-dimensional analogue is shown in Fig. 16.4. Due to the closed packing, the potential about each atom is only weakly anisotropic and to a first, good approximation one may therefore consider it being spherically symmetric inside each of the non-overlapping, atom-centered, so-called muffin-tin spheres that are shown in Fig. 16.5.

On the other hand, outside all spheres, in the so-called interstitial region, the potential is approximately constant. In total this gives the following approximate form for the potential (Slater, 1937):

$$V(\vec{r}) = \begin{cases} V_{\vec{R}}(|\vec{r} - \vec{R}|) & \text{for } \vec{r} \text{ inside sphere at } \vec{R} \\ V_0 & \text{for } \vec{r} \text{ in interstitial region.} \end{cases} \tag{16.21}$$

Figure 16.4 Schematic representation of a closely packed material

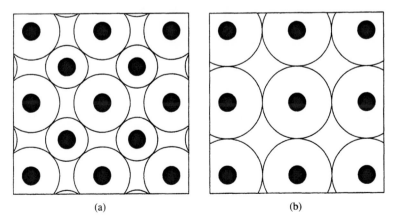

Figure 16.5 Two examples of the separation of space into muffin-tin spheres and the interstitial region

As a first approximation we will seek the exact solutions to the Kohn–Sham equations for this potential,

$$\left[-\tfrac{1}{2}\nabla^2 + V(\vec{r})\right]\psi(\vec{r}) = \varepsilon\psi(\vec{r}). \tag{16.22}$$

Inside any sphere, the potential is spherically symmetric, and Eq. (16.22) is a one-dimensional second-order differential equation that can be solved numerically once the eigenvalue ε is known. This leads to a certain eigenfunction

$$\psi(\vec{r}) = \phi_{l,m,\vec{R}}(\varepsilon; |\vec{r} - \vec{R}|)Y_{lm}\left(\frac{\vec{r} - \vec{R}}{|\vec{r} - \vec{R}|}\right), \tag{16.23}$$

where we have specified that the function depends on the (unknown) eigenvalue ε as well as on the atom $(\vec{R})$ and that it can be factorized in a radial dependence (ϕ) times an angular dependence where the latter is given as a harmonic function Y_{lm}.

In the interstitial region, the Kohn–Sham equations are

$$\left[-\tfrac{1}{2}\nabla^2 + V_0\right]\psi(\vec{r}) = \varepsilon\psi(\vec{r}). \tag{16.24}$$

This equation has, e.g., spherical waves as its solutions:

$$\psi(\vec{r}) = h_l^{(1)}(\kappa|\vec{r} - \vec{R}|)Y_{lm}\left(\frac{\vec{r} - \vec{R}}{|\vec{r} - \vec{R}|}\right), \tag{16.25}$$

where $h_l^{(1)}$ is a spherical Hankel function and where

$$\kappa^2 = 2(\varepsilon - V_0), \tag{16.26}$$

i.e., negative for bound states.

We can now *choose* a certain eigenvalue ε, construct the wavefunctions of Eqs. (16.23) and (16.25) and use these as basis functions in setting up the secular equation

$$\underline{\underline{H}} \cdot \underline{c} = \varepsilon \cdot \underline{\underline{O}} \cdot \underline{c}. \tag{16.27}$$

However, since the basis functions depend on ε, too, this equation becomes highly non-linear in ε and is therefore very complicated to solve.

There exist other formulations of this so-called KKR method (Korringa, 1947; Kohn and Rostoker, 1954), but we shall not discuss those here. Instead we mention that the great simplification to those came with the introduction of the so-called linearized methods (Andersen, 1975). Instead of letting ε above be undetermined, Andersen chose fixed values in the construction of the basis functions. These fixed values may be different for different atoms and l values and were labelled $\varepsilon_{v;l,m,\vec{R}}$. Inside the spheres this leads thus to functions

$$\phi_{l,m,\vec{R}}(\varepsilon_{v;l,m,\vec{R}}; \vec{r}) \equiv \phi_{l,m,\vec{R}}(\varepsilon_{v;l,m,\vec{R}}; |\vec{r} - \vec{R}|)Y_{lm}\left(\frac{\vec{r} - \vec{R}}{|\vec{r} - \vec{R}|}\right). \tag{16.28}$$

Moreover, he introduced

$$\dot{\phi}_{l,m,\vec{R}}(\varepsilon_{v;l,m,\vec{R}};\vec{r}) \equiv \frac{\partial\phi_{l,m,\vec{R}}(\varepsilon_{v;l,m,\vec{R}};\vec{r})}{\partial\varepsilon_{v;l,m,\vec{R}}}. \tag{16.29}$$

Also in the interstitial region he used fixed basis functions, often determined by setting $\kappa = 0$ in Eq. (16.26), whereby the functions of Eq. (16.25) decay as $|\vec{r} - \vec{R}|^{-l-1}$. Initially, this function is defined throughout the complete space (i.e., both inside the muffin-tin spheres and in the interstitial region), but subsequently it is replaced (i.e., augmented) inside any sphere (placed, e.g., at $\vec{R}'$) with linear combinations of the type

$$\sum_{l',m'} S_{l,m,l',m'}(\vec{R},\vec{R}')c_{l',m',\vec{R}'}\left[\phi_{l',m',\vec{R}'}(\varepsilon_{v;l',m',\vec{R}'};\vec{r}) + \omega_{l',m',\vec{R}'}\dot{\phi}_{l',m',\vec{R}'}(\varepsilon_{v;l',m',\vec{R}'};\vec{r})\right],$$
$$\tag{16.30}$$

where $c_{l',m',\vec{R}'}$ and $\omega_{l',m',\vec{R}'}$ are constants that are fixed by the requirement that the function of Eq. (16.25) and that of Eq. (16.30) is continuously and differentiably joined on the boundary of the muffin-tin spheres. Finally, $S_{l,m,l',m'}(\vec{R},\vec{R}')$ are the so-called structure constants for which closed analytical expressions exist and that do not depend on the potential but only on the structure of the system.

Andersen showed that one obtains in effect a method that can be considered linear in energy. This means that the errors due to the replacement of the exact energy ε by the fixed ε_v are of second order in their difference. The method is the so-called linearized muffin-tin orbital (LMTO) method.

A further simplification is obtained when the so-called atomic-sphere approximation (ASA) is introduced (Andersen, 1975). Then, the muffin-tin spheres are made larger so that they slightly overlap and that their total volume equals that of the system. Thus, the volume of the sphere overlaps equals that of the interstitial region. By neglecting both parts the method becomes highly efficient. Subsequently, the errors due to the neglected sphere overlaps and interstitial region can be partly removed using perturbation theory (the so-called combined-correction terms), but this requires that the errors are not too large. For systems that are less closed packed it has therefore become practice to introduce artificial, empty atoms in the interstitial regions, as shown in Fig. 16.6.

For $\kappa = 0$, the LMTOs decay as $|\vec{r} - \vec{R}|^{-l-1}$ and are accordingly long ranged, which may represent a problem for studies of crystalline materials. One may, however, apply a unitary transformation of the LMTOs so that they decay more rapidly (Andersen et al., 1985). For example, for two LMTOs centred on two neighbouring sites and each decaying as $|\vec{r} - \vec{R}|^{-1}$, a certain linear combination will decay more rapidly. In the more general case, one can construct linear combinations of LMTOs centred on several different neighbouring sites and thereby obtain highly localized basis functions. This has led to the so-called tight-binding LMTO (TB-LMTO) method.

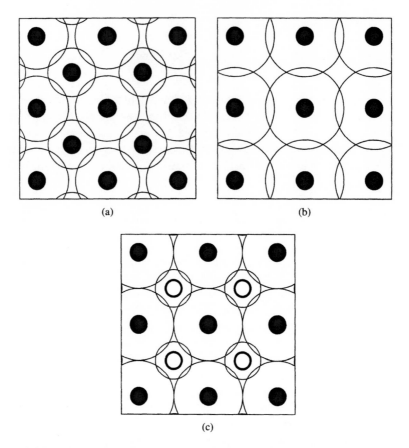

Figure 16.6 The atomic spheres for (a) the system of Fig. 16.5(a) and (b,c) that of Fig. 16.5(b). In (c), so-called empty spheres (with the open circles marking their centers) have also been introduced

As an alternative to the spherical waves of Eq. (16.25), one may also use plane waves,

$$e^{i\vec{k}\cdot\vec{r}}, \tag{16.31}$$

in the interstitial region. All arguments can be carried through as above, and in the linear version one arrives at the so-called linearized augmented plane-wave (LAPW) method.

Finally, the functions we have used are constructed solely from the muffin-tin potential which is an approximation to the potential. One may actually also include the full potential (i.e., deviations from spherical symmetry inside the spheres and from a constant in the interstitial region) and thus consider the LMTOs or LAPWs as nothing but a set of basis functions that are well adapted to the problem at hand. However, the calculation of the required matrix elements

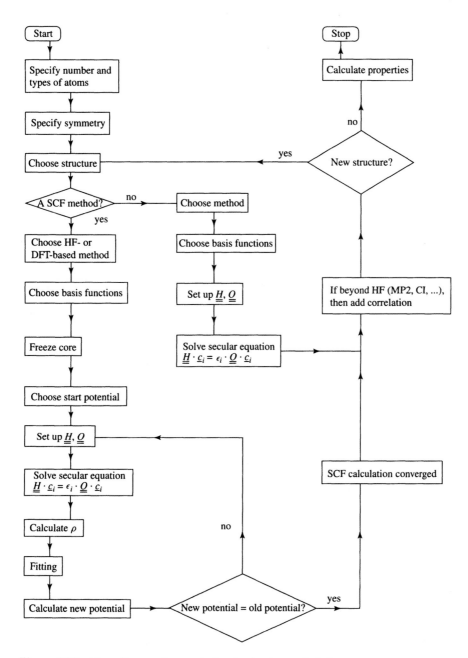

Figure 16.7 Flow diagram for an electronic-structure calculation

becomes easily quite involved (see, e.g., Springborg and Andersen, 1987), and the resulting methods are the FP-LMTO and FP-LAPW methods (FP: full-potential).

16.4 HOW TO CARRY A CALCULATION THROUGH

In Fig. 16.7 we have sketched a flow diagram for how to perform an electronic-structure calculation for a given system.

Once the system is specified, one has to choose some structure. This may be known from experiment or from information about related systems. Subsequently, one has to choose which kind of method is to be applied, whereby the first important question is whether a non-self-consistent method of a self-consistent method shall be used.

In the first case, one specifies then the method (which actually then is a semiempirical one), chooses the basis functions, sets up Hamilton and overlap matrices, and solves the secular equation

$$\underline{\underline{H}} \cdot \underline{c} = \varepsilon \cdot \underline{\underline{O}} \cdot \underline{c}. \tag{16.32}$$

In some cases also correlation effects are to be added subsequently.

If a new structure is to be studied, one determines this (e.g., using the methods to be discussed later), otherwise one may use the solutions (i.e., the eigenvalues ε and the eigenvectors $\underline{c}$ as well as the corrections due to correlation effects when such are included) in calculating the properties of interest.

For the self-consistent methods one has to choose between either Hartree-Fock-based or density-functional-based methods. In both cases, the basis set has to be defined, and one may choose to freeze the core electrons. Furthermore, one has to define an initial potential and from all this information one calculates the Hamilton and overlap matrices, which leads to the secular equation (16.32). The solutions to this defines an electron density

$$\rho(\vec{r}) = \sum_{i=1}^{N} |\psi_i(\vec{r})|^2. \tag{16.33}$$

This density, as well as the individual orbitals ψ_i, can be used in generating a new potential — in some cases fitting procedures are applied, too. If the new potential differs from the previous one with a difference that is larger than a pre-selected threshold, the iterative process of solving the single-particle equations (Hartree–Fock or Kohn–Sham equations) is continued; otherwise it is considered converged. A special case is that of CI calculations for which the correlation effects are added in a non-iterative way at the end of the iterative process of solving the Hartree–Fock equations. But independently of the method, also here one may repeat the whole calculation for a new structure if desired and, ultimately, one calculates the properties of interest.

17 Green's Function

17.1 GENERAL PROPERTIES

Very often we are confronted with the problem of determining the properties of a system that somehow deviates from an idealized behaviour, i.e., without certain perturbations it would have been considerably easier to study its electronic properties. For instance, we want to know the properties of an impurity in an otherwise perfect, periodic crystal or of an interface between two crystalline materials. Other examples include those of determining the time evolution of a system that is suddenly perturbed by some external interaction.

When it is known that the effects of the perturbations are small, perturbation theory (for example only to first order) may provide the sought information, but often this is not the case. Thus, for an impurity in a perfect, ideal crystal the overall properties of the crystalline material hardly change when incorporating the single impurity. But when one is explicitly interested in the properties of the impurity it is not sufficient to apply a low-order perturbation theory.

Green's function techniques offer an alternative way of treating such situations that is not restricted to the case of weak perturbations. In addition, Green's functions themselves possess many interesting properties that make them worth while studying *per se*. In this section we shall give a brief introduction to the general field of Green's functions.

Green's functions are not restricted to electronic-structure calculations. In fact, their origin goes back to before the development of modern quantum theory. In this subsection we shall consider a general example in order to demonstrate how Green's function are introduced and applied, and in Section 17.4 we shall explicitly consider their properties when applying Green's functions in electronic-structure calculations. In order to simplify the presentation we shall in this Section consider a one-dimensional case, but add that the generalization to more dimensions is straightforward.

We consider a general differential operator of the form

$$\hat{A} = \frac{d^n}{dx^n} + f_{n-1}(x)\frac{d^{n-1}}{dx^{n-1}} + \cdots + f_1(x)\frac{d}{dx} + f_0(x). \tag{17.1}$$

In order to relate it to a Hamilton operator, we write the Schrödinger equation (in the one-dimensional case) in the form

$$\frac{d^2}{dx^2}\psi(x) + 2[E - V(x)]\psi(x) = 0. \tag{17.2}$$

Then, the left-hand side has the form

$$\hat{A}\psi = 0 \tag{17.3}$$

with [cf. Eq. (17.1)]

$$n = 2$$
$$f_1(x) = 0$$
$$f_0(x) = 2[E - V(x)]. \tag{17.4}$$

$\hat{A}$ is, as defined in Eq. (17.1), a linear operator, i.e.,

$$\hat{A}[c_1 g_1(x) + c_2 g_2(x)] = c_1\hat{A}g_1(x) + c_2\hat{A}g_2(x). \tag{17.5}$$

Let us now assume that we want to solve the equation

$$\hat{A}y(x) = s(x). \tag{17.6}$$

Let us also assume that the boundary conditions are additive (most often, they are homogenous, i.e., the function and/or some of its derivatives are assumed to vanish on some of the boundaries). This means that if we can solve

$$\hat{A}y_1(x) = s_1(x) \tag{17.7}$$

and

$$\hat{A}y_2(x) = s_2(x) \tag{17.8}$$

with

$$s(x) = s_1(x) + s_2(x), \tag{17.9}$$

then

$$y(x) = y_1(x) + y_2(x) \tag{17.10}$$

is the solution to Eq. (17.6) with its boundary conditions [e.g., if $y_1(x_0) = y_{10}$ and $y_2(x_0) = y_{20}$ are the boundary conditions for Eqs. (17.7) and (17.8), respectively, then $y(x_0) = y_0$ with $y_0 = y_{10} + y_{20}$ is that of Eq. (17.6)]. This is the superposition principle.

In the extreme case, we can decompose $s(x)$ into an infinite set of small parts and solve the corresponding differential equation for each part separately. And this is exactly the principle behind the Green's function. To this end we notice that

$$s(x) = \int s(x')\delta(x' - x)\,dx' \tag{17.11}$$

and seek the solution to the equation

$$\hat{A}G(x, x') = -\delta(x' - x) \tag{17.12}$$

(the minus sign is a convention).

Having determined $G(x, x')$ we apply the superposition principle from above in determining

$$y(x) = -\int G(x, x')s(x')\,dx'. \tag{17.13}$$

$G(x, x')$ is the Green's function. It depends parametrically on x', so that for each value of x' we have to solve equation (17.12).

The right-hand side of Eq. (17.12) seems to cause problems. However, it is fairly easy to treat this. We assume that $G(x, x')$ as a function of x is continuous everywhere and that so are all its derivatives up to, but excluding, order $n - 1$, whereas we will allow the $(n - 1)$th derivative to be discontinuous at x'. We insert then the explicit form of $\hat{A}$ into Eq. (17.12) giving

$$\hat{A}G(x, x') = \left[\frac{d^n}{dx^n} + f_{n-1}(x)\frac{d^{n-1}}{dx^{n-1}} + \cdots + f_1(x)\frac{d}{dx} + f_0(x) \right] G(x, x')$$

$$= \frac{d^n}{dx^n}G(x, x') + f_{n-1}(x)\frac{d^{n-1}}{dx^{n-1}}G(x, x') + \cdots$$

$$+ f_1(x)\frac{d}{dx}G(x, x') + f_0(x)G(x, x')$$

$$\equiv -\delta(x' - x). \tag{17.14}$$

We integrate over an infinitesimally small region around x' and use the fact that all functions except for $(d^n G(x, x'))/(dx^n)$ are smooth in that interval so that the integrals over the infinitesimally small interval vanish. This gives

$$\int_{x'-}^{x'+} \frac{d^n}{dx^n}G(x, x')\,dx + \int_{x'-}^{x'+} f_{n-1}(x)\frac{d^{n-1}}{dx^{n-1}}G(x, x')\,dx + \cdots$$

$$+ \int_{x'-}^{x'+} f_1(x)\frac{d}{dx}G(x, x')\,dx + \int_{x'-}^{x'+} f_0(x)G(x, x')\,dx$$

$$= \frac{d^{n-1}G(x, x')}{dx^{n-1}}\bigg|_{x=x'+} - \frac{d^{n-1}G(x, x')}{dx^{n-1}}\bigg|_{x=x'-} \equiv -1. \tag{17.15}$$

On each side of $x = x'$, $G(x, x')$ is the solution to the simpler equation

$$\hat{A}G(x, x') = 0 \tag{17.16}$$

with the appropriate boundary conditions. Eq. (17.15) together with the requirement that the lower-order derivatives are continuous at $x = x'$ then provide the scheme for matching the two solutions at $x = x'$.

Let us now assume that $\hat{A}$ is a Hermitian and linear operator, and that we — somehow — have obtained the complete set of eigenvalues and orthonormal eigenfunctions,

$$\hat{A}\phi_i = a_i\phi_i. \tag{17.17}$$

We consider then the equation

$$[\hat{A} - a]G(x, x') = -\delta(x' - x). \tag{17.18}$$

x' is *not* a variable but instead a parameter. Equation (17.18) is therefore a differential equation where the dependence of G on x is to be determined, and where x' is to be treated as a kind of constant. This means that, for a given value of x', we can expand $G(x, x')$ in the complete set of functions $\{\phi_i\}$,

$$G(x, x') = \sum_i c_i(x')\phi_i(x), \tag{17.19}$$

where we have indicated that the expansion coefficients c_i will be different for different values of x', i.e., they depend on x'.

We insert Eq. (17.19) into Eq. (17.18),

$$\begin{aligned}
-\delta(x' - x) &= [\hat{A} - a]G(x, x') \\
&= [\hat{A} - a]\sum_i c_i(x')\phi_i(x) \\
&= \sum_i c_i(x')[\hat{A} - a]\phi_i(x) \\
&= \sum_i c_i(x')(a_i - a)\phi_i(x).
\end{aligned} \tag{17.20}$$

We multiply each side by $\phi_j^*(x)$ (i.e., any of the eigenfunctions), integrate over x, and use the fact that the functions ϕ_i are orthonormal. This gives

$$-\int \delta(x' - x)\phi_j^*(x)\,dx = \sum_i c_i(x')(a_i - a)\int \phi_j^*(x)\phi_i(x)\,dx, \tag{17.21}$$

or

$$-\phi_j^*(x') = c_j(x') \cdot (a_i - a), \tag{17.22}$$

i.e.,

$$c_j(x') = \frac{\phi_j^*(x')}{a - a_i}. \tag{17.23}$$

In total, therefore, [cf. Eq. (17.19)]

$$G(x, x') = \sum_i \frac{\phi_i^*(x')\phi_i(x)}{a - a_i}. \tag{17.24}$$

We have not specified the constant a in any way and may actually vary it freely. Therefore, it is useful to specify explicitly that G also depends on this parameter [cf. Eq. (17.18)],

$$G(x, x') \rightarrow G(x, x', a) \tag{17.25}$$

where G is a solution to Eq. (17.18).

Since $\hat{A}$ is assumed to be Hermitian and linear, the eigenvalues a_i are real. Therefore,

$$G(x, x', a) \rightarrow \infty \quad \text{for} \quad a \rightarrow a_i. \tag{17.26}$$

This problem can be circumvented by introducing a small imaginary constant $\pm i\delta$ and defining

$$G^{\pm}(x, x', a) = G(x, x', a \pm i\delta) = \sum_i \frac{\phi_i^*(x')\phi_i(x)}{a - a_i \pm i\delta}. \tag{17.27}$$

17.2 TWO EXAMPLES

In order to illustrate the concepts of the preceding subsection we shall consider two simple examples. We first study the second-order differential equation

$$\frac{d^2 y}{dx^2} = A \cdot \cos(\alpha x) \tag{17.28}$$

with the (homogeneous) boundary conditions

$$y(-l) = y(l) = 0. \tag{17.29}$$

The corresponding Green's function is calculated from the equation

$$\frac{d^2}{dx^2} G(x, x') = -\delta(x' - x). \tag{17.30}$$

For $x \neq x'$, the right-hand side vanishes and, accordingly, we can find the general solution to Eq. (17.30) easily:

$$G(x, x') = \begin{cases} b_- x + c_- & \text{for } x < x' \\ b_+ x + c_+ & \text{for } x > x'. \end{cases} \tag{17.31}$$

Since the function is to vanish for $x = \pm l$, we obtain

$$G(x, x') = \begin{cases} b_- \cdot (x + l) & \text{for } x < x' \\ b_+ \cdot (x - l) & \text{for } x > x'. \end{cases} \tag{17.32}$$

Moreover, continuity at $x = x'$ gives

$$b_+ = b \cdot (x' + l)$$
$$b_- = b \cdot (x' - l) \tag{17.33}$$

with the new parameter b. Finally, the first-order derivatives are discontinuous at $x = x'$,

$$\left. \frac{dG(x, x')}{dx} \right|_{x=x'+} - \left. \frac{dG(x, x')}{dx} \right|_{x=x'-} \equiv -1 \tag{17.34}$$

giving

$$b = -\frac{1}{2l}. \tag{17.35}$$

On the other hand, the general solution to Eq. (17.28) can easily be calculated,

$$y = d \cdot x + e - \frac{A}{\alpha^2} \cos(\alpha x), \tag{17.36}$$

and by using the boundary conditions (17.29) we obtain

$$y = \frac{A}{\alpha^2} \left[\cos(\alpha l) - \cos(\alpha x) \right]. \tag{17.37}$$

Alternatively, this solution can be calculated using the Green's function,

$$y = -\int G(x, x') A \cos(\alpha x') \, dx'$$
$$= -\int_x^l \frac{-1}{2l} (x' - l)(x + l) A \cos(\alpha x') \, dx'$$
$$- \int_{-l}^x \frac{-1}{2l} (x' + l)(x - l) A \cos(\alpha x') \, dx'. \tag{17.38}$$

Evaluating this involves only the calculation of some standard integrals and will therefore not be done here. One does, however, arrive at the result of Eq. (17.37)

As the second example we consider the operator

$$\frac{d^2}{dx^2} \tag{17.39}$$

together with the boundary conditions

$$y(0) = y(l) = 0 \tag{17.40}$$

The eigenvalue equation for this operator is

$$\frac{d^2 y_n(x)}{dx^2} = a_n y_n(x) \tag{17.41}$$

which has the solutions

$$y_n(x) = \sin(\alpha_n x) \qquad (17.42)$$

with

$$\alpha_n = n \cdot \frac{\pi}{l} \qquad (17.43)$$

and the eigenvalues

$$a_n = -\alpha_n^2. \qquad (17.44)$$

For this, we define the Green's function through

$$\left[\frac{d^2}{dx^2} - a\right] G(x, x') = -\delta(x' - x). \qquad (17.45)$$

This has the general solution

$$G(x, x') = \begin{cases} c_1 \sin(bx) + c_3 \cos(bx) & \text{for } x < x' \\ c_2 \sin(bx) + c_4 \cos(bx) & \text{for } x > x'. \end{cases} \qquad (17.46)$$

Here, we have introduced

$$b = \sqrt{-a}. \qquad (17.47)$$

The boundary conditions give

$$c_3 = 0$$

$$c_4 = -\tan(bl) \cdot c_2. \qquad (17.48)$$

Finally, the fact that $G(x, x')$ is continuous at $x = x'$, whereas the first-order derivatives are discontinuous (with a difference of -1) gives

$$\sin(bx')c_1 + \left[\tan(bl)\cos(bx') - \sin(bx')\right]c_2 = 0$$

$$\cos(bx')c_1 + \left[-\tan(bl)\sin(bx') - \cos(bx')\right]c_2 = b. \qquad (17.49)$$

The determinant for this 2×2 set of linear equations is

$$-\tan(bl), \qquad (17.50)$$

which vanishes exactly when a is one of the eigenvalues above in Eq. (17.44), making it impossible to calculate the Green's function in that case (i.e., the Green's function diverges as it should). Otherwise, it can be calculated with no problem, leading to an expression that at first appears different from that Eq. (17.24), but ultimately they will be identical.

17.3 RESIDUE THEORY

Through the introduction above of the small imaginary constant $\pm i\delta$, the Green's function becomes a function of a complex variable $a \pm i\delta$. We shall use some theorems for functions of complex variables in order to arrive at some useful results for the Green's function. We shall therefore start by recalling some of these results for functions of complex variables.

We consider a function f of a complex variable z. The function f is assumed to be 'reasonably well behaved', but it may have some poles $z_1, z_2, \ldots, z_N$, where the function diverges. N may be either finite or infinite, but we assume that the positions of the poles are well separated, i.e., they do not form a continuum.

For a function that is smooth (i.e., does not have any poles) we can expand it in a Taylor series,

$$f(z) = \sum_{n=0}^{\infty} f_n \cdot (z - z_0)^n, \tag{17.51}$$

where f_n is related to the nth derivative of f. We assume equivalently that we near any of the poles above, z_j, can write

$$f(z) = \sum_{n=-\infty}^{\infty} f_{n,j} \cdot (z - z_j)^n, \tag{17.52}$$

where we in contrast to Eq. (17.51) also have negative exponents. The series (17.52) is called a Laurent series.

Figure 17.1 shows an example of the positions of the poles of such a function $f(z)$. We have here five poles, numbered 1–5. If we now calculate the integral

$$\oint_{\delta C} f(z) \, dz, \tag{17.53}$$

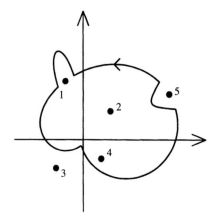

Figure 17.1 The positions of the poles of a function of a complex variable plus a curve over which the function is integrated

where δC is the curve in Fig. 17.1, this integral equals

$$\oint_{\delta C} f(z)\, dz = \sum_{j \in C} i\theta_j f_{-1,j}. \tag{17.54}$$

This (Cauchy's residue theorem) is one of the fundamental theorems of the theory of complex functions.

In Eq. (17.54), the summation goes over only those poles that are inside the curve δC. Furthermore, θ_j is the angle with which this is circumscribed (with sign!). In the example of Fig. 17.1 we have $\theta_1 = \theta_2 = \theta_4 = 2\pi$ and $\theta_3 = \theta_5 = 0$.

Only points z_j that are lying directly on the curve δC will have $\theta_j \neq 0, 2\pi$, and in that case the value of θ_j depends on how one chooses to circumvent z_j. Most important is the case of an integral along the real axis where the integrand has some poles. In Fig. 17.2 we show the case of a single pole, where one may choose to circumvent it as shown in either of the two parts of the figure. In the limit that the radius of the small semicircle circumventing the pole approaches zero one approaches the case of integrating through the pole. In the upper case, the contribution from the pole is calculated using Eq. (17.54) with $\theta_j = \pi$, in the lower case $\theta_j = -\pi$. We stress that, in contrast to the Cauchy's residue theorem above [Eq. (17.53)], which is valid independently of how we circumscribe the poles, the theorem here is only valid when the radii of the semicircles $\rightarrow 0$. Moreover, θ_j may then actually take other values depending on the angle with which the pole is circumscribed (but still only as long as the radius $\rightarrow 0$).

As an illustration we consider the example of Fig. 17.2. Here, we can write the integral along the upper curve as

$$\oint_{\delta C} f(z)\, dz = \oint_{\delta C}' f(z)\, dz + i\pi f_{i-1} \equiv 2\pi i f_{-1}, \tag{17.55}$$

where the prime on the integrand indicates that we exclude the small part through the pole in the limit that this part goes to 0. For the lower curve in Fig. 17.2 we find equivalently

$$\oint_{\delta C} f(z)\, dz = \oint_{\delta C}' f(z)\, dz - i\pi f_{-1} \equiv 0. \tag{17.56}$$

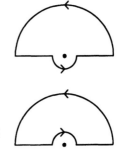

Figure 17.2 Two ways of considering the limit of integrating through a pole

It is easily recognized that these two expressions are mutually consistent. In the limit that we integrate over the pole we obtain $i\pi f_{-1}$ in accordance with Eqs. (17.55) and (17.56).

The quantities $f_{-1,j}$ in Eq. (17.54) are the coefficients in the Laurent series to the terms $(z - z_j)^{-1}$. These are the so-called residues.

This theorem is highly relevant when calculating integrals like

$$\int_{-\infty}^{\infty} f(x)\,dx \qquad (17.57)$$

where $f(x)$ now is a function of a real variable x, that may diverge at some points on the real axis. If we can generalize the function so that it becomes a function of a complex variable z, we can calculate the integral by considering the curve of Fig. 17.3. Assuming that $f(z)$ vanishes on the upper semicircle in the limit that its radius goes to ∞, the integral (17.57) can easily be calculated using the residues along the x axis. Of course, the large semicircle in Fig. 17.3 may just as well be replaced by one that lies below the real axis, and (some of) the semicircles about the poles may also be replaced by ones that lie above the real axis. Ultimately, the same result is obtained, as it should be!

For the Green's function

$$G^{\pm}(x, x', a) = G(x, x', a \pm i\delta) = \sum_i \frac{\phi_i^*(x')\phi_i(x)}{a - a_i \pm i\delta} \qquad (17.58)$$

the poles are at $a_i \pm i\delta$ and the residues equals $\phi_i^*(x')\phi_i(x)$.

Therefore, in the limit $\delta \to 0$, we have

$$\sum_i \phi_i^*(x')\phi_i(x) = \frac{1}{\pi} \lim_{\delta \to 0} \mathrm{Im} \int_{-\infty}^{\infty} G^-(x, x', a)\,da. \qquad (17.59)$$

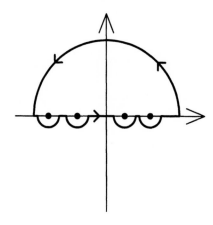

Figure 17.3 An example of the application of the residue theorem for the calculating of an integral along the real axis

Furthermore, the following *formal* identity,

$$\lim_{\varepsilon \to 0} \text{Im} \frac{1}{x \pm i\varepsilon} = \mp \pi \delta(x) \tag{17.60}$$

gives

$$\frac{1}{\pi} \lim_{\delta \to 0} \text{Im} G^-(x, x', a) = \sum_i \phi_i^*(x') \phi_i(x) \delta(a - a_i). \tag{17.61}$$

Subsequently, we may set $x = x'$, and integrate over x, resulting in the so-called density of states

$$D(a) = \frac{1}{\pi} \lim_{\delta \to 0} \text{Im} \int G^-(x, x, a) \, dx. \tag{17.62}$$

Comparing with Eq. (17.61), it is obvious that $D(a)$ equals 0 if there is no eigenfunction ϕ_j with an eigenvalue $a_j = a$. On the other hand, if one or more such eigenfunctions exist, $D(a)$ is a sum of δ-functions whose strength equals their number assuming that the eigenfunctions ϕ_i are normalized. For infinite systems, the eigenvalues a_j form one or more continua, and $D(a) \, da$ then gives the number of eigenfunctions with eigenvalues in the interval $[a, a + da]$.

Equation (17.60) can be proved as follows. We consider any function that is smooth at $x = 0$. Then residue theory gives

$$\lim_{\delta \to 0} \int \frac{f(x)}{x + i\delta} \, dx = i\pi \lim_{\delta \to 0} f(0). \tag{17.63}$$

Simultaneously

$$\int f(x) \delta(x) \, dx = f(0), \tag{17.64}$$

and identifying these two expressions leads immediately to one of the expressions in Eq. (17.60). The other one can be derived in an equivalent way.

17.4   GREEN'S FUNCTION AND ELECTRONIC STRUCTURE

So far our presentation of Green's function has been general. It may already have been guessed that ultimately we will let the operator $\hat{A}$ be the Hamilton operator, so that the eigenvalues a_i become the energy eigenvalues E_i. Alternatively, we may let $\hat{A}$ be the Fock operator if using the Hartree–Fock approach, or the effective Hamilton operator if using a density-functional approach within the formulation of Kohn and Sham. Then the eigenvalues a_i become the single-particle energies ε_i.

x is moreover no longer a one-dimensional variable, but for the single-particle models it will have the coordinates of a single particle (three position-space and one spin coordinate).

Equation (17.60) then defines the density of states as a function of energy,

$$D(E) = \frac{1}{\pi} \lim_{\delta \to 0} \text{Im} \int G^-(\vec{x}, \vec{x}, E) \, d\vec{x}, \tag{17.65}$$

where for simplicity we shall use E irrespective of whether we use a single-particle model or not.

For a single-particle model we define the Fermi energy ε_F so that

$$\int_{-\infty}^{\varepsilon_F} D(E) \, dE = N, \tag{17.66}$$

with N being the total number of particles.

The electron density is then given by

$$\rho(\vec{x}, \vec{x}) = \frac{1}{\pi} \lim_{\delta \to 0} \text{Im} \int_{-\infty}^{\varepsilon_F} G^-(\vec{x}, \vec{x}, E) \, dE. \tag{17.67}$$

Accordingly, the Green's function contains most of the information that is needed in describing the properties of the system of interest.

17.5 DYSON'S EQUATION

Our ultimate goal is to apply the Green's function in including extra effects for which we give some examples below. We assume that we have solved the Schrödinger equation for some unperturbed system,

$$\hat{H}_0 \Psi_0(\vec{x}) = E_0 \Psi_0(\vec{x}). \tag{17.68}$$

$\vec{x}$ is a shorthand notation for all coordinates.

Subsequently, we introduce the extra effect. This can, e.g., be that due to adding, removing, or exciting an electron, or that due to changing the structure by adding, removing, or substituting one or more atoms, or that due to displacing one or more atoms, or that due to surfaces or interfaces, or that due to some external interaction that furthermore can be time-dependent, and so on.

Denoting the potential due to the extra effect by $V(\vec{x}, t)$, the (time-dependent) Schrödinger equation becomes

$$\left(-i\frac{\partial}{\partial t} + \hat{H}_0 \right) \Psi(\vec{x}, t) = -V(\vec{x}, t)\Psi(\vec{x}, t). \tag{17.69}$$

In the spirit of Section 17.1 we shall replace the right-hand side of Eq. (17.69) with δ-functions in order to determine the Green's function. To this end it useful to Fourier transform the Green's function so that the energy dependence is replaced by a time dependence, i.e., we consider

$$G^{\pm}(\vec{x}, \vec{x}', t) = \frac{1}{2\pi} \int_{-\infty}^{\infty} e^{-iEt} G^{\pm}(\vec{x}, \vec{x}', E) \, dE. \tag{17.70}$$

Notice that we shall distinguish between the two Green's functions only through their last argument.

Instead of Eq. (17.69) we consider accordingly

$$\left(-i\frac{\partial}{\partial t} + \hat{H}_0\right) G_0^\pm(\vec{x}, \vec{x}', t) = -\delta(\vec{x}' - \vec{x})\delta(t). \qquad (17.71)$$

Since $\Psi_0(\vec{x})e^{-Et}$ is the solution to Eq. (17.69) with vanishing right-hand side, the equivalent of Eq. (17.13) gives the solution to Eq. (17.69) in the form

$$\Psi(\vec{x}, t) = \Psi_0(\vec{x})e^{-iEt} + \int\int G_0^+(\vec{x}, \vec{x}'', t - t')V(\vec{x}'', t')\Psi(\vec{x}'', t')\,d\vec{x}''dt'. \qquad (17.72)$$

Notice, that since the right-hand side of Eq. (17.69) contains the unknown wavefunction, so does the integrand of Eq. (17.72).

Alternatively, one may also work entirely with Green's functions. This has the advantage that the electron density and the density of states then can be directly extracted. In addition to G_0 defined above, we also introduce the Green's function for the full problem:

$$\left[-i\frac{\partial}{\partial t} + \hat{H}_0 + V(\vec{x}, t)\right] G^\pm(\vec{x}, \vec{x}'', t) = -\delta(\vec{x}'' - \vec{x})\delta(t). \qquad (17.73)$$

One can then show (this will not be done here) that the Green's functions are related through

$$G^\pm(\vec{x}, \vec{x}', t) = G_0^\pm(\vec{x}, \vec{x}', t)$$

$$+ \int\int G_0^\pm(\vec{x}, \vec{x}'', t' - t)V(\vec{x}'', t')G^\pm(\vec{x}'', \vec{x}', t')\,d\vec{x}''\,dt'. \qquad (17.74)$$

This is one form of the so-called Dyson's equation (Dyson, 1949).

For time-independent perturbations, i.e., when

$$V(\vec{x}, t) = V(\vec{x}), \qquad (17.75)$$

it is useful to return to the energy-dependent Green's functions by Fourier transforming Eq. (17.74). This leads to

$$G^\pm(\vec{x}, \vec{x}', E) = G_0^\pm(\vec{x}, \vec{x}', E) + \int G_0^\pm(\vec{x}, \vec{x}'', E)V(\vec{x}'')G^\pm(\vec{x}'', \vec{x}', E)\,d\vec{x}''. \qquad (17.76)$$

We have not attempted to derive these equations, since we consider this beyond the scope of this presentation. Here, we want to emphasize the usefulness of the concept of Green's functions and to present the main equation (Dyson's equation) according to which one — in principle — can calculate the Green's function. Once the Green's function has been calculated we can apply the methods of the preceding subsection in calculating other properties of the system of interest.

Since almost all electronic-structure methods apply some set of predefined basis functions, we shall now discuss Dyson's equation in terms of such functions.

17.6 BASIS FUNCTIONS

We consider the single-particle Schrödinger equation

$$\hat{H}_0 \psi_{0,k}(\vec{r}) = \varepsilon_{0,k} \psi_{0,k}(\vec{r}). \tag{17.77}$$

We assume that the eigenfunctions have been calculated (most likely only approximately through the application of the variational principle) by expanding them in some basis functions,

$$\psi_{0,k}(\vec{r}) = \sum_j c_{kj} \chi_j(\vec{r}). \tag{17.78}$$

The corresponding Green's function is

$$
\begin{aligned}
G_0^\pm(\vec{r}, \vec{r}', \varepsilon) &= \sum_k \frac{\psi_{0,k}^*(\vec{r}')\psi_{0,k}(\vec{r})}{\varepsilon - \varepsilon_k \pm i\delta} \\
&= \sum_{m,n} \chi_m^*(\vec{r}')\chi_n(\vec{r}) \sum_k \frac{c_{km}^* c_{kn}}{\varepsilon - \varepsilon_k \pm i\delta} \\
&\equiv \sum_{m,n} g_{0,nm}(\varepsilon) \chi_m^*(\vec{r}')\chi_n(\vec{r}).
\end{aligned} \tag{17.79}
$$

We assume now that the system is perturbed locally and that this perturbation can be described with the extra potential $V(\vec{r})$. For the new Green's function we seek an expansion like that of Eq. (17.79),

$$G^\pm(\vec{r}, \vec{r}', \varepsilon) = \sum_{m,n} g_{nm}(\varepsilon) \chi_m^*(\vec{r}')\chi_n(\vec{r}). \tag{17.80}$$

Dyson's equation (17.75) then gives

$$
\begin{aligned}
\sum_{mn} g_{nm}(\varepsilon)\chi_m^*(\vec{r}')\chi_n(\vec{r}) &= \sum_{mn} g_{0,nm}(\varepsilon)\chi_m^*(\vec{r}')\chi_n(\vec{r}) \\
&+ \sum_{mn} \sum_{m_1 n_1} g_{0,mn_1}(\varepsilon) g_{m_1 n}(\varepsilon)\chi_m^*(\vec{r}')\chi_n(\vec{r}) \\
&\times \int \chi_{m_1}^*(\vec{r}'')V(\vec{r}'')\chi_{n_1}(\vec{r}'')\, d\vec{r}''.
\end{aligned} \tag{17.81}
$$

This (matrix-)equation can, in principle, be solved for those coefficients g_{mn} that are supposed changing due to the perturbation V. It should be stressed that the equation (and therefore the solutions) depends on the energy ε and therefore has to be solved for a complete set of energies. However, once it has been solved, one does not need to reformulate the Green's function in terms of single-particle wavefunctions but can calculate electron densities, densities of states, etc., using the methods of Section 17.4.

We multiply Eq. (17.81) by $\chi_k(\vec{r}')\chi_l^*(\vec{r})$ and integrate over $\vec{r}$ and $\vec{r}'$ which results in

$$\sum_{mn} g_{nm}(\varepsilon)O_{mk}O_{ln} = \sum_{mn} g_{0,nm}(\varepsilon)O_{mk}O_{ln} + \sum_{mn}\sum_{m_1n_1} g_{0,n_1m}(\varepsilon)g_{nm_1}O_{mk}O_{ln}V_{m_1n_1} \tag{17.82}$$

with

$$O_{kl} = \langle \chi_k | \chi_l \rangle$$
$$V_{kl} = \langle \chi_k | V | \chi_l \rangle. \tag{17.83}$$

Equation (17.82) can be written in matrix form

$$\underline{\underline{O}} \cdot \underline{\underline{G}}(\varepsilon) \cdot \underline{\underline{O}} = \underline{\underline{O}} \cdot \underline{\underline{G}}_0(\varepsilon) \cdot \underline{\underline{O}} + \underline{\underline{O}} \cdot \underline{\underline{G}}(\varepsilon) \cdot \underline{\underline{V}} \cdot \underline{\underline{G}}_0(\varepsilon) \cdot \underline{\underline{O}}, \tag{17.84}$$

or, by multiplying by $\underline{\underline{O}}^{-1}$ both to the left and the right,

$$\underline{\underline{G}}(\varepsilon) = \underline{\underline{G}}_0(\varepsilon) + \underline{\underline{G}}(\varepsilon) \cdot \underline{\underline{V}} \cdot \underline{\underline{G}}_0(\varepsilon), \tag{17.85}$$

or

$$\underline{\underline{G}}(\varepsilon) \cdot \left[\underline{\underline{1}} - \underline{\underline{V}} \cdot \underline{\underline{G}}_0(\varepsilon) \right] = \underline{\underline{G}}_0(\varepsilon). \tag{17.86}$$

This is a matrix form for the Dyson equation.

The perturbation $V(\vec{r})$ may be spatially localized. This means that the matrix elements V_{kl} are only then non-vanishing when the basis functions χ_k and χ_l are centered in a certain finite region about the perturbation. Then, $\underline{\underline{V}}$ is non-zero only inside a finite block, and Eq. (17.86) becomes a finite-sized matrix equation whose size is determined by the size of the perturbation. And this equation can then be solved.

In a practical calculation one calculates the Green's function for the unperturbed system using some standard method we have described above. This results in $\underline{\underline{G}}_0(\varepsilon)$. Subsequently, the perturbation is introduced (e.g., one atom in an infinite crystal is replaced by another), and the finite region inside which the effects of the perturbation will be studied is defined. The Green's function of the perturbed system, i.e., $\underline{\underline{G}}(\varepsilon)$, is calculated from Eq. (17.86). From this, the electron density, etc., can be calculated, and when, e.g., a density-functional method is applied, the changes in the electron density may result in a changed perturbation potential $V(\vec{r})$ so that Dyson's equation (17.86) has to be solved once more. This procedure is continued until self-consistency is reached, in which case the electronic properties wanted can be extracted.

Part III
SPECIAL PROPERTIES

Having described the basic methods in detail, in this and the last part we shall turn to various applications. Therefore, we shall not describe the methods in such detail as hitherto. The idea is rather that the reader will obtain a feeling for the problems and their solutions that is specific for the various applications. The description of the theoretical approaches is enhanced by various examples that illustrate what kind of problems need to be tackled and what kind of results are obtained, but we shall not discuss in detail the specific problems and questions related to the individual example. In the present part we concentrate on the study of different properties.

18 Acidity and Basicity; Hardness and Softness

18.1 HARDNESS AND SOFTNESS

In the present section we shall study some issues that are often discussed in chemistry but that through the development of density-functional theory have obtained a firmer foundation (Parr *et al.*, 1978) and, furthermore, been extended with the possibility of studying the systems and properties in more detail. The quantities that we shall study are first of all relevant when describing chemical reactions and were originally introduced by Pearson (1963).

Within the density-functional formalism we considered a system of electrons moving in a given external potential (typically that of the nuclei). By studying a model system of non-interacting particles, the total electron density could be written as a sum over occupied (Kohn–Sham) orbitals,

$$\rho(\vec{r}) = \sum_{i=1}^{N} |\phi(\vec{r})|^2 = \sum_i n_i \phi_i^*(\vec{r}) \phi_i(\vec{r}), \qquad (18.1)$$

where we have introduced the orbital occupancies n_i that for the ground state equal 0 or 1.

The orbitals ϕ_i above are those of the model system (i.e., the one that was used in deriving the Kohn–Sham equations), but since experience has shown that they are very similar to those of the real system, we shall here treat them as being identical to those. This means that there may be smaller inaccuracies in our results, but since we in the present section rather are interested in qualitative aspects, this will not be important.

In Chapter 15 we showed that the total electronic energy is a functional of the electron density of Eq. (18.1), i.e.

$$E_e = E_e[\rho(\vec{r})], \qquad (18.2)$$

and we also showed that for a given system, i.e., for a given total number of electrons

$$N = \int \rho(\vec{r}) \, d\vec{r} \tag{18.3}$$

and given external potential

$$V_{ext}(\vec{r}), \tag{18.4}$$

E_e as a functional of all possible densities that satisfy Eq. (18.3) is at a minimum for the true ground-state density. This led to the variational principle,

$$\frac{\delta}{\delta \rho(\vec{r})} \left\{ E_e[\rho(\vec{r})] - \mu \left[\int \rho(\vec{r}) \, d\vec{r} - N \right] \right\} = 0. \tag{18.5}$$

Here, the Lagrange multiplier μ that takes care of the constraint (18.3) was found to be the chemical potential for the electrons,

$$\mu = \frac{\delta E_e[\rho(\vec{r})]}{\delta \rho(\vec{r})} = \left(\frac{\partial E_e}{\partial N} \right)_{V_{ext}}, \tag{18.6}$$

where in the second identity we have used the fact that the chemical potential is the energy required for adding a particle keeping everything else constant, in particular keeping the external potential constant.

We can actually study in more detail what happens when changing the number of electrons. This process is very important when studying the interactions between two systems, and our analysis is thus relevant when trying to explore how and whether two systems will interact and whether electrons will be transferred between them.

When changing the total number of electrons, the total electronic energy will change. The remaining part of the total energy, i.e., the electrostatic energy of the nucleus–nucleus interactions, will not change when assuming that the nuclei do not move. Our analysis focuses thus on the electrons. We can expand the total electronic energy in a Taylor series,

$$\Delta E_e \equiv E_e(N + \Delta N) - E_e(N)$$

$$= \left(\frac{\partial E_e}{\partial N} \right)_{V_{ext}} \Delta N + \frac{1}{2} \left(\frac{\partial^2 E_e}{\partial N^2} \right)_{V_{ext}} (\Delta N)^2 + \cdots$$

$$= \mu \Delta N + \eta (\Delta N)^2 + \cdots. \tag{18.7}$$

Here,

$$\eta \equiv \frac{1}{2} \left(\frac{\partial^2 E_e}{\partial N^2} \right)_{V_{ext}} = \frac{1}{2} \left(\frac{\partial \mu}{\partial N} \right)_{V_{ext}} \tag{18.8}$$

is the chemical *hardness*. It is seen to be directly related to the changes in the total electronic energy when the number of electrons is changed. It is large (the system is hard) when it costs much energy to change the number of electrons.

Equivalent to the hardness, also a *softness* can be defined,

$$S = \frac{1}{2\eta} = \left(\frac{\partial N}{\partial \mu}\right)_{V_{ext}}. \tag{18.9}$$

We may imagine changing the number of electrons by changing the orbital occupancies above in Eq. (18.1). This gives equivalently to Eq. (18.7)

$$\Delta E_e = \sum_i \left(\frac{\partial E_e}{\partial n_i}\right)_{V_{ext}, n_k \neq n_i} \Delta n_i + \sum_{i,j} \left(\frac{\partial^2 E_e}{\partial n_i \partial n_j}\right)_{V_{ext}, n_k \neq n_i, n_j} \Delta n_i \Delta n_j + \cdots$$

$$= -\sum_i \chi_i \Delta n_i + \sum_{i,j} \eta_{ij} \Delta n_i \Delta n_j + \cdots. \tag{18.10}$$

Here,

$$\chi_i = -\left(\frac{\partial E_e}{\partial n_i}\right)_{V_{ext}, n_k \neq n_i} \tag{18.11}$$

are called the *orbital electronegativities* and

$$\eta_{ij} = \left(\frac{\partial^2 E_e}{\partial n_i \partial n_j}\right)_{V_{ext}, n_k \neq n_i, n_j} \tag{18.12}$$

form the *orbital hardness matrix*. These quantities describe how sensitive the orbitals are to changes in the occupancies.

The information that can be gained from the orbital hardness matrix is of the following type. Imagine that we change the occupancy of one orbital through some chemical reaction. The orbital whose occupancy is changed may be defined through its spatial structure. For instance, the two interacting systems may approach each other in some well-defined way (cf. Fig. 18.1). We may imagine that when the two systems are sufficiently close, some electron transfer will take place, i.e., one or more of the occupancies of either of the two systems will change. Due to this change, the other orbitals and, eventually, also their occupancies will change. The orbital electronegativities and orbital hardness matrix describe in detail these changes. We notice that these quantities are specific for a single system, and, accordingly, that they can be calculated without specifying with which other system the system of interest ultimately may react.

One may also introduce the *Fukui index*,

$$f_i = \left(\frac{\partial N}{\partial n_i}\right)_{V_{ext}, n_k \neq n_i} \tag{18.13}$$

so that Eq. (18.10) can be written in the form

$$\Delta E_e = -\sum_i \chi_i f_i \Delta N + \sum_{i,j} \eta_{ij} f_i f_j (\Delta N)^2 + \cdots. \tag{18.14}$$

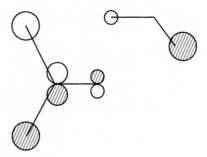

Figure 18.1 Schematic picture of an orbital of one system that will change occupancy when interacting with an orbital of another system

Finally, we may even define functions in position space instead of only numbers. For example,

$$
\eta_{ij} = \frac{1}{2} \left(\frac{\partial^2 E_e}{\partial n_i \partial n_j} \right)_{V_{\text{ext}}, n_k \neq n_i, n_j}
$$

$$
= \frac{1}{2} \int \int \left(\frac{\delta^2 E_e[\rho(\vec{r})]}{\delta \rho_i(\vec{r}_1) \delta \rho_j(\vec{r}_2)} \right)_{V_{\text{ext}}, n_k \neq n_i, n_j} d\vec{r}_1 \, d\vec{r}_2
$$

$$
= \frac{1}{2} \int \int \eta_{ij}(\vec{r}_1, \vec{r}_2) \, d\vec{r}_1 \, d\vec{r}_2, \tag{18.15}
$$

where

$$
\rho_i(\vec{r}) = \phi_i^*(\vec{r})\phi_i(\vec{r}) \tag{18.16}
$$

and where $\eta_{ij}(\vec{r}_1, \vec{r}_2)$ is an orbital hardness function. This function gives detailed (i.e., spatially resolved) information about how the orbitals change when the number of electrons are changed and are thus of direct relevance in studies of chemical interactions.

Studies of $\eta_{ij}(\vec{r}_1, \vec{r}_2)$ become, however, very easily very complicated since the number of functions (in six-dimensional position space) scales quadratically with the number of orbitals. It is simpler to study the total hardness function:

$$
\eta = \frac{1}{2} \left(\frac{\partial^2 E_e}{\partial N^2} \right)_{V_{\text{ext}}}
$$

$$
= \frac{1}{2} \int \int \left(\frac{\delta^2 E_e[\rho(\vec{r})]}{\delta \rho(\vec{r}_1) \delta \rho(\vec{r}_2)} \right)_{V_{\text{ext}}} d\vec{r}_1 \, d\vec{r}_2
$$

$$
= \frac{1}{2} \int \int \eta(\vec{r}_1, \vec{r}_2) \, d\vec{r}_1 \, d\vec{r}_2. \tag{18.17}
$$

Most convenient to study is the Fukui function. This can be defined by

$$f(\vec{r}) = \left(\frac{\delta N}{\delta \rho(\vec{r})} \right)_{V_{\text{ext}}}, \tag{18.18}$$

and it tells thus where electrons will be added or removed when changing the total number of electrons. That is, it will give information on where it is easier and where it is more difficult to add or remove electrons. Accordingly, it is hoped that calculating this function for a given molecular system will give information about how this molecule will interact with other molecules.

These definitions have been made possible with the introduction of the density-functional theory (Parr *et al.*, 1978). They are, however, related to older chemical definitions. Thus, from the ionization potential

$$I = E(N - 1) - E(N) \tag{18.19}$$

and the electron affinity

$$A = E(N) - E(N + 1) \tag{18.20}$$

one can obtain

$$\chi = \frac{I + A}{2}$$

$$\eta = \frac{I - A}{2}$$

$$S = \frac{1}{I - A} \tag{18.21}$$

which correspond to the simplest finite-difference approximations to the derivatives above.

Finally, for a typical molecular system with N electrons, an additional electron will occupy the $(N + 1)$st orbital [the, lowest unoccupied molecular orbital (LUMO)], whereas an electron will be removed from the Nth orbital [the highest occupied molecular orbital (HOMO)]. This means that there will be differences in the quantities we have discussed here, depending on whether we consider addition or removal of electrons. In order to distinguish between the two, one may put superindices on the quantities of interest, e.g., leading to $f^+(\vec{r})$ and $f^-(\vec{r})$ for the Fukui function of Eq. (18.18).

18.2   HARD AND SOFT ACIDS AND BASES PRINCIPLE

Studies of the softness, hardness, and Fukui functions are not yet common practice. They have, however, found one important application and this is in the studies of acids and bases.

Pearson proposed in 1963 the so-called hard and soft acids and bases (HSAB) principle. It states that 'hard acids prefer to coordinate to hard bases and soft acids to soft bases.' Although an apparently simple statement, it has been difficult to put it on a firm theoretical footing.

This has changed with the definitions above of softness and hardness, whereby it has become possible to argue quantitatively for the validity of this principle. By extending the analysis to the functions in position space, even more detailed information has been available than just contained in the original formulation of the HSAB principle.

This presentation of hardness and softness and related quantities will show that the density-functional formalism has allowed new approaches to be put on a firm theoretical basis and opened new lines of research. The quantities that have been defined are currently used in describing and quantifying molecular properties that are relevant for chemical reactions. Ultimately, once a detailed understanding of them has been developed, they may become useful in predicting chemical reactivities. Most likely, the future will bring many further discussions of these quantities, whereby first of all the understanding of reactivities and reactions will prosper.

19 Periodicity and Band Structures

19.1 HÜCKEL-LIKE MODEL FOR RING SYSTEMS

The Hückel and extended Hückel models represent some of the absolutely simplest approaches for studying the electronic structures of materials. They are useful in giving insights into some of the properties but are not always capable of providing accurate information on all properties. They are first of all constructed for calculating the single-particle energies, which — according to Koopmans' theorem — are approximations to ionization potentials and electron affinities. We shall here use models similar to the (extended) Hückel model as convenient tools for studying electronic orbitals and their energies in periodic systems. We shall thereby show how applying group theory, as described in Chapter 7, makes the calculations possible also for very large systems. We shall, however, stress that the basic principles are not restricted to the Hückel-like methods but apply to any single-particle (e.g., Hartree–Fock or Kohn–Sham) equation with appropriate modifications.

The Hartree–Fock equations are

$$\hat{F}\phi_i = \varepsilon_i\phi_i, \tag{19.1}$$

where

$$\hat{F} = \hat{h}_1 + \sum_j \left[\hat{J}_j - \hat{K}_j\right] \tag{19.2}$$

is the Fock operator. $\hat{h}_1$ contains the operator for the kinetic energy, $-\frac{1}{2}\nabla^2$, as well as the (external) Coulomb potential from the nuclei. $\hat{J}_j$ is the Coulomb operator from the jth electronic orbital, and $\hat{K}_j$ is the equivalent exchange operator.

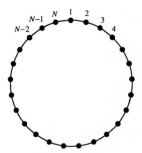

Figure 19.1   A (hypothetic) ring molecule consisting of N identical atoms

When we consider a ring molecule like that of Fig. 19.1 we see that this has a number of symmetry elements. Here, we shall only be concerned with one of these, i.e., the C_N axis. This symmetry element describes the fact that any rotation of $n \cdot (2\pi/N)$ (n being an integer) about the axis passing through the centre of the ring and perpendicular to the plane of the molecule maps the molecule onto itself. Therefore, the Fock operator is unchanged under this symmetry operator.

That this is so is easily seen by considering the individual terms of the operator. The kinetic-energy term does not depend on the nuclear coordinates (but on the electronic coordinates), so interchanging the nuclear coordinates does not change that term. Similar arguments apply to the electronic Coulomb and exchange operators. Only the Coulomb potential from the nuclei may change under the symmetry operations. However, these symmetry operations do nothing but map the individual nuclei onto equivalent ones, so that the Coulomb potential does not change either.

Within density-functional theory one solves the Kohn–Sham equations

$$\hat{h}_{\text{eff}}\phi_i = \varepsilon_i\phi_i. \tag{19.3}$$

Here,

$$\hat{h}_{\text{eff}} = \hat{h}_1 + V_{\text{C}} + V_{\text{xc}}. \tag{19.4}$$

As above, also this effective single-particle operator is invariant under the symmetry operations.

For the Hückel or extended-Hückel methods we solve also single-particle equations

$$\hat{h}\phi_i = \varepsilon_i\phi_i. \tag{19.5}$$

In contrast to the Hartree–Fock and the density-functional methods, the precise form of the operator is in this case not specified. Nevertheless, it must possess the same symmetry properties as those above; otherwise we cannot consider this method as a good approximation!

In *all* cases we solve some single-particle equations. The Hamilton-like single-particle operators possess the full symmetry of the system of interest, and we can therefore classify the solutions ϕ_i according to the irreducible representations of the group of the symmetry operations.

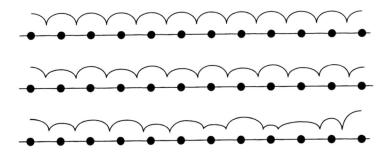

Figure 19.2 Schematic representation of the electron density for (upper curve) the case that it has the same symmetry properties as the underlying lattice, (middle curve) it has a lower symmetry, and (lower curve) it is incommensurable with the underlying lattice

We should, however, add that there are cases where this is not fulfilled, but where instead the electron density has a lower symmetry than that of the underlying lattice. We give some schematic representations of this situation in Fig. 19.2. In the first case, the electron density does possess the full symmetry of the chain, whereas in the second case the electron density shows an alternation. In the third case the electron density is not even commensurate with the lattice. The second and third case lead to the occurrence of so-called charge- or spin-density waves which are interesting in themselves. They may — but need not — be combined with distortions of the lattice. Here we shall, however, not consider such cases but assume that the electron density has the symmetry of the lattice as in the first example in Fig. 19.2.

Let us illustrate how the application of group theory works for the system of Fig. 19.1 and let us consider a Hückel-like model. We assume, accordingly, that the solutions to Eq. (19.5) can be written as a sum of functions centred on the different atoms. We will assume that we have only one function per atom, so that for the system of Fig. 19.1 we have N functions, one on each atom. By denoting the function on the jth atom by χ_j (with $j = 1, 2, \ldots, N$), we have

$$\psi_i = \sum_{j=1}^{N} \chi_j c_{ji}, \tag{19.6}$$

where the expansion coefficients c_{ji} are to be determined.

The coefficients c_{ji} are determined by solving the matrix eigenvalue equation (the secular equation)

$$\underline{\underline{H}} \cdot \underline{c}_i = \varepsilon_i \cdot \underline{\underline{O}} \cdot \underline{c}_i, \tag{19.7}$$

where the matrix elements of the Hamilton matrix are

$$H_{kl} = \langle \chi_k | \hat{h} | \chi_l \rangle \tag{19.8}$$

and those of the overlap matrix are

$$O_{kl} = \langle \chi_k | \chi_l \rangle. \tag{19.9}$$

In order to illustrate this for a simple model we shall assume that

$$O_{kl} = \delta_{k,l}, \tag{19.10}$$

i.e., the atomic (basis) functions are assumed to be orthonormal. This is actually the standard approximation within (extended) Hückel models.

For the Hamilton matrix elements we will assume

$$H_{kl} = \begin{cases} -e^{-\alpha d_{kl}} & \text{for } d_{kl} \le 3 \\ 0 & \text{for } d_{kl} > 3. \end{cases} \tag{19.11}$$

Here, d_{kl} is the shortest distance between atom k and l along the ring. For $N = 135$, for instance, $d_{17,35} = 18$, $d_{83,135} = 52$, $d_{1,135} = 1$, and $d_{8,124} = 19$. By the choice (19.11) the symmetry properties of the system are not violated. Furthermore, we have assumed that the interactions vanish beyond third-nearest neighbours. We stress that this is not meant as representing any real system but as being useful in demonstrating the complications in solving the secular equation.

$N = 10$ is not too large. It should therefore be possible to set up the matrix equation (19.7). It becomes

$$\begin{pmatrix} -1 & -e^{-\alpha} & -e^{-2\alpha} & -e^{-3\alpha} & 0 & 0 & 0 & -e^{-3\alpha} & -e^{-2\alpha} & -e^{-\alpha} \\ -e^{-\alpha} & -1 & -e^{-\alpha} & -e^{-2\alpha} & -e^{-3\alpha} & 0 & 0 & 0 & -e^{-3\alpha} & -e^{-2\alpha} \\ -e^{-2\alpha} & -e^{-\alpha} & -1 & -e^{-\alpha} & -e^{-2\alpha} & -e^{-3\alpha} & 0 & 0 & 0 & -e^{-3\alpha} \\ -e^{-3\alpha} & -e^{-2\alpha} & -e^{-\alpha} & -1 & -e^{-\alpha} & -e^{-2\alpha} & -e^{-3\alpha} & 0 & 0 & 0 \\ 0 & -e^{-3\alpha} & -e^{-2\alpha} & -e^{-\alpha} & -1 & -e^{-\alpha} & -e^{-2\alpha} & -e^{-3\alpha} & 0 & 0 \\ 0 & 0 & -e^{-3\alpha} & -e^{-2\alpha} & -e^{-\alpha} & -1 & -e^{-\alpha} & -e^{-2\alpha} & -e^{-3\alpha} & 0 \\ 0 & 0 & 0 & -e^{-3\alpha} & -e^{-2\alpha} & -e^{-\alpha} & -1 & -e^{-\alpha} & -e^{-2\alpha} & -e^{-3\alpha} \\ -e^{-3\alpha} & 0 & 0 & 0 & -e^{-3\alpha} & -e^{-2\alpha} & -e^{-\alpha} & -1 & -e^{-\alpha} & -e^{-2\alpha} \\ -e^{-2\alpha} & -e^{-3\alpha} & 0 & 0 & 0 & -e^{-3\alpha} & -e^{-2\alpha} & -e^{-\alpha} & -1 & -e^{-\alpha} \\ -e^{-\alpha} & -e^{-2\alpha} & -e^{-3\alpha} & 0 & 0 & 0 & -e^{-3\alpha} & -e^{-2\alpha} & -e^{-\alpha} & -1 \end{pmatrix}$$

$$\times \begin{pmatrix} c_{1i} \\ c_{2i} \\ c_{3i} \\ c_{4i} \\ c_{5i} \\ c_{6i} \\ c_{7i} \\ c_{8i} \\ c_{9i} \\ c_{10i} \end{pmatrix} = \varepsilon_i \cdot \begin{pmatrix} 1 & 0 & 0 & 0 & 0 & 0 & 0 & 0 & 0 & 0 \\ 0 & 1 & 0 & 0 & 0 & 0 & 0 & 0 & 0 & 0 \\ 0 & 0 & 1 & 0 & 0 & 0 & 0 & 0 & 0 & 0 \\ 0 & 0 & 0 & 1 & 0 & 0 & 0 & 0 & 0 & 0 \\ 0 & 0 & 0 & 0 & 1 & 0 & 0 & 0 & 0 & 0 \\ 0 & 0 & 0 & 0 & 0 & 1 & 0 & 0 & 0 & 0 \\ 0 & 0 & 0 & 0 & 0 & 0 & 1 & 0 & 0 & 0 \\ 0 & 0 & 0 & 0 & 0 & 0 & 0 & 1 & 0 & 0 \\ 0 & 0 & 0 & 0 & 0 & 0 & 0 & 0 & 1 & 0 \\ 0 & 0 & 0 & 0 & 0 & 0 & 0 & 0 & 0 & 1 \end{pmatrix} \cdot \begin{pmatrix} c_{1i} \\ c_{2i} \\ c_{3i} \\ c_{4i} \\ c_{5i} \\ c_{6i} \\ c_{7i} \\ c_{8i} \\ c_{9i} \\ c_{10i} \end{pmatrix} \tag{19.12}$$

There should be no need to stress that it is far from trivial to solve this equation!

The problem of solving Eq. (19.12) can, however, be simplified considerably by using the symmetry in constructing new symmetry-adapted basis functions.

This means that from the basis functions $\{\chi_j\}$ we construct new ones

$$\chi^{\tilde{k}} = \sum_{j=1}^{N} \chi_j u_{j\tilde{k}}. \tag{19.13}$$

We shall here use indices $\tilde{k}$ in specifying the different irreducible representations.

The coefficients $u_{j\tilde{k}}$ can be determined using group theory (i.e., the projection operators for the different irreducible representations) as demonstrated in Section 7.2. We may, however, also directly calculate them as follows. To this end we need one argument from group theory: all the irreducible representations are one-dimensional. This means that applying any of the symmetry operations $\hat{S}_n$ (being a rotation of $n \cdot (2\pi/N)$) brings the function over onto itself except for a constant prefactor. In particular, for $n = 1$ we find

$$\hat{S}_1 \chi^{\tilde{k}} \equiv \lambda_{\tilde{k}} \chi^{\tilde{k}}, \tag{19.14}$$

where $\lambda_{\tilde{k}}$ is this prefactor (which simultaneously is the character for the irreducible representation in the language of group theory).

By inserting Eq. (19.13) into Eq. (19.14), the latter can be rewritten as

$$\hat{S}_1 \chi^{\tilde{k}} = \hat{S}_1 \sum_{j=1}^{N} \chi_j u_{j\tilde{k}}$$

$$= \sum_{j=1}^{N} \left[\hat{S}_1 \chi_j \right] u_{j\tilde{k}}$$

$$= \sum_{j=1}^{N} \chi_{j-1} u_{j\tilde{k}}$$

$$= \sum_{j=1}^{N} \chi_j u_{j+1,\tilde{k}}, \tag{19.15}$$

where we have used the fact that by rotating the molecule, the function χ_j is brought over onto the function χ_{j-1} (with $\chi_0 \equiv \chi_N$).

Equation (19.14) gives also that the expression of Eq. (19.15) equals

$$\sum_{j=1}^{N} \lambda_{\tilde{k}} \chi_j u_{j\tilde{k}}, \tag{19.16}$$

whereby

$$u_{j+1,\tilde{k}} = \lambda_{\tilde{k}} u_{j\tilde{k}}, \tag{19.17}$$

or

$$u_{j\tilde{k}} = (\lambda_{\tilde{k}})^j u_{0\tilde{k}}. \tag{19.18}$$

By repeating the rotation N times, we have completed a full rotation, and the system is mapped identically onto itself. This means that

$$\left[\hat{S}_1\right]^N \chi^{\tilde{k}} = (\lambda_{\tilde{k}})^N \chi^{\tilde{k}} \equiv \chi^{\tilde{k}}, \qquad (19.19)$$

or

$$(\lambda_{\tilde{k}})^N = 1. \qquad (19.20)$$

This equation is valid only for

$$\lambda_{\tilde{k}} = e^{i\tilde{k}} \qquad (19.21)$$

with

$$\tilde{k} = 0, \pm\frac{2\pi}{N}, \pm\frac{4\pi}{N}, \ldots, \begin{cases} \pm\dfrac{(N-3)\pi}{N}, \pm\dfrac{(N-1)\pi}{N} & \text{for } N \text{ odd} \\ \pm\dfrac{(N-2)\pi}{N}, \pi & \text{for } N \text{ even} \end{cases} . \qquad (19.22)$$

It is possible to consider other values of $\tilde{k}$, although only differing from those above by a multiple of 2π, but any two values of $\tilde{k}$ differing by a multiple of 2π are equivalent, so those of Eq. (19.22) are the only inequivalent ones.

In total,

$$\chi^{\tilde{k}} = \frac{1}{\sqrt{N}} \sum_{j=1}^{N} e^{i(j-1)\tilde{k}} \chi_j, \qquad (19.23)$$

where $\tilde{k}$ is found from Eq. (19.22), and where we have *chosen* the constant

$$u_{0\tilde{k}} = \frac{1}{\sqrt{N}}. \qquad (19.24)$$

The functions $\chi^{\tilde{k}}$ are the symmetry-adapted basis functions. We *know* then from group theory that functions belonging to different irreducible representations (in this case, this means functions with different $\tilde{k}$) do not mix, i.e., the overlap and Hamilton matrix elements between them vanish.

This means that

$$\langle \chi^{\tilde{k}_1} | \chi^{\tilde{k}_2} \rangle = 0 \quad \text{for} \quad \tilde{k}_1 \neq \tilde{k}_2$$

$$\langle \chi^{\tilde{k}_1} | \hat{h} | \chi^{\tilde{k}_2} \rangle = 0 \quad \text{for} \quad \tilde{k}_1 \neq \tilde{k}_2. \qquad (19.25)$$

For $\tilde{k}_1 = \tilde{k}_2 \equiv \tilde{k}$ we proceed as follows:

$$\langle \chi^{\tilde{k}} | \chi^{\tilde{k}} \rangle = \left\langle \sum_{m=1}^{N} \chi_m u_{m\tilde{k}} \middle| \sum_{n=1}^{N} \chi_n u_{n\tilde{k}} \right\rangle$$

$$= \sum_{m,n=1}^{N} u_{m\tilde{k}}^* u_{n\tilde{k}} \langle \chi_m | \chi_n \rangle. \qquad (19.26)$$

Independently of the model (i.e., independent of whether the functions χ_i are orthonormal and of whether we apply the Hartree–Fock approach, a density-functional approach, a semi-empirical approach, or any other), the overlap $\langle \chi_m | \chi_n \rangle$ is that between two identical functions sitting on different sites, m and n. It is clear that since

$$\langle \chi_m | \chi_n \rangle = \langle \chi_{m+1} | \chi_{n+1} \rangle = \langle \chi_{m+2} | \chi_{n+2} \rangle = \cdots \tag{19.27}$$

we can write

$$\langle \chi_m | \chi_n \rangle = \langle \chi_1 | \chi_{n-m+1} \rangle, \tag{19.28}$$

i.e., the overlap depends only on the number of sites separating site m and site n. [In Eq. 19.28) we will let $n - m + 1$ be equal to $N + n - m + 1$, when it otherwise will be non-positive — this corresponds simply to stating that the matrix elements depend only on the number of units (including the sign) separating the two functions].

By inserting $u_{j\bar{k}}$ from Eqs. (19.18) and (19.24) we find then

$$\langle \chi^{\bar{k}} | \chi^{\bar{k}} \rangle = \sum_{m,n=1}^{N} \frac{1}{\sqrt{N}} e^{-im\bar{k}} \frac{1}{\sqrt{N}} e^{in\bar{k}} \langle \chi_m | \chi_n \rangle$$

$$= \frac{1}{N} \sum_{m,n=1}^{N} e^{i\bar{k}(n-m)} \langle \chi_m | \chi_n \rangle$$

$$= \frac{1}{N} \sum_{m,n=1}^{N} e^{i\bar{k}(n-m)} \langle \chi_1 | \chi_{n-m+1} \rangle. \tag{19.29}$$

Here,

$$e^{i\bar{k}(n-m)} = e^{i\bar{k}(N+n-m)} \tag{19.30}$$

since $\bar{k} \cdot N$ is an integer times 2π [cf. Eq. (19.22)]. Therefore, also the exponential in Eq. (19.29) depends only on the number of sites separating site m and site n. As a consequence, we can replace the double summation (over m and n) by a single summation (e.g., over n), when we simultaneously hold the other index fixed (e.g., $m = 1$) and multiply by N. In total, therefore,

$$\langle \chi^{\bar{k}} | \chi^{\bar{k}} \rangle = \sum_{n=1}^{N} e^{i\bar{k}(n-1)} \langle \chi_1 | \chi_n \rangle. \tag{19.31}$$

This formula is independent of the type of approach. In our example above it becomes even simpler, i.e.

$$\langle \chi^{\bar{k}} | \chi^{\bar{k}} \rangle = \sum_{n=1}^{N} e^{i\bar{k}(n-1)} \delta_{n,1} = 1 \tag{19.32}$$

due to Eq. (19.10).

The calculation of the Hamilton matrix elements proceeds exactly as above. Also here there is a high symmetry, so that — equivalent to Eq. (19.28) — we have

$$\langle \chi_m | \hat{h} | \chi_n \rangle = \langle \chi_1 | \hat{h} | \chi_{n-m+1} \rangle, \tag{19.33}$$

once again independently of the approach.

Then,

$$
\begin{aligned}
\langle \chi^{\tilde{k}} | \hat{h} | \chi^{\tilde{k}} \rangle &= \left\langle \sum_{m=1}^{N} \chi_m u_{m\tilde{k}} \middle| \hat{h} \middle| \sum_{n=1}^{N} \chi_n u_{n\tilde{k}} \right\rangle \\
&= \sum_{m,n=1}^{N} u_{m\tilde{k}}^* u_{n\tilde{k}} \langle \chi_m | \hat{h} | \chi_n \rangle \\
&= \sum_{m,n=1}^{N} \frac{1}{\sqrt{N}} e^{-im\tilde{k}} \frac{1}{\sqrt{N}} e^{in\tilde{k}} \langle \chi_m | \hat{h} | \chi_n \rangle \\
&= \frac{1}{N} \sum_{m,n=1}^{N} e^{i\tilde{k}(n-m)} \langle \chi_m | \hat{h} | \chi_n \rangle \\
&= \frac{1}{N} \sum_{m,n=1}^{N} e^{i\tilde{k}(n-m)} \langle \chi_1 | \hat{h} | \chi_{n-m+1} \rangle \\
&= \sum_{n=1}^{N} e^{i\tilde{k}(n-1)} \langle \chi_1 | \hat{h} | \chi_n \rangle. \tag{19.34}
\end{aligned}
$$

This expression is general and not restricted to the present model.

In our case, with $N = 10$ and the matrix elements given by Eq. (19.11), we find

$$\langle \chi^{\tilde{k}} | \hat{h} | \chi^{\tilde{k}} \rangle \equiv e_{\tilde{k}} = -1 - 2e^{-\alpha} \cos(\tilde{k}) - 2e^{-2\alpha} \cos(2\tilde{k}) - 2e^{-3\alpha} \cos(3\tilde{k}) \tag{19.35}$$

after some simple manipulations.

With this new basis set, the matrix equation (19.12) takes the form

$$
\begin{pmatrix}
e_{\tilde{k}_1} & 0 & 0 & 0 & 0 & 0 & 0 & 0 & 0 & 0 \\
0 & e_{\tilde{k}_2} & 0 & 0 & 0 & 0 & 0 & 0 & 0 & 0 \\
0 & 0 & e_{\tilde{k}_3} & 0 & 0 & 0 & 0 & 0 & 0 & 0 \\
0 & 0 & 0 & e_{\tilde{k}_4} & 0 & 0 & 0 & 0 & 0 & 0 \\
0 & 0 & 0 & 0 & e_{\tilde{k}_5} & 0 & 0 & 0 & 0 & 0 \\
0 & 0 & 0 & 0 & 0 & e_{\tilde{k}_6} & 0 & 0 & 0 & 0 \\
0 & 0 & 0 & 0 & 0 & 0 & e_{\tilde{k}_7} & 0 & 0 & 0 \\
0 & 0 & 0 & 0 & 0 & 0 & 0 & e_{\tilde{k}_8} & 0 & 0 \\
0 & 0 & 0 & 0 & 0 & 0 & 0 & 0 & e_{\tilde{k}_9} & 0 \\
0 & 0 & 0 & 0 & 0 & 0 & 0 & 0 & 0 & e_{\tilde{k}_{10}}
\end{pmatrix}
\cdot
\begin{pmatrix}
c_i^1 \\
c_i^2 \\
c_i^3 \\
c_i^4 \\
c_i^5 \\
c_i^6 \\
c_i^7 \\
c_i^8 \\
c_i^9 \\
c_i^{10}
\end{pmatrix}
$$

$$= \varepsilon_i \cdot \begin{pmatrix} 1 & 0 & 0 & 0 & 0 & 0 & 0 & 0 & 0 & 0 \\ 0 & 1 & 0 & 0 & 0 & 0 & 0 & 0 & 0 & 0 \\ 0 & 0 & 1 & 0 & 0 & 0 & 0 & 0 & 0 & 0 \\ 0 & 0 & 0 & 1 & 0 & 0 & 0 & 0 & 0 & 0 \\ 0 & 0 & 0 & 0 & 1 & 0 & 0 & 0 & 0 & 0 \\ 0 & 0 & 0 & 0 & 0 & 1 & 0 & 0 & 0 & 0 \\ 0 & 0 & 0 & 0 & 0 & 0 & 1 & 0 & 0 & 0 \\ 0 & 0 & 0 & 0 & 0 & 0 & 0 & 1 & 0 & 0 \\ 0 & 0 & 0 & 0 & 0 & 0 & 0 & 0 & 1 & 0 \\ 0 & 0 & 0 & 0 & 0 & 0 & 0 & 0 & 0 & 1 \end{pmatrix} \cdot \begin{pmatrix} c_i^1 \\ c_i^2 \\ c_i^3 \\ c_i^4 \\ c_i^5 \\ c_i^6 \\ c_i^7 \\ c_i^8 \\ c_i^9 \\ c_i^{10} \end{pmatrix}. \qquad (19.36)$$

Here, we have simply labelled the 10 different $\tilde{k}$ values of Eq. (19.22) by the numbers $1, 2, \ldots, 10$ and written the eigenfunctions as

$$\psi_i = \sum_{\tilde{k}=1}^{N} \chi^{\tilde{k}} c_i^{\tilde{k}} \qquad (19.37)$$

replacing Eq. (19.6).

It is clear that in this case $e_{\tilde{k}}$ of Eq. (19.35) *is* the eigenvalue and $\chi^{\tilde{k}}$ of Eq. (19.23) the eigenfunction.

For our model system we find the energy levels shown in Fig. 19.3, where we not only show those for $N = 10$ but also those for larger values of N (18, 50, and 450). It is clearly seen that at least for the larger system the levels become so dense that very little can be read off from the figure.

Alternatively, we may plot the levels as a function of the $\tilde{k}$ that describes the irreducible representation. This is done in Fig. 19.4. At first, this is nothing but another presentation of the results of Fig. 19.3, but as we shall show below for crystalline materials (in which case $\tilde{k}$ changes into a vector, $\vec{k}$), $\tilde{k}$ can be given a physical interpretation beyond that of being purely a group-theoretical classification label.

For $N \rightarrow \infty$, the discrete set of energies as a function of $\tilde{k}$ becomes infinite, and ε_i becomes a smooth function of $\tilde{k}$, $\varepsilon(\tilde{k})$. This function defines a so-called (energy) band.

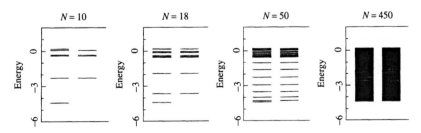

Figure 19.3 Energy levels for our model system for $N = 10$, 18, 50, and 450. λ of Eq. (19.11) was chosen equal to $\lambda = 0.3$

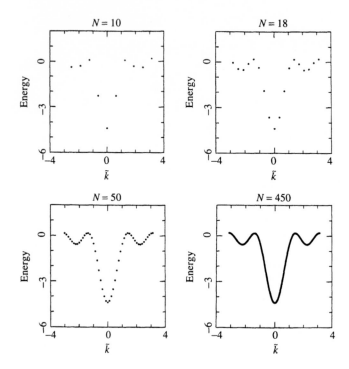

Figure 19.4   As Fig. 19.3, but shown as functions of $\tilde{k}$

19.2   BORN–VON KÁRMÁN ZONES

For the ring molecule (with N atoms) of the preceding section we required that the symmetry-adapted orbitals should be mapped onto themselves when we rotated the system one complete turn [Eq. (19.19)]. This means also that the values of the wavefunctions at the $(N + 1)$st site (which here is identical to the first site) are identical to those at the first site (these are the so-called periodic boundary conditions).

Exactly the same would have been obtained if we considered an infinite linear chain like that of Fig. 19.5 containing an infinite set of equivalent atoms. By requiring that a symmetry-adapted wavefunction is mapped identically onto itself when the system is translated by N units (this is once again the periodic-boundary

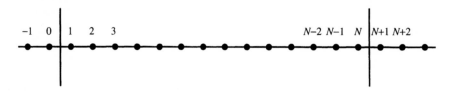

Figure 19.5   An infinite linear chain

conditions), we have the same construction as we applied for the ring molecule of N atoms. There is a difference in that for the ring molecule the symmetry operation $[\hat{S}_1]^N$ brought any atom onto exactly itself, whereas the equivalent operation for the infinite chain (i.e., N translations, each of one unit) brings any atom onto another, equivalent one. However, from a formal point of view these differences are unimportant.

Therefore, the ring molecule may be considered a finite approximation to an infinite linear chain. The size of the ring molecule defines a fragment of the infinite linear chain, as shown in Fig. 19.5. This fragment is known as the Born von Kármán zone (Born and von Kármán, 1912).

19.3 BAND STRUCTURES IN ONE DIMENSION

In the preceding two subsections we have seen how the concept of the band structures is natural for an infinite, periodic, one-dimensional system. For a ring system consisting of N identical units, each having *one* basis function, the construction of symmetry-adapted basis functions led to a natural classification of the eigenfunctions to the Schrödinger equation according to the irreducible representation of the group of symmetry operations. This irreducible representation could in turn be specified by a parameter $\tilde{k}$ that could take any out of a set of N values between $-\pi$ and π. In the limit $N \to \infty$, we have a whole continuum of $\tilde{k}$ values, and the single-particle eigenvalue ε becomes a function of $\tilde{k}$,

$$\varepsilon = \varepsilon(\tilde{k}). \tag{19.38}$$

We demonstrated this result for a very simple model that is closely related to the simple (extended) Hückel method. However, the general features behind the introduction of the band and the representation of the single-particle eigenvalue as in Eq. (19.38) as a function of a $\tilde{k}$ that describes the irreducible representation, remains true also when using other methods such as the Hartree–Fock or density-functional methods. But the simple model has the great advantage that very many of the calculations can be carried through analytically, whereby many features very easily can be demonstrated without the complexity related to the more exact approaches.

So far we have discussed only the simplest possible case of a periodic system with one basis function per repeated unit. In the present subsection we shall generalize this approach to include more functions per unit. First, we shall continue the general analysis by studying very simple models, and, subsequently, we shall discuss the band structures for some real systems as obtained with a density-functional method.

As the first and simplest extension we consider the system of Fig. 19.6. Compared with the system of Fig. 19.1 the only difference is that we now assume that we have two instead of one basis function per unit. We have represented

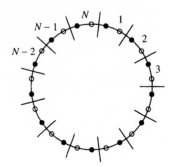

Figure 19.6 A ring molecule consisting of N repeated units each containing two basis functions

this as a ring molecule or chain (remember, the two are equivalent, as discussed in Section 19.2) consisting of alternating atoms A and B. We will assume that we have one basis function on each atom, i.e., in the nth unit we have the basis functions $\chi_{A,n}$ and $\chi_{B,n}$. As in Section 19.1 we utilize the symmetry of the system in constructing symmetry-adapted basis functions, but this time we will for each $\tilde{k}$ have two functions, i.e., one for the functions on the A sites,

$$\chi_A^{\tilde{k}} = \frac{1}{\sqrt{N}} \sum_{n=1}^{N} e^{ink} \chi_{A,n} \tag{19.39}$$

and one for those on the B sites,

$$\chi_B^{\tilde{k}} = \frac{1}{\sqrt{N}} \sum_{n=1}^{N} e^{ink} \chi_{B,n}. \tag{19.40}$$

As in Section 19.1 we use the fact that the Hamilton and overlap matrix elements vanish unless the functions belong to the same irreducible representation, i.e.

$$\langle \chi_A^{\tilde{k}_1} | \chi_A^{\tilde{k}_2} \rangle = \langle \chi_B^{\tilde{k}_1} | \chi_B^{\tilde{k}_2} \rangle = \langle \chi_A^{\tilde{k}_1} | \chi_B^{\tilde{k}_2} \rangle = \langle \chi_B^{\tilde{k}_1} | \chi_A^{\tilde{k}_2} \rangle = 0 \quad \text{for} \quad \tilde{k}_1 \neq \tilde{k}_2 \tag{19.41}$$

and

$$\langle \chi_A^{\tilde{k}_1} | \hat{h} | \chi_A^{\tilde{k}_2} \rangle = \langle \chi_B^{\tilde{k}_1} | \hat{h} | \chi_B^{\tilde{k}_2} \rangle = \langle \chi_A^{\tilde{k}_1} | \hat{h} | \chi_B^{\tilde{k}_2} \rangle = \langle \chi_B^{\tilde{k}_1} | \hat{h} | \chi_A^{\tilde{k}_2} \rangle = 0 \quad \text{for} \quad \tilde{k}_1 \neq \tilde{k}_2. \tag{19.42}$$

The only matrix elements that may be non-vanishing are those for functions of the same $\tilde{k}$. For these we find, completely analogously to Eqs. (19.31) and (19.34),

$$\langle \chi_A^{\tilde{k}} | \chi_A^{\tilde{k}} \rangle = \sum_{n=1}^{N} e^{i\tilde{k}(n-1)} \langle \chi_{A,1} | \chi_{A,n} \rangle$$

$$\langle \chi_B^{\tilde{k}} | \chi_A^{\tilde{k}} \rangle = \sum_{n=1}^{N} e^{i\tilde{k}(n-1)} \langle \chi_{B,1} | \chi_{A,n} \rangle$$

$$\langle \chi_A^{\tilde{k}} | \chi_B^{\tilde{k}} \rangle = \sum_{n=1}^{N} e^{i\tilde{k}(n-1)} \langle \chi_{A,1} | \chi_{B,n} \rangle$$

$$\langle \chi_B^{\tilde{k}} | \chi_B^{\tilde{k}} \rangle = \sum_{n=1}^{N} e^{i\tilde{k}(n-1)} \langle \chi_{B,1} | \chi_{B,n} \rangle$$

$$\langle \chi_A^{\tilde{k}} | \hat{h} | \chi_A^{\tilde{k}} \rangle = \sum_{n=1}^{N} e^{i\tilde{k}(n-1)} \langle \chi_{A,1} | \hat{h} | \chi_{A,n} \rangle$$

$$\langle \chi_B^{\tilde{k}} | \hat{h} | \chi_A^{\tilde{k}} \rangle = \sum_{n=1}^{N} e^{i\tilde{k}(n-1)} \langle \chi_{B,1} | \hat{h} | \chi_{A,n} \rangle$$

$$\langle \chi_A^{\tilde{k}} | \hat{h} | \chi_B^{\tilde{k}} \rangle = \sum_{n=1}^{N} e^{i\tilde{k}(n-1)} \langle \chi_{A,1} | \hat{h} | \chi_{B,n} \rangle$$

$$\langle \chi_B^{\tilde{k}} | \hat{h} | \chi_B^{\tilde{k}} \rangle = \sum_{n=1}^{N} e^{i\tilde{k}(n-1)} \langle \chi_{B,1} | \hat{h} | \chi_{B,n} \rangle. \tag{19.43}$$

This result is completely general for any periodic system with two basis functions per unit. In this case the Hamilton and overlap matrices, when using the symmetry-adapted functions as basis functions, will contain N blocks, each defining a 2×2 eigenvalue problem

$$\begin{pmatrix} \langle \chi_A^{\tilde{k}} | \hat{h} | \chi_A^{\tilde{k}} \rangle & \langle \chi_B^{\tilde{k}} | \hat{h} | \chi_A^{\tilde{k}} \rangle \\ \langle \chi_A^{\tilde{k}} | \hat{h} | \chi_B^{\tilde{k}} \rangle & \langle \chi_B^{\tilde{k}} | \hat{h} | \chi_B^{\tilde{k}} \rangle \end{pmatrix} \cdot \begin{pmatrix} c_{A,i}(\tilde{k}) \\ c_{B,i}(\tilde{k}) \end{pmatrix}$$
$$= \varepsilon_i(\tilde{k}) \begin{pmatrix} \langle \chi_A^{\tilde{k}} | \chi_A^{\tilde{k}} \rangle & \langle \chi_B^{\tilde{k}} | \chi_A^{\tilde{k}} \rangle \\ \langle \chi_A^{\tilde{k}} | \chi_B^{\tilde{k}} \rangle & \langle \chi_B^{\tilde{k}} | \chi_B^{\tilde{k}} \rangle \end{pmatrix} \cdot \begin{pmatrix} c_{A,i}(\tilde{k}) \\ c_{B,i}(\tilde{k}) \end{pmatrix}, \tag{19.44}$$

where we have emphasized that the coefficients and the eigenvalues will depend on $\tilde{k}$. Furthermore, the fact that the problem is a 2×2 eigenvalue problem means that for each $\tilde{k}$ we will have two different solutions; the index i distinguishes these.

These equations are to be solved for all $\tilde{k}$ values of Eq. (19.22). For $N \to \infty$ this would, in principle, mean for an infinite number of $\tilde{k}$ values, but since the solutions usually depend only slowly on $\tilde{k}$ it is often sufficient to consider a smaller finite set and interpolate between the solutions thereby obtained in order to obtain the results for a general $\tilde{k}$.

In order to illustrate the results of such calculations we shall consider a simple model for which

$$\langle \chi_{A,n} | \chi_{A,m} \rangle = \delta_{n,m}$$
$$\langle \chi_{B,n} | \chi_{A,m} \rangle = \langle \chi_{A,n} | \chi_{B,m} \rangle = 0$$
$$\langle \chi_{B,n} | \chi_{B,m} \rangle = \delta_{n,m} \tag{19.45}$$

for the overlap matrix elements (i.e., the atom-centred functions are assumed to be orthonormal), and

$$\langle \chi_{A,n} | \hat{h} | \chi_{A,m} \rangle = \varepsilon_A \delta_{n,m}$$

$$\langle \chi_{B,n} | \hat{h} | \chi_{A,m} \rangle = \langle \chi_{A,n} | \hat{h} | \chi_{B,m} \rangle = \begin{cases} t_{AB} & \text{for nearest neighbours} \\ 0 & \text{otherwise} \end{cases}$$

$$\langle \chi_{B,n} | \hat{h} | \chi_{B,m} \rangle = \varepsilon_B \delta_{n,m} \tag{19.46}$$

for the Hamilton matrix elements. Eq. (19.46) corresponds to assuming that the basis functions interact only with nearest neighbours, i.e., an A functions only with itself as well as its two nearest B neighbours, and a B function only with itself as well as its two nearest A neighbours.

In that case the eigenvalue problem becomes [using Eq. (19.43)]

$$\begin{pmatrix} \varepsilon_A & t_{AB}(1 + e^{i\tilde{k}}) \\ t_{AB}(1 + e^{-i\tilde{k}}) & \varepsilon_B \end{pmatrix} \cdot \begin{pmatrix} c_{A,i}(\tilde{k}) \\ c_{B,i}(\tilde{k}) \end{pmatrix}$$

$$= \varepsilon_i(\tilde{k}) \cdot \begin{pmatrix} 1 & 0 \\ 0 & 1 \end{pmatrix} \cdot \begin{pmatrix} c_{A,i}(\tilde{k}) \\ c_{B,i}(\tilde{k}) \end{pmatrix}. \tag{19.47}$$

In Fig. 19.7 we show the band structures for the four cases of Table 19.1. ε_A and ε_B give the energies for the functions in the case there is no interaction between the atoms. On the other hand, t_{AB} gives the strength of their interaction. Therefore, in case 1 in Fig. 19.7, the atomic levels are well separated (by eight energy units) compared with the strength of the interactions (that is only two energy units). Since $\varepsilon_A = -4$, whereas $\varepsilon_B = +4$, the energetically lowest band

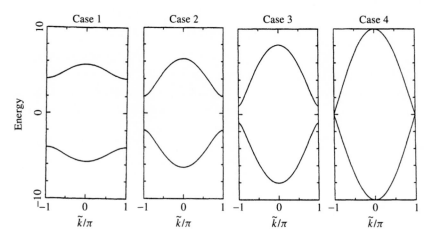

Figure 19.7 The band structures for the model of Fig. 19.6 and for the four cases of Table 19.1

Table 19.1 The four different sets of parameter values applied in studying the model of Fig. 19.6 and giving the results of Fig. 19.7

	Case 1	Case 2	Case 3	Case 4
ε_A	−4	−2	−1	0
ε_B	4	2	1	0
t_{AB}	2	3	4	5

of the two will be formed mainly by the $\chi_A^{\tilde{k}}$ functions whereas the highest band will be formed mainly by the $\chi_B^{\tilde{k}}$ functions. By solving Eq. (19.47) this can easily be verified. We find

$$\varepsilon_i(\tilde{k}) = \frac{\varepsilon_A + \varepsilon_B}{2} \pm \left[\frac{1}{4}(\varepsilon_A - \varepsilon_B)^2 + 2t_{AB}^2(1 + \cos \tilde{k}) \right]^{1/2} \qquad (19.48)$$

and

$$\frac{c_{B,i}(\tilde{k})}{c_{A,i}(\tilde{k})} = \frac{\varepsilon_i(\tilde{k}) - \varepsilon_A}{t_{AB}(1 + e^{i\tilde{k}})}, \qquad (19.49)$$

i.e., for $\varepsilon_i(\tilde{k}) \simeq \varepsilon_A$, $c_{B,i} \simeq 0$.

In case 2 we have reduced the energy difference between the A and B functions and simultaneously increased their interactions. This is continued for case 3, and in case 4 we have the special case that the two functions are completely equivalent (i.e., they have the same matrix elements). In that case we have actually not made full use of the symmetry. The black and white atoms of Fig. 19.6 are in this case equivalent (as far as the band structures are concerned), and we could have treated the system as having not two but only one atom per repeated unit. In that case we would return to the case of Section 19.1 and we would only have one band. This can partly be recognized in Fig. 19.7 for case 4: the upper and lower bands meet at $\tilde{k} = \pm\pi$, and it is possible to draw a single continuous band by, e.g., taking the upper band and 'unfolding it'. This process is shown schematically in Fig. 19.8 and will below be discussed further.

This example can be used as an example of a more general result. That is, if all bands (also when there are more than just two) meet pairwise at the so-called zone boundary ($\tilde{k} = \pm\pi$), then the system possesses a higher symmetry and it is possible to reduce the size of the repeated unit.

The example shows also that when the difference in the so-called on-site energies, i.e., $|\varepsilon_A - \varepsilon_B|$, gets smaller compared with the hopping integrals, i.e., t_{AB}, then the bands possess more curvature (i.e., dispersion) and the splittings are reduced.

As a second example we study that of Fig. 19.9. As for the example of Fig. 19.6, we assume that each site contains exactly one basis function and we will treat the system exactly as that, i.e., by considering a ring molecule of N identical units and then letting $N \to \infty$. And similarly to that example we will

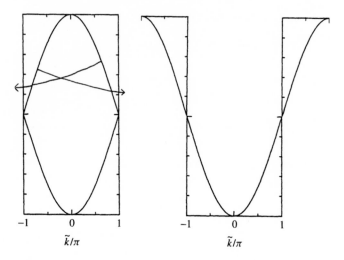

$\tilde{k}/\pi$ $\tilde{k}/\pi$

Figure 19.8   The process of unfolding the bands of case 4 in Fig. 19.7

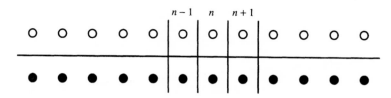

Figure 19.9   A chain consisting of repeated units each containing two basis functions

once again assume that the basis functions are orthonormal, and that they interact only with nearest neighbours. This means, equivalently to Eq. (19.45),

$$\langle \chi_{A,n} | \chi_{A,m} \rangle = \delta_{n,m}$$

$$\langle \chi_{B,n} | \chi_{A,m} \rangle = \langle \chi_{A,n} | \chi_{B,m} \rangle = 0$$

$$\langle \chi_{B,n} | \chi_{B,m} \rangle = \delta_{n,m} \tag{19.50}$$

for the overlap matrix elements, and, equivalent to Eq. (19.46),

$$\langle \chi_{A,n} | \hat{h} | \chi_{A,m} \rangle = \begin{cases} \varepsilon_A & \text{for } n = m \\ t_{AA} & \text{for } n = m \pm 1 \\ 0 & \text{otherwise} \end{cases}$$

$$\langle \chi_{B,n} | \hat{h} | \chi_{A,m} \rangle = \langle \chi_{A,n} | \hat{h} | \chi_{B,m} \rangle = \begin{cases} t_{AB} & \text{for } n = m \\ 0 & \text{otherwise} \end{cases}$$

$$\langle \chi_{B,n} | \hat{h} | \chi_{B,m} \rangle = \begin{cases} \varepsilon_B & \text{for } n = m \\ t_{BB} & \text{for } n = m \pm 1 \\ 0 & \text{otherwise.} \end{cases} \tag{19.51}$$

The secular equation, analogous to Eq. (19.47), is in this case

$$
\begin{pmatrix} \varepsilon_A + 2t_{AA}\cos(\tilde{k}) & t_{AB} \\ t_{AB} & \varepsilon_B + 2t_{BB}\cos(\tilde{k}) \end{pmatrix} \cdot \begin{pmatrix} c_{A,i}(\tilde{k}) \\ c_{B,i}(\tilde{k}) \end{pmatrix}
$$
$$
= \varepsilon_i(\tilde{k}) \cdot \begin{pmatrix} 1 & 0 \\ 0 & 1 \end{pmatrix} \cdot \begin{pmatrix} c_{A,i}(\tilde{k}) \\ c_{B,i}(\tilde{k}) \end{pmatrix}, \tag{19.52}
$$

which has the solutions

$$
\varepsilon_i(\tilde{k}) = \frac{\varepsilon_A + \varepsilon_B}{2} + (t_{AA} + t_{BB})\cos(\tilde{k})
$$
$$
\pm \left[\left(\frac{\varepsilon_A - \varepsilon_B}{2} + (t_{AA} - t_{BB})\cos(\tilde{k}) \right)^2 + t_{AB}^2 \right]^{1/2}. \tag{19.53}
$$

When the A and B systems are not interacting (i.e., $t_{AB} = 0$), then the solutions become

$$
\varepsilon_i(\tilde{k}) = \begin{cases} \varepsilon_A + 2t_{AA}\cos(\tilde{k}) \\ \varepsilon_B + 2t_{BB}\cos(\tilde{k}). \end{cases} \tag{19.54}
$$

There may exist certain $\tilde{k}$ for which the two solutions of Eq. (19.54) are identical, i.e.,

$$
\varepsilon_A + 2t_{AA}\cos(\tilde{k}) = \varepsilon_B + 2t_{BB}(\tilde{k}). \tag{19.55}
$$

Including the interactions between the A and B systems, i.e., setting $t_{AB} \neq 0$, then leads for this value of $\tilde{k}$ to a removal of the degeneracy,

$$
\varepsilon_i(\tilde{k}) = \frac{\varepsilon_A + \varepsilon_B}{2} + (t_{AA} + t_{BB})\cos(\tilde{k}) \pm t_{AB}. \tag{19.56}
$$

Also for other values of $\tilde{k}$ in the neighbourhood of the degeneracy of Eq. (19.55), the two solutions have a larger separation when including t_{AB}, i.e., a so-called avoided crossing is introduced.

We study the four cases of Table 19.2, which results in the band structures of Fig. 19.10. For the first case we have set $t_{AB} = 0$ which means that the A functions do not interact with the B functions and we thus end up with

Table 19.2 The five different sets of parameter values used in studying the model of Fig. 19.9 and giving the results of Fig. 19.10

	Case 1	Case 2	Case 3	Case 4	Case 5
ε_A	−6	0	0	0	0
ε_B	6	0	0	0	0
t_{AB}	0	0	0.1	1	4
t_{AA}	2	4	4	4	4
t_{BB}	−2	−4	−4	−4	−4

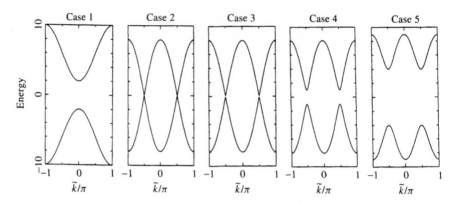

Figure 19.10 The band structures for the model of Fig. 19.9 and for the five cases of Table 19.2

two bands, of which one is completely formed by the A functions (the lower one) and the other is completely formed by the B functions. The two bands have a noteworthy difference. That is, one of them has the maximum at $\tilde{k} = 0$, whereas the other has a minimum there. For $\tilde{k} = 0$ the symmetry-adapted basis functions are simply the sum of the equivalent basis functions. Therefore, if neighbouring equivalent basis functions have an antibonding interaction, the corresponding band has a maximum for $\tilde{k} = 0$, whereas a bonding interaction leads to a minimum for $\tilde{k} = 0$. This point is illustrated further in Fig. 19.11.

Passing to case 2, the interactions between the two sets of functions is kept equal to 0. Therefore, we still have two bands, of which one is formed by A functions and the other by B functions, although in this case the bands are made to overlap (by reducing the energy difference $|\varepsilon_A - \varepsilon_b|$ compared with $|t_{AA}|$ and $|t_{BB}|$). That the functions would not interact could for instance be the case when the two sets of functions possess different symmetry properties according to another symmetry operation than that of translation. For instance, one set of functions could be symmetric and the other antisymmetric with respect to reflection in the plane of the nuclei for a planar system.

For case 3 we have included a very small interaction, and we see that the only consequence is that the two bands no longer cross, but rather that we have two bands with a small gap in between. The existence of non-vanishing matrix elements is general for orbitals of the same symmetry representation, and, therefore, bands that belong to the same symmetry representation will in general not cross but show the avoided crossing. Despite the occurrence of the small gaps, the wavefunctions with the lowest energy for $\tilde{k} = 0$ and those of the highest energy for $\tilde{k} = \pm \pi$ are formed mainly by A functions with, however, a small admixture of B functions. One can see this by drawing the hypothetic bands that would occur when neglecting the gaps (the avoided crossings — this would

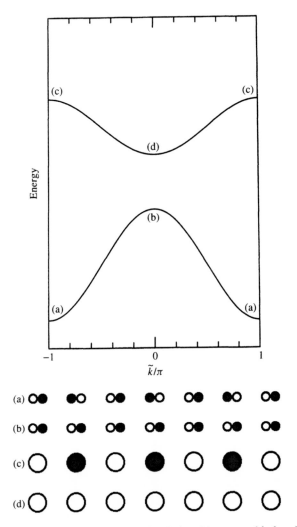

Figure 19.11 Qualitative band structures of a chain with one s orbital and one p orbital per unit

actually lead to the bands of case 2) and subsequently introduce the small gaps that mainly perturb the band structures and the orbitals close to those $\tilde{k}$ points where the avoided crossings occur.

For case 4 and case 5 the interactions between the two sets of functions have been increased and we see that this leads to increasingly larger gaps between the two bands. Nevertheless, the 'original' bands as found for case 2 can still, with some goodwill, be recognized.

Having studied one-dimensional systems with one function per unit in Section 19.1 and so far in this subsection with two functions per unit we now turn to some real one-dimensional systems. In that case the number of basis functions per repeated unit may be large, but we still construct symmetry-adapted linear combinations from equivalent ones of different units (so-called Bloch waves),

$$\chi_m^{\tilde{k}} = \frac{1}{\sqrt{N}} \sum_{n=1}^{N} \chi_{m,n} e^{ikn}. \tag{19.57}$$

Here, $m = 1, 2, \ldots, N_b$ denotes the different basis functions per unit, and n is a unit index. In order to treat the infinite, periodic system we will consider the limit $N \to \infty$.

Equivalent to the results above we have

$$\langle \chi_{m_1}^{\tilde{k}_1} | \chi_{m_2}^{\tilde{k}_2} \rangle = \delta_{\tilde{k}_1, \tilde{k}_2} \sum_{n=1}^{N} e^{i\tilde{k}_1(n-1)} \langle \chi_{m_1,1} | \chi_{m_2,n} \rangle \tag{19.58}$$

and

$$\langle \chi_{m_1}^{\tilde{k}_1} | \hat{h} | \chi_{m_2}^{\tilde{k}_2} \rangle = \delta_{\tilde{k}_1, \tilde{k}_2} \sum_{n=1}^{N} e^{i\tilde{k}_1(n-1)} \langle \chi_{m_1,1} | \hat{h} | \chi_{m_2,n} \rangle. \tag{19.59}$$

This result is independent of the method, i.e., it applies equally well to Hückel-like models, Hartree–Fock methods, and density-functional methods, as well as to any other single-particle method.

The eigenvalue problem (the secular equation) becomes a $\tilde{k}$-dependent eigenvalue equation

$$\underline{\underline{H}}(\tilde{k}) \cdot \underline{c}_i(\tilde{k}) = \varepsilon_i(\tilde{k}) \cdot \underline{\underline{O}}(\tilde{k}) \cdot \underline{c}_i(\tilde{k}), \tag{19.60}$$

where the overlap and Hamilton matrices have the size $N_b \times N_b$ with N_b being the number of basis functions per unit.

The solutions to this equation define the single-particle wavefunctions that become $\tilde{k}$-dependent,

$$\psi_i^{\tilde{k}} = \sum_{m=1}^{N_b} c_{i,m}(\tilde{k}) \chi_m^{\tilde{k}}. \tag{19.61}$$

Furthermore, the eigenvalues $\varepsilon_i(\tilde{k})$ define in total N_b bands. For each band we have N orbitals, which means that each band can accommodate one electron per unit. For many systems, the spin-up and spin-down wavefunctions are energetically degenerate and then each band can contain two electrons per unit; one for each spin direction.

The bands are now filled so that the correct number of orbitals are found. Let us consider the examples of Fig. 19.10, and let us assume that each band can contain two electrons per unit, by assuming that the bands for spin-up and spin-down orbitals are identical. Furthermore, let us assume that we have two electrons per unit. This means that for case 1 the energetically lowest band will be completely filled and the upper one completely empty. There will be an energy gap between the two and the Fermi energy, i.e., the energy separating empty and occupied bands, will be placed exactly in the middle of this gap. For case 2 each band will be exactly half filled, and the Fermi energy will lie at the energy where the bands cross, i.e., at the energy 0. For the last three cases we have again the situation from the first case, i.e., the Fermi energy will lie in the middle of the gap between the lowest and the highest band.

Systems for which the Fermi level passes through one or more bands are metals. When it lies in a gap separating occupied and empty orbitals, it is a semiconductor when the gap is not too large (up to about 3 eV), and otherwise an insulator. As we shall see below, the precise situation for a given system may depend sensitively on the structure of the system. It is very often found that the systems tend to have structures where the Fermi energy lies in a gap or at least where it cuts as few bands as possible. Alternatively formulated, systems or structures where the Fermi energy cuts many bands are most likely not stable and will possess some kind of distortion. This distortion might be a structural distortion where the atoms change position, or it could be the generation of some charge- or spin-density wave (cf. Fig. 19.2) as well as the occurrence of superconductivity or magnetism.

Before turning to the real systems we observe the following. If $\psi_i^{\tilde{k}}$ is an eigen-function for $\hat{h}$ with the eigenvalue $\varepsilon_i(\tilde{k})$, then $[\psi_i^{\tilde{k}}]^* = \psi_i^{-\tilde{k}}$ is an eigenfunction for $\hat{h}$ with the same eigenvalue $\varepsilon_i(\tilde{k}) = \varepsilon_i(-\tilde{k})$, i.e.,

$$[\psi_i^{-\tilde{k}}]^* = \psi_i^{\tilde{k}} \tag{19.62}$$

and

$$\varepsilon_i(-\tilde{k}) = \varepsilon_i(\tilde{k}). \tag{19.63}$$

This can be seen by realizing that for *any* symmetry-adapted function $\chi_m^{\tilde{k}}$ we have

$$\left[\chi_m^{-\tilde{k}}\right]^* = \left[\frac{1}{\sqrt{N}} \sum_{n=1}^{N} \chi_{m,n} e^{-i\tilde{k}n}\right]^*$$

$$= \frac{1}{\sqrt{N}} \sum_{n=1}^{N} \chi_{m,n}^* e^{i\tilde{k}n}, \tag{19.64}$$

i.e., its complex conjugate is a symmetry-adapted function for $-\tilde{k}$. In addition we use that $\varepsilon_i(\tilde{k})$ is real, whereby Eqs. (19.62) and (19.63) directly follow.

In total, this means that in depicting $\varepsilon(\tilde{k})$ we do not need the full interval $[-\pi, \pi]$ but only $[0, \pi]$.

The first real system we shall consider is the five different forms of poly-acetylene shown in Fig. 19.12. These forms can all be described as consisting of repeated C_2H_2 units, where for the three forms (c)–(e) we thereby have to make use of the zigzag symmetry. The band structures of all are shown in Fig. 19.13.

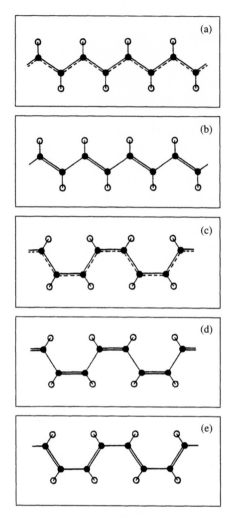

Figure 19.12 Five different structures of polyacetylene. The black and white circles represent carbon and hydrogen atoms, respectively

For the structure of Fig. 19.12(a) the bands [Fig. 19.13(a)] meet pairwise at $\tilde{k} = \pi$. As mentioned above, this indicates that the system has a higher symmetry. It is easily seen that this in fact is the case: it possesses a zigzag symmetry whereby one unit contains only one (and not two) CH unit. Unfolding the bands as demonstrated above in Fig. 19.8 leads to bands like those of Fig. 19.14 [notice that these differ slightly from those of Fig. 19.13(a) for reasons that are unimportant here].

Here we recognize three bands of which two shows an avoided crossing near -13 eV. Neglecting this avoided crossing, we end up with one broad band

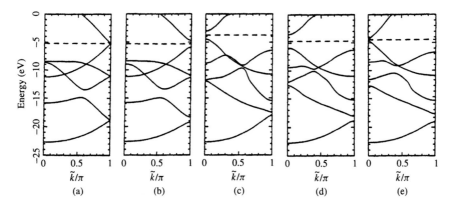

Figure 19.13 The band structures of the five different structures of polyacetylene of Fig. 19.12. The dashed lines mark the Fermi level, i.e., separate occupied and empty bands

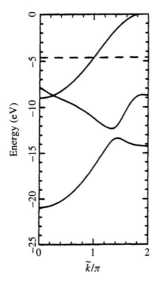

Figure 19.14 The unfolded band structures for the structure of Fig. 19.12(a)

from -21 to $-9\,\mathrm{eV}$ and a narrower one from -14 to $-8\,\mathrm{eV}$. By analysing the wavefunctions (these are not shown here), one can realize that the first of these is essentially due to σ bonds from the sp^2 hybrids between the carbon atoms, whereas the second is due to σ bonds between the carbon atoms and the hydrogen atoms, i.e., formed by sp^2 hybrids on the carbon atoms and $1s$ functions on the hydrogen atoms.

All these orbitals have the same symmetry, i.e., they are symmetric with respect to reflection in the plane of the nuclei. The last band, which in Fig. 19.14 is half filled, is due to functions that are antisymmetric with respect to this reflection. This is thus due to π functions, i.e., carbon p functions perpendicular to the plane of the nuclei. Since this belongs to another symmetry representation, the corresponding wavefunctions have vanishing matrix elements with those of the other bands and the band may cross the other ones. And so it does.

When passing from the system of Fig. 19.12(a) to that of Fig. 19.12(b) the symmetry is reduced, i.e., the zigzag symmetry is no longer present. Therefore, the degeneracies at $\tilde{k} = \pi$ are lifted as seen in Fig. 19.13(b). Nevertheless, the bands of Figs. 19.13(a) and 19.13(b) are very similar, and we shall not discuss this point further. However, the fact that the former system is metallic, whereas the latter is not, can be used in explaining why the system prefers the latter structure.

From the band structures of the three last systems of Fig. 19.12 we see how the changed structure leads to some changes in the band structures although many similarities with those of the other structures are recognized.

As the next system we consider polycarbontrile of Fig. 19.15. It resembles very much the two forms of polyacetylene of Figs. 19.12(a) and 19.12(b), except that here we have an alternation in the type of atoms and, therefore, there is no higher (zigzag) symmetry also for the structure with non-alternating bond lengths.

The band structures, shown in Fig. 19.16 together with those of polyacetylene with the structure of Fig. 19.12(b), are similar to those of polyacetylene. However, the near-degeneracies at $\tilde{k} = \pi$ are clearly removed. For polyacetylene we could imagine forming one broad band from the σ_1 band plus parts of the σ_2 and σ_3 bands (i.e., neglecting the avoided crossing between the two latter). As discussed above, this band would be due to the σ orbitals from the sp^2 hybrids between the carbon atoms. For polycarbonitrile this construction is much less valid, and instead we have one lower band (the σ_1 band) largely due to s and p functions on the nitrogen atoms and an energetically higher 'band' from the σ_2 and σ_3 bands due to those on the carbon atoms. Also for the energetically higher bands this separation is found.

In total we see that both structure and composition can influence the band structures significantly, but also that it is possible to recognize general trends.

As a last example in this Section we consider polyethylene of Fig. 19.17. Also this polymer has a mirror symmetry in the plane of the carbon atoms, and we can accordingly split the orbitals into those that are symmetric and those that are antisymmetric under this reflection. Furthermore, the system has also a zigzag symmetry, so when using the translational symmetry and having one C_2H_4 unit

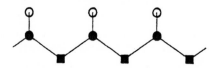

Figure 19.15 Structure of polycarbonitrile. Black squares, black circles, and white circles represent nitrogen atoms, carbon atoms, and hydrogen atoms, respectively

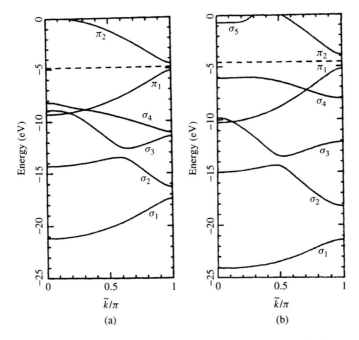

Figure 19.16 Bandstructures of (a) polyacetylene and (b) polycarbonitrile

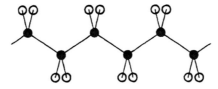

Figure 19.17 Structure of polyethylene. Black and white circles represent carbon and hydrogen atoms, respectively. The hydrogen atoms are lying symmetrically above and below the plane of the carbon atoms

per repeated unit we do not make full use of the symmetry and the bands meet pairwise at $\tilde{k} = \pi$ as seen in Fig. 19.18.

In Fig. 19.18 we see that also bands of σ symmetry cross. That this is not in conflict with what we have said above can be understood by noticing that when making full use of the zigzag symmetry and when unfolding the bands, the crossings will disappear.

On the other hand, there is an avoided crossing between the σ_3 and σ_4 bands. Neglecting this we can form one hypothetical band from different parts of the σ_1, σ_2, and σ_3 bands ranging from -23 to -7 eV, which is due to the σ bonds between the carbon atoms from the sp^3 hybrids. The other parts of the $\sigma_2 + \sigma_3$ bands plus the σ_4 band as well as the π bands are due to the bonds between the

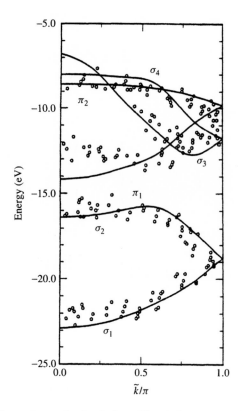

Figure 19.18 Band structures of polyethylene. The circles are experimental results from Seki *et al.* (1986)

carbon atoms and the hydrogen atoms. For each carbon atom there are two C−H bonds, and the σ bands originate from the symmetric (with respect to reflection in the plane of all carbon nuclei) combination of these bond orbitals, whereas the π bands originate from the antisymmetric combination.

19.4 BRILLOUIN ZONES

In Section 19.1 we saw how the translational symmetry naturally leads to the introduction of the symmetry-adapted basis functions. These were characterized by the number $\tilde{k}$ that simultaneously described the irreducible representation. When the operator for the translation of n units is called $\hat{T}_n$, a symmetry-adapted basis function $\chi^{\tilde{k}}$ obeys

$$\hat{T}_n \chi^{\tilde{k}} = e^{in\tilde{k}} \chi^{\tilde{k}}. \tag{19.65}$$

When the length of one unit is called a, then the operator $\hat{T}_n$ may also be described as translating the system by $n \cdot a$. Equivalent to Eq. (19.65) we may therefore also write

$$\hat{T}_n \chi^{\tilde{k}} = e^{in a \tilde{k}/a} \chi^{\tilde{k}},$$ (19.66)

or by replacing $\tilde{k}$ by

$$k = \tilde{k}/a,$$ (19.67)

we may reformulate Eq. (19.66) as

$$\hat{T}_n \chi^k = e^{in a k} \chi^k.$$ (19.68)

Instead of measuring the translation in numbers of units, we measure it thus in a length unit. k has therefore the dimension of length^{-1} (to be more precise, the argument of the exponential above will usually also contain a factor $\hbar^{-1}$, which in atomic units equals 1, so that k gets the dimension of momentum), whereas $\tilde{k}$ is dimensionless.

Equation (19.68) represents the traditional definition of the k number. It is seen to be closely related to the one we have pursued here using group-theoretical arguments. The introduction of k instead of $\tilde{k}$ has some advantages when giving k physical interpretations as we shall see below. On the other hand, for the systems where the primitive symmetry operation is a combined operation (for instance a translation and a rotation for the screw-axis symmetry), the transition from $\tilde{k}$ to k is less obvious. Here, we shall, nonetheless, use k instead of $\tilde{k}$.

Equation (19.65) may be written as

$$\hat{T}_n \chi^{\tilde{k}} = [e^{i\tilde{k}}]^n \chi^{\tilde{k}}.$$ (19.69)

This shows that for a given $\tilde{k}$, replacing

$$\tilde{k} \to \tilde{k} + 2\pi$$ (19.70)

leads to identically no changes. We may therefore *choose* to restrict $\tilde{k}$ to

$$-\pi < \tilde{k} \leq \pi.$$ (19.71)

This corresponds to

$$-\frac{\pi}{a} < k \leq \frac{\pi}{a}$$ (19.72)

and defines the so-called first Brillouin zone.

There is nothing wrong in choosing k values outside this region, but all information is contained within the first Brillouin zone. Independently of which set of k or $\tilde{k}$ values is chosen, there are only N inequivalent ones. And the construction of the Bloch waves is merely a unitary transformation of the original atom-centred functions. Therefore, no additional information is gained, neither is information lost, through the construction of the Bloch waves (i.e., the symmetry-adapted functions).

In the preceding section we discussed various examples of the bands for one-dimensional systems (or quasi-one-dimensional, since the atoms are not one-dimensional but real three-dimensional objects). We saw examples where higher symmetries could be used in unfolding the bands. When the higher symmetry allowed the size of the repeated unit to be halved, for instance, we see from Eq. (19.72) that then the size of the first Brillouin zone is doubled (a is changed to $\frac{a}{2}$). This is exactly what was used when unfolding the bands.

Some systems are periodic in two dimensions. These include surfaces, interfaces, and films. Also these can be treated using the symmetry properties. There are, however, no basic differences to the systems that are periodic in either one or three dimensions, and, therefore, we shall not consider the quasi-two-dimensional systems further here, but instead turn directly to materials with three-dimensional periodicity.

A crystalline material is supposed to be periodic in all three dimensions. This means that there exist three vectors $\vec{a}$, $\vec{b}$, and $\vec{c}$, so that any translation by

$$n_a\vec{a} + n_b\vec{b} + n_c\vec{c} \equiv \vec{R}_{n_a,n_b,n_c} \qquad (19.73)$$

maps the system onto itself. $\vec{a}$, $\vec{b}$, and $\vec{c}$ form the so-called basis and they need neither to have the same length, nor to be perpendicular, although they have to be linearly independent.

Just as in the one-dimensional case we will introduce the operator for the translation (19.73). Calling it $\hat{T}_{n_a,n_b,n_c}$ we can, equivalently to Eq. (19.68), introduce symmetry-adapted basis functions (characterized by the three-dimensional vector $\vec{k}$) through

$$\hat{T}_{n_a,n_b,n_c}\chi^{\vec{k}} = e^{i\vec{R}_{n_a,n_b,n_c}\cdot\vec{k}}\chi^{\vec{k}}. \qquad (19.74)$$

$\vec{k}$ hereby introduced is the conventional $\vec{k}$ vector, well-known from solid-state theory. Therefore, it will not be discussed in any detail here. We shall only mention that just as it was sufficient to consider k obeying Eq. (19.72) for the one-dimensional case, i.e., to consider only the first Brillouin zone, one can also restrict $\vec{k}$ to lie inside the first Brillouin zone in the three-dimensional case. The Brillouin zones are usually not simply cubes in the three-dimensional $\vec{k}$-space, but are some more complex polyhedra, which are characteristic for the individual crystal structures.

In Section 19.3 we saw that k and $-k$ were equivalent in the one-dimensional case. This stays valid also in the three-dimensional case [one may use the same arguments as in Eqs. (19.62)–(19.64)], and in some cases further symmetries may reduce the number of inequivalent $\vec{k}$ points even further. Whether this is the case depends on the symmetry properties of the crystalline material. But when it is the case, it can be of very great help in reducing the computational efforts in an actual calculation (since a typical calculation scales linearly with the number of $\vec{k}$ points).

Finally we mention that also in the three-dimensional case the calculation of the matrix elements follows as in Eqs. (19.53) and (19.54). Thus, assuming that

$$\vec{n} = (n_a, n_b, n_c) \tag{19.75}$$

describes the positions of the different repeated unit cells, we obtain

$$\langle \chi_{m_1}^{\vec{k}_1} | \chi_{m_2}^{\vec{k}_2} \rangle = \delta_{\vec{k}_1, \vec{k}_2} \sum_{\vec{n}} e^{i\vec{k}\cdot\vec{R}_{\vec{n}}} \langle \chi_{m_1, \vec{0}} | \chi_{m_2, \vec{n}} \rangle \tag{19.76}$$

and

$$\langle \chi_{m_1}^{\vec{k}_1} | \hat{h} | \chi_{m_2}^{\vec{k}_2} \rangle = \delta_{\vec{k}_1, \vec{k}_2} \sum_{\vec{n}} e^{i\vec{k}\cdot\vec{R}_{\vec{n}}} \langle \chi_{m_1, \vec{0}} | \hat{h} | \chi_{m_2, \vec{n}} \rangle. \tag{19.77}$$

These results are generally valid, independently of the kind of computational method applied (there may, however, be some deviations when $\hat{h}$ is orbital-dependent like, e.g., for the Fock operator or the self-interaction-corrected Kohn–Sham operator). Furthermore, the summations run over all unit cells, whereby it is assumed that for sufficiently distant unit cells the matrix elements can be discarded.

Somewhat different formulas apply when using plane waves as basis functions. These are not centred on individual atoms but are per construction symmetry-adapted and delocalized. Then, different sets of plane waves are used for different $\vec{k}$ points. We shall not, however, discuss this point further.

19.5 BAND STRUCTURES IN THREE DIMENSIONS

For a given crystalline material an electronic-structure calculation proceeds just as for a molecule with only some smaller methodological differences. It is of ultimate importance to construct the symmetry-adapted basis functions characterized by the vector $\vec{k}$. Thereby, the problem of solving the $\infty \times \infty$ secular equation

$$\underline{\underline{H}} \cdot \underline{c}_i = \varepsilon_i \cdot \underline{\underline{O}} \cdot \underline{c}_i \tag{19.78}$$

is transformed into finite ones:

$$\underline{\underline{H}}(\vec{k}) \cdot \underline{c}_i(\vec{k}) = \varepsilon_i(\vec{k}) \cdot \underline{\underline{O}}(\vec{k}) \cdot \underline{c}_i(\vec{k}). \tag{19.79}$$

These have only the dimension $N_b \times N_b$ with N_b being the number of basis functions per unit cell. On the other hand, this equation has in principle to be solved for *any* $\vec{k}$-vector in the first Brillouin zone which, of course, is impossible. Instead, one solves them for a finite set of $\vec{k}$ points and then interpolates the results in order to estimate the results for the infinite set of $\vec{k}$ points.

There are then two different approaches one may apply in selecting the finite set of discrete $\vec{k}$ points. For semiconductors or insulators (i.e., for systems where all bands are either filled or empty and the Fermi level lies in an energy gap) it can be of advantage to use the so-called special $\vec{k}$ points (Baldereschi, 1973; Monkhorst and Pack, 1976). The strategy behind their construction is that when adding the densities (or other $\vec{k}$-dependent quantities) for all $\vec{k}$ points in the first Brillouin zone for any band, the resulting density is particularly smooth and can to a good approximation be described by using the orbitals of the few special $\vec{k}$ points. That is,

$$\rho_i(\vec{r}) = \sum_{\vec{k}} |\psi_i^{\vec{k}}(\vec{r})|^2 \simeq \sum_{\vec{k} \in SS} w(\vec{k}) |\psi_i^{\vec{k}}(\vec{r})|^2, \qquad (19.80)$$

where SS denotes the set of special $\vec{k}$ points, i is a band index, and $w(\vec{k})$ is a weight factor. The first $\vec{k}$ summation in Eq. (19.80) is a summation over the infinite, continuous set, whereas the second one is over only the small, finite set. A very important point is that the special $\vec{k}$ points do not depend on the material or the bands (i), but only on the crystal structure.

For metallic systems this approach is not useful. Then, there will be partly filled bands, i.e. for some i above in Eq. (19.80), the $\vec{k}$ summation will not be over the complete first Brillouin zone but only over parts of it. Instead, one may use the so-called tetrahedron method (Jepsen and Andersen, 1971). Thereby, a regular set of equidistant $\vec{k}$ points is constructed,

$$\vec{k}_{(n_1, n_2, n_3)} = n_1 \cdot \vec{k}_1 + n_2 \cdot \vec{k}_2 + n_3 \cdot \vec{k}_3, \qquad (19.81)$$

where n_1, n_2, n_3 are integers, and $\vec{k}_1, \vec{k}_2, \vec{k}_3$ are certain vectors so defined that for

$$n_i = -N_i, -N_i + 1, \ldots, N_i - 1, N_i, \qquad (19.82)$$

$\vec{k}_{(n_1, n_2, n_3)}$ lies inside the first Brillouin zone. N_1, N_2, and N_3 are chosen parameters that characterize the density of the $\vec{k}$ points.

Subsequently, inside any so-called microcell, defined by the eight points (n_1, n_2, n_3), $(n_1 + 1, n_2, n_3)$, $(n_1, n_2 + 1, n_3)$, $(n_1, n_2, n_3 + 1)$, $(n_1 + 1, n_2 + 1, n_3)$, $(n_1 + 1, n_2, n_3 + 1)$, $(n_1, n_2 + 1, n_3 + 1)$, and $(n_1 + 1, n_2 + 1, n_3 + 1)$ one constructs four (non-overlapping) tetrahedra, each having four of these points as corners. Finally, from the eigenvalues and wavefunctions at these four corners of each tetrahedron one makes a linear interpolation in order to estimate these quantities at any $\vec{k}$ point inside the tetrahedron.

In Sections 19.1 and 19.3 we presented the band structures for some one-dimensional systems, i.e., we presented $\varepsilon_i(k)$ for *all* inequivalent k points. Similar plots can not be made for the three-dimensional systems for the simple reason that $\varepsilon_i(\vec{k})$ is a function of a three-dimensional variable $\vec{k}$. Instead, it has become practice to show the bands along some few lines in the first Brillouin zone. We shall here illustrate it through one simple example.

A number of semiconductors crystallize in the so-called zincblende structure shown in Fig. 19.19. This structure is closely related to the diamond structure

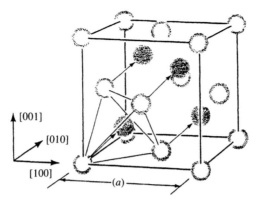

[001]

[010]

[100]

(a)

Figure 19.19 The zincblende structure. For the diamond structure, the dark and light atoms are identical. Reproduced with permission from Harrison (1980)

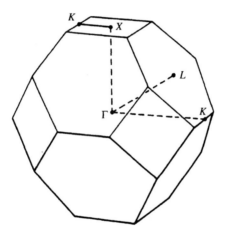

Figure 19.20 The first Brillouin zone and some high-symmetry points for the zincblende structure of Fig. 19.19. Reproduced with permission from Harrison (1980)

that can be obtained from the zincblende structure by letting the two different types of atoms become identical.

The first Brillouin zone of the zincblende structure is shown in Fig. 19.20, where we also show the names of some of the so-called high-symmetry $\vec{k}$ points. That these are placed at points of high symmetry should be obvious from the figure. Their names are chosen according to tradition. Most important is it that for *any* crystal structure, Γ is the $\vec{k} = (0, 0, 0)$ point.

In Fig. 19.21 we show the band structures for various semiconductors with this crystal structure. The bands are so shifted that the highest occupied orbital appears at $\varepsilon_i(\vec{k}) = 0$. This may at a first sight seem a surprising procedure, but one has to remember that usually one defines an energy axis so that the energy zero occurs infinitely far away from the system. For an infinite crystal that occupies the whole

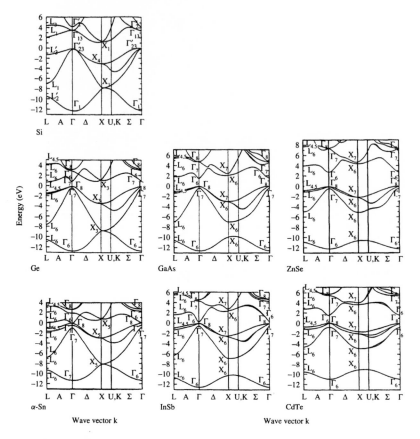

Figure 19.21 The band structures for various semiconductors with the zincblende crystal structure. Reproduced with permission from Harrison (1980)

space, there is no 'infinitely far away', and one can therefore arbitrarily define the zero of the energy axis. In Fig. 19.21 only the (occupied) valence bands and the lowest (unoccupied) conduction bands are shown. The core bands, which are very flat, appear close to the energies of the free-atom core orbitals — shifted according to the above-mentioned shift of the energy zero.

For silicon we see four occupied bands. That there are four bands can be seen, e.g., along the line $K - \Gamma$, whereas some of them become degenerate along, e.g., the line $L - \Gamma$. Each band can accommodate two electrons (due to spin). One unit cell contains two silicon atoms, and the bands can be interpreted as being due to the sp^3 hybrids. A more careful analysis reveals that the energetically lowest orbitals are mainly from $3s$ functions whereas the highest occupied ones are from $3p$ functions.

Passing to germanium only minor changes occur, which is also the case when passing to tin. Here, however, there are some splittings of the bands for instance along the line $L - \Gamma$, as also can be seen for germanium, although to a lesser extent. These are due to spin–orbit couplings (i.e., relativistic effects), and will not be discussed further here.

For the series Si–Ge–Sn the gap between occupied and unoccupied bands decreases, and is actually vanishing for Sn. When, on the other hand, increasing the ionicity (e.g., passing from Ge via GaAs to ZnSe), this band gap increases, and the bands become flatter. The reason for this is identical to what we discussed in Section 19.3 for passing from polyacetylene to polycarbonitrile. Therefore, we will not consider this in more detail here.

For some crystals, in particular systems based on $3d$ or $4f$ atoms, the bands for α spin differ from those for β spin and the system may be magnetic. In effect, one has then two sets of bands superposed, but except for being more complicated, there is principally no differences to what we have discussed here. In Section 22.5 we shall present one example of this case.

19.6 BLOCH'S FORMULATION

We have seen that the symmetry of the system can be used with advantage in constructing symmetry-adapted basis functions from the equivalent atom-centered ones of different unit cells. Therefore, the wavefunctions will be written in the form

$$\psi_j^{\vec{k}}(\vec{r}) = \sum_{m=1}^{N_b} c_{j,m}(\vec{k}) \chi_m^{\vec{k}}(\vec{r}), \tag{19.83}$$

where

$$\chi_m^{\vec{k}}(\vec{r}) = \frac{1}{\sqrt{N}} \sum_{\vec{n}} e^{i\vec{k}\cdot\vec{R}_{\vec{n}}} \chi_{m,\vec{n}}(\vec{r}), \tag{19.84}$$

and where $\vec{R}_{\vec{n}}$ is a lattice vector of Eq. (19.73). $N \to \infty$ is the number of unit cells. Often, the functions of Eqs. (19.83) and (19.84) are called Bloch waves or Bloch functions.

The functions $\chi_{m,\vec{n}}(\vec{r})$ obey

$$\chi_{m,\vec{n}}(\vec{r}) = \chi_{m,\vec{0}}(\vec{r} - \vec{R}_{\vec{n}}). \tag{19.85}$$

This formulation is closely related to having atom-centered basis functions, which is actually very often the case in practical electronic-structure calculations except when using plane waves as basis functions. Bloch suggested originally another formulation (Bloch, 1929). He showed that any electronic eigenfunction could be written in the form

$$\psi_j^{\vec{k}}(\vec{r}) = e^{i\vec{k}\cdot\vec{r}} u_j^{\vec{k}}(\vec{r}), \tag{19.86}$$

where $u_j^{\vec{k}}(\vec{r})$ is a periodic function,

$$u_j^{\vec{k}}(\vec{r} + \vec{R}_{\vec{n}}) = u_j^{\vec{k}}(\vec{r}), \tag{19.87}$$

although it depends on $\vec{k}$.

It is easily seen that $\psi_j^{\vec{k}}(\vec{r})$ transforms according to the irreducible representation $\vec{k}$,

$$
\begin{aligned}
\psi_j^{\vec{k}}(\vec{r} + \vec{R}_{\tilde{n}}) &= e^{i\vec{k}\cdot(\vec{r}+\vec{R}_{\tilde{n}})} u_j^{\vec{k}}(\vec{r} + \vec{R}_{\tilde{n}}) \\
&= e^{i\vec{k}\cdot\vec{R}_{\tilde{n}}} e^{i\vec{k}\cdot\vec{r}} u_j^{\vec{k}}(\vec{r} + \vec{R}_{\tilde{n}}) \\
&= e^{i\vec{k}\cdot\vec{R}_{\tilde{n}}} e^{i\vec{k}\cdot\vec{r}} u_j^{\vec{k}}(\vec{r}) \\
&= e^{i\vec{k}\cdot\vec{R}_{\tilde{n}}} \psi_j^{\vec{k}}(\vec{r}),
\end{aligned}
\tag{19.88}
$$

due to Eqs. (19.86) and (19.87).

Combining Eqs. (19.83) and (19.84) with Eq. (19.86) we obtain

$$
\psi_j^{\vec{k}}(\vec{r}) = \frac{1}{\sqrt{N}} \sum_m \sum_{\tilde{n}} c_{j,m}(\vec{k}) \chi_{m,\tilde{n}}(\vec{r}) e^{i\vec{k}\cdot\vec{r}_{\tilde{n}}} \equiv e^{i\vec{k}\cdot\vec{r}} u_j^{\vec{k}}(\vec{r}),
\tag{19.89}
$$

or

$$
u_j^{\vec{k}}(\vec{r}) = \frac{1}{\sqrt{N}} \sum_m \sum_{\tilde{n}} c_{j,m}(\vec{k}) \chi_{m,\tilde{n}}(\vec{r}) e^{i\vec{k}\cdot(\vec{R}_{\tilde{n}}-\vec{r})}
\tag{19.90}
$$

is a periodic function. This equation shows first of all that there is a non-trivial relation between the function u and the atom-centred functions χ.

Let us consider a simple example, i.e., a linear chain with one Gaussian per unit. Then, the k-dependent wavefunction becomes

$$
\psi^k(x) = \frac{1}{\sqrt{N}} \sqrt{\frac{2\alpha}{\pi}} \sum_n e^{-\alpha(x-nd)^2} e^{iknd},
\tag{19.91}
$$

where d is the lattice constant (neglecting a normalization constant). This function can be written as

$$
\begin{aligned}
\psi^k(x) &= e^{ikx} \left[\frac{1}{\sqrt{N}} \sqrt{\frac{2\alpha}{\pi}} \sum_n e^{-\alpha(x-nd)^2} e^{ind)} \right] e^{-ikx} \\
&= e^{ikx} \left[\frac{1}{\sqrt{N}} \sqrt{\frac{2\alpha}{\pi}} \sum_n e^{-\alpha(x-nd)^2} e^{-ik(x-nd)} \right] \\
&\equiv e^{ikx} u^k(x).
\end{aligned}
\tag{19.92}
$$

Hereby, the periodic function $u^k(x)$ is defined.

19.7   CRYSTAL MOMENTUM

The vector $\vec{k}$ has been introduced as a tool from group theory that can be used in simplifying the electronic-structure calculation for a crystalline material. It has, however, not been given any physical interpretation. But it is possible to do so. To this end we consider Bloch's formulation,

$$\psi_j^{\vec{k}}(\vec{r}) = e^{i\vec{k}\cdot\vec{r}} u_j^{\vec{k}}(\vec{r}). \tag{19.93}$$

Applying the momentum operator

$$\hat{\vec{p}} = -i\vec{\nabla} \tag{19.94}$$

on this function gives

$$\begin{aligned}
\hat{\vec{p}}\psi_j^{\vec{k}}(\vec{r}) &= -i\vec{\nabla}\psi_j^{\vec{k}}(\vec{r}) \\
&= -i\vec{\nabla}\left[e^{i\vec{k}\cdot\vec{r}} u_j^{\vec{k}}(\vec{r}) \right] \\
&= \left[-i\vec{\nabla} e^{i\vec{k}\cdot\vec{r}} \right] u_j^{\vec{k}}(\vec{r}) + e^{i\vec{k}\cdot\vec{r}}\left[-i\vec{\nabla} u_j^{\vec{k}}(\vec{r}) \right] \\
&= \vec{k} e^{i\vec{k}\cdot\vec{r}} u_j^{\vec{k}}(\vec{r}) + e^{i\vec{k}\cdot\vec{r}}\left[-i\vec{\nabla} u_j^{\vec{k}}(\vec{r}) \right] \\
&= \vec{k}\psi_j^{\vec{k}}(\vec{r}) + e^{i\vec{k}\cdot\vec{r}}\left[-i\vec{\nabla} u_j^{\vec{k}}(\vec{r}) \right].
\end{aligned} \tag{19.95}$$

For a completely free electron (such electrons are to a reasonable approximation found in simple metals), $u_j^{\vec{k}}(\vec{r})$ is a constant and the last term on the right-hand side of Eq. (19.95) vanishes. Then $\vec{k}$ is the momentum of the electron, and therefore it is called the crystal momentum. For fewer free electrons the momentum gets modified by effects due to the spatial variations in $u_j^{\vec{k}}(\vec{r})$.

We may, as a further illustration, consider the example above from Eq. (19.91). Then (using the fact that this model is assumed to be one-dimensional),

$$\begin{aligned}
\hat{p}\psi^k(x) &= -i\frac{d}{dx}\psi^k(x) \\
&= -i\left[ik \cdot e^{ikx} u^k(x) + e^{ikx}\frac{1}{\sqrt{N}}\sqrt{\frac{2\alpha}{\pi}}\frac{d}{dx}\sum_n e^{-\alpha(x-nd)^2} e^{-ik(x-nd)} \right] \\
&= k \cdot \psi^k(x) - ie^{ikx}\frac{1}{\sqrt{N}}\sqrt{\frac{2\alpha}{\pi}}\sum_n e^{-\alpha(x-nd)^2 - ik(x-nd)}\left[-2\alpha(x-nd) - ik \right] \\
&= k \cdot \psi^k(x) + \frac{1}{\sqrt{N}}\sqrt{\frac{2\alpha}{\pi}}\sum_n e^{-\alpha(x-nd)^2 + iknd}[2i\alpha(x-nd) - k].
\end{aligned} \tag{19.96}$$

Here, the second term represents a modification to k compared to the simplest interpretation of k as the momentum.

19.8 WANNIER FUNCTIONS

In Section 19.1 we studied one-dimensional systems with one basis function per unit cell. We saw how we could construct symmetry-adapted basis functions that were delocalized over the complete system and how these gave rise to the band. Any of the wavefunctions were thus spread out all over the system, but due to its construction we *knew* that originally the band was due to some basis functions that were localized at the individual units.

When increasing the complexity as was done in Section 19.3, where we first studied the slightly more complicated systems with two basis functions per unit cell, it was still possible — through careful analysis — to recognize the localized basis functions that were responsible for the bands. As the number of basis functions and of bands increases it becomes less obvious how to interpret the band structures as being due to localized functions and one may ask the question whether it is possible, in the general case, to construct such localized functions. The answer is 'yes, in principle', although this is only very rarely done. These localized functions are the so-called Wannier functions (Wannier, 1937, 1962).

In order to illustrate their properties we shall here consider only one-dimensional systems. The extension to more dimensions is straightforward. We consider accordingly a ring molecule of N repeated units, and for each we have N_b basis functions. We construct the symmetry-adapted basis functions as in Section 19.1,

$$\chi_m^{\tilde{k}} = \frac{1}{\sqrt{N}} \sum_{n=1}^{N} e^{ikn} \chi_{m,n} \tag{19.97}$$

and the solutions to the single-particle equations

$$\hat{h}\psi_j^{\tilde{k}} = \varepsilon_j(\tilde{k})\psi_j^{\tilde{k}} \tag{19.98}$$

become

$$\psi_j^{\tilde{k}} = \sum_{m=1}^{N_b} c_{j,m}(\tilde{k})\chi_m^{\tilde{k}}, \tag{19.99}$$

where $\tilde{k}$ is given by Eq. (19.22).

Before proceeding we observe that, if $\psi_j^{\tilde{k}}$ is a normalized solution to Eq. (19.98), then so is

$$e^{i\phi_j(\tilde{k})}\psi_j^{\tilde{k}}. \tag{19.100}$$

That is, we can multiply any solution with any phase factor without changing *any* physical observable.

A Wannier function is now constructed by summing all the functions (19.100) over one band, i.e., over all $\tilde{k}$,

$$w_{j,p}(\vec{r}) = \frac{1}{\sqrt{N}} \sum_{\tilde{k}} e^{i\phi_j(\tilde{k})}\psi_j^{\tilde{k}}(\vec{r}) e^{-i\tilde{k}p}, \tag{19.101}$$

where p is an integer.

By using the fact that

$$\langle \psi_{j_1}^{\tilde{k}_1} | \psi_{j_2}^{\tilde{k}_2} \rangle = \delta_{j_1,j_2} \delta_{\tilde{k}_1,\tilde{k}_2} \tag{19.102}$$

(the first δ is due to the fact that orbitals of the same $\tilde{k}$ but from different bands are orthogonal, whereas the latter is due to the orthogonality of functions of different irreducible representations), we can easily show that also the Wannier functions are orthonormal:

$$\langle w_{j_1,p_1} | w_{j_2,p_2} \rangle = \frac{1}{N} \sum_{\tilde{k}_1,\tilde{k}_2} e^{i[\phi_{j_2}(\tilde{k}_2)-\phi_{j_1}(\tilde{k}_1)]} e^{i(\tilde{k}_1 p_1 - \tilde{k}_2 p_2)} \langle \psi_{j_1}^{\tilde{k}_1} | \psi_{j_2}^{\tilde{k}_2} \rangle$$

$$= \frac{1}{N} \sum_{\tilde{k}_1,\tilde{k}_2} e^{i[\phi_{j_2}(\tilde{k}_2)-\phi_{j_1}(\tilde{k}_1)]} e^{i(\tilde{k}_1 p_1 - \tilde{k}_2 p_2)} \delta_{j_1,j_2} \delta_{\tilde{k}_1,\tilde{k}_2}$$

$$= \frac{1}{N} \sum_{\tilde{k}} e^{i[\phi_{j_2}(\tilde{k})-\phi_{j_1}(\tilde{k})]} e^{i(\tilde{k} p_1 - \tilde{k} p_2)} \delta_{j_1,j_2}$$

$$= \frac{1}{N} \sum_{\tilde{k}} e^{i[\phi_{j_1}(\tilde{k})-\phi_{j_1}(\tilde{k})]} e^{i(\tilde{k} p_1 - \tilde{k} p_2)} \delta_{j_1,j_2}$$

$$= \frac{1}{N} \sum_{\tilde{k}} e^{i(\tilde{k} p_1 - \tilde{k} p_2)} \delta_{j_1,j_2}$$

$$= \frac{1}{N} \sum_{\tilde{k}} e^{i\tilde{k}(p_1 - p_2)} \delta_{j_1,j_2}. \tag{19.103}$$

The $\tilde{k}$ are those of Eq. (19.22), i.e., they can be written as

$$\tilde{k} = \tilde{k}_{\min} + l \cdot \frac{2\pi}{N} \tag{19.104}$$

with

$$l = 0, 1, 2, \ldots, N-1 \tag{19.105}$$

and with $\tilde{k}_{\min}$ being the smallest one.

Therefore,

$$\sum_{\tilde{k}} e^{i\tilde{k}p} = e^{i\tilde{k}_{\min}p} \sum_{l=0}^{N-1} e^{il\,p2\pi/N} = e^{i\tilde{k}_{\min}p} \sum_{l=0}^{N-1} [e^{ip2\pi/N}]^l = e^{i\tilde{k}_{\min}p} \frac{1 - e^{i2\pi p}}{1 - e^{ip2\pi/N}}. \tag{19.106}$$

We have here used the standard expression

$$\sum_{n=0}^{N-1} x^n = \frac{1 - x^N}{1 - x}. \tag{19.107}$$

Since p is an integer, the expression in Eq. (19.106) vanishes unless $p = 0$. In that case, the sum is that of N times 1, i.e., N. This gives, inserted in Eq. (19.103),

$$\langle w_{j_1, p_1} | w_{j_2, p_2} \rangle = \delta_{j_1, j_2} \delta_{p_1, p_2}, \tag{19.108}$$

i.e., the Wannier functions are orthonormal.

The construction of the Wannier functions is nothing but a unitary transformation of the eigenfunctions $\psi_j^{\tilde{k}}$. Therefore, we have as many Wannier functions as we have eigenfunctions, i.e., N per band. They differ in the choice of p, which may take any of the values

$$p = 1, 2, \ldots, N. \tag{19.109}$$

So far we have not verified that these functions in any way resemble localized atom-centred functions. In fact, it is far from clear that this is the case, but in order to approach this question let us consider the simple model of Section 19.1, i.e. a system with one basis function per unit cell. For this system the Wannier functions are [according to Eq. (19.101)]

$$
\begin{aligned}
w_p(\vec{r}) &= \frac{1}{\sqrt{N}} \sum_{\tilde{k}} e^{i\phi(\tilde{k})} \psi^{\tilde{k}}(\vec{r}) e^{-i\tilde{k}p} \\
&= \frac{1}{\sqrt{N}} \sum_{\tilde{k}} e^{i\phi(\tilde{k})} \chi^{\tilde{k}}(\vec{r}) e^{-i\tilde{k}p} \\
&= \frac{1}{\sqrt{N}} \sum_{\tilde{k}} e^{i\phi(\tilde{k})} \left[\frac{1}{\sqrt{N}} \sum_{n=1}^{N} \chi_n(\vec{r}) e^{i\tilde{k}n} \right] e^{-i\tilde{k}p} \\
&= \frac{1}{N} \sum_{\tilde{k}} \sum_{n=1}^{N} e^{i\phi(\tilde{k})} \chi_n(\vec{r}) e^{i\tilde{k}(n-p)}, \tag{19.110}
\end{aligned}
$$

where we have omitted the index j (since we only have one basis function per unit cell).

We now *choose* $\phi(\tilde{k}) = 0$, and apply Eq. (19.107) in obtaining

$$
\begin{aligned}
w_p(\vec{r}) &= \frac{1}{N} \sum_{\tilde{k}} \sum_{n=1}^{N} \chi_n(\vec{r}) e^{i\tilde{k}(n-p)} \\
&= \sum_{n=1}^{N} \chi_n(\vec{r}) \left[\frac{1}{N} \sum_{\tilde{k}} e^{i\tilde{k}(n-p)} \right] \\
&= \sum_{n=1}^{N} \chi_n(\vec{r}) \delta_{n,p}, \tag{19.111}
\end{aligned}
$$

i.e., the Wannier function w_p is in fact the original, atom-centred basis function centered at the pth unit cell, i.e., χ_p.

This gives first of all a meaning of the parameter p. Second, it tells that the change

$$p \rightarrow p + N \qquad (19.112)$$

results in the same Wannier function, i.e., p is only specified within an integral multiple of N. Third, it shows that other choices of the phase factors may lead to other Wannier functions. For example, choosing

$$\phi(\tilde{k}) = a \cdot \tilde{k} \qquad (19.113)$$

(notice that this choice is not motivated by physical reasons but chosen more or less randomly) gives

$$w_p(\vec{r}) = \sum_{n=1}^{N} \chi_n(\vec{r}) \left[\frac{1}{N} e^{i\tilde{k}_{\min}(n-p+a)} \frac{1 - e^{ia2\pi}}{1 - e^{i(n-p+a)2\pi/N}} \right], \qquad (19.114)$$

i.e., a function that for a non-integral has non-vanishing components over the complete system. Therefore, the spatial extent of the Wannier functions depends sensitively on the phase factors. It can, however, be shown (des Cloizeaux, 1964) that for $N \rightarrow \infty$ a Wannier function decays — in principle — exponentially for $|\vec{r}| \rightarrow \infty$, although it is not clear whether this behaviour is also found at shorter ranges. But this behaviour is the reason for using Wannier functions whereby it is hoped that the exponential decay is reached already at a short distance from the centre of the function.

We shall study two further properties of the Wannier functions. First, replacing p by $p + 1$ gives

$$w_{j,p+1}(\vec{r}) = \frac{1}{\sqrt{N}} \sum_{\tilde{k}} e^{i\phi_j(\tilde{k})} \psi_j^{\tilde{k}}(\vec{r}) e^{-i\tilde{k}(p+1)}$$

$$= \frac{1}{\sqrt{N}} \sum_{\tilde{k}} e^{i\phi_j(\tilde{k})} \sum_{m=1}^{N_b} c_{j,m}(\tilde{k}) \chi_m^{\tilde{k}}(\vec{r}) e^{-i\tilde{k}(p+1)}$$

$$= \frac{1}{\sqrt{N}} \sum_{\tilde{k}} e^{i\phi_j(\tilde{k})} \sum_{m=1}^{N_b} c_{j,m}(\tilde{k}) \sum_{n=1}^{N} \chi_{m,n}(\vec{r}) e^{i\tilde{k}n} e^{-i\tilde{k}(p+1)}$$

$$= \sum_{m,n} \left\{ \frac{1}{\sqrt{N}} \sum_{\tilde{k}} e^{i\phi_j(\tilde{k})} c_{j,m}(\tilde{k}) e^{i\tilde{k}n} e^{-i\tilde{k}(p+1)} \right\} \chi_{m,n}(\vec{r})$$

$$= \sum_{m,n} \left\{ \frac{1}{\sqrt{N}} \sum_{\tilde{k}} e^{i\phi_j(\tilde{k})} c_{j,m}(\tilde{k}) e^{i\tilde{k}(n-p-1)} \right\} \chi_{m,n}(\vec{r})$$

$$= \sum_{m,n} \left\{ \frac{1}{\sqrt{N}} \sum_{\tilde{k}} e^{i\phi_j(\tilde{k})} c_{j,m}(\tilde{k}) e^{i\tilde{k}[(n-1)-p]} \right\} \chi_{m,n}(\vec{r})$$

$$= \sum_{m,q(=n-1)} \left\{ \frac{1}{\sqrt{N}} \sum_{\tilde{k}} e^{i\phi_j(\tilde{k})} c_{j,m}(\tilde{k}) e^{i\tilde{k}(q-p)} \right\} \chi_{q+1,n}(\vec{r}). \quad (19.115)$$

This is exactly the expression for the Wannier function $w_{j,p}(\vec{r})$ translated by one unit cell. Thus, two Wannier functions only differing in the p index are identical but translated relatively to each other.

Since the Wannier functions are nothing but a unitary transform of the Bloch waves, they result in the same total electron density. This means that

$$\sum_{p=1}^{N} |w_{j,p}(\vec{r})|^2 = \sum_{\tilde{k}} |\psi_j^{\tilde{k}}(\vec{r})|^2. \quad (19.116)$$

This identity may also be shown rigorously by inserting the definition of the Wannier functions into the right-hand side of Eq. (19.116),

$$\sum_{p=1}^{N} |w_{j,p}(\vec{r})|^2 = \sum_{p=1}^{N} \left| \frac{1}{\sqrt{N}} \sum_{\tilde{k}} e^{i\phi_j(\tilde{k})} \psi_j^{\tilde{k}}(\vec{r}) e^{-i\tilde{k}p} \right|^2$$

$$= \frac{1}{N} \sum_{p=1}^{N} \sum_{\tilde{k}_1,\tilde{k}_2} e^{i[\phi_j(\tilde{k}_2)-\phi_j(\tilde{k}_1)]} e^{i(\tilde{k}_1-\tilde{k}_2)p} \psi_j^{\tilde{k}_2*}(\vec{r}) \psi_j^{\tilde{k}_1}(\vec{r}). \quad (19.117)$$

The p summation gives a non-zero contribution only for $\tilde{k}_1 = \tilde{k}_2$, and in that case it becomes N. Substituting $\tilde{k}_1 = \tilde{k}_2 \equiv \tilde{k}$ then gives

$$\sum_{p=1}^{N} |w_{j,p}(\vec{r})|^2 = \sum_{\tilde{k}} \psi_j^{\tilde{k}*}(\vec{r}) \psi_j^{\tilde{k}}(\vec{r}), \quad (19.118)$$

which is identical to Eq. (19.116).

The advantage of the Wannier functions is that they permit the construction of functions that — in principle — are localized, and where there is only one function per unit cell and band. Thus, when large basis sets are used the number of functions can be reduced by constructing the Wannier functions without losing accuracy. This means that constructing the Wannier functions may be a means of obtaining a good starting point for including perturbations in the calculations that otherwise would be very difficult to include with larger basis sets or with basis sets of less localized functions. However, the Wannier functions are not automatically localized, but by adjusting the phases $\phi_j(\tilde{k})$ one may hope to obtain localized functions. Nevertheless, the applications and construction of Wannier functions in electronic-structure calculations for specific systems is still in its infancy.

19.9 DENSITY OF STATES

For a ring molecule consisting of N identical units, each having one basis function, we can construct N symmetry-adapted Bloch waves characterized by $\tilde{k} = a \cdot k$ with a being the length of one unit cell. These Bloch waves define in turn a single band with $\varepsilon = \varepsilon(k)$. The N basis functions can at most accommodate $2N$ electrons (including spin) and, accordingly, so can the band. Alternatively stated, the band can be filled with two electrons per unit.

Having N_b basis functions per unit rather than just one leads to N_b bands which can still each contain two electrons per unit. The same arguments can be carried through for structures that are periodic in two or three dimensions.

The first Brillouin zone spans the interval

$$-\pi/a < k \leq \pi/a \qquad (19.119)$$

and has accordingly the length $\frac{2\pi}{a}$. Thus, in k space we can have two electrons per band and per $\frac{2\pi}{a}$.

In Fig. 19.22 we show the band structures for a system like that of Fig. 19.1, for which we assume that

$$\langle \chi_n | \chi_m \rangle = \delta_{n,m}$$

$$\langle \chi_n | \hat{h} | \chi_m \rangle = \begin{cases} -t & \text{for } n, m \text{ neighbours} \\ \varepsilon_0 & n = m \\ 0 & \text{otherwise.} \end{cases} \qquad (19.120)$$

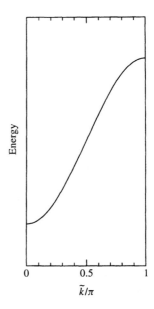

Figure 19.22 The band structures for the model of Eqs. (19.120) and (19.121)

For this model,

$$\varepsilon(k) = \varepsilon_0 - 2t\cos(ak). \tag{19.121}$$

We see in the figure that the energy band spans the interval

$$\varepsilon_0 - 2t \le \varepsilon(k) \le \varepsilon_0 + 2t. \tag{19.122}$$

We may now ask how many states we have per energy interval. For the model of Fig. 19.22 we have for ε of Eq. (19.122) that this number is proportional to $1/|d\varepsilon(k)/dk|$ evaluated for the energy of interest, i.e., the more flat the band is, the more states per energy interval. Since the length of the first Brillouin zone equals $\frac{2\pi}{a}$, the proportionality constant equals $1/\frac{2\pi}{a}$.

The number of states per energy interval is called the density of states $D(\varepsilon)$. Accordingly $D(\varepsilon)\,d\varepsilon$ is the number of states in the energy interval $[\varepsilon; \varepsilon + d\varepsilon]$.

For the model above we have in total

$$D(\varepsilon) = \frac{1}{\left(\dfrac{2\pi}{a}\right)} \int_{-\pi/a}^{\pi/a} \frac{1}{\left|\dfrac{d\varepsilon(k)}{dk}\right|} \delta(\varepsilon(k) - \varepsilon)\,dk$$

$$= \frac{a}{2\pi} \int_{-\pi/a}^{\pi/a} \frac{1}{\left|\dfrac{d\varepsilon(k)}{dk}\right|} \delta(\varepsilon(k) - \varepsilon)\,dk. \tag{19.123}$$

For the model above we see immediately that for a given ε obeying Eq. (19.122) we have two k points (differing in sign) for which $\varepsilon(k) = \varepsilon$. Thus, the density of states becomes

$$\frac{2a}{2\pi} \frac{1}{|a \cdot 2t\sin(ak)|}. \tag{19.124}$$

Using Eq. (19.121) we find

$$|a \cdot 2t\sin(ak)| = 2at\sqrt{1 - \cos^2(ak)}$$

$$= 2at\sqrt{1 - \frac{\varepsilon(k) - \varepsilon_0}{2t}^2}$$

$$= a\sqrt{(2t)^2 - [\varepsilon(k) - \varepsilon_0]^2} \tag{19.125}$$

so that

$$D(\varepsilon) = \frac{1}{2\pi} \frac{1}{\sqrt{(2t)^2 - [\varepsilon(k) - \varepsilon_0]^2}}. \tag{19.126}$$

This function diverges for

$$\varepsilon(k) = \varepsilon_0 \pm 2t, \tag{19.127}$$

i.e., at the band edges where (see Fig. 19.119) the bands are flat. The density of states is shown in Fig. 19.120.

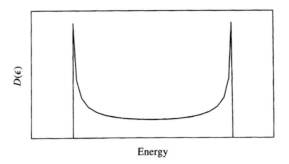

Figure 19.23 The density of states for the model of Eqs. (19.120) and (19.121)

When we have more bands, i.e., $\varepsilon(k)$ is changed into $\varepsilon_i(k)$ with i labelling the different bands, we have to include an additional summation over the different bands. Thereby Eq. (19.123) is modified into

$$D(\varepsilon) = \frac{a}{2\pi} \sum_i \int_{-\pi/a}^{\pi/a} \frac{1}{\left|\dfrac{d\varepsilon(k)}{dk}\right|} \delta(\varepsilon_i(k) - \varepsilon)\,dk \tag{19.128}$$

and, finally, for a crystalline material that is periodic in three dimensions, the integration is over the first Brillouin zone (labelled 1BZ) and the prefactor is the volume of this (V_{1BZ}). Then,

$$D(\varepsilon) = \frac{1}{V_{1BZ}} \sum_i \int_{1BZ} \frac{1}{\left|\nabla_{\vec{k}}\varepsilon_i(\vec{k})\right|} \delta(\varepsilon_i(\vec{k}) - \varepsilon)\,d\vec{k}. \tag{19.129}$$

The singularity in $D(\varepsilon)$ of Eq. (19.126) is a true singularity, i.e., the function diverges. It is called a van Hove singularity (van Hove, 1954). For systems that are periodic in two or three dimensions similar points exist where the density of states shows an irregular behaviour but does not diverge. Also in that case one talks about a van Hove singularity.

20 Structure and Forces

20.1 HELLMANN–FEYNMAN THEOREM

In solving the electronic Schrödinger equation

$$\hat{H}_e \Psi_e = E_e \Psi_e \qquad (20.1)$$

or the equivalent density-functional equation, one does not calculate the total energy E of the system but only the electronic part of it. However, as shown earlier in this manuscript (Chapter 8), E and E_e are related by

$$E = E_e + \frac{1}{2} \sum_{k \neq l=1}^{M} \frac{Z_k Z_l}{|\vec{R}_k - \vec{R}_l|}, \qquad (20.2)$$

where Z_k and $\vec{R}_k$ is the charge and position of the kth nucleus, respectively.

The theoretically optimized structure is that for which E is lowest. Finding the optimal structure theoretically corresponds accordingly to determine that set of positions $\{\vec{R}_k\}$ for which E has its minimum.

For a given system one may optimize the structure by starting out with some set of positions and in a more or less qualified way change these until the minimum of E is found. In this process, knowledge of the forces

$$\vec{F}_k = -\vec{\nabla}_{\vec{R}_k} E \qquad (20.3)$$

would be of immense help. Therefore it would be very desirable to be able to calculate these. In this chapter we shall discuss how this can be done with current electronic-structure methods. First, however, we shall discuss the Hellmann–Feynman theorem (Hellmann, 1937; Feynman, 1939. This theorem is general and the calculation of the forces represents just one example of its application (another application was presented in Section 15.11).

We consider the Schrödinger equation

$$\hat{H} \psi = E \psi \qquad (20.4)$$

and assume that all involved quantities depend on some parameter λ. Later in this section we will let λ be a nuclear coordinate, and the statement is thus that both the Hamilton operator, the wavefunction, and the eigenvalue depends on this nuclear coordinate.

In calculating the forces we need the derivatives (20.3). Assuming that we have solved Eq. (20.4) *exactly*, and that ψ is normalized

$$\langle \psi | \psi \rangle = 1 \tag{20.5}$$

we have

$$
\begin{aligned}
\frac{d}{d\lambda} E &= \frac{d}{d\lambda} \langle \psi | \hat{H} | \psi \rangle \\
&= \frac{d}{d\lambda} \int \psi^*(\vec{x}) \hat{H} \psi(\vec{x}) \, d\vec{x} \\
&= \int \frac{d\psi^*(\vec{x})}{d\lambda} \hat{H} \psi(\vec{x}) \, d\vec{x} + \int \psi^*(\vec{x}) \frac{d\hat{H}}{d\lambda} \psi(\vec{x}) \, d\vec{x} + \int \psi^*(\vec{x}) \hat{H} \frac{d\psi(\vec{x})}{d\lambda} \, d\vec{x} \\
&= \int \frac{d\psi^*(\vec{x})}{d\lambda} E \psi(\vec{x}) \, d\vec{x} + \int \psi^*(\vec{x}) \frac{d\hat{H}}{d\lambda} \psi(\vec{x}) \, d\vec{x} + \left[\int \frac{d\psi^*(\vec{x})}{d\lambda} \hat{H} \psi(\vec{x}) \, d\vec{x} \right]^* \\
&= E \int \frac{d\psi^*(\vec{x})}{d\lambda} \psi(\vec{x}) \, d\vec{x} + \int \psi^*(\vec{x}) \frac{d\hat{H}}{d\lambda} \psi(\vec{x}) \, d\vec{x} + \left[\int \frac{d\psi^*(\vec{x})}{d\lambda} E \psi(\vec{x}) \, d\vec{x} \right]^* \\
&= E \int \frac{d\psi^*(\vec{x})}{d\lambda} \psi(\vec{x}) \, d\vec{x} + \int \psi^*(\vec{x}) \frac{d\hat{H}}{d\lambda} \psi(\vec{x}) \, d\vec{x} + E \left[\int \frac{d\psi^*(\vec{x})}{d\lambda} \psi(\vec{x}) \, d\vec{x} \right]^* \\
&= E \int \frac{d\psi^*(\vec{x})}{d\lambda} \psi(\vec{x}) \, d\vec{x} + E \left[\int \frac{d\psi^*(\vec{x})}{d\lambda} \psi(\vec{x}) \, d\vec{x} \right]^* + \int \psi^*(\vec{x}) \frac{d\hat{H}}{d\lambda} \psi(\vec{x}) \, d\vec{x} \\
&= E \int \frac{d\psi^*(\vec{x})}{d\lambda} \psi(\vec{x}) \, d\vec{x} + E \int \frac{d\psi(\vec{x})}{d\lambda} \psi^*(\vec{x}) \, d\vec{x} + \int \psi^*(\vec{x}) \frac{d\hat{H}}{d\lambda} \psi(\vec{x}) \, d\vec{x} \\
&= E \left[\int \frac{d\psi^*(\vec{x})}{d\lambda} \psi(\vec{x}) \, d\vec{x} + \int \frac{d\psi(\vec{x})}{d\lambda} \psi^*(\vec{x}) \, d\vec{x} \right] + \int \psi^*(\vec{x}) \frac{d\hat{H}}{d\lambda} \psi(\vec{x}) \, d\vec{x} \\
&= E \frac{d}{d\lambda} \int \psi^*(\vec{x}) \psi(\vec{x}) \, d\vec{x} + \int \psi^*(\vec{x}) \frac{d\hat{H}}{d\lambda} \psi(\vec{x}) \, d\vec{x} \\
&= E \frac{d}{d\lambda} 1 + \int \psi^*(\vec{x}) \frac{d\hat{H}}{d\lambda} \psi(\vec{x}) \, d\vec{x} \\
&= \int \psi^*(\vec{x}) \frac{d\hat{H}}{d\lambda} \psi(\vec{x}) \, d\vec{x}. \tag{20.6}
\end{aligned}
$$

This is the Hellmann–Feynman theorem. It states that the derivative of the energy with respect to the *any* parameter equals the expectation value of the derivative of the Hamilton operator for the state of interest. In deriving it we have used that the Hamilton operator is Hermitian, that the energy is real, and that the wavefunction is normalized.

The Hellmann–Feynman theorem is actually nothing but a special application of first-order perturbation theory. To see this we write

$$\frac{d}{d\lambda}E = \lim_{\Delta\lambda\to 0}\left\{ \frac{1}{\Delta\lambda}\left[\langle\psi(\lambda+\Delta\lambda)|\hat{H}(\lambda+\Delta\lambda)|\psi(\lambda+\Delta\lambda)\rangle \right.\right.$$

$$\left.\left. - \langle\psi(\lambda)|\hat{H}(\lambda)|\psi(\lambda)\rangle \right] \right\}, \tag{20.7}$$

where we have indicated that the eigenfunction ψ depends on the parameter λ.

Keeping only first-order terms in $\Delta\lambda$ (since it ultimately has to $\to 0$), we have

$$\hat{H}(\lambda+\Delta\lambda) \simeq \hat{H}(\lambda) + \Delta\lambda\cdot\frac{d\hat{H}}{d\lambda}, \tag{20.8}$$

where $\Delta\lambda$ is small. Therefore, first order perturbation theory gives that

$$\langle\psi(\lambda+\Delta\lambda)|\hat{H}(\lambda+\Delta\lambda)|\psi(\lambda+\Delta\lambda)\rangle \simeq \left\langle \psi(\lambda)|\hat{H}(\lambda) + \Delta\lambda\cdot\frac{d\hat{H}}{d\lambda}|\psi(\lambda) \right\rangle$$

$$= \langle\phi(\lambda)|\hat{H}(\lambda)|\phi(\lambda)\rangle + \Delta\lambda\left\langle \phi(\lambda)|\frac{d\hat{H}}{d\lambda}|\phi(\lambda) \right\rangle. \tag{20.9}$$

Combining this with Eqs. (20.8) and (20.7) leads immediately to the Hellmann–Feynman theorem.

We have already used this theorem above in Section 15.11 when discussing the adiabatic connection formula. In that case λ was a parameter describing the strength of the electron–electron interactions. Here, we shall let λ be a nuclear coordinate.

Unfortunately, the practical application of the Hellmann–Feynman theorem is somewhat limited. In deriving the theorem we have assumed that we, in every point of space, have solved the Schrödinger equation (20.4) exactly. In a practical calculation, however, we obtain an approximate wavefunction by minimizing

$$\frac{\langle\phi|\hat{H}|\phi\rangle}{\langle\phi|\phi\rangle} \tag{20.10}$$

so that Eq. (20.4) is *not* identically satisfied. Therefore, we cannot use the Hellmann–Feynman theorem directly in calculating forces but have to modify it. This will be discussed below.

20.2 FORCES

In calculating the forces on the nuclei we will let λ above be one of the components of the position vector of any of the nuclei, i.e., the αth component of $\vec{R}_k$ ($\alpha = x$, y, or z). We will denote this $R_{k,\alpha}$ and seek accordingly

$$F_{k,\alpha} = -\frac{dE}{dR_{k,\alpha}}. \tag{20.11}$$

In a typical electronic-structure calculation, we do not solve the electronic Schrödinger equation

$$\hat{H}_e \Psi_e = E_e \Psi_e \tag{20.12}$$

in every point, but instead approximate the solution

$$\Psi_e \simeq \Phi \tag{20.13}$$

by applying the variational method. Therefore, we cannot directly use the Hellmann–Feynman theorem above. Instead, we have [cf. Eq. (20.6)]

$$\frac{dE_e}{dR_{k,\alpha}} = \langle \Phi | \frac{\partial \hat{H}_e}{\partial R_{k,\alpha}} | \Phi \rangle + \left\langle \frac{\partial \Phi}{\partial R_{k,\alpha}} \middle| \hat{H}_e \middle| \Phi \right\rangle + \left\langle \Phi \middle| \hat{H}_e \middle| \frac{\partial \Phi}{\partial R_{k,\alpha}} \right\rangle. \tag{20.14}$$

By using that $\hat{H}_e$ is Hermitian, this may be written as

$$\frac{dE_e}{dR_{k,\alpha}} = \langle \Phi | \frac{\partial \hat{H}_e}{\partial R_{k,\alpha}} | \Phi \rangle + \left\langle \frac{\partial \Phi}{\partial R_{k,\alpha}} \middle| \hat{H}_e \middle| \Phi \right\rangle + \left\langle \frac{\partial \Phi}{\partial R_{k,\alpha}} \middle| \hat{H}_e \middle| \Phi \right\rangle^*. \tag{20.15}$$

This formula is actually independent of whether we use a Hartree–Fock approach (or an improvement thereof including correlation effects) or a density-functional approach. We have not assumed anything about $\hat{H}_e$, except that the eigenvalue problem (20.12) is solved more or less accurately according to Eq. (20.13). In order to reduce the complications we shall now, however, assume that Φ is a single Slater determinant. Accordingly, we restrict ourselves to either the Hartree–Fock method or a density-functional method within the formulation of Kohn and Sham. The formulas may be extended so that they also apply to, e.g., CI methods, but this is beyond the scope of this presentation.

From the Hellmann–Feynman theorem we know that the last two terms on the right-hand side of Eq. (20.15) vanish when the Schrödinger, Hartree–Fock, or Kohn–Sham equations are solved exactly. Therefore, when applying approximate solutions obtained from the variational principle, we will expect that these two terms are small. However, as they are written in Eq. (20.15), it is not obvious that they are small. But to see that this is the case we proceed as follows.

Independently of whether we use Hartree–Fock or density-functional methods, we approximate the total electronic energy as some function E_e. This is calculated using the approximate solutions to some single-particle equations,

$$\hat{h}\phi_i \simeq \varepsilon_i\phi_i, \tag{20.16}$$

where $\hat{h}$ is either the Fock operator or the effective single-particle operator in the Kohn–Sham equations. In Eq. (20.16) we have emphasized that the functions

$$\phi_i = \sum_{m=1}^{N_b} \chi_m c_{mi} \tag{20.17}$$

are only approximate solutions to Eq. (20.16), since the set of basis functions $\{\chi_m\}$ in general is finite and not complete.

E_e will then depend on the basis functions χ_m, on the expansion coefficients c_{mi}, and, furthermore, on the nuclear coordinates, including $R_{k,\alpha}$. That is,

$$E_e = E_e(\{\chi_m\}, \{c_{mi}\}, R_{k,\alpha}), \tag{20.18}$$

where we include explicitly only the single nuclear coordinate of our interest.

The coefficients c_{mi} are determined by minimizing E_e under the constraint that the single-particle functions ϕ_i are orthonormal. This means minimizing

$$G = E_e(\{\chi_m\}, \{c_{mi}\}, R_{k,\alpha}) - \sum_i \varepsilon_i(\langle\phi_i|\phi_i\rangle - 1) \tag{20.19}$$

with respect to varying c_{mi} and c_{mi}^*. [In Eq. (20.19) we have used G and not F as earlier in order to avoid confusing it with the forces.] This leads to the well-known eigenvalue equation (or secular equation)

$$\frac{\partial E_e}{\partial c_{mi}^*} = \varepsilon_i \sum_l O_{ml} c_{li}$$

$$\frac{\partial E_e}{\partial c_{mi}} = \varepsilon_i \sum_l O_{lm} c_{li}^*, \tag{20.20}$$

where

$$O_{ml} = \langle\chi_m|\chi_l\rangle = \langle\chi_l|\chi_m\rangle^* = O_{lm}^* \tag{20.21}$$

is an overlap-matrix element and where the left-hand sides of Eq. (20.20) are written in a somewhat unusual form, which will be useful below. Moreover, we have

$$\frac{\partial E_e}{\partial c_{mi}^*} = \sum_l \langle\chi_m|\hat{h}|\chi_l\rangle c_{li}$$

$$\frac{\partial E_e}{\partial c_{mi}} = \sum_l \langle\chi_l|\hat{h}|\chi_m\rangle c_{li}^* \tag{20.22}$$

which is nothing but one part of the secular equation.

When varying the nuclear coordinate $R_{k,\alpha}$, E_e will change. This change has a number of origins. First, E_e depends on $R_{k,\alpha}$ directly. Second, any of the basis functions χ_m as well as its complex conjugate χ_m^* may vary in any point of space when varying $R_{k,\alpha}$ (this is, e.g., the case when the basis function is centred on the kth atom). And third, any of the expansion coefficients c_{mi} as well as its complex conjugate c_{mi}^* may vary. This means that

$$\frac{dE_e}{dR_{k,\alpha}} = \frac{\partial E_e}{\partial R_{k,\alpha}} + \sum_m \int \left[\frac{\delta E_e}{\delta \chi_m} \frac{d\chi_m}{dR_{k,\alpha}} + \frac{\delta E_e}{\delta \chi_m^*} \frac{d\chi_m^*}{dR_{k,\alpha}} \right] d\vec{r}$$

$$+ \sum_{im} \left[\frac{\partial E_e}{\partial c_{mi}} \frac{dc_{mi}}{dR_{k,\alpha}} + \frac{\partial E_e}{\partial c_{mi}^*} \frac{dc_{mi}^*}{dR_{k,\alpha}} \right]. \tag{20.23}$$

Here, the first term takes care of the explicit dependence of E_e on the nuclear coordinate, and this is in fact the term that is contained within the Hellmann–Feynman theorem. The second term is an integral over all space involving functional derivatives, since *any* basis function may change in *any* point of space. Finally, the last term is that due to the expansion coefficients.

From the single-particle equations (20.20) we have that

$$\sum_{im} \left[\frac{\partial E_e}{\partial c_{mi}} \frac{dc_{mi}}{dR_{k,\alpha}} + \frac{\partial E_e}{\partial c_{mi}^*} \frac{dc_{mi}^*}{dR_{k,\alpha}} \right]$$

$$= \sum_i \varepsilon_i \sum_{lm} \left[O_{lm}^* c_{li}^* \frac{dc_{mi}}{dR_{k,\alpha}} + O_{ml} c_{li} \frac{dc_{mi}^*}{dR_{k,\alpha}} \right]. \tag{20.24}$$

From

$$1 = \langle \phi_i | \phi_i \rangle = \sum_{lm} c_{li}^* c_{mi} O_{lm} \tag{20.25}$$

we obtain upon differentiation

$$0 = \sum_{lm} \left[\frac{dc_{li}^*}{dR_{k,\alpha}} c_{mi} O_{lm} + c_{li}^* \frac{dc_{mi}}{dR_{k,\alpha}} O_{lm} + c_{li}^* c_{mi} \frac{dO_{lm}}{dR_{k,\alpha}} \right]$$

$$= \sum_{lm} \left[\frac{dc_{mi}^*}{dR_{k,\alpha}} c_{li} O_{ml} + c_{li}^* \frac{dc_{mi}}{dR_{k,\alpha}} O_{lm} + c_{li}^* c_{mi} \frac{dO_{lm}}{dR_{k,\alpha}} \right]. \tag{20.26}$$

Inserting this into Eq. (20.24) gives

$$\sum_{im} \left[\frac{\partial E_e}{\partial c_{mi}} \frac{dc_{mi}}{dR_{k,\alpha}} + \frac{\partial E_e}{\partial c_{mi}^*} \frac{dc_{mi}^*}{dR_{k,\alpha}} \right]$$

$$= -\sum_i \varepsilon_i \sum_{ml} c_{li}^* c_{mi} \frac{dO_{lm}}{dR_{k,\alpha}}$$

$$= -\sum_i \varepsilon_i \sum_{ml} c_{li}^* c_{mi} \int \left[\frac{d\chi_l^*(\vec{r})}{dR_{k,\alpha}} \chi_m(\vec{r}) + \chi_l^*(\vec{r}) \frac{d\chi_m(\vec{r})}{dR_{k,\alpha}} \right] d\vec{r}. \tag{20.27}$$

The last identity is obtained from the definition of the overlap-matrix elements.

Equation (20.27) gives the last term of the right-hand side of Eq. (20.23). For the second-last one we need to specify how E_e depends on the basis functions. Within density-functional theory we have

$$E_e = \sum_i \langle \phi_i | \hat{h}_1 | \phi_i \rangle + \frac{1}{2} \int V_C(\vec{r}) \rho(\vec{r}) \, \mathrm{d}\vec{r} + E_{xc}[\rho(\vec{r})], \qquad (20.28)$$

where $\hat{h}_1$ is the one-electron operator that contains the kinetic-energy contribution and the contribution from the electrostatic potential of the nuclei. V_C is the Coulomb potential of the electron density and E_{xc} is the exchange-correlation energy. Both these last two quantities depend on the electron density $\rho(\vec{r})$, which in turn depends on the eigenfunctions ϕ_i that depend on the basis functions.

Similarly, with the Hartree–Fock approximation one gets

$$E_e = \sum_i \langle \phi_i | \hat{h}_1 | \phi_i \rangle + \frac{1}{2} \sum_{i,j} \left[\langle \phi_i \phi_j | \hat{h}_2 | \phi_i \phi_j \rangle - \langle \phi_j \phi_i | \hat{h}_2 | \phi_i \phi_j \rangle \right], \qquad (20.29)$$

where $\hat{h}_2$ is the electron–electron interaction operator,

$$\hat{h}_2 = \frac{1}{|\vec{r}_1 - \vec{r}_2|}. \qquad (20.30)$$

In both cases we can write E_e as a functional of the basis functions χ_m. We can then take the functional derivatives $\delta E_e / \delta \chi_m$ and $\delta E_e / \delta \chi_m^*$ by considering changing one single basis function χ_m or χ_m^* in one single point. The procedure shall not be carried through in detail here since it is completely equivalent to what was done in Section 9.2 when deriving the Hartree–Fock equations. The result is

$$\frac{\delta E_e}{\delta \chi_m(\vec{r})} = \sum_i \sum_l c_{mi} c_{li}^* \hat{h} \chi_l^*(\vec{r})$$

$$\frac{\delta E_e}{\delta \chi_m^*(\vec{r})} = \sum_i \sum_l c_{mi}^* c_{li} \hat{h} \chi_l(\vec{r}), \qquad (20.31)$$

where $\hat{h}$ is the operator of Eq. (20.16), i.e., the Fock operator for the Hartree–Fock–Roothaan equations and the effective Hamilton operator for the Kohn–Sham equations.

Combining Eqs. (20.23), (20.27), and (20.31) gives

$$\frac{\mathrm{d}E_e}{\mathrm{d}R_{k,\alpha}} = \frac{\partial E_e}{\partial R_{k,\alpha}} + \sum_i \sum_{ml} c_{li}^* c_{mi} \int \frac{\mathrm{d}\chi_l^*(\vec{r})}{\mathrm{d}R_{k,\alpha}} [\hat{h} - \varepsilon_i] \chi_m(\vec{r}) \, \mathrm{d}\vec{r}$$

$$+ \sum_i \sum_{ml} c_{li} c_{mi}^* \int \frac{\mathrm{d}\chi_l(\vec{r})}{\mathrm{d}R_{k,\alpha}} [\hat{h} - \varepsilon_i] \chi_m^*(\vec{r}) \, \mathrm{d}\vec{r}. \qquad (20.32)$$

The first term is the term from the Hellmann–Feynman theorem that is usually called the Hellmann–Feynman force. The second and third term vanish *if* we solve the single-particle equations (20.16) exactly or *if* the basis functions are independent of the nuclear coordinates as is the case for plane waves, but otherwise it is the correction due to the approximations in the single-particle wavefunctions. It is due to Pulay (1969) and is therefore called the Pulay force.

In order to see that the Pulay force vanishes when we have solved the single-particle equations exactly we use the fact that

$$\sum_m c_{mi} \chi_m(\vec{r}) = \phi_i(\vec{r}), \tag{20.33}$$

so that

$$\sum_m c_{mi}[\hat{h} - \varepsilon_i]\chi_m(\vec{r}) = [\hat{h} - \varepsilon_i]\phi_i(\vec{r}) = \hat{h}\phi_i(\vec{r}) - \varepsilon_i\phi_i(\vec{r}). \tag{20.34}$$

The Pulay force takes care of the fact that the single-particle equations are solved only approximately through application of the variational principle. When further approximations in solving the density-functional, Hartree–Fock, ... , equations are introduced (for instance, a frozen-core approximation, approximate fits of densities or potentials, etc.), also these would have to be taken into account when calculating the forces. This, however, will not be discussed further here.

20.3 STRUCTURE OPTIMIZATION

Within the Born–Oppenheimer approximation the electronic properties of a given system is calculated by solving the electronic Schrödinger equation or the equivalent density-function equation for fixed, chosen nuclear positions, $\vec{R}_1, \vec{R}_2, \ldots, \vec{R}_M$. The total energy

$$E = E_e + \frac{1}{2} \sum_{k \neq l = 1}^{M} \frac{Z_k Z_l}{|\vec{R}_k - \vec{R}_l|} \tag{20.35}$$

is then a function of the nuclear positions,

$$E = E(\vec{R}_1, \vec{R}_2, \ldots, \vec{R}_M). \tag{20.36}$$

In Eq. (20.35), E_e is the electronic energy, which is usually calculated as the minimum of an expectation value

$$E_e = \frac{\langle \Phi | \hat{H}_e | \Phi \rangle}{\langle \Phi | \Phi \rangle}. \tag{20.37}$$

The theoretically predicted structure is that for which E has its minimum, i.e.,

$$\vec{F}_1 = \vec{F}_2 = \ldots = \vec{F}_M = \vec{0}, \tag{20.38}$$

where

$$\vec{F}_k = -\vec{\nabla}_{\vec{R}_k} E \tag{20.39}$$

is the force on the kth nucleus. The α component on this is calculated as

$$F_{k,\alpha} = -\frac{dE}{dR_{k,\alpha}} = -\frac{dE_e}{dR_{k,\alpha}} + \frac{1}{2} \sum_{l=1(l \neq k)}^{M} \frac{Z_k Z_l}{|\vec{R}_k - \vec{R}_l|^2} (R_{k,\alpha} - R_{l,\alpha}). \tag{20.40}$$

Here, the second term is that due to the nucleus–nucleus interaction, whereas the first term can be calculated using the methods of the preceding two subsections. This involves first of all the Hellmann–Feynman force, but in very many cases (the only exceptions are those cases where the basis functions either are plane waves or form a complete set) also the Pulay forces. As discussed above, also other terms may in some cases occur when additional approximations are invoked.

By calculating the forces analytically so that they are the derivatives of the total energy, it is possible to optimize a structure automatically; i.e., to determine that structure for which Eq. (20.38) is satisfied. We start out with a given structure characterized by the nuclear coordinates $\vec{R}_1^0, \vec{R}_2^0, \ldots, \vec{R}_M^0$, where the upper index indicates that this set is the initial set of coordinates. $F_{k,\alpha}^0$ for *that* set of nuclear coordinates (therefore the upper index) tells how to shift $R_{k,\alpha}$ in order to reduce the total energy, so the simplest possible approach is to change the nuclear coordinates according to

$$\vec{R}_k^0 \to \vec{R}_k^1 = \vec{R}_k^0 + \gamma \cdot \vec{F}_k^0. \tag{20.41}$$

This scheme can be continued:

$$\vec{R}_k^p \to \vec{R}_k^{p+1} = \vec{R}_k^p + \gamma \cdot \vec{F}_k^p \tag{20.42}$$

until the structure does not change any further within a predefined tolerance, at which point the optimization of the structure is considered completed. γ is some chosen positive parameter that has the dimension of time squared and p is an index that enumerates the displacement steps.

The above scheme (the *steepest-descend* method) is the absolutely simplest one that can be constructed through knowledge of the forces. It is not the most intelligent one, first of all since it in many cases requires very many geometry steps before converging. Alternatives, e.g., the so-called conjugate-gradient method, are therefore often used, but describing these in detail is beyond the scope of this presentation (see, e.g., Press *et al.*, 1992). Instead we shall consider a slightly more advanced scheme that simultaneously makes a connection to classical mechanics. It is due to Verlet (1967) and works as follows.

According to Newton's law, the force $\vec{F}_k$ equals the mass of that particle (this is essentially the nuclear mass, since the electrons are so much lighter) times its acceleration,

$$\vec{F}_k = M_k \cdot \vec{a}_k. \tag{20.43}$$

Let us assume that we consider times $t - \Delta t$, t, and $t + \Delta t$ (this corresponds to displacement steps with $t_p = t_0 + p \cdot \Delta t$), where Δt is so small that the acceleration is roughly constant in the complete interval $[t - \Delta t, t + \Delta t]$. Then, truncating after the second-order terms in the Taylor series

$$\vec{R}_k(t + \Delta t) = \vec{R}_k(t) + \vec{v}_k(t) \cdot \Delta t + \tfrac{1}{2}\vec{a}_k(t) \cdot (\Delta t)^2 + \cdots$$
$$\vec{R}_k(t - \Delta t) = \vec{R}_k(t) - \vec{v}_k(t) \cdot \Delta t + \tfrac{1}{2}\vec{a}_k(t) \cdot (\Delta t)^2 + \cdots \tag{20.44}$$

is a good approximation.

Adding the two equations of Eq. (20.44) then gives

$$\vec{R}_k(t + \Delta t) \simeq 2\vec{R}_k(t) - \vec{R}_k(t - \Delta t) + \vec{a}_k(t) \cdot (\Delta t)^2$$
$$= 2\vec{R}_k(t) - \vec{R}_k(t - \Delta t) + \frac{1}{M_k}\vec{F}_k(t) \cdot (\Delta t)^2, \tag{20.45}$$

where we have used Eq. (20.43). Since the ultimate goal is to determine the structure for which the forces vanishes, it is less important how the masses M_k are chosen but one may, e.g., simply use the true atomic masses or set them all equal to some constant.

Equation (20.45) shows that from knowledge of the positions at two succeeding time steps and of the forces, it is possible to calculate the positions at the following time step. This is the Verlet method.

All methods we have discussed so far are based on using information from one or two previous structures as well as the forces from one previous structure in calculating a new structure. Thus, very much information that has been calculated for other structures is simply abandoned, although it could be useful in getting more insight into how the total energy depends on the structure, and thus also in locating the total-energy minimum. More advanced methods like that of Broyden (1965) are based on the strategy that one shall use as much information as is available. These methods will, however, not be discussed further here.

20.4 THE CLASSICAL LAGRANGIAN

The procedure for optimizing the structure that we have applied so far is as follows: we consider a given set of nuclear coordinates. For this we calculate self-consistently the electronic properties, including the forces. Subsequently, the nuclei are shifted and the whole procedure is repeated. Accordingly, for *each* structure we optimize the electronic degrees of freedom, whereas the nuclear degrees of freedom are optimized only once.

Car and Parrinello (1985) suggested an alternative method according to which electronic and nuclear degrees of freedom are optimized simultaneously. Above, with the method of Verlet, the nuclei are treated classically and by introducing a time coordinate the motion of the nuclei is treated using Newtonian classical mechanics. Also the method of Car and Parrinello is based on classical mechanics but in the formulation of Lagrange. In order to understand their method we shall, therefore, need some tools from classical mechanics.

We assume that we have a number of particles, $i = 1, 2, \ldots, N$. They have a certain total kinetic energy

$$T = \sum_{i=1}^{N} \frac{1}{2} m_i |\vec{v}_i|^2. \tag{20.46}$$

The particles move in a potential that defines the forces acting on the particles,

$$\vec{F}_i = -\vec{\nabla}_i V, \tag{20.47}$$

where

$$\vec{\nabla}_i V \equiv \vec{\nabla}_{\vec{r}_i} V. \tag{20.48}$$

Here, the positions of the N particles are described by the vectors $\vec{r}_1, \vec{r}_2, \ldots, \vec{r}_N$.

We will now assume that the positions are functions of some other generalized coordinates as well as of the time,

$$\vec{r}_i = \vec{r}_i(q_1, q_2, \ldots, q_n, t). \tag{20.49}$$

The number n may be smaller than or equal to $3N$, dependent on whether the system is exposed to some constraint. A very simple example is that of particles moving on a sphere. Then $\vec{r}_i$ is a three-dimensional position vector, but by letting the q_j be the polar coordinates ϕ and θ we need only two independent ones for each particle since the last polar coordinate r is given through the constraint that the particles move on the sphere.

Just as we can use $\vec{r}_i$ and the time-derivatives $\dot{\vec{r}}_i$ as independent variables, we may alternatively use the parameters q_j and $\dot{q}_j$. Then the equations of motion

$$\vec{F}_i = \dot{\vec{p}}_i \tag{20.50}$$

can be rewritten as

$$\sum_i (\vec{F}_i - \dot{\vec{p}}_i) \cdot \delta \vec{r}_i = 0 \tag{20.51}$$

with $\delta \vec{r}_i$ being arbitrary. This is the so-called D'Alembert principle. Equation (20.51) can be formulated with the help of the new variables as

$$\frac{d}{dt} \left(\frac{\partial L}{\partial \dot{q}_j} \right) - \frac{\partial L}{\partial q_j} = 0 \tag{20.52}$$

where
$$L = T - V \tag{20.53}$$

(T being the total kinetic energy) is the so-called Lagrangian.

Deriving Eq. (20.52) requires some steps that have not been reproduced here. The interested reader can find it in any textbook on classical mechanics (e.g., Goldstein, 1950), but here we just want to stress that the formulation exists, that it is essentially Newton's law, and that it is based on the Lagrangian of Eq. (20.53). Finally, we notice that we have here not specified the parameters q_j and their time derivatives in any way. They are just a set of parameters that specify completely the distribution and time dependence of our system.

20.5   THE CAR–PARRINELLO METHOD

Car and Parrinello applied the formalism above but for a non-classical system. They assumed that the Born–Oppenheimer approximation is valid and considered thus the electrons as particles that will be treated with quantum theory whereas the nuclei are considered as being classical particles. The total energy is thus

$$E = E_e + \frac{1}{2} \sum_{\substack{k \neq l = 1}}^{M} \frac{Z_k Z_l}{|\vec{R}_k - \vec{R}_l|}, \tag{20.54}$$

where the second term is the classical electrostatic energy of the nucleus–nucleus interactions and E_e is the electronic energy that is obtained either from the Schrödinger equation

$$\hat{H}_e \Psi_e = E_e \Psi_e \tag{20.55}$$

or from the equivalent density-functional expression.

Independently of whether a density-functional approach or one based on the Schrödinger equation is applied, E_e is written as a functional of some single-particle wavefunctions (e.g., the Kohn–Sham orbitals or the Hartree–Fock orbitals) as well as on the positions of all the nuclei. That is,

$$E_e = E_e(\{\phi_i\}, \{\vec{R}_k\}). \tag{20.56}$$

We introduce now a time coordinate and assume that the wavefunctions and the nuclear positions depend on this time. Moreover, each single-particle wavefunction ϕ_i depends on the position-space coordinates when neglecting the spin dependences. In total, therefore,

$$E = E(\{\phi_i(\vec{r}, t)\}, \{\vec{R}_k(t)\}). \tag{20.57}$$

Car and Parrinello now treat this as if it were the potential energy for some classical system, i.e., it takes the role of V above in Section 20.4. As for the classical system, E depends on some generalized coordinates $\{q_i\}$, but in this case we have not a finite but an infinite set. This is so, since we will consider the value of *any* of the wavefunctions ϕ_i in *any* position-space point $\vec{r}$ as a parameter.

We also need to define a kinetic energy. This is the sum of two terms, one from the nuclei that simply becomes

$$T_n = \frac{1}{2} \sum_k M_k |\dot{\vec{R}}_k|^2, \qquad (20.58)$$

and one from the electrons. For the latter, Car and Parrinello use

$$T_e = \mu \sum_i \int |\dot{\phi}_i(\vec{r})|^2 \, d\vec{r}, \qquad (20.59)$$

where μ is some fictitious mass ascribed to the electronic wavefunctions.

Car and Parrinello define accordingly the following Lagrangian:

$$L = T_e + T_n - E. \qquad (20.60)$$

For this they write down the Lagrange equation (20.52), with the additional constraint that the wavefunctions have to be orthonormal at any time,

$$\int \phi_i^*(\vec{r}, t)\phi_j(\vec{r}, t) \, d\vec{r} = \delta_{i,j}. \qquad (20.61)$$

These constraints are included via Lagrange multipliers by modifying E,

$$E \to E - \sum_{i,j} \lambda_{i,j} \left[\int \phi_i^*(\vec{r}, t)\phi_j(\vec{r}, t) \, d\vec{r} - \delta_{i,j} \right]. \qquad (20.62)$$

From Eq. (20.52) we then obtain two different types of equations, of which one is found for the case that the variable q_j equals a nuclear coordinate. In that case,

$$M_k \ddot{\vec{R}}_k = -\vec{\nabla}_{\vec{R}_k} E. \qquad (20.63)$$

This is simply the classical equations of motion for the nuclei that were also used in Section 20.3.

For the electronic degrees of freedom, we have to change the simple derivatives by functional derivative (since our parameters now are functions in a continuous position space). Then we find

$$\mu \ddot{\phi}_i(\vec{r}, t) = -\frac{\delta E}{\delta \phi_i^*(\vec{r}, t)} + \sum_j \lambda_{i,j} \phi_j(\vec{r}, t). \qquad (20.64)$$

Here, we have taken the functional derivative with respect to ϕ_i^* — the equation for the derivative with respect to ϕ_i is essentially only the complex conjugate of Eq. (20.64).

Within Hartree–Fock theory,

$$\frac{\delta E}{\delta \phi_i^*(\vec{r}, t)} = \hat{F}(t)\phi_i(\vec{r}, t), \qquad (20.65)$$

where $\hat{F}$ is the Fock operator (which in the present case becomes time-dependent), whereas within density-functional theory,

$$\frac{\delta E}{\delta \phi_i^*(\vec{r}, t)} = \hat{h}_{\text{eff}}(t)\phi_i(\vec{r}, t), \qquad (20.66)$$

with $\hat{h}_{\text{eff}}$ being the effective single-particle operator of the Kohn–Sham equations. (Here, the functional derivatives are calculated exactly as done in Section 9.2 when we derived the Hartree–Fock equations) In both cases we have, accordingly, that, in the time-independent limit, Eq. (20.64) reduces to the normal single-particle equations.

In their implementation, Car and Parrinello used the density-functional expression for the total energy. Furthermore, they used plane waves as basis functions and treated infinite, periodic systems. The equations were solved simultaneously for the nuclear and electronic degrees of freedom, and the 'mass' μ was chosen considerably larger than that of real electrons.

The kinetic energy of the nuclei defines a real temperature. This means that one may perform a calculation at a well-defined temperature that furthermore may be changed during the calculation. Keeping the temperature constant will result in (predicted) results that should be relevant for that temperature.

This is illustrated in Fig. 20.1 for the case of liquid and amorphous selenium. In principle, such systems are not periodic and one would have to treat the complete macroscopic system. This is, however, not possible of computational reasons and one approximates the systems as being periodic. For the systems of Fig. 20.1 one has accordingly considered periodically repeated units of 64 atoms. In order to avoid looking at the positions of each individual atom, the autocorrelation function $g(r)$ is a useful quantity to explore, which furthermore is accessible experimentally. $g(r)dr$ gives the average number of neighbours with distances between r and $r + dr$ from any atom. The function $g(r)$ may be generalized for systems containing different types of atoms so that $g_{AB}(r)dr$ gives the average number of B neighbours at a distance between r and $r + dr$ around an A atom.

An alternative application of the Car–Parrinello method amounts to starting a calculation at a high temperature. Then, the nuclei move rapidly around in large parts of space, whereby many different types of structures may occur. Subsequently, one may gradually reduce the temperature and thereby hope that

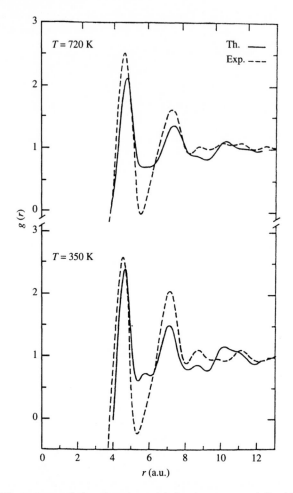

Figure 20.1 The autocorrelation function $g(r)$ for (upper curve) liquid selenium and (lower curve) amorphous selenium. Reproduced with permission from Hohl and Jones (1991). Copyright 1991 by the American Physical Society

the structure gets trapped in that of the global total-energy minimum and not only in a local one. This method is known as *simulated annealing*.

Figure 20.2 shows an example of the application of this approach for smaller, finite Se_N clusters. Also here it is assumed that the system is periodic (in order to be able to use plane waves as basis functions), but here one unit cell is made so large that neighbouring molecules in effect do not interact.

Since a real time is introduced into the calculations, they may also be used in exploring time-dependent properties. This could, e.g., be the vibrations of the atoms, whereby the frequencies with which the atoms oscillate are those of the vibrations. Figure 20.3 shows an example of this where the phonons in crystalline

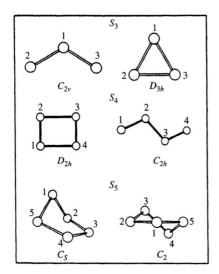
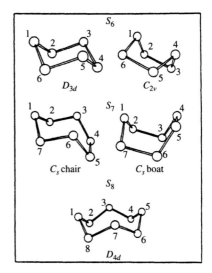

Figure 20.2 Optimized structures of different Se_N clusters. Reproduced by permission of the American Institute of Physics from Hohl *et al.* (1988)

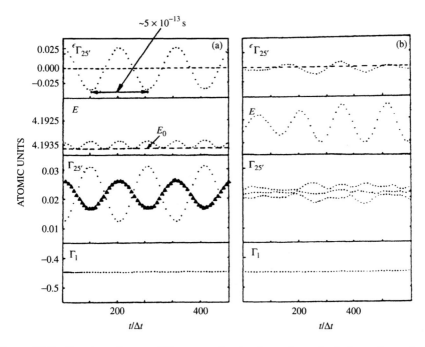

Figure 20.3 Results of two different molecular-dynamics calculations devoted to studying the vibrations in silicon. The upper curves and the third curves from the top are relative atomic positions, whereas the other curves are potential energy per atom. Reproduced with permission from Car and Parrinello (1985). Copyright 1985 by the American Physical Society

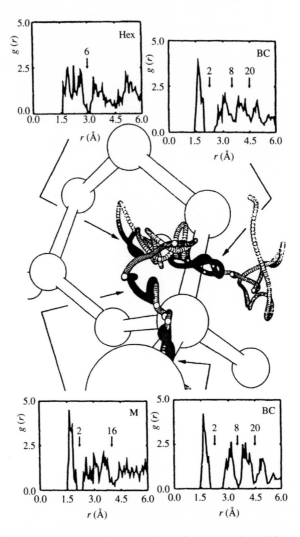

Figure 20.4 The time evolution of the position of a proton that diffuses in crystalline silicon. The inserts show the pair correlation functions (i.e., the number of neighbours as a function of distance) for the number of silicon atoms around the proton for the shown positions of the proton. Reproduced with permission from Buda *et al.* (1989). Copyright 1989 by the American Physical Society

silicon have been studied. In Fig. 20.4 we show an example of the diffusion of a proton in crystalline silicon. Here, the inserts show the pair-correlation functions for the number of silicon atoms around the proton for various positions of the proton.

But first of all, the method is used for obtaining structures — also as a function of temperature.

21 Vibrations

21.1 MOLECULAR VIBRATIONS AND DYNAMICAL MATRIX

Within the Born–Oppenheimer approximation we calculate the total electronic energy E_e (at least approximately) for fixed nuclear coordinates

$$\vec{X} = (\vec{R}_1, \vec{R}_2, \ldots, \vec{R}_M) = (R_{1,x}, R_{1,y}, R_{1,z}, R_{2,x}, R_{2,y}, R_{2,z}, \ldots, R_{M,x}, R_{M,y}, R_{M,z}) \tag{21.1}$$

(neglecting the nuclear spin).

The total energy

$$E = E_e + \frac{1}{2} \sum_{k \neq l = 1}^{M} \frac{Z_k Z_l}{|\vec{R}_k - \vec{R}_l|} \tag{21.2}$$

is a function of $\vec{X}$,

$$E = E(\vec{X}) = E(R_{1,x}, R_{1,y}, R_{1,z}, R_{2,x}, R_{2,y}, R_{2,z}, \ldots, R_{M,x}, R_{M,y}, R_{M,z}). \tag{21.3}$$

For the optimized structure $\vec{X}^e$, E has a minimum, and by expanding E to second order about the minimum (notice that the first-order derivatives vanish at the minimum) we find

$$E(\vec{X}) \simeq E(\vec{X}^e) + \frac{1}{2} \sum_{k_1, k_2 = 1}^{M} \sum_{\alpha_1, \alpha_2 = x, y, z} \frac{\partial^2 E(\vec{X}^e)}{\partial R_{k_1, \alpha_1} \partial R_{k_2, \alpha_2}}$$

$$\times (R_{k_1, \alpha_1} - R_{k_1, \alpha_1}^e)(R_{k_2, \alpha_2} - R_{k_2, \alpha_2}^e). \tag{21.4}$$

In principle, the series includes also higher-order terms but in most cases the approximation above (the so-called harmonic approximation) is a good one, although there exists cases where higher (anharmonic) terms are to be included in order to obtain an accurate description of the vibrational properties of the material of interest.

The matrix

$$\underline{\underline{H}} = \left(\frac{\partial^2 E(\vec{X}^e)}{\partial R_{k_1,\alpha_1} \partial R_{k_2,\alpha_2}} \right) \tag{21.5}$$

is the so-called Hessian. With some computer programs the optimized structure $\vec{X}^e$ is calculated automatically using the forces as discussed above in Chapter 20, i.e., $\vec{X}^e$ is obtained as a structure for which the forces $\partial E/\partial R_{k,\alpha}$ vanish. Actually, this is also the case for a saddle point, and in order to be absolutely sure that the optimized structure indeed belongs to a total-energy minimum one has to require that *all* eigenvalues of the Hessian are non-negative. Otherwise there would be certain linear combinations of $(R_{k,\alpha} - R_{k,\alpha}^e)$ that would lead to a reduction in E.

For some computational methods also the Hessian is calculated directly and analytically. In Chapter 20 we discussed how the forces, i.e., the first-order derivatives of E with respect to the nuclear coordinate, could be calculated and observed that due to the fact that we do not solve the Schrödinger, Hartree–Fock, or Kohn–Sham equations exactly, but only approximately through application of the variational method, we could not directly apply the Hellmann–Feynman theorem but had to include additional contributions to the forces. The calculation of the second-order derivatives as needed for the Hessian is certainly not easier, but the *basic* principles are as presented above for the forces.

As an alternative one may first determine the structure of the total-energy minimum. Subsequently, one may calculate the changes in the total energy when displacing the various atoms slightly by considering more (many) different displacements. The results of these calculations may finally be fitted with an expression like Eq. (21.4).

Having written the total energy in the form of Eq. (21.4) we can study vibrations. We will then treat the nuclei as quantum particles and consider accordingly the following nuclear Hamilton operator

$$\hat{H}_n = -\sum_{k=1}^{M} \frac{1}{2M_k} \nabla^2_{\vec{R}_k} + \frac{1}{2} \sum_{k_1,k_2=1}^{M} \sum_{\alpha_1,\alpha_2=x,y,z} \frac{\partial^2 E(\vec{X}^e)}{\partial R_{k_1,\alpha_1} \partial R_{k_2,\alpha_2}}$$
$$\times (R_{k_1,\alpha_1} - R_{k_1,\alpha_1}^e)(R_{k_2,\alpha_2} - R_{k_2,\alpha_2}^e). \tag{21.6}$$

The eigenvalues and eigenfunctions to the nuclear Schrödinger equation

$$\hat{H}_n \Psi_n = E_n \Psi_n \tag{21.7}$$

then define the energies and patterns (modes) of the vibrations of the system of interest.

In order to solve this equation it is of advantage to introduce the new coordinates

$$\vec{u}_k = \sqrt{M_k}(\vec{R}_k - \vec{R}_k^e), \tag{21.8}$$

that are the deviations from the equilibrium position but scaled with the square root of the masses. The nuclear Hamilton operator of Eq. (21.6) then takes the form

$$\hat{H}_n = -\sum_{k=1}^{M} \frac{1}{2}\nabla_{\tilde{u}_k}^2 + \frac{1}{2}\sum_{k_1,k_2=1}^{M} \sum_{\alpha_1,\alpha_2=x,y,z} \frac{1}{\sqrt{M_{k_1}M_{k_2}}} \frac{\partial^2 E(\vec{X}^e)}{\partial R_{k_1,\alpha_1}\partial R_{k_2,\alpha_2}} u_{k_1,\alpha_1} u_{k_2,\alpha_2}.$$

(21.9)

As for the Hessian, this leads naturally to the definition of a matrix

$$\underline{\underline{D}} = \left(\frac{1}{\sqrt{M_{k_1}M_{k_2}}} \frac{\partial^2 E(\vec{X}^e)}{\partial R_{k_1,\alpha_1}\partial R_{k_2,\alpha_2}} \right),$$

(21.10)

which is the so-called *dynamical matrix*.

Also in this case we should in principle solve the nuclear Schrödinger equation (21.7). The problem with this equation, however, is that all the different u mix, i.e., it is a true $3M$-dimensional problem. But if we diagonalize the dynamic matrix, i.e., write it as

$$\underline{\underline{D}} = \underline{\underline{U}}^\dagger \cdot \underline{\underline{\Lambda}} \cdot \underline{\underline{U}},$$

(21.11)

where $\underline{\underline{U}}$ is a unitary matrix and $\underline{\underline{\Lambda}}$ is a diagonal matrix containing the eigenvalues of $\underline{\underline{D}}$, then we can define new coordinates

$$\tilde{u}_i = \sum_{k=1}^{3M} U_{ik} u_k.$$

(21.12)

Here, we have simplified the notation by letting

$$u_1 \equiv u_{1,x}$$
$$u_2 \equiv u_{1,y}$$
$$\cdots$$
$$u_{3M} \equiv u_{M,z}.$$

(21.13)

These can be used in rewriting the nuclear Hamilton operator. First, we find for the kinetic-energy part,

$$\frac{\partial^2}{\partial u_i^2} = \frac{\partial}{\partial u_i}\left[\sum_k \frac{\partial \tilde{u}_k}{\partial u_i} \frac{\partial}{\partial \tilde{u}_k}\right]$$
$$= \frac{\partial}{\partial u_i}\left[\sum_k U_{ik} \frac{\partial}{\partial \tilde{u}_k}\right]$$

$$= \sum_k U_{ik} \sum_l \frac{\partial \tilde{u}_l}{\partial u_i} \frac{\partial^2}{\partial \tilde{u}_k \partial \tilde{u}_l}$$

$$= \sum_{k,l} U_{ik} U_{il} \frac{\partial^2}{\partial \tilde{u}_k \partial \tilde{u}_l}. \tag{21.14}$$

Then,

$$-\frac{1}{2} \sum_i \frac{\partial^2}{\partial u_i^2} = -\frac{1}{2} \sum_i \sum_{k,l} U_{ik} U_{il} \frac{\partial^2}{\partial \tilde{u}_k \partial \tilde{u}_l}$$

$$= -\frac{1}{2} \sum_{k,l} \frac{\partial^2}{\partial \tilde{u}_k \partial \tilde{u}_l} \sum_i U_{ik} U_{il}$$

$$= -\frac{1}{2} \sum_{k,l} \frac{\partial^2}{\partial \tilde{u}_k \partial \tilde{u}_l} \delta_{k,l}$$

$$= -\frac{1}{2} \sum_k \frac{\partial^2}{\partial \tilde{u}_k^2} \tag{21.15}$$

due to the unitarity of the matrix $\underline{\underline{U}}$.

Therefore, the nuclear Hamilton operator takes the form

$$\hat{H}_n = \sum_k \left[-\frac{1}{2} \frac{\partial^2}{\partial \tilde{u}_k^2} + \frac{1}{2} \Lambda_{k,k} \tilde{u}_k^2 \right], \tag{21.16}$$

i.e., there is no mixing between the different components (the different k). In this form, it is straightforward to calculate the eigenvalues and eigenfunctions. To this end we observe that the form of the nuclear Hamilton operator in Eq. (21.16) makes it possible to write the nuclear wavefunction as a product,

$$\Psi_n(\vec{R}_1, \vec{R}_2, \ldots, \vec{R}_M) = \psi_1(\tilde{u}_1) \cdot \psi_2(\tilde{u}_2) \ldots \psi_{3M}(\tilde{u}_{3M}). \tag{21.17}$$

Inserting this into the nuclear Schrödinger equation gives $3M$ independent equations of the form

$$\left[-\frac{1}{2} \frac{d^2}{d\tilde{u}_k^2} + \frac{1}{2} \Lambda_{k,k} \tilde{u}_k^2 \right] \psi_k(\tilde{u}_k) = E_k \psi_k(\tilde{u}_k). \tag{21.18}$$

The eigenvalues are then

$$E_k = \left(m_k + \tfrac{1}{2} \right) \sqrt{\Lambda_{k,k}}, \tag{21.19}$$

where the integers m_k distinguish the different eigenvalues and eigenfunctions.

Thus, $\tilde{u}_k$ describes how the different atoms move when the kth vibrational mode is excited and the eigenvalues $\Lambda_{k,k}$ of the dynamical matrix are the square roots of the vibration excitation energies.

There is one aspect that should be added. I.e., since we may translate or rotate the molecule without any energy costs, we will have 5 or 6 modes of zero energy, where 5 is valid for linear molecules and 6 for others.

21.2 PHONONS

For an infinite crystalline material the approach of the preceeding subsection can, in principle, also be applied. The main problem is that since the number of atoms is infinite, the dynamical matrix becomes infinitely large, and it is not straightforward to calculate its eigenvalues and eigenvector. But just as, for the case of electrons moving in an infinite, periodic crystalline material, we could use symmetry arguments in transforming the $\infty \times \infty$ Hamilton and overlap matrices into an infinite set of finite matrices that could be treated independently, we can also use symmetry in transforming the infinite dynamical matrix into an infinite set of finite dynamical matrices.

We let $R_{l,\vec{n},\alpha}$ be the α component (x, y, or z) of the position of the lth atom in the unit cell described by

$$\vec{T}_{\vec{n}} = n_a \cdot \vec{a} + n_b \cdot \vec{b} + n_c \cdot \vec{c}, \tag{21.20}$$

where n_a, n_b, and n_c are integers and $\vec{a}$, $\vec{b}$, and $\vec{c}$ are the basis vectors describing the translation-symmetry properties of the crystal.

Subsequently, we introduce

$$u_{l,\vec{n},\alpha} = R_{l,\vec{n},\alpha} - R^e_{l,\vec{n},\alpha}. \tag{21.21}$$

The dynamical matrix then contains the quantities

$$\frac{1}{\sqrt{M_{l_1}M_{l_2}}} \frac{\partial^2 E_n}{\partial u_{l_1,\vec{n}_1,\alpha_1} \partial u_{l_2,\vec{n}_2,\alpha_2}}. \tag{21.22}$$

Related to the case of electrons in an infinite, periodic crystal we apply group theory in forming symmetry-adapted displacements:

$$u^{\vec{k}}_{l,\alpha} = \sum_{\vec{n}} e^{i\vec{k}\cdot\vec{T}_{\vec{n}}} u_{l,\vec{n},\alpha}. \tag{21.23}$$

With these, the total energy per unit cell is written as (to second order in the displacements)

$$\frac{E}{N} = \frac{E_e}{N} + \frac{1}{2N} \sum_{\vec{k}_1,\vec{k}_2} \sum_{l_1,l_2} \sum_{\alpha_1,\alpha_2=x,y,z} \frac{\partial^2 E(\vec{X}^e)}{\partial u^{\vec{k}_1*}_{l_1,\alpha_1} \partial u^{\vec{k}_2}_{l_2,\alpha_2}} u^{\vec{k}_1*}_{l_1,\alpha_1} u^{\vec{k}_2}_{l_2,\alpha_2}, \tag{21.24}$$

with $N \to \infty$ being the number of unit cells and where we have used the complex conjugated on one of the derivatives since we have introduced complex displacement vectors.

We see that for the dynamical matrix we need the derivatives

$$\frac{1}{N} \frac{\partial^2 E(\vec{X}^e)}{\partial u_{l_1,\alpha_1}^{\vec{k}_1*} \partial u_{l_2,\alpha_2}^{\vec{k}_2}}. \tag{21.25}$$

This quantity is the change in the total energy under two infinitesimal displacements of all nuclei. E is completely invariant under any translation, i.e., it corresponds to the irreducible representation with $\vec{k} = \vec{0}$. On the other hand, $u_{l_1,\alpha_1}^{\vec{k}_1*}$ belongs to the irreducible representation of $\vec{k}_1^*$, whereas $u_{l_2,\alpha_2}^{\vec{k}_2}$ belongs to $\vec{k}_2$. The derivative of Eq. (21.25) transforms therefore according to the irreducible representations of

$$\vec{k}_1^* \otimes \vec{0} \otimes \vec{k}_2 = (-\vec{k}_1) \otimes \vec{0} \otimes \vec{k}_2. \tag{21.26}$$

The derivative is only then non-vanishing if this direct product contains the representation $\vec{0}$ and, as discussed in Chapter 7, this is only the case if

$$\vec{k}_1 = \vec{k}_2. \tag{21.27}$$

Another way to see this is to calculate the derivatives (21.25) directly:

$$\frac{1}{N} \frac{\partial^2 E(\vec{X}^e)}{\partial u_{l_1,\alpha_1}^{\vec{k}_1*} \partial u_{l_2,\alpha_2}^{\vec{k}_2}} = \frac{1}{N} \sum_{\vec{n}_1,\vec{n}_2} \frac{\partial^2 E(\vec{X}^e)}{\partial u_{l_1,\vec{n}_1,\alpha_1} \partial u_{l_2,\vec{n}_2,\alpha_2}} \frac{\partial u_{l_1,\vec{n}_1,\alpha_1}}{\partial u_{l_1,\alpha_1}^{\vec{k}_1*}} \frac{\partial u_{l_2,\vec{n}_2,\alpha_2}}{\partial u_{l_2,\alpha_2}^{\vec{k}_2}}$$

$$= \frac{1}{N} \sum_{\vec{n}_1,\vec{n}_2} \frac{\partial^2 E(\vec{X}^e)}{\partial u_{l_1,\vec{n}_1,\alpha_1} \partial u_{l_2,\vec{n}_2,\alpha_2}} \left(\frac{\partial u_{l_1,\alpha_1}^{\vec{k}_1*}}{\partial u_{l_1,\vec{n}_1,\alpha_1}} \right)^{-1} \left(\frac{\partial u_{l_2,\alpha_2}^{\vec{k}_2}}{\partial u_{l_2,\vec{n}_2,\alpha_2}} \right)^{-1}$$

$$= \frac{1}{N} \sum_{\vec{n}_1,\vec{n}_2} \frac{\partial^2 E(\vec{X}^e)}{\partial u_{l_1,\vec{n}_1,\alpha_1} \partial u_{l_2,\vec{n}_2,\alpha_2}} e^{i\vec{k}_1 \cdot \vec{T}_{\vec{n}_1}} e^{-i\vec{k}_2 \cdot \vec{T}_{\vec{n}_2}}. \tag{21.28}$$

This is a double summation over all unit cells. The first factor, i.e., the second-order derivative, depends only on the relative position of the two unit cells, i.e., on

$$\vec{n}_1 - \vec{n}_2 \equiv \vec{n}. \tag{21.29}$$

For a given $\vec{n}$, the second factor

$$e^{i\vec{k}_1 \cdot \vec{T}_{\vec{n}_1}} e^{-i\vec{k}_2 \cdot \vec{T}_{\vec{n}_2}} = e^{i(\vec{k}_1 \cdot \vec{T}_{\vec{n}_1} - \vec{k}_2 \cdot \vec{T}_{\vec{n}_2})} = e^{i\vec{k}_1 \cdot \vec{T}_{\vec{n}}} e^{i(\vec{k}_1 - \vec{k}_2) \cdot T_{\vec{n}_2}}, \tag{21.30}$$

when summed over all $\vec{n}_2$, vanishes unless $\vec{k}_1 - \vec{k}_2 = 0$, i.e., unless Eq. (21.27) is satisfied. That this is so can, e.g., be seen by observing that in the expression

$$\lim_{N \to \infty} \frac{1}{N} \sum_{\vec{n}} e^{i\vec{q} \cdot \vec{T}_{\vec{n}}} \tag{21.31}$$

the sum oscillates (and accordingly vanishes when being divided by $N \to \infty$) except when $\vec{q} = \vec{0}$.

In total, the matrix elements of the dynamical matrix vanish unless the two u belong to the same $\vec{k}$. This means that we can split the dynamical matrix into blocks belonging to different $\vec{k}$ and each having the dimension $3M_0 \times 3M_0$, where M_0 is the number of atoms per unit cell.

We need accordingly to study the total energy per unit cell for $\vec{k}_1 = \vec{k}_2$, i.e.,

$$\frac{1}{N} \frac{\partial^2 E(\vec{X}^e)}{\partial u_{l_1,\alpha_1}^{\vec{k}*} \partial u_{l_2,\alpha_2}^{\vec{k}}} = \sum_{\vec{n}} \frac{\partial^2 E(\vec{X}^e)}{\partial u_{l_1,\vec{n},\alpha_1} \partial u_{l_2,\vec{0},\alpha_2}} e^{i\vec{k}\cdot\vec{T}_{\vec{n}}} \tag{21.32}$$

which has been obtained using the fact that the different contributions in Eq. (21.24) depend only on the difference $\vec{n}$ [Eqs. (21.28)–(21.29)].

With this formulation we see that the $\infty \times \infty$ dynamical matrix can be split into an infinite set of finite dynamical matrices, each characterized by a specific $\vec{k}$. Therefore, one may now calculate the vibrational frequencies as a function of $\vec{k}$, and ultimately plot these as a function of $\vec{k}$, just as we did it for the electronic single-particle energies in Chapter 19 when discussing electronic band structures. For the vibrations we obtain thereby the phonon dispersion curves. Below as well as in Section 21.3 we shall see examples of such curves as obtained with density-functional methods.

A calculation may now proceed as follows. We consider a specific $\vec{k}$ for which we want to calculate the phonon energies. Starting with the structure for the lowest total energy we subsequently consider all the independent displacements of the atoms for this specific $\vec{k}$. This is a far from trivial procedure, since the atoms may move differently in different unit cells, cf. Fig. 21.1. This means that one may have to use very large unit cells in order to calculate the total-energy changes accompanying the phonon displacements. For example, for the

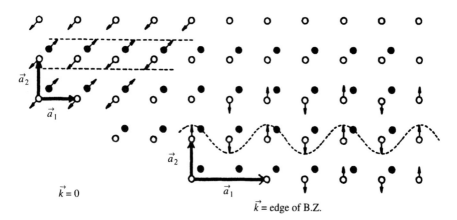

Figure 21.1 Schematic representation of the atomic displacements for a two-dimensional system with two atoms per unit cell. In the left part is shown a phonon for $\vec{k} = 0$, whereas $\vec{k}$ lies at the edge of the first Brillouin zone in the right part. Reproduced by permission of Plenum Press from Kunc and Martin (1983)

two types of displacement of Fig. 21.1 we have in one case ($\vec{k} = 0$) to use the standard unit cell of two atoms, whereas for the other case we have to use a doubled unit cell with four atoms. For even more complex $\vec{k}$ the size of the unit that is repeated periodically may contain even (many) more unit cells. Therefore, practical calculations applying this approach are limited to the phonons of so-called high-symmetry $\vec{k}$ points, i.e., to those for which the periodically repeated unit is not much larger than the original unit cell.

For the different independent displacement patterns we calculate the changes in the total energy and fit the results subsequently with the form of Eq. (21.24) (with $\vec{k}_1 = \vec{k}_2 = \vec{k}$),

$$\frac{E}{N} = \frac{E_e}{N} + \frac{1}{2N} \sum_{l_1,l_2} \sum_{\alpha_1,\alpha_2=x,y,z} \frac{\partial^2 E(\vec{X}^e)}{\partial u_{l_1,\alpha_1}^{\vec{k}*} \partial u_{l_2,\alpha_2}^{\vec{k}}} u_{l_1,\alpha_1}^{\vec{k}*} u_{l_2,\alpha_2}^{\vec{k}}, \qquad (21.33)$$

whereby the dynamical matrix for this $\vec{k}$ point can be calculated. Finally, the phonon energies and displacement patterns can be obtained upon diagonalizing the dynamical matrix.

This approach is known as the *frozen-phonon* approach, meaning that we simulate the displacements of the phonons by rigidly shifting the atoms as it is relevant for that specific $\vec{k}$ point. Even simpler is it if the displacement pattern

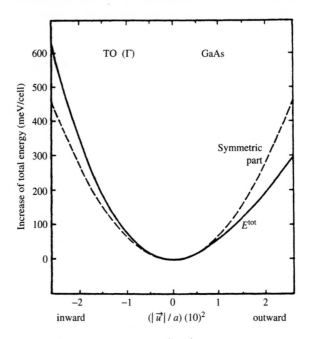

Figure 21.2 Total energy per GaAs unit for a $\vec{k} = \vec{0}$ displacement for GaAs together with the harmonic approximation ('symmetric part'). Reproduced by permission of Plenum Press from Kunc and Martin (1983)

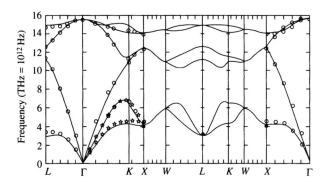

Figure 21.3 Phonon energies (phonon dispersions) of crystalline Si along some high-symmetry $\vec{k}$ lines. The open circles and the stars show experimental results. Reproduced with permission from Wei and Chou (1992). Copyright 1992 by the American Physical Society

is known or can be guessed in advance, whereby only that specific displacement pattern needs to be considered in order to determine the energy of that specific phonon. An example of the results of a such calculation is shown in Fig. 21.2 for GaAs for a Γ phonon (i.e., $\vec{k} = \vec{0}$). Here we also see that the harmonic approximation fails for larger displacements.

A slightly more complicated approach is to consider different displacement patterns for different $\vec{k}$. Subsequently, the variations in the total energy per unit cell can be fitted with a form related to that of Eq. (21.33), with the second-order derivatives of the total energy given by Eq. (21.32). By assuming that the (in principle) infinite series in Eq. (21.32) can be truncated after a finite number of terms (i.e., that only not too distant neighbours interact), one can then determine those constants $\partial^2 E(\vec{X}^e)/\partial u_{l_1,\vec{n},\alpha_1} \partial u_{l_2,\vec{0},\alpha_2}$ that are assumed to be non-zero. Subsequently, the phonon energies at any $\vec{k}$ point can be calculated. A result of such a calculation is shown in Fig. 21.3. A main problem with this approach is that for polar materials (i.e., for crystals with more than one type of atoms between which a charge transfer takes place) long-range Coulomb interactions may make the series in Eq. (21.31) slowly convergent.

21.3 LINEAR-RESPONSE THEORY

Within the harmonic approximation the phonon energies are determined through the second-order expansion of the total energy as a function of nuclear displacements away from their equilibrium positions. In Chapter 20 we demonstrated that the Hellmann–Feynman theorem could be used in determining forces, but that in many practical calculations this theorem is useful only when the so-called Pulay forces could be neglected. This latter was the case either when the basis set was complete or when the basis functions did not depend on the nuclear positions, e.g., when plane waves are used as basis functions. The method we shall develop in the present subsection was originally developed by Giannozzi *et al.* (1991) for this latter case and we shall therefore restrict ourselves to that. In other cases one would have to include further complications but we consider that beyond the aim of this presentation. Furthermore, we shall here exclusively consider density-functional methods.

We consider the general case that our calculations include a set of parameters $\{\lambda_u\}$. Later we will let these be the nuclear coordinates. The Hellmann–Feynman theorem — which is assumed valid — states then that

$$\frac{dE_e}{d\lambda_i} = \langle \Phi | \frac{d\hat{H}}{d\lambda_i} | \Phi \rangle = \int \rho(\vec{r}) \frac{dV(\vec{r})}{d\lambda_i} \, d\vec{r}. \tag{21.34}$$

Here, Φ is the N-particle Slater determinant constructed from the solutions to the Kohn–Sham equations

$$\left[-\frac{1}{2}\nabla^2 + V_{\text{eff}}(\vec{r}) \right] \psi_i(\vec{r}) = \varepsilon_i \psi_i(\vec{r}). \tag{21.35}$$

Moreover, in Eq. (21.34) we have used the fact that that part of the total electronic energy E_e that originates from the kinetic energy does not contribute to the Hellmann–Feynman forces, so that we only have a contribution from the potentials.

As mentioned above, we are interested in the second-order derivatives of the total energy. Therefore, we differentiate Eq. (21.34) once more obtaining

$$\frac{d^2E_e}{d\lambda_i d\lambda_j} = \int \left[\frac{d\rho(\vec{r})}{d\lambda_j} \frac{dV(\vec{r})}{d\lambda_i} + \rho(\vec{r}) \frac{d^2V(\vec{r})}{d\lambda_i d\lambda_j} \right] d\vec{r}. \tag{21.36}$$

In addition we need the derivatives of the electrostatic energy of the nucleus-nucleus interactions,

$$\frac{d^2E_n}{d\lambda_i d\lambda_j} = \frac{\partial^2}{\partial\lambda_i \partial\lambda_j} \left[\frac{1}{2} \sum_{k \neq m=1}^{M} \frac{Z_k Z_m}{|\vec{R}_k - \vec{R}_m|} \right]. \tag{21.37}$$

We now let the parameters λ_i and λ_j be symmetry-adapted displacements of the nuclei, characterized by $\vec{k}$,

$$\lambda_i = u_{l_1,\alpha_1}^{\vec{k}*}$$

$$\lambda_j = u_{l_2,\alpha_2}^{\vec{k}} \tag{21.38}$$

with

$$u_{l,\alpha}^{\vec{k}} = \sum_{\vec{n}} e^{i\vec{k}\cdot\vec{T}_{\vec{n}}} u_{l,\vec{n},\alpha}, \tag{21.39}$$

where $u_{l,\vec{n},\alpha}$ is the αth component of the displacement of the lth atom of the $\vec{n}$th unit cell away from its equilibrium position,

$$\vec{u}_{l,\vec{n}} = (u_{l,\vec{n},x}, u_{l,\vec{n},y}, u_{l,\vec{n},z}) = (\vec{R}_{l,\vec{n}} - \vec{R}_{l,\vec{n}}^e). \tag{21.40}$$

Then, the second-order derivatives of the nuclear energy, Eq. (21.37), are easily calculated. What we need is the second-order derivatives of the electronic energy, Eq. (21.36). One of the terms involves the second-order derivatives of the potential

$$\int \rho(\vec{r}) \frac{\mathrm{d}^2 V(\vec{r})}{\mathrm{d}u_{l_1,\alpha_1}^{\vec{k}*} \, \mathrm{d}u_{l_2,\alpha_2}^{\vec{k}}} \, \mathrm{d}\vec{r} = \int \rho(\vec{r}) \sum_{\vec{n}_1,\vec{n}_2} \frac{\mathrm{d}^2 V(\vec{r})}{\mathrm{d}u_{l_1,\vec{n}_1,\alpha_1} \, \mathrm{d}u_{l_2,\vec{n}_2,\alpha_2}} \, \mathrm{e}^{-\mathrm{i}\vec{k}\cdot(\vec{T}_{\vec{n}_2}-\vec{T}_{\vec{n}_1})}. \quad (21.41)$$

Since $V(\vec{r})$ is periodic, $\mathrm{d}^2 V(\vec{r})/\mathrm{d}u_{l_1,\vec{n}_1,\alpha_1} \, \mathrm{d}u_{l_2,\vec{n}_2,\alpha_2}$ depends only on $\vec{T}_{\vec{n}_2} - \vec{T}_{\vec{n}_1}$ and the term in Eq. (21.41) vanishes unless $\vec{k} = \vec{0}$. In order to be able to ignore this term completely, one may consider any other value of $\vec{k}$, arbitrarily close to $\vec{k} = \vec{0}$, and subsequently use the fact that the phonon energies are continuous functions of $\vec{k}$, also at $\vec{k} = 0$. Alternatively, one may explicitly study $\vec{k} = \vec{0}$ (e.g., using a frozen-phonon approach), but we will not do so here.

Instead, we turn to the other term of Eq. (21.36),

$$\int \frac{\mathrm{d}\rho(\vec{r})}{\mathrm{d}u_{l_1,\lambda_1}^{\vec{k}*}} \frac{\mathrm{d}V(\vec{r})}{\mathrm{d}u_{l_2,\lambda_2}^{\vec{k}}} \, \mathrm{d}\vec{r}. \quad (21.42)$$

Here, $\mathrm{d}V(\vec{r})/\mathrm{d}u_{l_2,\lambda_2}^{\vec{k}}$ involves the first-order changes in $V(\vec{r})$ when the system is perturbed as described by $u_{l_2,\lambda_2}^{\vec{k}}$. First of all, $V(\vec{r})$ will change since the potential generated by the nuclei will change (since the nuclei move). This term is to be included. Second, the change in $V(\vec{r})$ will lead to a change in the electron distribution $\rho(\vec{r})$, which in turn will change $V(\vec{r})$. This is, however, a second-order term, and, therefore, for the first-order derivative above it is irrelevant. In total, $\mathrm{d}V(\vec{r})/\mathrm{d}u_{l_2,\lambda_2}^{\vec{k}}$ can be calculated by considering *solely* the changes in the nuclear potential.

The other quantity of Eq. (21.42) is the change in the electron density due to the perturbation from the nuclear displacements. That is, it is the (linear) response of the electron density to displacing the nuclei. Calculating this implies calculating the changes in the eigenfunctions of Eq. (21.35).

Displacing the nuclei will lead to a change in the electron density,

$$\rho(\vec{r}) \rightarrow \rho(\vec{r}) + \Delta\rho(\vec{r}). \quad (21.43)$$

Thereby, the effective potential of Eq. (21.35),

$$V_{\mathrm{eff}}(\vec{r}) = V_{\mathrm{n}}(\vec{r}) + V_{\mathrm{C}}(\vec{r}) + V_{\mathrm{xc}}(\vec{r}) \quad (21.44)$$

will change. In Eq. (21.44), V_{n} is the electrostatic potential of the nuclei,

$$V_{\mathrm{C}}(\vec{r}) = \frac{1}{2} \int \frac{\rho(\vec{r}\,')}{|\vec{r} - \vec{r}\,'|} \, \mathrm{d}\vec{r}\,' \quad (21.45)$$

is the electrostatic (Coulomb) potential of the electrons, and V_{xc} is the exchange-correlation potential. Due to the displacements of the nuclei and the accompanying electron-density changes [Eq. (21.43)], $V_{eff}(\vec{r})$ changes according to

$$V_{eff}(\vec{r}) \rightarrow V_{eff}(\vec{r}) + \Delta V_{eff}(\vec{r}) \tag{21.46}$$

with

$$\Delta V_{eff}(\vec{r}) = \Delta V_n(\vec{r}) + \frac{1}{2} \int \frac{\Delta\rho(\vec{r}')}{|\vec{r} - \vec{r}'|} \, d\vec{r}' + \int \Delta\rho(\vec{r}') \cdot \frac{\delta V_{xc}(\vec{r})}{\delta\rho(\vec{r}')} \, d\vec{r}'. \tag{21.47}$$

On the other hand,

$$\Delta\rho(\vec{r}) = \sum_i \left[|\psi_i(\vec{r}) + \Delta\psi_i(\vec{r})|^2 - |\psi_i(\vec{r})|^2 \right] \simeq 2 \sum_i |\psi_i^*(\vec{r}) \cdot \Delta\psi_i(\vec{r})|, \tag{21.48}$$

where $\Delta\psi_i(\vec{r})$ is the change in the eigenfunction to Eq. (21.35) due to the extra potential of Eq. (21.47). This may now be calculated either using first-order perturbation theory (only first order, since we need only the first-order derivatives), or, alternatively, using a Green's function approach. In either case, Eqs. (21.47) and (21.48) have to be solved self-consistently: the changes in the electron density due to the the changes in the potential shall equal those that cause the changes in the potential. The formulas for the perturbation theory are the simpler ones but of computational reasons it may be better to apply the Green's function method. We shall, however, not give the formulas here, since the derivation is complicated.

That these methods work well and, in addition, can give the phonon energies at *any* $\vec{k}$ point is illustrated in Fig. 21.4. The calculations were done — for a given $\vec{k}$ point — by calculating the second-order derivatives of the total energy with respect to the nuclear displacements. To this end, the expression of Eq. (21.36) or Eq. (21.42) has to be calculated. The derivative of the potential is easily calculated, as mentioned above, and in order to determine the derivative of the electron density, the linear-response method is employed. Subsequently, the dynamical matrix (which is $\vec{k}$ dependent) is set up, and its eigenvalues give the phonon energies.

21.4   WHAT IS RESPONSE THEORY?

Above we stated that linear-response theory should be applied without discussing the precise meaning behind this. The basic principle is that disturbing a given system somehow will lead to changes in the system properties. When focusing on those that depend linearly on the strength of the perturbation, then one considers the linear response.

This definition is very vague and general, but should make it clear that one has to be careful in defining the properties whose linear dependence on the perturbation is of interest. One has also to make sure that the perturbation is

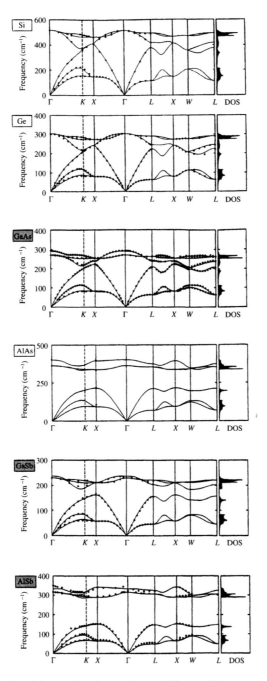

Figure 21.4 Calculated (curves) and experimental (diamonds) phonon dispersion curves for various semiconductors. The 'DOS' curves on the right-hand sides are the phonon density of states. Reproduced with permission from Giannozzi *et al.* (1991). Copyright 1991 by the American Physical Society

so weak that the linear response is sufficiently accurate. Let us therefore first consider a simple example.

The ground-state energy of the harmonic oscillator,

$$-\frac{1}{2}\frac{d^2\psi}{dx^2} + \frac{1}{2}kx^2\psi = \varepsilon\psi \qquad (21.49)$$

is

$$\varepsilon = \frac{1}{2}\sqrt{k}. \qquad (21.50)$$

We will add a perturbation of the form

$$\Delta V = \frac{1}{2}\Delta k \cdot (x - x_0)^2 \qquad (21.51)$$

and consider Δk as describing the strength of this. x_0 is a constant.

The total potential then becomes

$$\frac{1}{2}kx^2 + \frac{1}{2}\Delta k \cdot (x - x_0)^2 = \frac{1}{2}(k + \Delta k)\left(x - \frac{\Delta k}{k + \Delta k}x_0\right)^2 + \frac{1}{2}\frac{k\Delta k}{k + \Delta k}x_0^2. \qquad (21.52)$$

Thus, the ground-state energy changes into

$$\varepsilon = \frac{1}{2}\sqrt{k + \Delta k} + \frac{1}{2}\frac{k\Delta k}{k + \Delta k}x_0^2, \qquad (21.53)$$

which, however, to first order in Δk, becomes

$$\varepsilon = \frac{1}{2}\sqrt{k} + \left(\frac{1}{4\sqrt{k}} + \frac{1}{2}x_0^2\right)\Delta k. \qquad (21.54)$$

This corresponds to the linear-response result, which, actually, also could have been obtained with perturbation theory.

But linear-response theory may be applied at many other situations. Thus, for a certain system with some electron density $\rho(\vec{r})$, the electron density may be changed for one reason or another,

$$\rho(\vec{r}) \to \rho(\vec{r}) + \Delta\rho(\vec{r}). \qquad (21.55)$$

This change will lead to a change in the potential and, subsequently, the complete electron density will change. Thus, in any point in space there will be a change in the potential due to the change in that in any other point in space,

$$\Delta V(\vec{r}) = \int \frac{\delta V(\vec{r})}{\delta\rho(\vec{r}')}\Delta\rho(\vec{r}')\,d\vec{r}'. \qquad (21.56)$$

Here, we have, as above, only considered the linear response — this time of the potential — to the changes in the electron density.

Formulas like that of Eq. (21.56) will become important later in this text.

22 Electronic Excitations

22.1 EIGENVALUE SPECTRUM AND DENSITY OF STATES

We consider a system for which we assume that we can solve the stationary Schrödinger equation

$$\hat{H}_0 \Psi_n = E_n \Psi_n \tag{22.1}$$

exactly, so that the set of eigenfunctions $\{\Psi_n\}$ forms a complete set of orthonormal functions.

We assume that at $t = 0$ the system is in its ground state $(n = 0)$ and at that time an extra time-dependent perturbation is turned on. We will denote this by $\hat{H}_1(t)$, which, as indicated, is time-dependent. Due to this extra perturbation the system is no longer stationary, and the wavefunction Ψ may change as a function of time. This means that the time-dependent wavefunction can be written as a superposition of the form (since the Ψ_n form a complete set of functions)

$$\Psi = \sum_n a_n(t) \Psi_n \, e^{-iE_n t}. \tag{22.2}$$

The occupation $|a_n(t)|^2$ of the state Ψ_n depends on time, and according to Fermi's golden rule (corresponding to time-dependent perturbation theory), the transition rate

$$W_{n,0}(t) = \frac{d}{dt} |a_n(t)|^2 \tag{22.3}$$

(here, the indices n and 0 characterize the final and initial states) is

$$W_{n,0}(t) = 2\pi |\langle \Psi_n | \hat{H}_1(t) | \Psi_0 \rangle|^2. \tag{22.4}$$

Here, $\langle \Psi_n | \hat{H}_1(t) | \Psi_0 \rangle$ depends on the perturbation as well as on the wavefunctions of the initial and final states. This may therefore vary for different experimental situations. Equation (22.3) is only a first-order approximation. When the matrix element $\langle \Psi_n | \hat{H}_1(t) | \Psi_0 \rangle$ vanishes, $W_{n,0}(t)$ may still be non-zero due to higher-order effects, but is in that case most often only small.

Very many experiments include an energy discrimination, i.e., one studies the intensity as a function of time and energy. Denoting the measured energy E, the quantity of interest is then

$$I(E, t) = \sum_n W_{n,0}(t)\delta[E - (E_n - E_0)]$$

$$= 2\pi \sum_n |\langle \Psi_n | \hat{H}_1(t) | \Psi_0 \rangle|^2 D(E_n)D(E_0)\delta[E - (E_n - E_0)]$$

$$= 2\pi \sum_n |\langle \Psi_n | \hat{H}_1(t) | \Psi_0 \rangle|^2 D(E - E_0)D(E_0)\delta[E - (E_n - E_0)]. \quad (22.5)$$

Here, $D(E_i)$ is the so-called density of states, which gives the number of states with the energy E_i and which was discussed in Section 19.9. The expression of Eq. (22.4) is mainly the density of states $D(E - E_0)$ multiplied by some matrix elements. Therefore, it is very useful to study the density of states itself, and only if desired also include the matrix-element effects.

Let us now be a little more specific. We assume that the system of our interest is described within a single-particle model (e.g., with a Hartree–Fock or density-functional method) so that the ground-state corresponds to occupying the N energetically lowest single-particle orbitals. In Fig. 22.1 we show an example of a system with $N = 4$ electrons where we furthermore assume that we have in total only four orbitals (for any real system, there are infinitely many orbitals).

Exciting only one electron leads to configurations like those of the four right panels of Fig. 22.1. Assuming that Koopmans' theorem is valid (i.e., neglecting relaxation effects caused by charge transfers or redistributions by structural

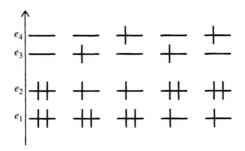

Figure 22.1 Schematic representation of the energy levels of a four-electron system and their occupancies for (left panel) the ground state and for those states obtained by exciting one electron

changes, which may lead to changes in the orbitals and their energies) the four excited states in Fig. 22.1 have the excitation energies $e_3 - e_2$, $e_4 - e_2$, $e_3 - e_1$, and $e_4 - e_1$. Therefore, the quantity of Eq. (22.4) can only then be non-vanishing, when E equals one of these energies.

Instead of the presentation of Fig. 22.1, that of Fig. 22.2 is more useful. Here, we have only presented the energies of the different single-particle orbitals together with the number of electrons they can accommodate [i.e., the density of states $D(e)$]. For the ground-state, all levels up to the Fermi level, e_F, are occupied, whereas those for higher energies are empty. The possible excitations are then *any* that excite one electron from below the Fermi level to above the Fermi level. The energy of the excitation is simply the difference of the two orbital energies. Therefore, Fig. 22.2 contains all the information that is relevant for the spectrum of Eq. (22.4), as long as we assume that Koopmans' theorem is valid, and when neglecting matrix-element effects. Therefore, the density of states as that of Fig. 22.2 is the one we shall study.

In Fig. 22.2 we see that the smallest possible excitation energy equals $e_3 - e_2$. This energy is called the energy gap. For metals it vanishes whereas for semiconductors it is non-vanishing and up to about 3 eV, and even larger for insulators. The orbital with the energy e_2 in Fig. 22.2 is called the highest occupied molecular orbital (HOMO) for molecular systems and the highest occupied crystal orbital (HOCO) for crystalline systems. That with the energy e_3 is equivalently called either the lowest unoccupied molecular orbital (LUMO) or the lowest unoccupied crystal orbital (LUCO). In crystalline materials, the bands below the Fermi level are called valence bands and those above are called conduction bands.

For infinite, periodic materials we saw in Chapter 19 that the discrete energy levels are broadened into continuous bands. Therefore, the density of states for the electronic orbitals will no longer be a function with discrete peaks, but consists

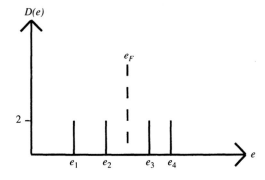

Figure 22.2 The positions of the single-particle energies of the example of Fig. 22.1 as a function of energy with the abscissa being the number of states (density of states). For the ground state, the Fermi level (marked e_F) separates occupied and empty orbitals

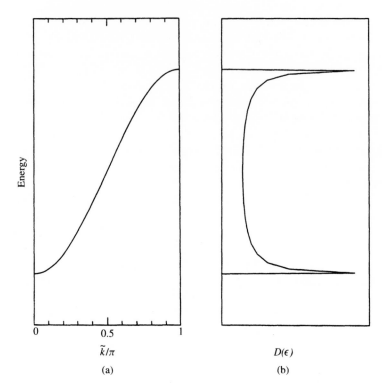

Figure 22.3 (a) A single band for a one-dimensional chain; (b) the corresponding density of states. The latter has been rotated by 90° in order to make the presentation clearer

of more continuous parts; cf. Section 19.9. In Fig. 22.3 we show how the density of states becomes when considering a single band for a one-dimensional system. We see that at the bottom and top of the band, where the band is flat, the density of states diverges. This is a so-called van Hove singularity.

In principle, there is for an infinite, periodic system an infinitely large density of states in those energy intervals where the bands occur. However, as we showed in Chapter 19, each complete band can accommodate two electrons (when including spin) per repeated unit. Therefore, it is more convenient to study the density of states per unit cell. It also means that when integrating the density of states over a complete band (e.g., when integrating the curve of the right-hand part of Fig. 22.3 over the whole energy interval of the figure) one obtains the result 2.

In Fig. 22.4 we show a somewhat more complicated case, i.e., the band structures and the density of states for polyacetylene. It is easily seen how the density of states has sharp peaks where the bands are flat, which may be either at the band energies at $\tilde{k} = 0$ or $\tilde{k} = \pi$ or at energies where the bands show avoided crossings.

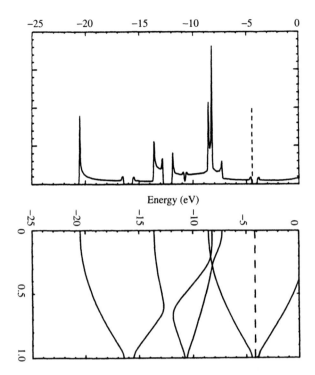

Figure 22.4 The density of states (upper part) and the band structures (rotated by 90°, lower part) of polyacetylene

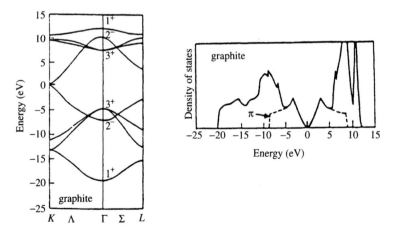

Figure 22.5 The band structures (left part) and density of states (right part) of a single layer of graphite. The Fermi energy has been set at 0. Reproduced with permission from Robertson and O'Reilly (1987). Copyright 1987 by the American Physical Society

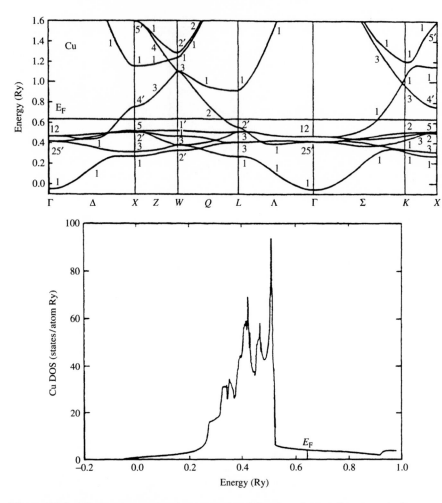

Figure 22.6 The band structures (upper part) and density of states (lower part) of copper. The energies are given in rydbergs (1 rydberg = 13.605 eV). Reproduced with permission from Jepsen *et al.* (1981). Copyright 1981 by the American Physical Society

For systems that are periodic in more dimensions the singularities in the density of states get less pronounced. This can be seen, e.g., in Fig. 22.5, where we show the band structures and density of states of a single layer of graphite. This system is periodic in two dimensions, so that the $\vec{k}$ vector is a two-dimensional vector and the first Brillouin zone is two-dimensional, too.

This effect is even more pronounced in three dimensions, cf. Fig. 22.6, where we show the band structures and density of states for crystalline copper.

22.2 SINGLE-PARTICLE EXCITATIONS

As a first approximation, the density of states of the single-particle orbitals provides the relevant information when studying single-particle excitations. This approximation amounts to assuming that the orbitals do not change when one of the electrons is excited. Therefore, we will expect that it is the better fulfilled the larger the system is, since then the perturbations due to exciting a single electron may become less pronounced. Furthermore, in studying the density of states we assume also that matrix-element effects can be ignored. In many cases selection rules make this approximation less valid, so that peaks that should appear according to the density of states may be (almost) suppressed due to matrix-element effects, i.e., the corresponding transitions are (almost) forbidden.

Since the density of states is closely linked to the single-particle energies, the first question one may ask is therefore whether the single-particle energies are calculated accurately. We shall accordingly first address this question. We shall therefore compare Hartree–Fock-based and density-functional-based methods since these are the ones that most directly give a set of single-particle energies.

According to the basic approach of Kohn and Sham, the single-particle energies, ε_i, as calculated from the Kohn–Sham equations,

$$\hat{h}_{\text{eff}}\psi_i(\vec{r}) = \varepsilon_i\psi_i(\vec{r}), \qquad (22.6)$$

have in principle no physical meaning. It can be shown (Almbladh and von Barth, 1985; see also Section 15.8) that the energy of the highest occupied orbital should be the ionization potential (in the case that the exact effective Hamilton operator $\hat{h}_{\text{eff}}$ is known), but for the others there is no theoretical justification for relating them to a physically observable density of states. Nevertheless, it has turned out that neglecting this formal inconsistency is a good approximation in very many cases and so we shall do here.

On the other hand, the single-particle energies ε_i from the Hartree–Fock equations

$$\hat{F}\psi_i(\vec{r}) = \varepsilon_i\psi_i(\vec{r}) \qquad (22.7)$$

can be related to ionization potentials and electron affinities according to Koopmans' theorem (Koopmans 1933; see also Section 9.3).

The first system we shall study is a single hydrogen atom. For this, the exact single-particle energy is -0.5 a.u., which also is the energy obtained with an accurate Hartree–Fock calculation. On the other hand, a density-functional calculation with a local-density approximation gives only -0.25 a.u. The error is due to the self-interaction, that was discussed above in Section 15.10, i.e., to the fact that we include the Coulomb interaction of the electron with itself, that in the exact theory shall be cancelled by the exchange interaction of the electron with itself, but that with the local-density approximation is not cancelled exactly. Actually, the finding that the density-functional eigenvalues of the occupied orbitals are too high is general, and often this error is as large as a couple of eV.

As a next system we study an isolated Rn atom. This is a closed-shell atom, so the Hartree–Fock approximation is assumed to be good (cf. the discussion

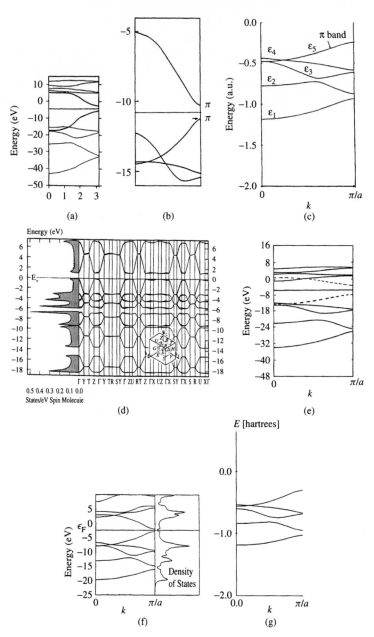

in Section 13.6). Neglecting relativistic effects (although these may be large, as discussed below in Chapter 23), we obtain the results of Table 22.1. Here we see that the Hartree–Fock energies span a broader energy region as those of the density-functional studies. This is a general finding that holds also for larger and extended systems, and, in general, it turns out that for larger systems the density-functional eigenvalues of the uppermost occupied orbitals are in better agreement with experimental information than are the Hartree–Fock eigenvalues. The physical reason is that correlation effects that tend to screen the interactions between different atoms are, by definition, absent in the Hartree–Fock calculations, so that more distant electronic orbitals interact too strongly within the Hartree–Fock picture.

In order to compare a number of different methods on the same system we show in Fig. 22.7 the band structures from different theoretical studies on polyacetylene. There are many differences and in particular the results from semiempirical methods may deviate markedly from those of parameter-free methods. Furthermore, as above we see that the bands from Hartree–Fock-based

Table 22.1 Single-particle energies in Hartree (27.21 eV) for a single Rn atom as calculated with Hartree–Fock (HF) and density-functional (DFT) methods. (Relativistic effects are neglected)

Orbital	HF	DFT
$1s$	−3230.31	−3204.76
$2s$	−556.913	−546.589
$2p$	−536.677	−527.543
$3s$	−138.422	−133.382
$3p$	−128.671	−124.186
$3d$	−110.701	−106.958
$4s$	−33.9206	−31.2453
$4p$	−29.4911	−27.1236
$4d$	−21.3312	−19.4645
$4f$	−10.10753	−8.96787
$5s$	−6.90574	−5.90564
$5p$	−5.22513	−4.42469
$5d$	−2.32624	−1.92753
$6s$	−0.873967	−0.643459
$6p$	−0.427981	−0.309910

Figure 22.7 Band structures of polyacetylene as obtained with different methods. (a) semiempirical MNDO calculations (from Young et al., 1979), (b) extended Hückel calculations (from Whangbo et al., 1979); (c) parametrized Hartree–Fock calculations (from Brédas et al., 1981), (d) density-functional calculations on crystalline polyacetylene (from Grant and Batra, 1979), (e) semiempirical CNDO/S calculations (from Ford et al., 1982), (f) density-functional calculations (from Mintmire and White, 1983a,b), and (g) parameter-free Hartree–Fock calculations (from Karpfen and Petkov, 1979). The results were reproduced by permission of (a,c,e) American Institute of Physics, (b) The Royal Society of London, (d,g) Pergamon Press, and (f) The American Physical Society

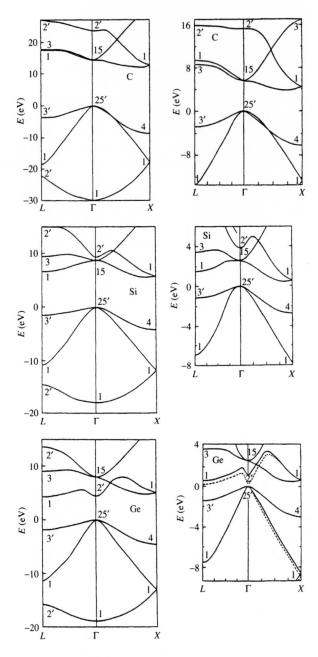

Figure 22.8 Band structures of some crystalline semiconductors from (left side; from Svane, 1987) Hartree–Fock calculations and (right side; from Glötzel *et al.*, 1980) density-functional calculations. The lowest valence bands are not shown for the density-functional studies. The Hartree–Fock results are reproduced by permission of The American Physical Society, and the density-functional results by permission of Pergamon Press

methods (essentially all of the figure except for those from density-functional methods) are broader than those of the density-functional methods, which in turn are known from experience to provide reasonably accurate results for the band structures and density of states of the occupied levels. On the other hand, the density-functional energies lie typically at overall too high energies, as mentioned above.

Finally, we show in Fig. 22.8 a similar comparison between Hartree–Fock and density-functional band structures for some simple crystalline semiconductors, supporting the conclusions from above.

The density of states is useful in studying excitations where one need not bother about the matrix-element effects. The occupied parts of the density of states can then be used in simulating the spectrum that is obtained when ionizing the system of interest, i.e., when the final state can be considered as being identical to the initial state except that one electron occupies a free-electron orbital.

In other situations, the excited electron does not enter a free-electron orbital, but one of the higher-lying (so-called conduction) bands, i.e., it is still bound to the system. Then, one electron is removed from an orbital below the Fermi level and added to an orbital above the Fermi level. In that case, matrix-element effects very often have very pronounced effects on the spectra. Neglecting these, the spectrum would simply be the convolution of the density of states of the occupied orbitals with those of the empty orbitals. But with the matrix-element effects the spectrum may be strongly modified.

Before turning to a more accurate description of these excitation energies, we briefly sketch the experimental conditions.

Group theory is a powerful tool for studying when electronic excitations are possible and when not. Both the initial and the final state (Ψ_0 and Ψ_n, respectively) belongs to certain irreducible representations of the group of symmetry operations of the system, say Ψ_0 belong to R_0 and Ψ_n^* to R_n, respectively. Similarly, the interaction operator $\hat{H}_1$ (i.e., the operator that describes the excitation and which depends on details of the experiment) can be decomposed into various contributions from one or more irreducible representations, $R_1^H \oplus R_2^H \oplus \cdots \oplus R_P^H$. In many cases this sum contains only one or just some few terms. Only when the direct product

$$R_n \otimes (R_1^H \oplus R_2^H \oplus \cdots \oplus R_P^H) \otimes R_0 \tag{22.8}$$

contains the 'trivial' irreducible representation, the matrix element

$$\langle \Psi_n | \hat{H}_1(t) | \Psi_0 \rangle \tag{22.9}$$

may be non-vanishing.

A simple example of this is that of photons interacting with the electrons of a crystal. Then, $\vec{k}$ can be used in characterizing the irreducible representation. The interaction operator in this case belongs to $\vec{k} \simeq \vec{0}$, i.e., $\vec{k} = \vec{0}$ is a good approximation. When exciting an electron from an orbital with $\vec{k}_1$ to one with

$\vec{k}_2$, we thus require that

$$(-\vec{k}_2) \otimes \vec{0} \otimes \vec{k}_1 \equiv \vec{0} \oplus \cdots, \qquad (22.10)$$

or that

$$\vec{k}_2 = \vec{k}_1. \qquad (22.11)$$

This means that only so-called *vertical* excitations are allowed, i.e., that

$$\Delta \vec{k} = \vec{k}_2 - \vec{k}_1 = \vec{0}. \qquad (22.12)$$

Therefore, one has strong peaks when there are occupied and empty bands that are parallel (but not necessarily flat).

In other cases it is possible to study other excitations. In particular, it can be possible (by using electrons as the projectiles instead of photons) to select other values of $\Delta \vec{k} \neq \vec{0}$. By choosing many different values of $\Delta \vec{k}$ it is then almost possible to determine the band structure experimentally.

Another type of experiment involves exciting a core electron to an empty orbital that remains bound to the system. Since the core electron is strongly localized to a specific atom, and since the core electrons of different types of atoms have different excitation energies, it is thereby possible to study the local behaviour of the empty orbitals. On the other hand, the removal of core electrons has in many cases strong effects on all electrons and it is necessary to include the effects due to the relaxations of all the other electrons following the excitation.

Finally, we recall of the concept of Dyson orbitals that we introduced in Section 14.5. These orbitals are designed for specifically determining how the electronic distribution is modified upon exciting a single electron into a free-electron orbital.

22.3   DIELECTRIC MATRIX

The band structures and density of states for the uppermost occupied bands (the valence bands) are — according to experience — in most (but not all) cases accurately described by Hartree–Fock or density-functional methods with the bands being somewhat (about 20–30%) too wide for the Hartree–Fock case. The situation is different for the unoccupied bands (the conduction bands). Here, first of all the band gap separating the valence and conduction bands (the so-called optical gap) is a critical quantity. For a typical semiconductor with a gap of 1–2 eV, the density-functional calculations will give a gap of up to 1 eV, i.e., grossly underestimate its size, whereas according to Hartree–Fock calculations it will be above 5 eV, i.e., grossly overestimated. The Hartree–Fock overestimate may be understood as follows. Within the Hartree–Fock approximation we calculate self-consistently the electronic eigenvalues for the ground state. That is, the potential felt by each of the N electrons in the ground state is that due to the $N - 1$ other electrons. On the other hand, for unoccupied orbitals, the potential is that due to the N occupied orbitals, so these orbitals experience in effect a potential from one more electron. Therefore, their energies are higher.

A pragmatic approach to the so-called band-gap problem is that provided by what is known as the scissor operator (Baraff and Schlüter, 1984). This amounts to rigidly shifting *all* conduction bands so much that the gap obtains the correct size, but without making *any* other adjustment of the bands. In many cases (e.g., for the density-functional bands of crystalline silicon) it is a good approximation, but it is certainly not always the case.

A better approach is to calculate directly the response of the system to the excitation of one electron from a valence orbital to a conduction orbital. Keeping only linear responses (corresponding to only including first-order terms in pertur-bation theory) the dielectric matrix provides the relevant information. We shall here review the properties of the dielectric matrix and subsequently discuss how it can be calculated following Hybertsen and Louie (1987). In the next subsection we shall sketch its application to the problem of calculating the excitation ener-gies for crystalline materials. We stress at this point that this way of obtaining the conduction bands is much more complicated than that of using the scissor operator, but also much more accurate.

We shall start out by introducing some of the general terms that are used in describing the responses to the changes.

Exciting an electron corresponds to changing the charge distribution inside the material. The relevant question we shall address is therefore, how does the material respond to this change? We shall keep only linear terms, and accordingly assume that the response to two changes is the sum of the response to each change treated individually. Therefore, we study first the general case of how the system responds to an extra point charge at a given position. Already at this point we recognize that we have to distinguish between two different cases: if the added charge is due to an electron, this electron will subsequently be indistinguishable from all the other electrons of the material, whereas adding another type of particle will not lead to this.

The point charge $\delta\rho_{ext}(\vec{r})$ gives rise to an extra Coulomb potential $\delta V_{ext}(\vec{r})$, which in turn induces an electron density $\delta\rho_{ind}(\vec{r})$. The two former are related through Poison's equation,

$$-\nabla^2 \delta V_{ext}(\vec{r}) = 4\pi \delta \rho_{ext}(\vec{r}). \qquad (22.13)$$

Simultaneously, the total electron density

$$\rho(\vec{r}) = \rho_0(\vec{r}) + \delta\rho_{ext}(\vec{r}) + \delta\rho_{ind}(\vec{r}) \qquad (22.14)$$

contains one term from the unperturbed system, one term from the extra point charge, and one term from the induced electron density.

The total Coulomb potential $V_C(\vec{r})$ is a solution to Poison's equation for the complete electron density

$$-\nabla^2 V_C(\vec{r}) = 4\pi \rho(\vec{r}). \qquad (22.15)$$

As long as the potential of the added charge $\delta V_{ext}(\vec{r})$ is small compared with the complete potential $V_{tot}(\vec{r})$, one may assume that the former depends only linearly on the latter (this corresponds to the approximation of linear response, as we shall see below). This means that we write

$$\delta V_{ext}(\vec{r}) = \int \varepsilon(\vec{r}, \vec{r}\,')\delta V_{tot}(\vec{r}\,')\,d\vec{r}\,'. \tag{22.16}$$

Here, $\varepsilon(\vec{r}, \vec{r}\,')$ is the so-called dielectric matrix. It is called a matrix due to its dependence on two parameters, $\vec{r}$ and $\vec{r}\,'$, but in contrast to the more well-known matrices, the parameters here are continuous and the matrix, equivalently, $\infty \times \infty$. One may also define its inverse through

$$\delta V_{tot}(\vec{r}) = \int \varepsilon^{-1}(\vec{r}, \vec{r}\,')\delta V_{ext}(\vec{r}\,')\,d\vec{r}\,', \tag{22.17}$$

where ε^{-1} is the inverse dielectric matrix, which does *not* equal $1/\varepsilon$! Instead, ε and ε^{-1} are related through an equation similar to Eq. (22.21) below.

It turns out that it is more useful to work with the so-called polarizability. Once again we consider only the linear response of the system to the extra potential, and study then the induced electron density $\delta\rho_{ind}(\vec{r})$ due to the extra potential $\delta V_{ext}(\vec{r})$. Since the former is assumed to depend linearly on the latter we have in the most general case

$$\delta\rho_{ind}(\vec{r}) = \int \chi(\vec{r}, \vec{r}\,')\delta V_{ext}(\vec{r}\,')\,d\vec{r}\,', \tag{22.18}$$

where $\chi(\vec{r}, \vec{r}\,')$ is the polarizability. χ describes how the change in the potential in one point $\vec{r}\,'$ leads to a change in the charge density in any other point $\vec{r}$.

Also for this, an inverse can be defined,

$$\delta V_{ext}(\vec{r}) = \int \chi^{-1}(\vec{r}, \vec{r}\,')\delta\rho_{ind}(\vec{r}\,')\,d\vec{r}\,'. \tag{22.19}$$

We may combine Eqs. (22.18) and (22.19) to obtain

$$\delta\rho_{ind}(\vec{r}) = \int \chi(\vec{r}, \vec{r}\,')\delta V_{ext}(\vec{r}\,')\,d\vec{r}\,'$$

$$= \int \chi(\vec{r}, \vec{r}\,') \int \chi^{-1}(\vec{r}\,', \vec{r}\,'')\delta\rho_{ind}(\vec{r}\,'')\,d\vec{r}\,''\,d\vec{r}\,'$$

$$= \int \left[\int \chi(\vec{r}, \vec{r}\,')\chi^{-1}(\vec{r}\,', \vec{r}\,'')\,d\vec{r}\,' \right] \delta\rho_{ind}(\vec{r}\,'')\,d\vec{r}\,'' \tag{22.20}$$

or

$$\int \chi(\vec{r}, \vec{r}\,')\chi^{-1}(\vec{r}\,', \vec{r}\,'')\,d\vec{r}\,' = \delta(\vec{r} - \vec{r}\,''). \tag{22.21}$$

We limit ourselves now to density-functional theory. We can then write

$$\delta\rho_{ind}(\vec{r}) = \int \chi_0(\vec{r}, \vec{r}\,')\delta V_{tot}(\vec{r}\,')\,d\vec{r}\,', \tag{22.22}$$

where $\chi_0(\vec{r}, \vec{r}')$ is the so-called independent-particle polarizability and where δV_{tot} is the total change in the potential, i.e., it consists of both the extra external potential and that induced by the perturbation:

$$\delta V_{tot}(\vec{r}) = \delta V_{ext}(\vec{r}) + \delta V_{ind}(\vec{r}). \tag{22.23}$$

δV_{ind} is mainly a screening potential that is built up by the system in order to reduce the effects of the external perturbation.

According to density-functional theory, the total potential consists of three terms,

$$V_{tot}(\vec{r}) = V_{ext}(\vec{r}) + V_C(\vec{r}) + V_{xc}(\vec{r}), \tag{22.24}$$

where $V_{ext}(\vec{r})$ is the total external potential, i.e., without the perturbation δV_{ext} it equals the Coulomb potential of the nuclei. Furthermore,

$$V_C(\vec{r}) = \int \frac{\rho(\vec{r}')}{|\vec{r} - \vec{r}'|} \, d\vec{r}' \tag{22.25}$$

is the Coulomb potential of the electron distribution, and

$$V_{xc}(\vec{r}) = \frac{\delta E_{xc}}{\delta \rho(\vec{r})} \tag{22.26}$$

is the exchange-correlation potential.

Changing

$$\rho(\vec{r}) \rightarrow \rho(\vec{r}) + \delta \rho(\vec{r}) \tag{22.27}$$

leads to changes in V_C and V_{xc}, but not in V_{ext}. We find then

$$\delta V_{ind}(\vec{r}) = \int \frac{\delta \rho(\vec{r}')}{|\vec{r} - \vec{r}'|} \, d\vec{r}' + \int \frac{\delta V_{xc}(\vec{r}')}{\delta \rho(\vec{r}')} \delta \rho(\vec{r}') \, d\vec{r}'. \tag{22.28}$$

We now combine the various expressions. This gives

$$\chi^{-1}(\vec{r}, \vec{r}') = \frac{\delta V_{ext}(\vec{r}')}{\delta \rho(\vec{r})} = \frac{\delta V_{tot}(\vec{r}')}{\delta \rho(\vec{r})} - \frac{\delta V_{ind}(\vec{r}')}{\delta \rho(\vec{r})}$$

$$= \chi_0^{-1}(\vec{r}, \vec{r}') - \frac{1}{|\vec{r} - \vec{r}'|} - K_{xc}(\vec{r}, \vec{r}'). \tag{22.29}$$

In deriving this, we have first used Eq. (22.19) as defining $\chi^{-1}(\vec{r}, \vec{r}')$. Subsequently, Eq. (22.23) is used in rewriting δV_{tot}, and Eqs. (22.22) and (22.28) lead to the last identity.

In Eq. (22.29) we have introduced the so-called exchange-correlation kernel,

$$K_{xc}(\vec{r}, \vec{r}') \equiv \frac{\delta^2 E_{xc}}{\delta \rho(\vec{r}) \delta \rho(\vec{r}')} = \frac{\delta V_{xc}(\vec{r})}{\delta \rho(\vec{r}')}, \tag{22.30}$$

which can be represented analytically once a local or a non-local density approximation has been chosen.

Then, from Eq. (22.29) χ can be calculated once χ_0 is known. Finally, we give the connection to the dielectric matrix,

$$\varepsilon(\vec{r}, \vec{r}') = \frac{\delta V_{\text{ext}}(\vec{r})}{\delta V_{\text{tot}}(\vec{r}')}, \tag{22.31}$$

where Eq. (22.16) has been used.

To this end we need to specify the extra charge. There are two principally different cases. Either the extra charge is different from the electrons of the system (it could, e.g., be a positron, a muon, etc.), or it is indistinguishable from the electrons and accordingly has to fulfill the quantum-mechanical anti-symmetry conditions. The latter is the case when an electron is excited. In the former case we find that

$$\varepsilon^{-1}(\vec{r}, \vec{r}') = \frac{\delta V_{\text{tot}}(\vec{r}')}{\delta V_{\text{ext}}(\vec{r})} = \frac{\delta V_{\text{ext}}(\vec{r}')}{\delta V_{\text{ext}}(\vec{r})} = \delta(\vec{r} - \vec{r}') + \frac{\delta V_{\text{ind}}(\vec{r}')}{\delta V_{\text{ext}}(\vec{r})}$$

$$= \delta(\vec{r} - \vec{r}') + \frac{\delta \rho(\vec{r}')}{|\vec{r} - \vec{r}'|} = \delta(\vec{r} - \vec{r}') + \int \frac{\chi(\vec{r}'', \vec{r})}{|\vec{r}'' - \vec{r}'|} \, d\vec{r}'', \tag{22.32}$$

by using Eqs. (22.23) and (15.56).

In the second case, also the exchange-correlation part will be included,

$$\varepsilon^{-1}(\vec{r}, \vec{r}') = \frac{\delta V_{\text{tot}}(\vec{r}')}{\delta V_{\text{ext}}(\vec{r})} = \delta(\vec{r} - \vec{r}') + \frac{\delta V_{\text{ind}}(\vec{r}')}{\delta V_{\text{ext}}(\vec{r})}$$

$$= \delta(\vec{r} - \vec{r}') + \int \chi(\vec{r}'', \vec{r}) \left[\frac{1}{|\vec{r}'' - \vec{r}'|} + K_{\text{xc}}(\vec{r}'', \vec{r}) \right] d\vec{r}''. \tag{22.33}$$

This is the case that is relevant here when studying electronic excitations.

We see that the central quantity to calculate is $\chi_0(\vec{r}, \vec{r}')$. Knowing this it is in principle possible to calculate any other quantity above [i.e., from χ_0 we can calculate χ using Eq. (22.29)]. χ_0 describes how the electron density changes (to first order!) when the potential changes infinitesimally [cf. Eq. (22.22)]. This means that we need the first-order changes in the eigenfunctions to the Kohn–Sham equations

$$\left[-\tfrac{1}{2} \nabla^2 + V_{\text{n}}(\vec{r}) + V_{\text{C}}(\vec{r}) + V_{\text{xc}}(\vec{r}) \right] \psi_i(\vec{r}) = \varepsilon_i \psi_i(\vec{r}). \tag{22.34}$$

Adding an extra potential δV_{tot} at the point $\vec{r}'$ gives that $\psi_i(\vec{r})$ changes according to (from perturbation theory)

$$\psi_i(\vec{r}) \rightarrow \psi_i(\vec{r}) + \sum_{j \neq i} \frac{\langle \psi_j | \delta V_{\text{tot}} | \psi_i \rangle}{\varepsilon_i - \varepsilon_j} \psi_j(\vec{r}). \tag{22.35}$$

All orbitals up to number N are occupied before turning on the perturbation. These are in addition orthonormal. Requiring that this remains the case (to first order in the perturbation) also after turning on the perturbation this can only be the case when the summation is changed into one over only the unoccupied orbitals (cf. Section 6.1). That is, we rewrite Eq. (22.35) as

$$\psi_i(\vec{r}) \rightarrow \psi_i(\vec{r}) + \sum_{j>N} \frac{\langle \psi_j | \delta V_{tot} | \psi_i \rangle}{\varepsilon_i - \varepsilon_j} \psi_j(\vec{r}). \tag{22.36}$$

The first-order change in the electron density is then easily calculated leading to the following expression for χ_0:

$$\chi_0(\vec{r}, \vec{r}') = \frac{\delta \rho_{ind}(\vec{r})}{\delta V_{tot}(\vec{r}')} = \frac{1}{\delta V_{tot}(\vec{r}')} \sum_{i=1}^{N} \sum_{j>N} \frac{\langle \psi_j | \delta V_{tot} | \psi_i \rangle}{\varepsilon_i - \varepsilon_j} \psi_j^*(\vec{r}) \psi_j(\vec{r})$$

$$= \sum_{i=1}^{N} \sum_{j=N+1}^{\infty} \frac{1}{\varepsilon_i - \varepsilon_j} \left[\psi_i^*(\vec{r}) \psi_j(\vec{r}) \psi_i(\vec{r}') \psi_j^*(\vec{r}') \right.$$

$$\left. + \psi_i(\vec{r}) \psi_j^*(\vec{r}) \psi_i^*(\vec{r}') \psi_j(\vec{r}') \right]. \tag{22.37}$$

Equation (22.29) subsequently gives $\chi(\vec{r}, \vec{r}')$ and from Eq. (22.33) we obtain the dielectric matrix.

Equation (22.37) shows that χ_0 will be represented as an expansion in products of four basis functions, once the single-particle eigenfunctions ψ are expanded in some basis functions. The same will hold true also for χ and the dielectric matrix.

Once the dielectric matrix has been calculated, we may use it in studying the response of a given system on various perturbations. This is illustrated in Figs. 22.9 and 22.10 for the case of an extra electrostatic field and for an extra point charge, respectively.

The dielectric matrix we have introduced is the stationary case of the more general time-dependent one that relates internal and external time-dependent fields,

$$\vec{E}(\vec{r}, t) = \int \int \varepsilon^{-1}(\vec{r}, \vec{r}', t - t') \vec{D}(\vec{r}', t') \, dt' \, d\vec{r}'. \tag{22.38}$$

Often, it may be assumed that $\varepsilon^{-1}(\vec{r}, \vec{r}', t - t')$ depends only on the difference $\vec{r} - \vec{r}'$. Then, taking the Fourier transform (both for the position and for the time coordinates) of Eq. (22.38), one arrives at

$$\vec{E}(\vec{q}, \omega) = \frac{1}{\varepsilon(\vec{q}, \omega)} \vec{D}(\vec{q}, \omega) \tag{22.39}$$

in standard formulation (i.e., with $\vec{q}$ being the momentum coordinate, and ω being the energy coordinate).

$-\text{Im}[1/\varepsilon(\vec{q}, \omega)]$ is known as the loss function and it describes the probability that particles (e.g., photons or incoming electrons) with a given momentum ($\vec{q}$)

Figure 22.9   Valence electron density (left part) for diamond, silicon, and germanium (from top to bottom) together with the changes due to an electrostatic field in the vertical direction of the plots (right part). Reproduced with permission from Hybertsen and Louie (1987). Copyright 1987 by the American Physical Society

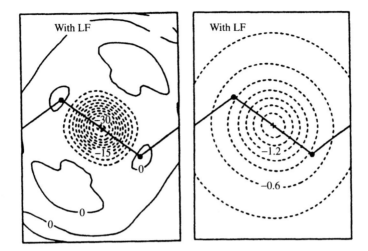

Figure 22.10 The induced change in the electron density (left part) and the screening potential (right part) for germanium for an extra charge at the centre of the Ge–Ge bond. Reproduced with permission from Hybertsen and Louie (1987). Copyright 1987 by the American Physical Society

and energy (ω) can interact with the material. The calculation of this follows in principle as that above with the extra complication of the additional time (or energy) dependence. The most important change is that the denominator in Eq. (22.37) will change according to

$$\varepsilon_i - \varepsilon_j \rightarrow \varepsilon_i - \varepsilon_j + \omega. \tag{22.40}$$

Results of such calculations are shown in Figs. 22.11 and 22.12 for two examples. Here, we separate the Fourier-transformed dielectric matrix into a real and an imaginary part,

$$\varepsilon(\vec{q}, \omega) = \varepsilon_1(\vec{q}, \omega) + i\varepsilon_2(\vec{q}, \omega). \tag{22.41}$$

For the loss function we use the identity (cf. Section 17.3)

$$\lim_{\eta \to 0} \operatorname{Im} \frac{1}{x + i\eta} = -\pi\delta(x), \tag{22.42}$$

and by comparing Eqs. (22.32), (22.33), (22.37), and (22.40) it can be realized that there is a particularly strong absorption when the energy ω equals the energy difference between the single-particle orbitals. However, in addition there are also other resonances. These correspond to the case that not a single electron but the complete collection of electrons is excited, i.e., to so-called plasmons.

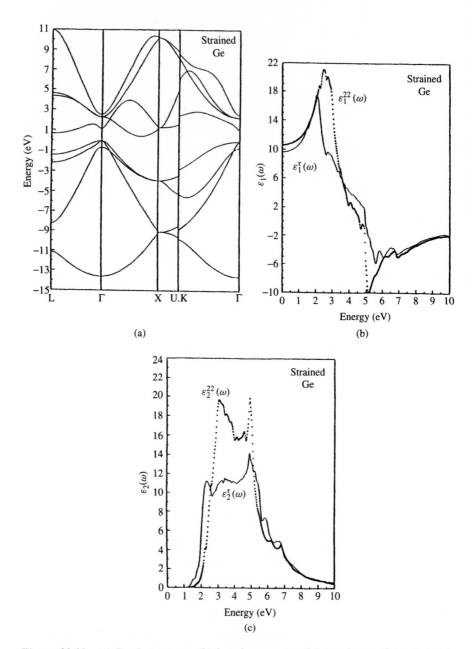

Figure 22.11 (a) Band structures, (b) imaginary part and (c) real part of the dielectric function for strained germanium. Reproduced with permission from Ghahramani *et al.* (1990). Copyright 1990 by the American Physical Society

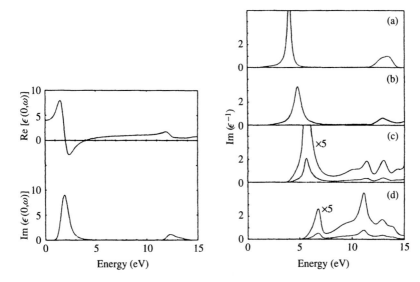

Figure 22.12 Calculated optical dielectric function (left part) and energy-loss function (right part) for polyacetylene. The different panels in the left part corresponds to different values of q, i.e., from top to bottom 0, 0.26, 0.52, and 0.78 Å^{-1}. Reproduced with permission from Mintmire and White (1983). Copyright 1983 by the American Physical Society

For the first example presented here (germanium with a strained structure) in Fig. 22.11, the band-structure effects are the dominating ones, whereas plasmons dominate for the second example in Fig. 22.12 (polyacetylene).

22.4 QUASI-PARTICLES

Having introduced the concept of the dielectric matrix we can now turn to the ultimate goal, i.e., to study the response of the system of interest to the excitation of a single electron. In particular, we want to calculate the energy that is required for this excitation.

In Chapter 17 we discussed Green's function. We saw that within a single-particle approach Green's function could be written as

$$G^{\pm}(\vec{r}, \vec{r}\,', \varepsilon) = \sum_{i} \frac{\psi_i^*(\vec{r}\,')\psi_i(\vec{r})}{\varepsilon - \varepsilon_i \pm \mathrm{i}\delta}, \tag{22.43}$$

where ψ_i and ε_i are the solutions to some single-particle Schrödinger-, Hartree–Fock-, or Kohn–Sham-like equations

$$\hat{h}\psi_i(\vec{r}) = \varepsilon_i\psi_i(\vec{r}). \tag{22.44}$$

The density of states could be obtained through

$$D(\varepsilon) = \frac{1}{\pi} \lim_{\delta \to 0} \mathrm{Im} \int G^-(\vec{r}, \vec{r}, \varepsilon) \, d\vec{r}, \qquad (22.45)$$

which actually is generally valid also without the explicit knowledge of the relation between G on the one side and ψ_i and ε_i on the other. Equation (22.43) is valid within a single-particle approximation, i.e., it is assumed that the individual particles (electrons) behave largely independently of all the other particles. Then, the density of states as obtained from the Green's function — irrespectively of how the latter has been calculated — is that of the individual single-particle orbitals. As discussed in Section 22.1, this density of states is highly relevant when discussing excitation energies. However, also when the particles interact strongly and the approximation of independent particles is not good Green's function can be calculated, although it is not obvious how it is related to single-particle orbitals and energies as in Eq. (22.43). Instead of talking about individual particles one talks about quasiparticles.

Within the context of the one-particle Green's function, the energies of these quasiparticles are associated with the peaks in the density of states (also called the spectral function). If the peak is sufficiently sharp, a well-defined quasiparticle energy can be associated with it. Our aim is thus to identify the peaks in the density of states. To this end we use the fact that for the electronic system the Green's function can be written in the form of Eq. (22.43). That is, we go the other way around by starting with the Green's function for the general case of interacting particles and subsequently stating (without proof!) that it can then be written in the form (22.43), where, however, we do not know the single-particle equation that defines ψ_i and ε_i.

It has been shown, on the other hand (and this proof is very far from trivial) that this equation can be written as (Hedin and Lundqvist, 1969)

$$\left[-\frac{1}{2}\nabla^2 + V_n(\vec{r}) + V_C(\vec{r}) \right] \psi_i(\vec{r}) + \int \Sigma(\vec{r}, \vec{r}\,', \varepsilon_i) \psi_i(\vec{r}\,') \, d\vec{r}\,' = \varepsilon_i \psi_i(\vec{r}),$$

$$(22.46)$$

where $\Sigma(\vec{r}, \vec{r}\,', \varepsilon_i)$ is the so-called electron self-energy operator that contains all the effects of exchange and correlation among the electrons.

The physical idea behind Eq. (22.46) is that of studying what happens when an electron is being excited. The main problem is that the electrons are charged. This means that when changing the wavefunction of one electron, all wavefunctions of all other electrons are changed, and then the first one changes again, and so on. This means that the potential in the point $\vec{r}$ felt by any particular electron (in orbital i) is specific for that orbital and, furthermore, depends not only on $\vec{r}$, but also on the electrons in all other points $\vec{r}\,'$ of space. All these effects are hidden in the self energy $\Sigma(\vec{r}, \vec{r}\,', \varepsilon_i)$, except for one part, i.e., the classical electrostatic potential of the electronic distribution, i.e., $V_C(\vec{r})$.

The Kohn–Sham equations with a local-density or generalized-gradient approximation are obtained by considering a particularly simple form of the self energy for a non-existing model system of non-interacting particles, i.e.,

$$\Sigma(\vec{r}, \vec{r}\,', \varepsilon_i) \simeq V_{xc}(\vec{r}) \cdot \delta(\vec{r} - \vec{r}\,'). \tag{22.47}$$

Here, on the other hand, we need to include explicitly that the particles are charged and that they interact with each other in a dynamical way.

In words, $\Sigma(\vec{r}, \vec{r}\,', \varepsilon_i)$ contains the effects of all the other electrons on a specific orbital ψ_i, except for the classical Coulomb potential. Thus, whenever *any* of the orbitals is changed, not only the Coulomb potential but also the self energy changes. Since the Green's function contains the information of how the system responds to perturbations, it seems reasonable that the self energy can be written with the help of the Green's function. Furthermore, the physics behind the self energy is the electrostatic interactions, and therefore we would expect that to enter the self energy, too. On the other hand, the electrostatic interactions are screened (as described by the dielectric matrix), and therefore one should modify the unscreened interaction (called V) accordingly. This gives some *qualitative* arguments for the so-called GW approximation (Hedin, 1965; Hedin and Lundqvist 1969), according to which (in a shorthand notation; notice that the quantities are to be interpreted as matrices)

$$\Sigma = GW \tag{22.48}$$

with G being the Green's function and

$$W = \varepsilon^{-1}V \tag{22.49}$$

being the screened Coulomb interaction and ε being the dielectric matrix.

We shall not discuss this approach any further, since it is far from trivial. Instead we hope that we have given a vague qualitative idea of how the quasiparticle energies can be calculated. It has been applied for some systems, and in Fig. 22.13 we show how the density-functional eigenvalues with a local-density approximation are modified when replacing them with the quasiparticle energies. For some materials, the modifications are almost two constant values, one for the levels below the Fermi energy (negative energies) and another for those above the Fermi energy. When this is the case, the scissor-operator approach is justified, otherwise not.

22.5   FERMI SURFACES

In Fig. 22.14 we show two examples of the band structures for two different materials. In both cases, the Fermi energy is placed at the energy zero and it

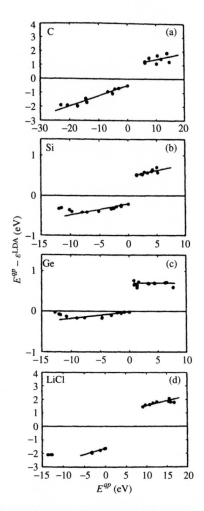

Figure 22.13 The difference between the quasiparticle energies (labelled E^{qp}) and the eigenvalues in the density-functional calculations with a local-density approximation, for states at several high-symmetry points in the first Brillouin zone. The straight lines are guides to the eye. The four panels correspond to (a) diamond, (b) Si, (c) Ge, and (d) LiCl. Reproduced with permission from Hybertsen and Louie (1986). Copyright 1986 by the American Physical Society

is seen that several bands cross it at different places in the first Brillouin zone (i.e., at different $\vec{k}$). In order to illustrate the properties of the materials a little further, we show in Fig. 22.15 the corresponding density of states decomposed into the two spin directions. For one of them, $SrRuO_3$, the two components are slightly different, i.e., the system possesses a spin-polarization and is accordingly

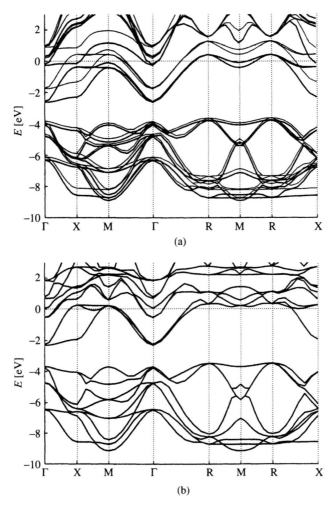

Figure 22.14   The band structures for (a) $SrRuO_3$ and (b) $CaRuO_3$ in an idealized cubic perovskite structure. For the first system, the spin-up and spin-down orbitals have different energies, and the bands are correspondingly shown as thick and thin lines for the two spin directions. Reproduced by permission of Institute of Physics Publishing from Santi and Jarlborg (1997)

magnetic. The computational method that has been used in obtaining the results for $SrRuO_3$ and $CaRuO_3$ (an LMTO method which employs atomic spheres; see Section 16.3) allows for a decomposition of the individual orbitals into atomic and angular components. Therefore, the density of states in Fig. 22.15 has been decomposed equivalently. Here, however, the details of this are irrelevant.

We may now locate those $\vec{k}$ points for which the bands cross the Fermi energy. Since the bands are continuous functions of $\vec{k}$, the set of these points defines surfaces in the three-dimensional $\vec{k}$ space, called Fermi surfaces. These are experimentally observables (e.g., in resonance experiments when the material is placed in strong magnetic fields as well as in momentum-spectroscopy experiments) and therefore of interest to calculate, too.

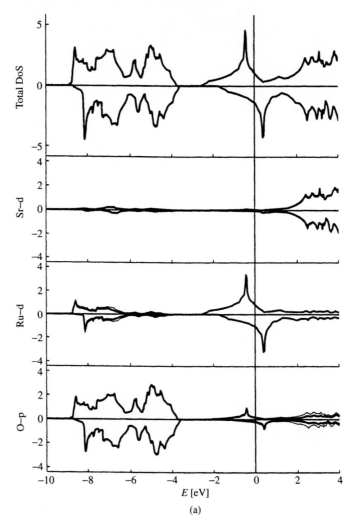

(a)

Figure 22.15   The density of states for the two systems of Fig. 22.14. For each system the two different curves (pointing in each direction) correspond to the two spin components. Furthermore, the density of states has been split into different atomic and angular components. Reproduced by permission of Institute of Physics Publishing from Santi and Jarlborg (1997)

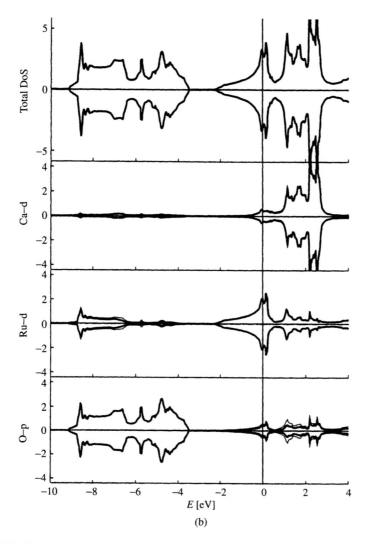

Figure 22.15   *(continued)*

For a system consisting of free electrons, the Fermi surface will be a sphere. When the effects of the crystal are small, the Fermi surface will still be close to a sphere, as is the case for, e.g., crystalline copper. However, in other cases the Fermi surface is complicated and may even contain more parts as is the case for those of the systems of Figs. 22.14 and 22.15, cf. Fig. 22.16.

The results above were obtained using a density-functional method. In principle, one might attempt to obtain the same results with a Hartree–Fock method. Here, however, we encounter a fundamental problem that is illustrated in

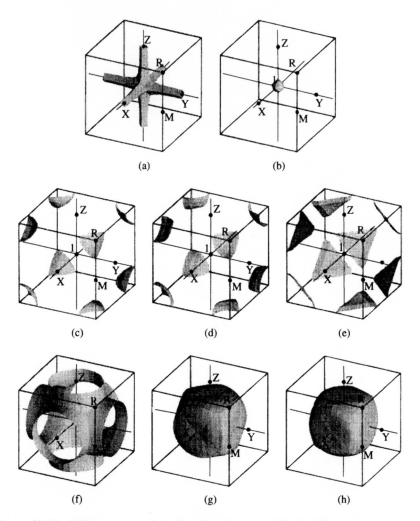

Figure 22.16 Different parts of the Fermi surfaces for $SrRuO_3$. The panels (a)–(e) are due to the one spin component, whereas those of (f)–(h) are due to the other spin component. Reproduced by permission of Institute of Physics Publishing from Santi and Jarlborg (1997)

Fig. 22.17 for the simple case of a linear chain of hydrogen atoms with constant bond lengths. This system has one electron per unit and should in principle have a half-filled band. By calculating the band structures for different values of k in the first Brillouin zone we obtain a curve like that of Fig. 22.17(a), i.e., a number of different energies that can be connected with a smooth curve. Calculating, however, the density of states, cf. Fig. 22.17(b), we discover that when the number of k points is made *very* large, the density of states will

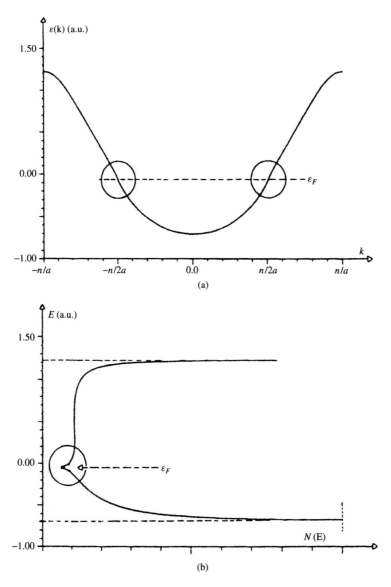

Figure 22.17 (a) The band structures and (b) the density of states for a linear chain of hydrogen atoms as obtained with a Hartree–Fock approach. Reproduced by permission of American Institute of Physics from Delhalle *et al.* (1988)

ultimately approach zero at the Fermi energy. This is a fundamental aspect of Hartree–Fock theory which one can choose to ignore, so that the results for a smaller set of k points are used in estimating the band structures and the Fermi surface. Nevertheless, it is a fundamental problem and one should be aware of its presence!

23 Relativistic Effects

23.1 THE DIRAC EQUATION

All of the discussion up to the present has been based on the Schrödinger equation. This equation, deduced by the correspondence principle from the Hamiltonian formalism of non-relativistic classical mechanics, has all the invariance properties of the Hamiltonian from which it derives. In particular, if the system is isolated, it is invariant under spatial rotations and translations. It can also be shown that it is invariant under Galilean transformations (i.e., transformations to other coordinate systems that move with a constant velocity with respect to the original one). However, the equation is not invariant to the more general Lorentz transformations that are relevant when considering velocities approaching that of light and, therefore, the predictions of the solutions to the Schrödinger equation are only accurate as long as relativistic effects can be neglected.

A further deficiency of the Schrödinger equation is that the relativistic equivalence of mass and energy results in that the number of particles in general not is conserved.

These problems have led to the formulation of relativistic quantum theory. This is a very large field that covers high-energy physics (i.e., the study of elementary particles), electromagnetic fields, etc. Here, we shall be concerned only with those consequences relativistic effects may have on electronic properties of atoms, molecules, or solids. We shall therefore assume that the energies of interest are so small that the number of particles is conserved (i.e., the energies that are involved in the creation or annihilation of electrons are many orders of magnitude larger than those of our interest).

The equation that replaces the Schrödinger equation is the Dirac equation (Dirac, 1928). This equation describes a single particle moving in some electromagnetic field. Furthermore, the particle is supposed to have a spin of $\frac{1}{2}$, as is the case for an electron. We shall not derive the equation here (this is certainly beyond the scope of this presentation), but simply present the final equation. But before doing so it is useful to recall the Schrödinger equation for such a particle.

In general the (time-dependent) wavefunction for a single electron has two components, one for the spin-α component and one for the spin-β component. We may, accordingly, formally write the wavefunction as a two-component vector

$$\Psi^{nr}(\vec{r}, t) = \begin{pmatrix} \psi_1^{nr}(\vec{r}, t) \\ \psi_2^{nr}(\vec{r}, t) \end{pmatrix} \tag{23.1}$$

with ψ_i^{nr} being the two components. The Schrödinger equation is an equation for both components, and it can be written as

$$\hat{H}^{nr}\Psi^{nr}(\vec{r}, t) = i\frac{\partial}{\partial t}\Psi^{nr}(\vec{r}, t), \tag{23.2}$$

i.e., we have a set of two equations. The equations are in many cases not coupled and in those cases we may equally well write them as two separate equations,

$$\hat{H}^{nr}\psi_i^{nr}(\vec{r}, t) = i\frac{\partial}{\partial t}\psi_i^{nr}(\vec{r}, t). \tag{23.3}$$

The position-space probability density is

$$\rho^{nr}(\vec{r}, t) = |\psi_1^{nr}(\vec{r}, t)|^2 + |\psi_2^{nr}(\vec{r}, t)|^2. \tag{23.4}$$

In the relativistic case we have not two but four components, i.e.,

$$\Psi(\vec{r}, t) = \begin{pmatrix} \psi_1(\vec{r}, t) \\ \psi_2(\vec{r}, t) \\ \psi_3(\vec{r}, t) \\ \psi_4(\vec{r}, t) \end{pmatrix}. \tag{23.5}$$

In this case the position-space probability density is

$$\rho(\vec{r}, t) = \sum_{i=1}^{4} |\psi_i(\vec{r}, t)|^2. \tag{23.6}$$

In the general case that the electron moves in an electromagnetic field, we can use the two potentials V and $\vec{A}$ in describing this field. V is the scalar potential that may be the electrostatic potential of a nucleus but which may also include potentials from external, applied electromagnetic fields. $\vec{A}$ is the vector potential that is required when magnetic fields are present. The Dirac equation can then be written in the compact form

$$\left[\left(i\frac{\partial}{\partial t} - V\right) - \vec{\alpha} \cdot (-ic\vec{\nabla} - \vec{A}) - \beta mc^2\right]\Psi = \vec{0}. \tag{23.7}$$

Here, it has to be remembered that Ψ is a four-component vector, so Eq. (23.7) represents in the general case four coupled first-order differential equations.

$\vec{\alpha}$ is a three-dimensional vector, each being a 4×4 matrix,

$$\vec{\alpha} = \left(\begin{pmatrix} 0 & 0 & 0 & 1 \\ 0 & 0 & 1 & 0 \\ 0 & 1 & 0 & 0 \\ 1 & 0 & 0 & 0 \end{pmatrix}, \begin{pmatrix} 0 & 0 & 0 & -i \\ 0 & 0 & i & 0 \\ 0 & -i & 0 & 0 \\ i & 0 & 0 & 0 \end{pmatrix}, \begin{pmatrix} 0 & 0 & 1 & 0 \\ 0 & 0 & 0 & -1 \\ 1 & 0 & 0 & 0 \\ 0 & -1 & 0 & 0 \end{pmatrix} \right) \tag{23.8}$$

and β is a single 4×4 matrix

$$\beta = \begin{pmatrix} 1 & 0 & 0 & 0 \\ 0 & 1 & 0 & 0 \\ 0 & 0 & -1 & 0 \\ 0 & 0 & 0 & -1 \end{pmatrix}. \tag{23.9}$$

The Dirac equation is complicated to solve due to its four components. Only in a few cases have attempts been made to solve it directly when dealing with electronic-structure calculations of atoms, molecules, or solids. Instead the equation is approximated, so that some of the relativistic effects are neglected whereas others are kept.

Furthermore, the Dirac equation is not exact. Thus, it assumes that the electro-static interactions between the charged particles are instantaneous which is not in accordance with the assumption that the speed of light sets an upper limit for the propagation of interactions. This deficiency can be recovered through an extra term to the Hamiltonian, the so-called Breit interaction (Breit, 1929, 1930, 1932). And the Dirac equation is not invariant under general Lorentz transformations (Bethe and Salpeter, 1957). A final problem is that the eigenvalues of the Dirac equations are not bound, i.e., there is no lowest eigenvalue, although for electrons one is interested only in those of positive energies.

23.2   THE SCHRÖDINGER EQUATION

One of the most popular methods is that of eliminating the so-called small components. Then it is most convenient to separate the four components of Ψ into two vectors, each of two components, i.e.,

$$\Psi = \begin{pmatrix} \Phi \\ \chi \end{pmatrix} \tag{23.10}$$

with

$$\Phi = \begin{pmatrix} \psi_1 \\ \psi_2 \end{pmatrix} \tag{23.11}$$

and

$$\chi = \begin{pmatrix} \psi_3 \\ \psi_4 \end{pmatrix}. \tag{23.12}$$

Subsequently, the four-component Dirac equation can be written as two coupled two-component equations,

$$\left[\vec{\sigma} \cdot (-ic\vec{\nabla} - \vec{A})\chi + (V + mc^2)\right]\Phi = i\frac{\partial}{\partial t}\Phi$$

$$\left[\vec{\sigma} \cdot (-ic\vec{\nabla} - \vec{A})\Phi + (V - mc^2)\right]\chi = i\frac{\partial}{\partial t}\chi. \tag{23.13}$$

Here,

$$\vec{\sigma} = \left(\begin{pmatrix} 0 & 1 \\ 1 & 0 \end{pmatrix}, \begin{pmatrix} 0 & -i \\ i & 0 \end{pmatrix}, \begin{pmatrix} 1 & 0 \\ 0 & -1 \end{pmatrix}\right). \tag{23.14}$$

We search for stationary solutions and, hence, replace

$$i\frac{\partial}{\partial t} \to E, \tag{23.15}$$

and assume that the wavefunctions Φ and χ are independent of time.

Subsequently, by solving the second of the equations in Eq. (23.13) for χ and then substituting this into the first of the equations in Eq. (23.13) we obtain first

$$\chi = \frac{1}{mc^2 + E - V}\vec{\sigma} \cdot (-ic\vec{\nabla} - \vec{A})\Phi \tag{23.16}$$

and subsequently

$$\left[\vec{\sigma} \cdot (-ic\vec{\nabla} - \vec{A})\frac{1}{mc^2 + E - V}\vec{\sigma} \cdot (-ic\vec{\nabla} - \vec{A}) + V\right]\Phi = (E - mc^2)\Phi. \tag{23.17}$$

We stress that this transformation is exact.

E is the total energy of the single electron, i.e., it includes the 'normal' energy and the rest-mass energy mc^2. Therefore, $E - mc^2$ of Eq. (23.17) is the 'normal' energy (e.g., the binding energy of an electron, etc.). Furthermore, in the cases of interest here, mc^2 is to be considered very large, so that [from Eq. (23.16)]

$$\chi \ll \Phi. \tag{23.18}$$

Therefore, χ is called the small component, and Φ the large component.

In order to remove a number of the complications due to the relativistic effects one may now proceed in different ways. The simplest one is to neglect the small component and to set the denominator in Eq. (23.17) equal to twice the electronic rest-mass energy, mc^2. Then, one ends up with the Schrödinger equation that we have treated so many times throughout this manuscript.

A slightly more complicated approximation is that of keeping terms up to second order in $1/c$. To this end, one can replace

$$\frac{1}{mc^2 + E - V} = \frac{1}{2mc^2 + (E - mc^2) - V} = \frac{1}{2mc^2} \frac{1}{1 + \frac{(E - mc^2) - V}{2mc^2}}$$

$$\simeq \frac{1}{2mc^2} \left[1 - \frac{(E - mc^2) - V}{2mc^2} + \left(\frac{(E - mc^2) - V}{2mc^2} \right)^2 + \cdots \right].$$

$$(23.19)$$

It turns out, however, that truncating after, e.g., the second-order term leads to problems, i.e., one can no longer neglect the small components, and solving these equations becomes rather complicated.

An alternative method has been proposed by Foldy and Wouthuysen (1950) whereby one can consistently truncate the solution to the complete set of equations (23.16) and (23.17) after any power in $1/c$. Considering systems without magnetic fields, one arrives thereby (when truncating after $1/c^2$ terms) at the following Schrödinger-like equation (the derivation will not be shown here!):

$$[\hat{H}_{\text{Schrödinger}} + \hat{H}_{\text{rel}}]\Psi = E'\Psi, \qquad (23.20)$$

where the first part of the Hamilton operator is the standard non-relativistic one,

$$E' = E - mc^2 \qquad (23.21)$$

is the 'normal' energy without rest-mass energy, and

$$\hat{H}_{\text{rel}} = -\frac{1}{8m^3c^2}\nabla^4 + \frac{1}{8m^2c^2}\nabla^2 V + \frac{1}{4m^2c^2}(-i\vec{\nabla}V \times \vec{\nabla}) \cdot \vec{s} \qquad (23.22)$$

is relativistic corrections. Then, $\hat{H}_{\text{Schrödinger}} + \hat{H}_{\text{rel}}$ of Eq. (23.20) is the so-called Pauli Hamiltonian.

The first term in Eq. (23.22) is the so-called mass-velocity term, whereas the second term is the Darwin term. Finally, the last one is the spin–orbit coupling term. For a central potential (i.e., a potential that is spherically symmetric), the last term can be rewritten as

$$\frac{1}{4m^2c^2}(-i\vec{\nabla}V \times \vec{\nabla}) \cdot \vec{s} = \frac{1}{4m^2c^2}\frac{1}{r}\frac{dV}{dr}(\vec{s} \cdot \vec{l}), \qquad (23.23)$$

which gives the justification for its name. Due to its dependence on the derivative of the potential, it has largest contributions from the regions where the potential changes most rapidly, i.e., close to the nuclei. And in these regions the potential is in fact close to spherically symmetric.

For a Coulombic potential from a nucleus (assuming that the nucleus is a point charge), the Darwin term reduces to a δ-function centred at the nucleus.

In many practical calculations where relativistic effects are included, Eqs. (23.20) and (23.22) are solved. This means that one includes only 'some extra terms' in the Schrödinger equation which will take care of the relativistic effects. This is the simplest approach and is in fact often sufficiently accurate.

There is one problem related with the spin–orbit coupling term (but not with the other ones). That is, the spin–orbit coupling term (also when including only the spherically symmetric parts of the potential in its evaluation) leads to a lowering of the symmetry. First of all, the two spin components will interact (i.e., they will mix so that the spin no longer is a good quantum number), but also the symmetry properties for the orbitals when studying the space-group symmetry properties of the system of interest may be different from the case where the spin–orbit interactions are neglected. Thus, with these interactions, the calculations easily become more involved. Therefore, in many cases the spin–orbit couplings are neglected, whereas the other terms of Eq. (23.22) are kept. This results in so-called scalar-relativistic calculations.

Nevertheless, the spin–orbit couplings are important. In Table 23.1 we show the electronic single-particle energies from a density-functional calculation on a single Rn atom both without and with the relativistic effects of Eq. (23.22). Without the spin–orbit couplings, all pairs of energies would be identical.

Furthermore, in Fig. 23.1 we show the band structures of a Bi zigzag chain both when including and when not including the spin–orbit couplings. In one calculation the spin–orbit couplings were completely neglected and the Kohn–Sham equations in the scalar-relativistic case solved self-consistently (i.e., solved iteratively until the input and output potentials were identical). In another calculation, one further iteration was carried through where the spin–orbit couplings were

Table 23.1 Single-particle energies in hartrees (1 hartree = 27.21 eV) for a single Rn atom as calculated with a density-functional method either without (NR) or with (R) relativistic effects. For all $l \neq 0$, the spin–orbit couplings lead to a splitting of the (n, l) shells into two. The 'NR' results are those of Table 22.1

Orbital	NR	R
$1s$	−3204.76	−3615.41
$2s$	−546.589	−657.751
$2p$	−527.543	−632.87, −531.85
$3s$	−133.382	−161.275
$3p$	−124.186	−149.993, −127.150
$3d$	−106.958	−108.883, −104.049
$4s$	−31.2453	−38.4174
$4p$	−27.1236	−33.4525, −27.7009
$4d$	−19.4645	−19.6667, −18.5955
$4f$	−8.96787	−8.09595, −7.83708
$5s$	−5.90564	−7.39933
$5p$	−4.42469	−5.59619, −4.38462
$5d$	−1.92753	−1.80685, −1.64361
$6s$	−0.643459	−0.825337
$6p$	−0.309910	−0.405451, −0.272136

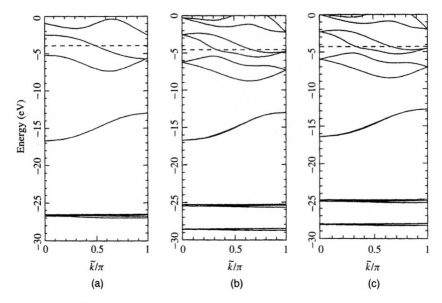

Figure 23.1   The band structures for a zigzag chain of Bi atoms with Bi−Bi bond lengths of 5.6 a.u. (a) has been obtained with a scalar relativistic calculation (without spin−orbit couplings), whereas the spin−orbit couplings have been included in (b) in one additional iteration. In (c) they have been included self-consistently

included, and in the last calculation, the spin−orbit couplings were included self-consistently. Once again, the importance of this effect is clearly recognized, but it is also seen that the spin−orbit couplings may be included perturbatively.

The idea behind the above approach has been to introduce some kind of expansion in Eq. (23.17). This expansion is not uniquely defined and there have accordingly been more different proposals. Whenever one introduces an expansion, it is important that the quantity in which one expands is small. For the Pauli Hamiltonian one expands in $(E − V)/(2c^2)$, which is not a small quantity close to the nuclei. Among the alternative approaches we mention here the *no-pair* approximation of Douglas and Kroll (1974) and Hess (1986) as well as the transformations of Heully *et al.* (1986) and Chang *et al.* (1986) and the more recent zero-order regular approximation (ZORA) and first-order regular approximation (FORA) of van Lenthe *et al.* (1993).

24 Molecules and Solids in Electromagnetic Fields

24.1 POLARIZABILITIES AND HYPERPOLARIZABILITIES

So far we have focused on properties of molecules or solids that are left isolated or at most perturbed by some static, external interactions from other parts of the complete system. We have studied structural properties, the electronic orbitals, changes when different systems are interacting, but only marginally changes induced by interactions between the system of interest and external electromagnetic fields. Only through the discussion of excitations we have indicated that the system of interest may respond to such fields. In the present section we shall consider the general response of the systems to electromagnetic fields, and in the present subsection we shall focus on electric fields.

When a molecule is exposed to a static (DC) field, its electronic and structural properties may change. In the general case we can quantify the changes through the changes of the dipole moment. Letting the field be described through the vector

$$\vec{E} = (E_x, E_y, E_z) \tag{24.1}$$

and the dipole moment through

$$\vec{\mu} = (\mu_x, \mu_y, \mu_z), \tag{24.2}$$

$\vec{\mu}$ may contain both linear and non-linear dependences on $\vec{E}$,

$$\mu_i = \mu_i^{(0)} + \sum_j \alpha_{ij} E_j + \sum_{jk} \beta_{ijk} E_j E_k + \sum_{jkl} \gamma_{ijkl} E_j E_k E_l + \cdots. \tag{24.3}$$

Here, $\vec{\mu}^{(0)}$ is the dipole moment of the system without external electrostatic fields, whereas α is the polarizability, and β, γ, ... are the first, second, ... hyperpolarizability. Furthermore, the indices i, j, ..., specify the different components (i.e., x, y, or z).

In the case that the DC field is replaced by an AC field, the quantities of Eq. (24.3) become frequency-dependent,

$$\mu_i(\omega_0) = \mu_i^{(0)}(\omega_0) + \sum_j \alpha_{ij}(\omega_0, \omega_j) E_j(\omega_j)$$

$$+ \sum_{jk} \beta_{ijk}(\omega_0, \omega_j, \omega_k) E_j(\omega_j) E_k(\omega_k)$$

$$+ \sum_{jkl} \gamma_{ijkl}(\omega_0, \omega_j, \omega_k, \omega_l) E_j(\omega_j) E_k(\omega_k) E_l(\omega_l) + \cdots . \quad (24.4)$$

For a general electromagnetic field, μ may also depend on the magnetic field, but this effect is much smaller than that of the electric field and will therefore be ignored here.

The different terms on the right-hand side of Eq. (24.4) can be interpreted as describing different processes. Let us assume that the system before the AC field is turned on is in a given state, e.g., the ground state, and let us choose its energy as the energy zero. $\vec{\mu}(\omega_i)$ can then be considered the dipole moment of the system when it is in an excited state with the energy ω_i. This state can now be thought of as reached through different processes. One possibility is that the system *is* already in it, and in that case we must have $\omega_i = 0$. This corresponds to the first term on the right-hand side, so we must have

$$\mu_i^{(0)}(\omega_0) = 0 \quad \text{for} \quad \omega_0 \neq 0. \quad (24.5)$$

Alternatively, a photon with the energy ω_j may be absorbed, which corresponds to the second term on the right-hand side of Eq. (24.4), and for which we hence must have

$$\alpha_{ij}(\omega_0, \omega_j) = 0 \quad \text{for} \quad -\omega_0 + \omega_j \neq 0. \quad (24.6)$$

Or two photons may be involved, giving

$$\beta_{ijk}(\omega_0, \omega_j, \omega_k) = 0 \quad \text{for} \quad -\omega_0 + \omega_j + \omega_k \neq 0. \quad (24.7)$$

The three-photon process leads to

$$\gamma_{ijkl}(\omega_0, \omega_j, \omega_k, \omega_l) = 0 \quad \text{for} \quad -\omega_0 + \omega_j + \omega_k + \omega_l \neq 0. \quad (24.8)$$

These identities can also be shown using the Hamilton operator of Eq. (24.16).

The polarizabilities and hyperpolarizabilities are defined for single, finite molecules for which the dipole moment is well defined (this will be discussed further below). For macroscopic systems, e.g., molecular crystals containing many weakly interacting, finite molecules, or 'real' solids, one considers instead of the dipole moment, the polarization $\vec{P}$. When the system is exposed to an (AC or DC) electromagnetic field, the polarization may change:

$$P_i(\omega_0) = P_i^{(0)}(\omega_0) + \sum_j \chi_{ij}^{(1)}(\omega_0, \omega_j) E_j(\omega_j)$$

$$+ \sum_{jk} \chi_{ijk}^{(2)}(\omega_0, \omega_j, \omega_k) E_j(\omega_j) E_k(\omega_k)$$

$$+ \sum_{jkl} \chi_{ijkl}^{(3)}(\omega_0, \omega_j, \omega_k, \omega_l) E_j(\omega_j) E_k(\omega_k) E_l(\omega_l) + \cdots, \quad (24.9)$$

where the χ are susceptibilities. For molecular crystals the (hyper-)polarizabilities and the susceptibilities are closely related, where the former describes the response of the system per molecule, whereas the latter describes that per volume unit. But also in the general case, the two sets of quantities are related, and we shall therefore not distinguish between them in discussing how to treat them theoretically. Finally, the restrictions of Eqs. (24.5)–(24.8) hold also for the susceptibilities.

Before discussing how they are calculated let us study some of the more fundamental properties. First, the dipole moment of a finite molecule with M nuclei and an electron density of $\rho(\vec{r})$ is given through

$$\vec{\mu} = \sum_{k=1}^{M} Z_k \vec{R}_k - \int \vec{r} \rho(\vec{r}) \, d\vec{r}. \quad (24.10)$$

Here, Z_k and $\vec{R}_k$ are the charge and position of the k nucleus, respectively.

For a neutral system,

$$\sum_k Z_k = \int \rho(\vec{r}) \, d\vec{r}, \quad (24.11)$$

and the dipole moment is independent of the choice of the origin of the coordinate system. Thus, shifting

$$\vec{R}_k \rightarrow \vec{R}_k + \vec{X}$$
$$\vec{r} \rightarrow \vec{r} + \vec{X}, \quad (24.12)$$

gives a new dipole moment of

$$\vec{\mu}' = \sum_{k=1}^{M} Z_k (\vec{R}_k + \vec{X}) - \int (\vec{r} + \vec{X}) \rho(\vec{r}) \, d\vec{r}$$

$$= \sum_{k=1}^{M} Z_k \vec{X} - \int \vec{X} \rho(\vec{r}) \, d\vec{r} + \sum_{k=1}^{M} Z_k \vec{R}_k - \int \vec{r} \rho(\vec{r}) \, d\vec{r}$$

$$= \vec{X} \left[\sum_{k=1}^{M} Z_k - \int \rho(\vec{r}) \, d\vec{r} \right] + \vec{\mu} = \vec{\mu}. \quad (24.13)$$

This is not, however, the case for a charged system, where the dipole moment accordingly, depends on the choice of the origin of the coordinate system.

Infinite, although periodic, systems pose an additional problem. To see this, we study the system of Fig. 24.1. We have here a linear chain of N units, with each unit containing two different types of atoms, A and B. We will let the A atom of the nth unit be placed at $n \cdot a$, and the B atom of that unit be placed at $n \cdot a + x$. Let us assume that the atoms are charged, $-q$ on the A atoms and $+q$ on the B atoms. The dipole moment per unit is then

$$\frac{\mu}{N} = \frac{1}{N} \sum_{n=1}^{N} [q \cdot (na + x) - q \cdot na] = qx. \qquad (24.14)$$

However, as shown in Fig. 24.1, we may — for the infinite system — also split the system into unit cells in another way. Then, the A atoms are placed at $(n + 1) \cdot a$ and the B atoms at $n \cdot a + x$. This means that

$$\frac{\mu}{N} = \frac{1}{N} \sum_{n=1}^{N} [q \cdot (na + x) - q \cdot (n + 1)a] = q \cdot (x - a). \qquad (24.15)$$

For the infinite system, where the separation into unit cells is arbitrary, this means that the dipole moment is *not* unique. This is a general problem that also exists for real crystalline materials. For the latter it can, however, be shown (Resta, 1994) that the polarization is defined to within an additive constant of $(2/\Omega)\vec{R}$, where Ω is the volume of one unit cell, and $\vec{R}$ is a lattice vector. It can also be shown that the polarization is a bulk effect (Vanderbilt and King-Smith, 1993), i.e., it is a property that is determined by the electrons in the interior of the system; Eqs. (24.14) and (24.15) may otherwise suggest that the total dipole moment is solely determined by the charges of and distances between the 'surfaces' (i.e., the A and B atoms of the 1st and Nth unit).

For a monochromatic radiation of frequency ω_0 the electromagnetic field leads to an extra term in the *total* Hamilton operator equal to

$$\Delta \hat{H} = \left[\sum_{k=1}^{M} Z_k \vec{R}_k \cdot \vec{E} - \sum_{i=1}^{N} \vec{r}_i \cdot \vec{E} \right] e^{-i\omega_0 t}. \qquad (24.16)$$

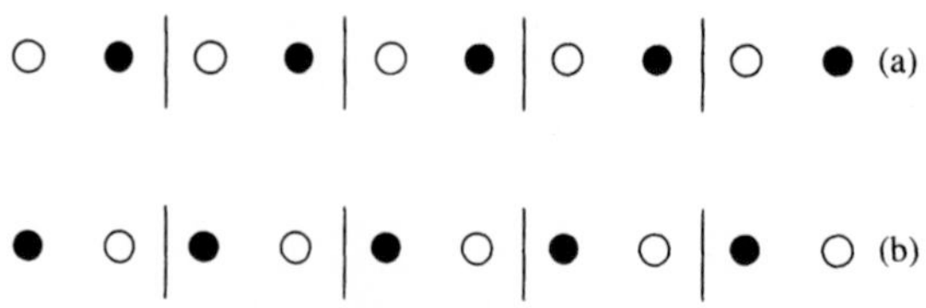

Figure 24.1 A part of an infinite linear chain consisting of two different types of atoms. The two panels correspond to two different ways of separating the system into unit cells

Here, the first term is the term for the nuclei, and the second is that for the electrons.

Let us consider a neutral system that furthermore has the special property that it is centrosymmetric, i.e., replacing

$$\vec{r} \rightarrow -\vec{r} \tag{24.17}$$

leaves the system unchanged. For this system the total dipole moment without the electromagnetic field vanishes (due to the symmetry). Turning on the field may lead to a non-vanishing dipole moment $\vec{\mu}$. Replacing

$$\vec{E} \rightarrow -\vec{E} \tag{24.18}$$

leads to

$$\vec{\mu} \rightarrow -\vec{\mu}. \tag{24.19}$$

Comparing with Eq. (24.4) shows, however, that this can be fulfilled only if

$$\beta_{ijk}(\omega_0, \omega_j, \omega_k) = 0, \tag{24.20}$$

which actually also holds for all higher-order, even hyperpolarizabilities.

Let us now turn to explicit calculations of the polarizabilities and the hyperpolarizabilities. Let us first consider the static DC case, i.e.,

$$\omega_0 = 0. \tag{24.21}$$

In that case, solving the Schrödinger equation (or the Hartree–Fock of Kohn–Sham equations) can be carried through as for the case without the static field (with some complications, however, that will be discussed below). The total energy will change according to

$$E = E_0 + \sum_i \mu_i^{(0)} E_i + \sum_{ij} \alpha_{ij} E_i E_j + \sum_{ijk} \beta_{ijk} E_i E_j E_k$$
$$+ \sum_{ijkl} \gamma_{ijkl} E_i E_j E_k E_l + \cdots, \tag{24.22}$$

where E_0 is the energy without the external field. That this is so can be seen by noting that the extra Hamilton operator (24.16) leads to an extra term to the nuclear energy equal to

$$\sum_{k=1}^{M} Z_k \vec{R}_k \cdot \vec{E}, \tag{24.23}$$

and to the electronic energy equal to

$$-\left\langle \Psi \left| \sum_{i=1}^{N} \vec{r}_i \cdot \vec{E} \right| \Psi \right\rangle = -\vec{E} \cdot \int \vec{r} \rho(\vec{r}) \, d\vec{r}. \tag{24.24}$$

Here, we have assumed that Ψ is the exact N-electron wavefunction for the full Hamilton operator, which includes the effects of the electrostatic field.

Since the extra Hamilton operator of Eq. (24.16) contains both a term for the nuclei and one for the electrons, both the nuclei and the electrons will respond to the presence of the electrostatic field. This means that the structure for the system without the extra field will most likely not be left unchanged, but the nuclei will move under the influence of the field. This effect will produce one contribution to the polarizability and hyperpolarizabilities, but will not be discussed further here. The other contribution is that of the electrons. For these, there are two principally different approaches for calculating them. One approach amounts to performing different calculations for different strengths and directions of the field $\vec{E}$, and subsequently fitting the results with an expression of the form of Eq. (24.3) or (24.22). This is a straightforward method but requires that many calculations are done, and it can, in principle, also be applied to determine the contribution from the nuclei. In that case, the complete structure will be optimized for *any* strength and *any* direction of the field $\vec{E}$, which of course is non-trivial.

The other approach recognizes that the expansion (24.22) resembles that of a perturbation expansion. Then, one can attempt to include the effects of the extra field perturbatively, and for the different polarizabilities or hyperpolarizabilities directly relate them with different terms in the perturbation expansion. As discussed in Section 6.1, higher than first-order perturbation (i.e., the determination of the hyperpolarizabilities) requires formally that *all* eigenfunctions to the unperturbed system are determined. In a given calculation where a finite basis set is applied this is never the case, and, furthermore, those that are determined

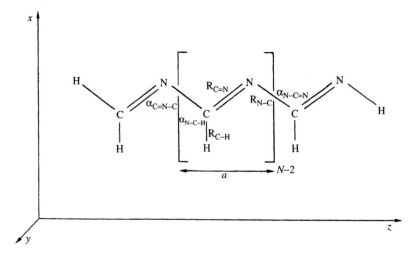

Figure 24.2 Structure of finite oligomers of polycarbonitrile. Reprinted with permission from Champagne *et al.* (1997). Copyright 1997 American Chemical Society

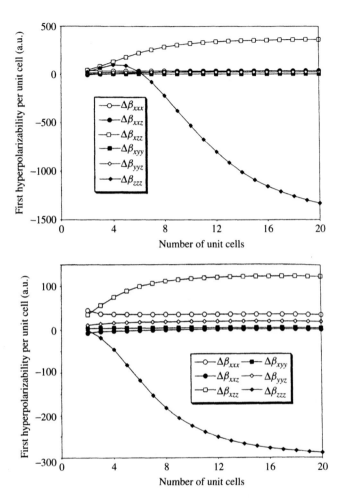

Figure 24.3 Evolution with chain length of the first hyperpolarizabilities per unit cell for polycarbonitrile. Reprinted with permission from Champagne *et al.* (1997). Copyright 1997 American Chemical Society

may in addition be more or less exact. Therefore such calculations require fairly large basis sets.

A special case is that of infinite systems. For these, the extra Hamilton operator of Eq. (24.16) breaks the translational symmetry, although it may exist without the field. This means that one can no longer construct symmetry-adapted Bloch waves characterized by a $\vec{k}$ from the equivalent basis functions of different unit cells, which could be used in simplifying the calculations. Instead, one would in principle have to treat the complete system. A further complication for the infinite system is that the Hamilton operator diverges for $|\vec{r}| \rightarrow \infty$, which means that the matrix elements diverge. On the other hand, calculations on large finite

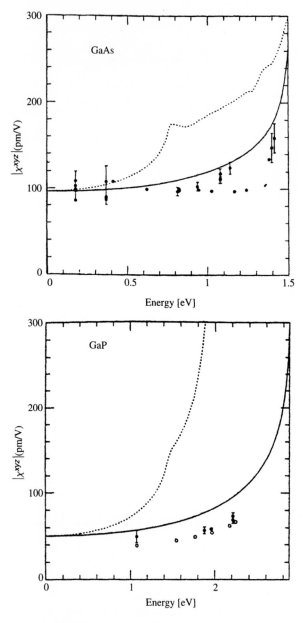

Figure 24.4 The second-order susceptibilities $\chi_{xyz}^{(2)}(-2\omega, \omega, \omega)$ (solid line) and $\chi_{xyz}^{(2)}(-\omega, \omega, 0)$ (dotted line) for crystalline GaAs and GaP as a function of ω. Experimental values are also included. $\chi_{xyz}^{(2)}(-2\omega, \omega, \omega)$ corresponds to the so-called second-harmonic generation, and $\chi_{xyz}^{(2)}(-\omega, \omega, 0)$ to the linear electro-optic effect. Reproduced with permission from Hughes and Sipe (1996). Copyright 1996 by the American Physical Society

systems have shown that the field is largely screened so that inside the material the potential does not diverge. But formally the problems remain.

One common approach to these problems is to consider finite systems of K repeated units and study the polarizabilities and hyperpolarizabilities per unit as functions of K. It is then hoped that these quantities will converge to their $K \to \infty$ values for not too large K.

In Fig. 24.3 we show an example of a such calculation for polycarbonitrile of Fig. 24.2. It is seen that the convergence as a function of K is slow, and it will be stressed that for the higher hyperpolarizabilities the convergence is usually even slower. The results of Fig. 24.3 were obtained with an *ab initio* Hartree–Fock method and by using the perturbation-series expression.

For the AC fields only perturbation-theory expressions are applied. These are quite complicated and will therefore not be reproduced here. In Section 22.4 we discussed how the quasi-particle energies could be calculated within a density-functional formalism. With these, accurate electron-excitation energies could be calculated. We showed, however, that it was necessary to include the full response of the complete system to the excitation of a single electron, i.e., to take care of the fact that the electron–electron interactions make all electrons respond to this excitation. Similarly, when calculating non-linear responses to the perturbation of the electrons through an electromagnetic field it is important to take these extra electron–electron–interaction effects into account. As one may understand, this is far from trivial, and we shall not attempt to give any detailed description of the approach here. Instead, we show just a single example of the results of such calculations. We show accordingly $\chi^{(2)}_{xyz}(-2\omega, \omega, \omega)$ and $\chi^{(2)}_{xyz}(-\omega, \omega, 0)$ for GaAs and GaP in Fig. 24.4. These were calculated with a density-functional method for crystalline materials using a perturbation-theory approach and taking some (but not all) of the electron–electron-interaction effects into account.

24.2 MAGNETIC RESONANCES

As discussed in Chapter 23, any electromagnetic field can be specified through the scalar potential $V(\vec{r})$ and the vector potential $\vec{A}(\vec{r})$. The electric field is then

$$\vec{E}(\vec{r}) = -\vec{\nabla} V(\vec{r}), \tag{24.25}$$

and the magnetic field is

$$\vec{H}(\vec{r}) = \vec{\nabla} \times \vec{A}(\vec{r}). \tag{24.26}$$

For the sake of simplicity we shall consider only a single, isolated atom, so that the potential becomes spherical symmetric. Furthermore, we shall only consider the case of a homogeneous magnetic field, i.e., we consider $\vec{H}$ to be constant within the atom. For N electrons the Hamilton operator then becomes (cf.

Section 23.2; i.e., we include relativistic effects up to $1/c^2$)

$$\hat{H} = \frac{1}{2}\sum_{i=1}^{N}\left\{\left[-i\vec{\nabla}_i - \frac{1}{c}\vec{A}(\vec{r}_i)\right]^2 + \frac{2}{c}\vec{s}_i \cdot \vec{H} + \frac{2}{c}\xi(r_i)\vec{r}_i \times \vec{A}(\vec{r}_i) \cdot \vec{s}_i + \xi(r_i)\vec{l}_i \times \vec{s}_i\right\}$$

$$+ \sum_{i=1}^{N}\frac{-Z}{|\vec{r}_i - \vec{R}|} + \frac{1}{2}\sum_{i \neq j=1}^{N}\frac{1}{|\vec{r}_i - \vec{r}_j|}. \tag{24.27}$$

Here, $\vec{s}_i$ and $\vec{l}_i$ are the spin and angular momentum, respectively, of the ith electron. Furthermore, Z and $\vec{R}$ are the number and position of the nucleus, respectively. Finally,

$$\xi(r) = -\frac{1}{2c^2}\frac{1}{r}\frac{dV(r)}{dr}, \tag{24.28}$$

where $V(r)$ is the spherically symmetric potential.

One may now write the total Hamilton operator of Eq. (24.27) as one magnetic-field-independent term plus a remainder that depends on the presence of the magnetic field:

$$\hat{H} = \hat{H}_0 + \Delta\hat{H}. \tag{24.29}$$

This gives

$$\Delta\hat{H} = \sum_{i=1}^{N}\left\{-i\frac{1}{c}\vec{A}(\vec{r}_i) \cdot \vec{\nabla}_i + \frac{1}{2c^2}|\vec{A}(\vec{r}_i)|^2 + \frac{1}{c}\vec{H} \cdot \vec{s}_i\right\}$$

$$= \sum_{i=1}^{N}\left\{-i\frac{1}{2c}\vec{H} \times \vec{r}_i \cdot \vec{\nabla}_i + \frac{1}{2c^2}|\vec{A}(\vec{r}_i)|^2 + \frac{1}{c}\vec{H} \cdot \vec{s}_i\right\}$$

$$= \frac{1}{2c}\vec{H} \cdot \sum_{i=1}^{N}(\vec{l}_i + 2\vec{s}_i) + \frac{1}{8c^2}\sum_{i=1}^{N}|\vec{H} \times \vec{r}_i|^2$$

$$= \frac{1}{2c}\vec{H} \cdot (\vec{L} + 2\vec{S}) + \frac{1}{8c^2}\sum_{i=1}^{N}|\vec{H} \times \vec{r}_i|^2 \tag{24.30}$$

with $\vec{L}$ and $\vec{S}$ being the total angular momentum and total spin, respectively of the electrons.

In deriving this equation we have used that.

$$\vec{A}(\vec{r}) = \tfrac{1}{2}\vec{H} \times \vec{r} \tag{24.31}$$

since $\vec{H}$ is considered constant [with this choice, Eq. (24.26) is satisfied].

We see that the magnetic field couples both to spin and to position-space degrees of freedom for the electrons.

Before proceeding we add that the approach above is based on perturbation theory. This means that it is assumed that the magnetic field is weak. When the magnetic field is strong, the situation may be markedly different, and strong, only poorly understood, effects may show up.

Furthermore, the system we have studied represents a simplification. First, we have considered only one isolated atom, whereby the potential becomes spherically symmetric and the spin–orbit coupling does take the form of a coupling between spin and orbital angular momenta. For molecular systems, the Hamilton operator will take a somewhat more complicated form, but for our discussion the only important point is that the magnetic field interacts with both spin and position-space components of the total wavefunction. Second, the Hamilton operator of Eq. (24.27) considers only the electronic degrees of freedom and accordingly neglects the nuclei. But also the nuclei may possess spins with which the magnetic field can interact. Also this is utilized in some experiments, as we shall see. Third, in most practical cases the magnetic-field effects are small. Experiments use this by applying static magnetic fields of a certain strength whereby the resonances are determined by applying an additional time-dependent magnetic field. The fields are chosen according to which part of the system is being studied (i.e., the nuclei or the electrons), and by varying the frequency of the time-dependent field one searches for resonances. The energies of these resonances are first of all determined by that part of the system that is studied (i.e., they take one value for a ^{1}H nucleus, another for a ^{13}C nucleus, a third value for an electron, etc.). But smaller shifts of the resonance energies occur due to the surroundings of the particle of interest, whereby important information about chemical compositions of a given compound can be obtained.

In the more general case of a molecule consisting of a number of nuclei and electrons all magnetic moments interact with the magnetic field and with each other. In addition, a magnetic field can act in two ways on the electrons: by rotating the electron clouds as rigid objects, and by distorting these clouds. The first interaction leads to so-called diamagnetic (shielding) effects, since the rotated electron clouds produce a secondary magnetic field which partially compensates the original one. The second interaction produces paramagnetic (deshielding) effects which create a magnetic field that aligns with, and hence increases, the original one. Furthermore, since the electron clouds can be asymmetric (e.g., non-spherical), the secondary magnetic fields may not be collinear with the original field.

Assuming that the external magnetic field $\vec{B}$ is weak, one may expand the total energy in powers of the magnetic field and the nuclear and total electronic magnetic moments (Beveridge, 1977),

$$E = E_0 - \vec{\gamma} \cdot \vec{B} - \frac{1}{2}\vec{B} \cdot \underline{\underline{\chi}} \cdot \vec{B} - \sum_M \vec{\mu}_M \cdot \vec{B} + \sum_M \vec{\mu}_M \cdot \underline{\underline{\sigma}}_M \cdot \vec{B}$$

$$+ \frac{1}{2}\sum_{M \neq N} \vec{\mu}_M \cdot \underline{\underline{K}}_{MN} \cdot \vec{\mu}_N - \vec{\mu}_e \cdot \vec{B} + \vec{\mu}_e \cdot \underline{\underline{\Delta g}} \cdot \vec{B} + \sum_M \vec{\mu}_e \cdot \underline{\underline{A}}_M \cdot \vec{\mu}_M.$$

$$(24.32)$$

Here, $\vec{\mu}_M = g_M \beta_M \vec{I}_M$ and $\vec{\mu}_e = -g\beta\vec{S}$ with g_M and g being the nuclear and electronic g values, β_M and β the nuclear and Bohr magnetons, and $\vec{I}_M$ and $\vec{S}$

the nuclear and the electronic spin momenta. Furthermore, $\vec{\gamma}$ is the permanent magnetic moment of the system, $\underline{\underline{\chi}}$ is the magnetic susceptibility tensor, $\underline{\underline{\sigma}}_M$ is the shielding tensor for the Mth nucleus, $\underline{\underline{K}}_{MN}$ is the so-called reduced nuclear spin–spin coupling tensor for the Mth and Nth nuclei, $\underline{\underline{\Delta g}}$ is the shift in the free-electron g-tensor, and $\underline{\underline{A}}_M$ is the electron-nuclear hyperfine tensor. All quantities are, in principle accessible with electronic-structure calculations.

As a first example we consider nuclear magnetic resonance NMR experiments. With this experimental technique one studies the spin of a given type of nucleus, e.g., ^{1}H, ^{3}H, ^{13}C, ^{19}F, ^{35}Cl, or ^{37}Cl. The total spin of these nuclei $\vec{I}$ is non-vanishing, and its z component can therefore take different values. In a static magnetic field the energy of the nucleus depends on the value of the component of the spin parallel to the magnetic field, and it is often a good approximation to write this as

$$E(m_I) = -\gamma B m_I, \tag{24.33}$$

where m_I is the component of $\vec{I}$ along the magnetic field, B is the strength of the magnetic field, and γ is some constant that depends on the nucleus of interest.

The resonance criterion is then.

$$\omega = E(m_I - 1) - E(m_I) = \gamma B. \tag{24.34}$$

Thus, varying ω for a given B and determining the resonances can be used in giving information on γ.

γ depends first of all on the type of nucleus that is being studied. But, in addition, smaller changes occur due to the couplings with the electrons in the neighbourhood and with the spins of the neighbouring nuclei. Therefore, NMR experiments give information on the chemical surroundings of given nuclei. From the electronic-structure-calculation point of view the relevant property is the electronic density at the site of the nucleus.

Another example is the electron spin resonance (ESR) technique, where one studies the spin of the electrons. This technique — like the NMR technique — can be applied only when the total spin is non-vanishing. That is, for molecular systems it is most often used in studying radicals with an odd number of electrons. The relevant formulas are very similar to those above for the NMR technique. Here, some compound-specific variations in the constant γ are due to couplings between the spin of different electrons and couplings between the electron spin and the nuclear spins. As one example we study the couplings between the electronic and the nuclear spins. These are known as hyperfine interactions. From an electronic-structure point of view their strength is determined by

$$\hat{H}_{\text{hf}} = \text{cst.} \times [\rho_\alpha(\vec{R}) - \rho_\beta(\vec{R})]\,\vec{s} \cdot \vec{I} = a \cdot \vec{s} \cdot \vec{I}. \tag{24.35}$$

Here, cst. is some constant that is unimportant here, $\rho_\alpha(\vec{R})$ and $\rho_\beta(\vec{R})$ are respectively the spin-up and spin-down electron density at the site of the nucleus ($\vec{R}$), and $\vec{s}$ and $\vec{I}$ are respectively the electron and nuclear spin.

Table 24.1 Calculated hyperfine coupling constants (in gauss) of different isolated atoms calculated with different types of basis sets together with experimental values. From Ishii and Shimizu (1995)

Basis	^{1}H	^{11}B	^{13}C	^{14}N	^{17}O	^{19}F
GTO		−5.9, −10.7	−2.2, −8.5	−0.1, 1.7	−0.8, −0.4	2.7
STO	518	0.7, 0.5	6.6, 7.0	4.9, 4.2	−16.4, −11.6	109
Exp.	508	4.1	7.0	3.7	−12.3	108

Table 24.2 As Table 24.1 but for the atoms of different small molecules of the form XH. From Ishii and Shimizu (1995)

XH	Basis	Nucleus	Value
CH	GTO	^{13}C	9
CH	STO	^{13}C	16.7, 17.1
CH	Exp.	^{13}C	16.7
CH	GTO	^{1}H	16
CH	STO	^{1}H	19.1, 19.7
CH	Exp.	^{1}H	20.6
NH	GTO	^{14}N	3
NH	STO	^{14}N	7.7, 7.3
NH	Exp.	^{14}N	6.9
NH	GTO	^{1}H	−18
NH	STO	^{1}H	−21.1
NH	Exp.	^{1}H	−23.6
OH	GTO	^{17}O	−6
OH	STO	^{17}O	−20.0, −16.8
OH	Exp.	^{17}O	−18.3
OH	GTO	^{1}H	−21
OH	STO	^{1}H	−23.1, −23.3
OH	Exp.	^{1}H	−26.1

The constant in Eq. (24.35) is well known and one can therefore attempt to calculate the hyperfine coupling constants a using electronic-structure methods. Table 24.1 provides one example for this obtained by a density-functional method. Here, we observe that the use of Gaussians as basis functions (i.e., the entries 'GTO') may be less accurate. This is understandable when noticing that the Gaussians have a wrong behaviour near the nuclei (i.e., they have flat tangents, there whereas Slater-type orbitals, STOs, have non-vanishing derivatives at the nuclei—the so-called cusp problem). On the other hand, this problem may be avoided by using large basis sets of Gaussians.

Table 24.2 shows results of similar calculations for systems where the atoms are parts of some molecules. Besides noticing the good agreement between theory and experiment we also see how the coupling constants depend on the chemical surroundings of the atom of interest, as mentioned above.

Part IV
SPECIAL SYSTEMS

In the last part of this manuscript we shall briefly describe a number of different systems that in one way or another differ from being finite molecules or infinite, periodic crystals. Therefore, the study of their electronic properties is often possible only when applying special methods that are specifically developed for that particular type of system. These methods as well as representative examples of their applications will be briefly described here without, however, going into too much detail. Thereby the examples will rather illustrate what kind of problems are to be tackled and what kind of results are obtained. The reader who is interested in more details is recommended to consult the original literature.

25 Impurities

25.1 THE ONE-DIMENSIONAL CASE

In order to approach the subject of impurities in extended systems, let us first consider the one-dimensional case in some detail. We start out with an infinite periodic chain like that of Fig. 25.1.

We saw in Chapter 19 that for the infinite, periodic chain it was very helpful to use the periodicity of the system in constructing symmetry-adapted basis functions (the so-called Bloch waves) that subsequently could be used in transforming the problem of solving the $\infty \times \infty$ secular matrix equation into that of solving an infinite set of finite secular equations. In effect this means that the problem of calculating the electronic properties becomes tractable. Furthermore, the k number (or $\bar{k}$ number) that classifies the Bloch waves could be used in defining band structures like that schematic shown in Fig. 25.2.

The band structures are thus ultimately linked to the periodicity of the system. This is, on the other hand, not the case for the density of states, for which we show a simple example in Fig. 25.3. The density of states could in principle also be calculated without constructing the Bloch waves, e.g., by studying systems of increasing size in the limit of infinite size. This is, however, computationally a much more complicated task compared to directly using the symmetry and Bloch waves.

When the chain contains an impurity like that of the lower part of Fig. 25.1, the periodicity is destroyed. This means that it is not possible to construct symmetry-adapted basis functions, and the concept of band structures loses its meaning, although a density of states still exists.

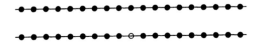

Figure 25.1 Schematic presentation of an infinite chain. In the upper part it is periodic, whereas an impurity destroys the periodicity in the lower part

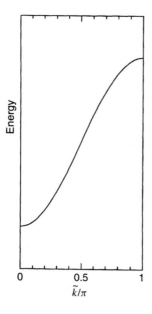

Figure 25.2 A single band for an infinite, periodic chain

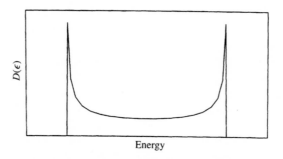

Figure 25.3 The density of states corresponding to the band of Fig. 25.2

An impurity changes the properties of the material. Often, there will be additional single-particle states that are localized to the region of the impurity. These may be either empty or occupied. When the concentration of impurities is very low, one may take the conservative point of view and simply ignore their presence. For many properties (e.g., structural and mechanical ones) this may be a good approximation, but for others it is precisely the presence of the impurities that is the interesting point, and a central goal is to understand the properties of the impurities. An example will illustrate this point.

Figure 25.4 shows different structures of polyacetylene. This material contains carbon and hydrogen atoms that are bonded first of all via σ bonds from carbon sp^2 hybrids and hydrogen $1s$ functions. In addition, one π electron per carbon atom participates in π bonds between the carbon atoms. As discussed

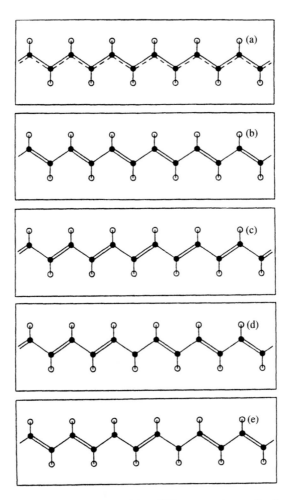

Figure 25.4 Schematic representation of five different structures of polyacetylene. Black and white circles correspond to carbon and hydrogen atoms, respectively. In (a) all C–C bond lengths are identical, whereas they alternate in (b) and (c). The two structures in (b) and (c) differ in this alternation, but are otherwise identical. (d) contains an impurity where the alternation changes from one pattern to the other, whereas (e) contains an impurity where the alternation locally is distorted

in Section 19.3, the lowest total energy is found for a structure where the C–C bond lengths alternate, like in Fig. 25.4(b) or Fig. 25.4(c). The two structures of these figures are energetically degenerate, and an 'impurity' like that of Fig. 25.4(d) may exist. This is rather a structural distortion, where the bond-length alternation changes from one pattern to the other. Or an 'impurity' (structural distortion) like that of Fig. 25.4(e) occurs, where the bond-length alternation is only locally distorted. Upon passing we mention that the distortion of Fig. 25.4(d) is called a soliton and that of Fig. 25.4(e) a polaron.

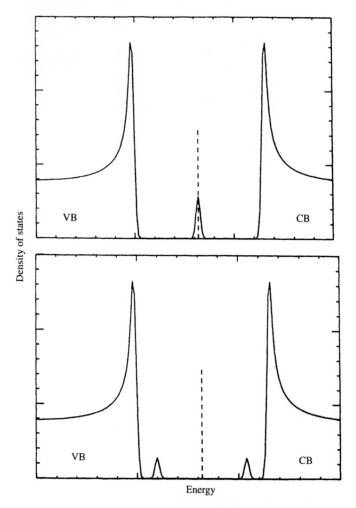

Figure 25.5 The density of states for a polyacetylene chain with (upper part) a soliton and (lower part) a polaron. The vertical dashed line marks the position of the Fermi level for the neutral system, and 'VB' and 'CB' mark the density of states for the π valence and conduction bands, respectively, for the system without the impurities

These impurities involve only redistributions of the π electrons and we will therefore focus on those. It turns out that the density of states for the two impurity-containing structures contains additional features in the gap separating occupied and empty orbitals, cf. Fig. 25.5. This means that when the material is being doped (i.e., electrons are added to or removed from the material), the extra charges will enter these orbitals. Thus, for charge transport processes the properties of these levels are of ultimate importance.

This is not only the case for this specific system. Thus, for many semiconductors, the dopants lead to exactly such extra levels in the gap between occupied and empty orbitals. Therefore, for many studies of the properties of semiconductors, studying the impurities is a very important part.

25.2 SUPERCELLS

In order to describe the concept of supercells it is useful to consider a two-dimensional model like that of Fig. 25.6(a). We have here a single impurity in an otherwise periodic lattice.

The presence of the impurity is felt in many ways by the lattice. First of all, the potential felt by the electrons will be different close to the impurity than elsewhere in the crystal. This may lead to extra impurity-induced levels that may, but need not, have energies outside the energy regions (bands) of the perfectly periodic crystal without the impurity.

Second, if the impurity is, e.g., a vacancy or a different type of atom or an extra atom, the lattice may change (i.e., relax) in the proximity of the impurity. It is not obvious exactly what will happen. Consider, e.g., the case of a vacancy, i.e., the impurity in Fig. 25.6(a) is to be considered as the absence of

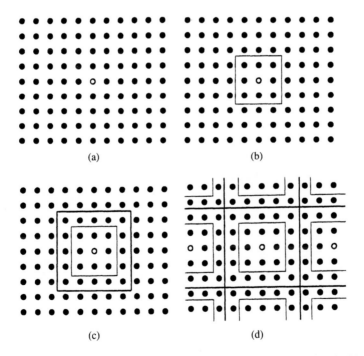

Figure 25.6 (a) A single impurity in a periodic lattice. (b) The region inside which the presence of the impurity can be felt. (c) The size of the (super)cell that is repeated periodically. (d) The periodically repeated supercells

an atom. One may argue that there will be extra space for the other atoms and that these may prefer to move in there and fill this space. Alternatively, one may argue that by considering the chemical bonds between the atoms as due to springs, the removal of the springs to the vacancy will make the neighbouring atoms move further away from the vacancy. What in any given situation actually does take place can only be determined through detailed studies, experimental or theoretical.

We may now assume that the effects of the presence of the impurity is felt only inside a finite region of the infinite material, as sketched in Fig. 25.6(b). Therefore, when considering a larger region, like that of Fig. 25.6(c), we include both atoms that are inside the defect region and atoms that are outside it. Ultimately, we may repeat the larger cell (the so-called supercell) periodically and arrive at a system like that of Fig. 25.6(d).

For this, the impurities are assumed to be so well separated (this has to be checked in the calculations) that they do not feel each others' presence. Furthermore, through this construction we have recovered a periodic system that can be treated with the methods outlined in Chapter 19, although the size of the repeated unit cell is considerably larger than that without the impurity (for the case of Fig. 25.6 its size changes from 1 to 25 atoms). On the other hand, the concentration of the impurities may be very much larger than what is realistic and one may have to modify the results accordingly when comparing with experimental results.

We shall present two examples of the application of this procedure. In the first one, defects in crystalline GaAs are studied. Here, the case that an As atom replaces a Ga atom at a certain site is studied. GaAs crystallizes in the zincblende structure, so that without the defects one may consider the structure as consisting of crossing zigzag chains in all three directions, each chain consisting of alternating As and Ga atoms. In the ideal case, one unit cell contains two atoms. In the present study, supercells with 18 atoms (i.e., cells of $3 \times 3 \times 1$ standard unit cells) were studied, and all structural coordinates were optimized. Two different arrangements were considered, one where the As atom simply replaces a Ga atom and where only minor lattice relaxations (i.e., structural changes upon this replacement) take place, and one where large lattice relaxations take place. In Fig. 25.7 we show the total electron density for the latter case together with the densities of the highest occupied and lowest unoccupied orbitals. The latter show, just as for the polyacetylene chain studied above, that the two orbitals closest to the Fermi level are well localized to the region of the defect, so that these orbitals are very important for the doped material.

The other example is that of an Se atom replacing an As atom or vice versa in As_2Se_3. The crystal structure (cf. Fig. 25.8) consists of unit cells with 20 atoms, and for the defect studies supercells with 40 atoms were considered. All structural degrees of freedom were relaxed (by using forces as discussed in Chapter 20). Different states may occur for the defect, differing both in the structure and in the

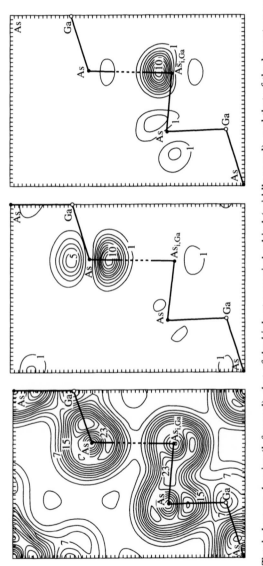

Figure 25.7 Total electron density (left panel), that of the highest occupied orbital (middle panel), and that of the lowest unoccupied orbital (right panel) for a GaAs crystal with an As atom placed at the site of a Ga atom followed by large lattice relaxations. Reproduced with permission from Chadi and Chang (1988). Copyright 1988 by the American Physical Society

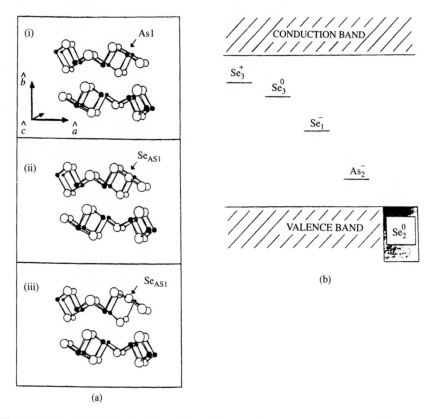

Figure 25.8 (a) Schematic presentation of (i) crystalline As_2Se_3, and (ii,iii) of two defect structures. (b) The energy levels due to defect for the different states and charges of the defect. Reproduced with permission from Tarnow *et al.* (1988). Copyright 1988 by the American Physical Society

charge of the defect, and in the right part of Fig. 25.8 we show how the defects induce extra levels that in many cases appear in the gap between the occupied valence bands and the empty conduction bands for the defect-free crystal.

25.3 GREEN'S FUNCTION AND IMPURITIES

In Chapter 17 we discussed the concept of Green's function and showed how this function could be useful in studying changes in the electronic structure when introducing a perturbation, e.g., due to an impurity. For the system without the impurity, the Schrödinger equation is

$$\hat{H}_0\Psi_0(\vec{x}) = E_0\Psi_0(\vec{x}), \tag{25.1}$$

or

$$[\hat{H}_0 - E_0]\Psi_0(\vec{x}) = 0. \tag{25.2}$$

The Green's function is obtained as the solution to the equation

$$[\hat{H}_0 - E]G_0(\vec{x}, \vec{x}', E) = -\delta(\vec{x} - \vec{x}'). \tag{25.3}$$

Knowing all solutions to the Schrödinger equation

$$\hat{H}_0\Psi_n(\vec{x}) = E_n\Psi_n(\vec{x}), \tag{25.4}$$

the Green's function could also be written as

$$G_0(\vec{x}, \vec{x}', E) = \sum_n \frac{\Psi_n^*(\vec{x}')\Psi_n(\vec{x})}{E - E_n}. \tag{25.5}$$

In a density-functional or Hartree–Fock calculation, Ψ_0 is written as a Slater determinant

$$\Psi_0(\vec{x}) = |\psi_1, \psi_2, \ldots, \psi_N|, \tag{25.6}$$

and the individual single-particle wavefunctions are expanded in all practical applications in a set of basis functions,

$$\psi_i(\vec{x}_n) = \sum_k \chi_k(\vec{x}_n)c_{ki}. \tag{25.7}$$

Then, the Green's function is that for the single particles, i.e., the Hamilton operator is to be replaced by the Fock or the Kohn–Sham operator, respectively, and the energies are the single-particle energies. Nevertheless, the main results remain and we shall therefore not discuss that case separately.

Assuming that the basis functions $\{\chi_k\}$ define a sufficiently complete set, the Green's function can be written in the form

$$G_0(\vec{x}, \vec{x}', E) = \sum_{k,l} \chi_k^*(\vec{x}')\chi_l(\vec{x})g_{0,lk}(E), \tag{25.8}$$

where the $g_{kl}^0(E)$ are coefficients that depend on the energy E, cf. Section 17.6.

Including the perturbation due to the impurity, the Schrödinger equation takes the form

$$[\hat{H}_0 + V(\vec{x}) - E]\Psi(\vec{x}) = 0 \tag{25.9}$$

with $V(\vec{x})$ being the perturbation.

Also here we introduce a Green's function,

$$[\hat{H}_0 + V(\vec{x}) - E]G(\vec{x}, \vec{x}', E) = -\delta(\vec{x} - \vec{x}') \tag{25.10}$$

and also this Green's function can be expanded as in Eq. (25.8):

$$G(\vec{x}, \vec{x}', E) = \sum_{k,l} \chi_k^*(\vec{x}')\chi_l(\vec{x})g_{lk}(E). \tag{25.11}$$

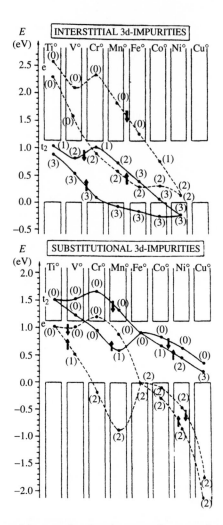

Figure 25.9 Single-particle energies for the ground states of neutral transition-metal impurities in crystalline silicon. A number in parentheses gives the occupancy of a localized gap state, or of a resonance in the valence (below 0.0 eV) or conduction (above 1.15 eV) band. The upper panel gives the case of an interstitial impurity, and the lower one that of a substitutional impurity. Reproduced with permission from Beeler *et al.* (1985). Copyright 1985 by the American Physical Society

We showed in Section 17.5 that G and G_0 are related through the so-called Dyson equation,

$$G(\vec{x}, \vec{x}', E) = G_0(\vec{x}, \vec{x}', E) + \int G_0(\vec{x}, \vec{x}'', E) V(\vec{x}'') G(\vec{x}'', \vec{x}', E) \, d\vec{x}''. \quad (25.12)$$

In this equation we insert the two expansions (25.8) and (25.11) where we use that the coefficients $g_{0,lk}$ are known from a calculation on the system without the impurity. We have accordingly,

$$\sum_{k,l} \chi_k^*(\vec{x}')\chi_l(\vec{x})g_{lk} = \sum_{k,l} \chi_k^*(\vec{x}')\chi_l(\vec{x})g_{0,lk}(E) + \sum_{k,l,m,n} \int \chi_k^*(\vec{x}'')\chi_l(\vec{x})$$

$$\times V(\vec{x}'')\chi_m^*(\vec{x}')\chi_n(\vec{x}'')\, d\vec{x}''\, g_{0,lk}(E)g_{nm}(E). \quad (25.13)$$

Subsequently, we multiply by $\chi_q(\vec{x}')\chi_p^*(\vec{x})$, where q and p are arbitrarily chosen, and integrate over $\vec{x}'$ and $\vec{x}$. This gives Dyson's equation in matrix form:

$$\underline{\underline{G}}(E) = \underline{\underline{G}}_0(E) + \underline{\underline{G}}_0(E) \cdot \underline{\underline{V}} \cdot \underline{\underline{G}}(E). \quad (25.14)$$

This is the (matrix) equation that will be solved as a function of E. One has to define the spatial region where the extra potential is assumed to be non-vanishing (so that the size of the matrices becomes finite). This corresponds to the region of Fig. 25.6(b). Furthermore, the potential has components both from the presence of the impurity and from the redistribution of the electrons and the lattice distortions. Accordingly, the whole procedure has to be done self-consistently: A given potential leads to a Green's function $G(\vec{x}, \vec{x}')$ from which a density of states and an electron density can be calculated. The latter defines a new potential that subsequently leads to a new Green's function, and the whole procedure is repeated until the input and output potentials and densities agree.

Having solved Dyson's equation, various quantities can be calculated directly from the Green's function as discussed in Chapter 20.

In Fig. 25.9 we show an example of the application of this method. Here, transition-metal impurities in crystalline silicon have been studied. These impurities were considered in two different cases, i.e., either as substituting a silicon atom or as occupying a so-called interstitial site that is not occupied by any atoms in the impurity-free case. The results of Fig. 25.9 show how some of the impurities lead to extra levels in the gap between empty and occupied bands whereas others do not.

25.4 TRANSFER MATRICES IN ONE DIMENSION

Transfer matrices have developed into a useful concept in studying systems that can be considered one-dimensional although to a lesser extent for parameter-free, electronic-structure methods than for semiempirical methods or phenomenological models. The one-dimensional materials are materials for which the properties along one direction are those of interest whereas those in the other directions are considered less interesting. Often the interactions in those two other directions are much weaker than those in the first direction.

Starting with the infinite, periodic chain of Fig. 25.10(a) we have a system that certainly can be characterized as quasi-one-dimensional. In addition, this system is periodic. This system can be treated with the band-structure methods we discussed in Chapter 19. Introducing a single impurity as in Fig. 25.10(b), the periodicity is destroyed and the band-structure concepts can no longer be applied. This is also the case for the system of Fig. 25.10(c) that contains irregularly placed impurities. Finally, the chain of Fig. 25.10(d) is again periodic.

The method we shall develop here is applicable only for single-particle models like the (extended) Hückel model. Thus, our aim is to solve the single-particle equation

$$\hat{h}\psi_i = \varepsilon_i\psi_i. \tag{25.15}$$

The eigenfunctions ψ_i are assumed to be expanded in a given set of basis functions,

$$\psi_i = \sum_j \chi_j c_{ji}, \tag{25.16}$$

where χ_j are the basis functions and c_{ji} the (unknown) expansion coefficients. The basis functions are assumed to be orthonormal, and the matrix elements

$$\langle \chi_j|\hat{h}|\chi_k \rangle = t_{jk} \tag{25.17}$$

are some constants that characterize the model.

At this point we notice that the approach we have taken actually may be used for the system of Fig. 25.11, too. Here, we show schematically a system that is non-periodic in only one dimension but periodic in the other (for a three-dimensional structure: in the other *two* dimensions). Perpendicular to the direction of the aperiodicity we may use the periodicity in constructing symmetry-adapted basis functions (Bloch waves) characterized by a $\vec{k}$ vector (this $\vec{k}$ vector is one-dimensional for the structure of Fig. 25.11, and two-dimensional for a three-dimensional system). For *any* $\vec{k}$ we may subsequently study the properties in the third direction (i.e., the direction of the aperiodicity) by applying the methods to be discussed here.

In order to illustrate the concept of transfer matrices we shall consider a specific example, which could be relevant for the systems of Fig. 25.10. We assume that

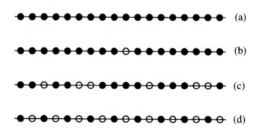

Figure 25.10 Different cases of an infinite, one-dimensional chain. That of (a) is periodic with one atom per unit cell. The chain of (b) contains a single impurity, whereas that of (c) contains many, irregularly placed impurities. Finally, the chain of (d) is again periodic, but with two atoms per unit cell

Figure 25.11   A two-dimensional lattice that is periodic in one dimension but irregular in the other

we have one basis function per site, and that the sites are labelled by the integers $n = \ldots, -1, 0, 1, \ldots$. The basis function of the nth atom will be labelled χ_n, and we shall call the black atoms of Fig. 25.10 A atoms and the other ones B atoms. Finally, we set

$$\langle \chi_n | \hat{h} | \chi_n \rangle = \begin{cases} \varepsilon_A & \text{for an A atom} \\ \varepsilon_B & \text{for a B atom} \end{cases} \tag{25.18}$$

and

$$\langle \chi_n | \hat{h} | \chi_m \rangle = \begin{cases} \varepsilon_n & \text{for } n = m \\ t & \text{for } n = m \pm 1 \\ 0 & \text{otherwise.} \end{cases} \tag{25.19}$$

Here, ε_n takes the two values of Eq. (25.18).

The approximations (25.18) and (25.19) correspond to assuming that the interactions are non-vanishing only between nearest neighbours, that the on-site terms take only two values depending on the type of atom, and that the nearest-neighbour hopping integrals are constant.

As discussed in Section 12.2, we may now write the Hamilton operator in the form

$$\hat{h} = \sum_n \left[\varepsilon_n \hat{a}_n^\dagger \hat{a}_n + t(\hat{a}_{n+1}^\dagger \hat{a}_n + \hat{a}_n^\dagger \hat{a}_{n+1}) \right]. \tag{25.20}$$

The operators $\hat{a}_n^\dagger$ and $\hat{a}_n$ are the creation and annihilation operators discussed in Section 12.2.

The general solution to the Schrödinger equation

$$\hat{h}\psi = \varepsilon\psi \tag{25.21}$$

can be written in the form

$$\psi = \sum_n c_n \chi_n. \tag{25.22}$$

We now use that

$$\hat{a}_m^\dagger \hat{a}_n \chi_k = \delta_{k,n} \chi_m. \tag{25.23}$$

Then, upon inserting Eq. (25.22) into Eq. (25.21) with $\hat{h}$ given by Eq. (25.20) we obtain

$$\sum_n [\varepsilon_n c_n \chi_n + t c_n \chi_{n+1} + t c_n \chi_{n-1}] = \varepsilon \sum_n c_n \chi_n. \tag{25.24}$$

Multiplying by χ_p^* and integrating (and using the orthonormality of the basis functions) we end up with

$$\varepsilon_p c_p + t c_{p-1} + t c_{p+1} = \varepsilon c_p. \tag{25.25}$$

We combine this equation with the trivial equation

$$c_p = c_p \tag{25.26}$$

in obtaining

$$
\begin{aligned}
\begin{pmatrix} c_{p+1} \\ c_p \end{pmatrix}
&= \begin{pmatrix} \dfrac{\varepsilon - \varepsilon_p}{t} & -1 \\ 1 & 0 \end{pmatrix} \cdot \begin{pmatrix} c_p \\ c_{p-1} \end{pmatrix} \\[2mm]
&= \begin{pmatrix} \dfrac{\varepsilon - \varepsilon_p}{t} & -1 \\ 1 & 0 \end{pmatrix} \cdot \begin{pmatrix} \dfrac{\varepsilon - \varepsilon_{p-1}}{t} & -1 \\ 1 & 0 \end{pmatrix} \cdot \begin{pmatrix} c_{p-1} \\ c_{p-2} \end{pmatrix} \\[2mm]
&= \begin{pmatrix} \dfrac{(\varepsilon - \varepsilon_p)(\varepsilon - \varepsilon_{p-1})}{t^2} - 1 & -\dfrac{\varepsilon - \varepsilon_p}{t} \\ \dfrac{\varepsilon - \varepsilon_{p-1}}{t} & -1 \end{pmatrix} \cdot \begin{pmatrix} c_{p-1} \\ c_{p-2} \end{pmatrix} \\[2mm]
&\equiv \underline{\underline{T}}_p \cdot \begin{pmatrix} c_{p-1} \\ c_{p-2} \end{pmatrix},
\end{aligned}
\tag{25.27}
$$

where $\underline{\underline{T}}_p$ is the transfer matrix.

For large systems one may use

$$\begin{pmatrix} c_{N+1} \\ c_N \end{pmatrix} = \underline{\underline{T}}_N \cdot \underline{\underline{T}}_{N-2} \cdots \cdots \underline{\underline{T}}_{-M} \cdot \begin{pmatrix} c_{-M-1} \\ c_{-M-2} \end{pmatrix} \tag{25.28}$$

(assuming that $N + M$ is even) in studying how the orbital changes throughout the system.

A very special case is that of Fig. 25.15(d). Here, the transfer matrix is independent of p and equals

$$\underline{\underline{T}} = \begin{pmatrix} \dfrac{(\varepsilon - \varepsilon_A)(\varepsilon - \varepsilon_B)}{t^2} - 1 & -\dfrac{\varepsilon - \varepsilon_A}{t} \\ \dfrac{\varepsilon - \varepsilon_B}{t} & -1 \end{pmatrix}. \tag{25.29}$$

Let us now diagonalize this matrix,

$$\underline{\underline{T}} = \underline{\underline{U}} \cdot \underline{\underline{\Lambda}} \cdot \underline{\underline{U}}^\dagger, \tag{25.30}$$

where $\underline{\underline{U}}$ is a unitary matrix, and where $\underline{\underline{\Lambda}}$ is a diagonal matrix containing the eigenvalues of $\underline{\underline{T}}$.

Since $\underline{\underline{U}}$ is unitary, we find immediately,

$$\begin{pmatrix} c_{N+1} \\ c_N \end{pmatrix} = \underline{\underline{U}} \cdot \underline{\underline{\Lambda}}^{(N+M)/2} \cdot \underline{\underline{U}}^\dagger \cdot \begin{pmatrix} c_{-M-1} \\ c_{-M-2} \end{pmatrix}. \tag{25.31}$$

The matrix $\underline{\underline{\Lambda}}^{N+M}$ contains the eigenvalues of $\underline{\underline{\Lambda}}$ to the power $N + M$. Thus, by requiring that the orbital for an infinite system does not diverge at any end (i.e., for $N, M \to \infty$) we have to have that at least one of the eigenvalues of the transfer matrix has the absolute value 1.

Let us therefore study the eigenvalues of the transfer matrix of Eq. (25.29). These are obtained from

$$\begin{vmatrix} \dfrac{(\varepsilon - \varepsilon_A)(\varepsilon - \varepsilon_B)}{t^2} - 1 - \lambda & -\dfrac{\varepsilon - \varepsilon_A}{t} \\[2ex] \dfrac{\varepsilon - \varepsilon_B}{t} & -1 - \lambda \end{vmatrix} = 0. \tag{25.32}$$

Inserting

$$\lambda = e^{i\theta} \tag{25.33}$$

(i.e., that the eigenvalue has the absolute value 1), we can now solve Eq. (25.32) for ε (i.e., *not* for λ). It is easily realized that this equation leads to the same energies as that obtained by diagonalizing the $\tilde{k}$-dependent Hamilton matrix

$$\begin{vmatrix} \varepsilon_A - \varepsilon & t e^{i\tilde{k}} \\ t e^{-i\tilde{k}} & \varepsilon_B - \varepsilon \end{vmatrix} = 0. \tag{25.34}$$

Thus, this method offers an alternative way of obtaining the band structures for a periodic system.

It can be shown (Springborg, 1994a,b) that this is generally valid for a single-particle model when it can be assumed that the hopping integrals $\langle \chi_m | \hat{h} | \chi_n \rangle$ vanish for sufficiently large $|n - m|$.

Finally, in the present model, $\theta = \tilde{k}$, as can be shown, and the method can also be applied to the study of non-periodic chains (where it has actually found its largest area of application, Springborg, 1994a,b).

When including impurities into the model one may distinguish between two different cases: a single, isolated impurity, or many impurities. Approaches based on transfer matrices have mainly been used in the latter case, for which one may once again distinguish between two cases: completely randomly placed impurities of a certain concentration or so-called quasiperiodic structures where the impurities are placed non-periodically but nevertheless deterministically. In both of these two cases the starting point is Eq. (25.28) where analysis of the transfer-matrix product $\underline{\underline{T}}_N \cdot \underline{\underline{T}}_{N-2} \cdots \cdots \underline{\underline{T}}_{-M}$ gives information about the density of states.

Also for individual, isolated impurities, the transfer matrices can be used in analysing the density of states, although this has been done in only a few cases.

26 Surfaces and Interfaces

26.1 GENERAL CONSIDERATIONS

As impurities, also surfaces and interfaces destroy the periodicity of an otherwise infinite, periodic, crystalline material. The surfaces are very important, first of all since they form the region of contact with the surroundings. But in addition, as we shall see below, they have properties that are specific for the surfaces, i.e., not shared by the atoms of the interior of the material.

Let us first consider the structures of Figs. 26.1(a) and 26.1(b). They correspond to the ideal case that the atoms near the surface have the same bond lengths and bond angles, etc., as in the bulk, i.e., there is no structural relaxation. For the infinite, periodic, three-dimensional crystal of each material separately [Fig. 26.1(a)], or of the single material [Fig. 26.1(b)], we have seen in Chapter 18 that the periodicity can be used in constructing symmetry-adapted basis functions characterized by a three-dimensional $\vec{k}$ vector. Then, the single-particle wavefunctions can also be classified according to this $\vec{k}$, i.e., any such wavefunction is constructed by the basis functions belonging to only one specific $\vec{k}$.

If we write

$$\vec{k} = (k_a, k_b, k_c), \tag{26.1}$$

we can assume that one of the components, say k_c, describes the periodicity perpendicular to the direction of the surface or interface in Figs. 26.1(a) and 26.1.(b). The other two components, k_a and k_b, describe the periodicity parallel to the surface or interface. When introducing the surface or interface, the periodicity parallel to this is kept (at least to a first approximation; see below), i.e., we may still use k_a and k_b in classifying the single-particle wavefunctions. But the last component, i.e., k_c, of $\vec{k}$, can no longer be used in classifying the wavefunctions, since this periodicity is destroyed. Therefore, we can only use a two-dimensional

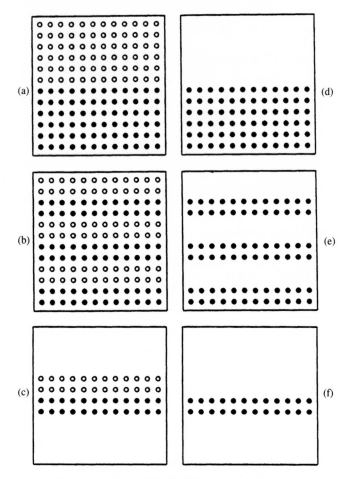

Figure 26.1 (a) Interface between two crystalline materials; (b) surface of one crystalline material; (c,d) periodic structures for studying the interfaces or surfaces; (e,f) thin films for studying the interfaces or surfaces

$\vec{k}$ vector in classifying the wavefunctions, and for any value of this

$$\vec{k} = (k_a, k_b) \tag{26.2}$$

we will have *all* wavefunctions for *any* value of k_c for the crystalline material. Therefore, the band structures we discussed in Section 18.5 for three-dimensional crystalline materials will no longer be nice narrow curves but will be broad bands.

A simple model demonstrates this. We consider a material with one atom per unit cell and one orbital per atom. We assume that the band structures can be described by using a simple Hückel-like model with nearest-neighbour hopping integrals $-t_a$ in the a direction, $-t_b$ in the b direction, and $-t_c$ in the c direction.

Moreover, we assume that these interactions are the only non-vanishing ones, and that each atom has two neighbours in each of the three directions. Then, for the crystalline material, the band structures are given by

$$\varepsilon(\vec{k}) = \varepsilon(k_a, k_b, k_c) = -2t_a \cos(ak_a) - 2t_b \cos(bk_b) - 2t_c \cos(ck_c). \quad (26.3)$$

If we now discard the translational symmetry along the c direction, we have for each fixed (k_a, k_b) *all* energy values for *all* different values of k_c, i.e., we have a broad band spanning the region

$$-2t_a \cos(ak_a) - 2t_b \cos(bk_b) - 2t_c \le \varepsilon(k_a, k_b)$$
$$\le -2t_a \cos(ak_a) - 2t_b \cos(bk_b) + 2t_c, \quad (26.4)$$

that for each (k_a, k_b) is $4t_c$ wide.

We illustrate this point further in Fig. 26.2 for the case of an interface between NiSi$_2$ and Si. Here, it is clearly seen how the bands become broad and how most of the individual bands can be traced back to the separate crystalline materials.

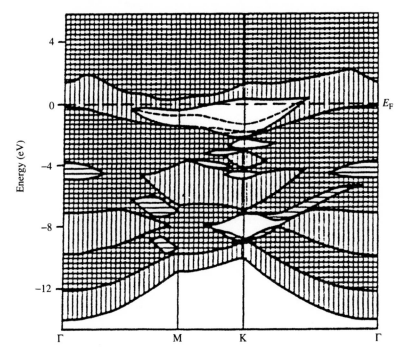

Figure 26.2 The two-dimensional band structures for an interface between NiSi$_2$ and Si. The band structures of the individual crystals are marked as vertically hatched regions for NiSi$_2$ and horizontally hatched regions for Si. The dashed bands correspond to orbitals that are localized to the region of the interface. Reproduced with permission from Das *et al.* (1989). Copyright 1989 by the American Physical Society

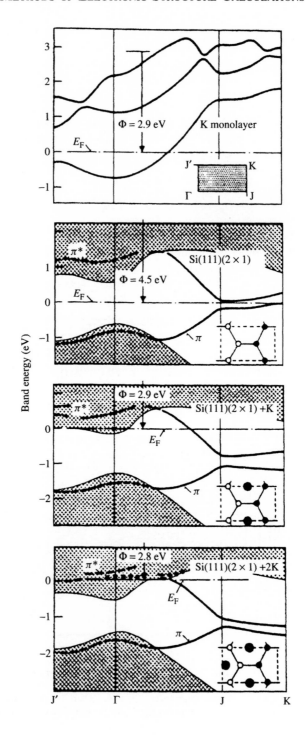

It is also recognized that there may still appear energy gaps for certain (k_a, k_b) where there are no allowed bands.

Starting out with a single crystal there are many ways of forming a surface. In the example of Fig. 26.2 the surface of the silicon crystal is the so-called (111) surface, meaning that its normal is the $(1, 1, 1)$ vector. In principle, the other material (in our case here, $NiSi_2$) has also very many different surfaces, but for the example of Fig. 26.2, only one is relevant (there are only two ways $NiSi_2$ and Si will bond together, and they are constructed from the same interface but differ in the relative orientation), and therefore (111) specifies completely the interface.

The example of Fig. 26.3 shows some further aspects. It is seen that a true two-dimensional material, i.e., the single layer of K atoms, possesses a band structures consisting of different narrow curves, cf. Fig. 26.3(a). On the other hand, for the pure Si(111) surface we see once again the broad bands, but in addition we have some thin bands. These correspond to wavefunctions well localized to the surface and do therefore not exist for the pure, infinite, periodic, three-dimensional material. Furthermore, the atoms of the surface do not sit exactly as in the bulk (this should not be a surprise since the chemical surroundings of the surface atoms are different from those of the bulk atoms). Actually, the symmetry is lowered, so that one needs two of the original unit cells in describing the unit cell of the relaxed system; the system possesses a so-called 2×1 reconstruction. This is indicated on the panels of Fig. 26.3.

Finally, the two last panels in Fig. 26.3 show the band structures when placing one or two layers of K atoms (whose bands are shown in the upper panel) on the Si surface.

26.2 SUPERCELLS

The surfaces and interfaces destroy the translational symmetry in one direction. This means that there is translational symmetry in only two directions and that the size of one unit cell is infinite in the third direction. Of course, it is not possible to treat such systems directly.

One approach to circumvent this problem is, as for the impurities, to construct artificial systems that are periodic in all three directions and simultaneously contain the surfaces or interfaces repeatedly. Figs. 26.1(c) and 26.1(d) show how such periodic structures may look. Each periodically repeated supercell is to be so large that the atoms that are most distant from the interface or surface do behave

---

Figure 26.3 (a) The band structures for a single layer of K atoms; (b) the band structures for the a pure Si (111) surface; (c) the band structures of a pure Si surface with one layer of K atoms; (d) The band structures of a pure Si surface with two layers of K atoms. The insert in the uppermost panel shows the two-dimensional Brillouin zone, and those in the other panels show the positions of the atoms with the smaller atoms being Si atoms and the larger ones being K atoms. Moreover, light and dark atoms correspond to different layers. Reproduced with permission from Ciraci and Batra (1988). Copyright 1988 by the American Physical Society

like bulk atoms and that the wavefunctions located at one surface or interface do not interact with those of any other.

In Fig. 26.2 we showed the band structures for an interface between NiSi$_2$ and Si. These were in fact calculated with a supercell method, where it was assured that the supercells were sufficiently large according to the criteria above. To ensure this, different calculations for different sizes of the supercells were carried through.

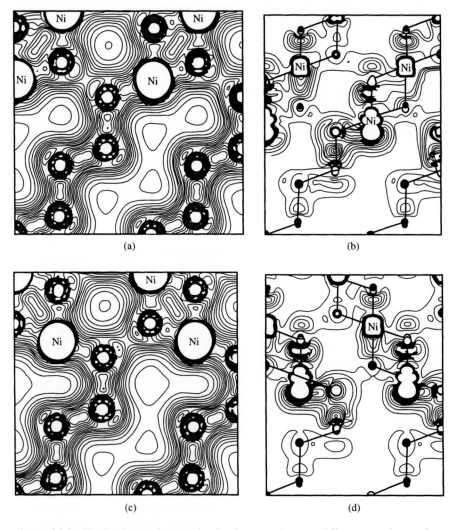

(a)   (b)

(c)   (d)

Figure 26.4 Total valence-electron density for (a,c) the two different interfaces of the NiSi$_2$+Si system. (b,d) The valence electron density of the interface states for the two interfaces. Reproduced with permission from Das *et al.* (1989). Copyright 1989 by the American Physical Society

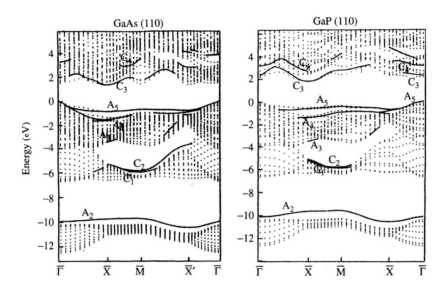

Figure 26.5 Surface band structures for the (110) surface of (left part) GaAs and (right part) GaP. The full lines indicate surface states. Reproduced with permission from Manghi *et al.* (1990). Copyright 1990 by the American Physical Society

In Fig. 26.4 we show the total valence-electron density for this system. As discussed above, one may construct two different types of interfaces, differing in the relative orientation of the two crystals, and, accordingly, Fig. 26.4 shows two different sets of valence electron densities. Furthermore, we saw in Fig. 26.5 that most of the bands could be derived from the band structures of each half-part individually. But in Fig. 26.5 we also recognize some extra bands (marked by dashed lines) that originate from orbitals localized to the region of the interface. Their densities are also shown in Fig. 26.4, where it is confirmed that they are interface states.

Another example is presented in Fig. 26.5 The results here were obtained using supercells of 15 atomic layers and 8 'vacuum layers'. Once again we recognize both bulk-like bands (actually, their number is finite due to the finite number of atomic layers whereas for the truly semiinfinite system the discrete bands merge into broader, continuous features) as well as surface states.

26.3 GREEN'S FUNCTIONS

In Section 25.3 we saw how Green's functions could be used in studying the electronic properties of impurities in crystals. The advantage with this approach is that Green's functions provide a natural way of incorporating additional effects on top of another calculation. Thereby, one has to solve Dyson's equation

$$G(\vec{x}, \vec{x}', E) = G_0(\vec{x}, \vec{x}', E) + \int G_0(\vec{x}, \vec{x}'', E) V(\vec{x}'') G(\vec{x}'', \vec{x}', E) \, d\vec{x}'', \quad (26.5)$$

where G_0 is the Green's function for the system without the additional effect, whereas G is that after the inclusion of this extra effect.

Within Hartree–Fock or density-functional methods, G_0 can be expressed in terms of the eigenfunctions to the single-particle equations for the unperturbed system,

$$G_0(\vec{x}, \vec{x}', E) = \sum_n \frac{\psi_n^*(\vec{x}')\psi_n(\vec{x})}{E - \varepsilon_n}. \quad (26.6)$$

For a system that is periodic in one or more dimensions, the eigenfunctions ψ_n can be classified according to their $\vec{k}$, and we can accordingly split the sum in Eq. (26.6) into different ones for different pairs of $\vec{k}$,

$$G_0(\vec{x}, \vec{x}', E) = \sum_{\vec{k}} G_0^{\vec{k}}(\vec{x}, \vec{x}', E), \quad (26.7)$$

with

$$G_0^{\vec{k}}(\vec{x}, \vec{x}', E) = \sum_n \frac{\psi_n^{\vec{k}*}(\vec{x}')\psi_n^{\vec{k}}(\vec{x})}{E - \varepsilon_n(\vec{k})}, \quad (26.8)$$

where the n summation now runs only over those wavefunctions that belong to that specific value of $\vec{k}$ and where we have used the fact that for the unperturbed system the eigenfunctions can be classified according to their $\vec{k}$.

Introducing the surface or interface, we can still classify the orbitals according to the two-dimensional $\vec{k}$ that describes the symmetry properties of the wavefunctions parallel to the surface or interface. This means that we first calculate the Green's function for the full three-dimensional, periodic crystalline material by performing an ordinary band-structure calculation. The eigenfunctions to this define the Green's function for each three-dimensional $\vec{k}$. That is, we end up with a set of Green's functions

$$G_0^{k_a,k_b,k_c} \quad (26.9)$$

with (k_a, k_b, k_c) given by Eq. (26.1). Subsequently, we define

$$G_0^{k_a,k_b} = \sum_{k_c} G_0^{k_a,k_b,k_c}. \quad (26.10)$$

This function is the Green's function for the infinite periodic crystal without a surface or interface.

Also the Green's function for the perturbed system can be classified according to the two-dimensional (k_a, k_b) vector similar to Eq. (26.10), since the eigenfunctions can be so classified. And since the potential $V(\vec{x}\,'')$ of Eq. (26.5) transforms according to $(k_a, k_b) = (0, 0)$, the Green's functions of different (k_a, k_b) do not mix.

For the system with a surface or interface we can construct two Green's function of the type of Eq. (26.10), one for each half-part. For the interface case we have accordingly one Green's function for each material that is joined through the interface, and for the surface case we have one Green's function for the crystalline material and one for the vacuum. Each of these two Green's functions is actually defined in the complete space. The first step is therefore to remove half of the space for each Green's function. And the second step is to turn on the interactions between the two different halves.

In a practical calculation this is done as follows. We assume that the effects of the interface or surface can be felt only within a finite (not too large) number of layers on each side. Further away each material behaves as in the bulk. In order to determine the Green's function for the semiinfinite crystal we need therefore only to determine it in a finite region close to the surface or interface. We calculate it by considering the Dyson's equation for a perturbing potential that equals all those terms that involve interactions between the two sides. Then, the perturbing potential takes care of removing the connection between the two halves and we end up with two non-interacting semiinfinite crystals. Secondly (for the interface), we add the potential due to the interactions between the two different materials once again through the Dyson's equation. For the surface this second step is not necessary.

In order to describe this approach in some further detail we consider the one-dimensional case of Fig. 26.6. We consider a single electron moving in some potential that we write as the sum of atom-centred potentials,

$$V(x) = \sum_n V_n(x - n \cdot a),\qquad(26.11)$$

where a is the lattice constant and n is an atom index. Through this construction we have split the potential into well-localized parts that can be ascribed to the individual units (atoms). In (a) we show the infinite, periodic chain which we

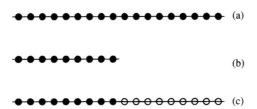

Figure 26.6 (a) An infinite periodic chain; (b) a one-dimensional chain with a surface; (c) an interface between two one-dimensional chains

assume to consist of A atoms. Since all atoms are identical, Eq. (26.11) takes the form

$$V(x) = \sum_n V_A(x - n \cdot a). \tag{26.12}$$

Removing all atoms for $n \geq 0$ we arrive at the situation of Fig. 26.6(b).
For this system, the potential is

$$V(x) = \sum_{n<0} V_A(x - n \cdot a) \tag{26.13}$$

whereas for the B part of the system it is

$$V(x) = \sum_{n \geq 0} V_B(x - n \cdot a). \tag{26.14}$$

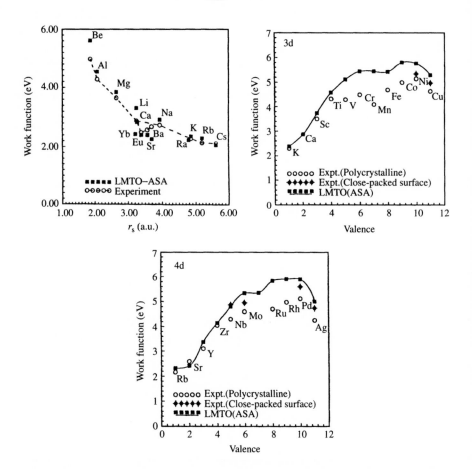

Figure 26.7 Work functions for different elemental crystals. The results marked LMTO are the theoretical ones. Reproduced with permission from Skriver and Rosengaard (1992). Copyright 1992 by the American Physical Society

We will now assume that this potential remains unchanged far away from the interface, whereas in a finite region $[-n_A; n_B]$ around the interface of Fig. 26.6(c) it changes due to charge redistributions. Thus the perturbing potential is

$$\Delta V(x) = \sum_{n=-n_A}^{n_B} \Delta V(x - n \cdot a). \qquad (26.15)$$

In a true electronic-structure calculation the definition of the various potentials is less obvious. However, the potential from the nuclei and the core electrons is easily separated into various atomic components. For the valence electrons one has to be careful in specifying the 'atomic' potentials. In a real system there will be further effects due to electronic re-distributions (which will change both the

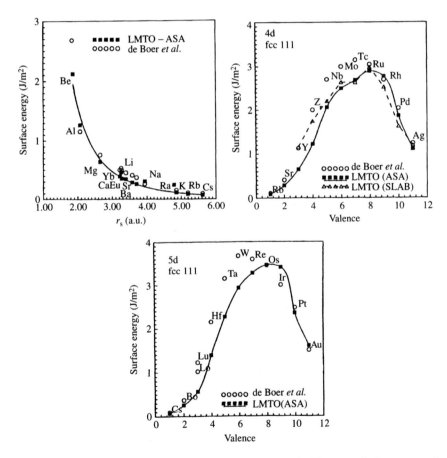

Figure 26.8 Surface energies for different elemental crystals. The open circles are experimental values and the results marked LMTO are the theoretical ones. Reproduced with permission from Skriver and Rosengaard (1992). Copyright 1992 by the American Physical Society

Coulomb and the exchange-correlation potential) and to atomic relaxations, but despite these problems it is often possible to determine a finite region to which the perturbation is confined. Only in that case is the Dyson equation in matrix form (see Sections 17.6 and 25.3) finite and soluble.

Such methods have been developed. They are, however, far less common than are those based on the supercell approach. Where comparisons are possible they seem to produce equally accurate results. In Fig. 26.7 we show an example of calculated work functions (i.e., the energy required to remove an electron from the bulk to infinitely far away from the crystal) for some different materials, and in Fig. 26.8 we show calculated surface energies (i.e., the binding energy per surface area) for the same systems. Both sets of results were obtained by a Green's function method.

26.4   RECONSTRUCTIONS

In Fig. 26.9 we show the (111) surface of the diamond structure (for instance of Si). We see that at the surface the atoms are 'lacking neighbours', i.e., the bonds towards the next layers of atoms are so-called dangling bonds.

The existence of these dangling bonds is an unfavourable situation for these atoms and the single electrons of each dangling bond may seek to find a partner with which it can form some kind of bond. The simplest way this can be achieved

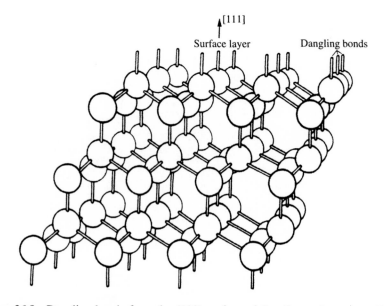

Figure 26.9   Dangling bonds from the (111) surface of the diamond structure. Reproduced by permission of Oxford University Press from Prutton (1994)

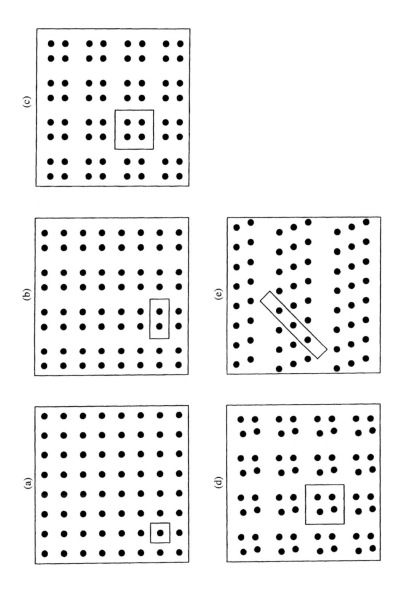

Figure 26.10 Schematic view of (a) a pure, unreconstructed surface as well as of (b) 2×1, (c) 2×2, (d) 2×2, and (e) $\sqrt{3} \times \sqrt{3}$ reconstructions. A two-dimensional unit cell is also shown in each panel

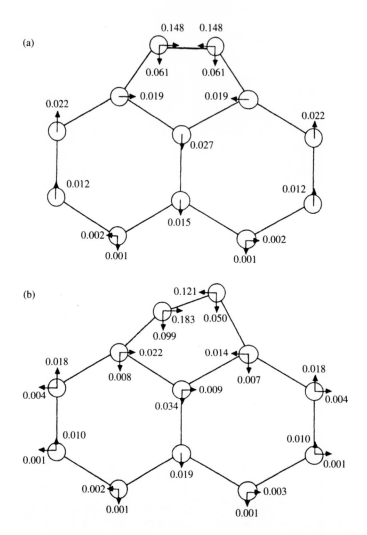

Figure 26.11 Section through the Si(100) surface showing the relaxations of the atoms for two different surface reconstruction patterns. The two topmost atoms in each panel are those directly at the surface. Reprinted from *Surface Science*, **236**, Roberts and Needs, Total energy calculation of dimer reconstructions on the silicon (001) surface, p. 112. Copyright 1990 with permission from Elsevier Science

is that the dangling bonds pairwise bend towards each other, whereby some structural relaxations of the surface may take place in order to support the creation of such pairs. This is one example of a so-called surface reconstruction.

In Fig. 26.3 we saw the outcome of exactly the reconstruction of the surface of Fig. 26.9 for the case of crystalline silicon. In that case the dangling bonds did pairwise form some weak bond at the surface and the surface relaxed so that the unit cell became doubled, i.e., we experienced a so-called 2×1 reconstruction.

The 2×1 reconstruction is one of the simplest ones. In Fig. 26.10 we show some few further examples of reconstruction patterns. As drawn in Fig. 26.10 only the atoms at the outermost layer change their positions, but it should be obvious that these shifts also affect the next atomic layers, although not so strongly as the dangling bonds affect the outermost layers. In total the relaxations can be felt by some few (typically 1–5) atomic layers.

In Fig. 26.11 we show one example of how the atoms can change positions due to surface relaxations. The example shown is that of an Si(100) surface and the two panels show the displacements of the atoms of the five outermost atomic layers. Two different relaxation patterns were considered. Moreover, the calculations were done using a supercell method.

One of the most complex surface reconstruction patterns is the 7×7 reconstruction of the Si(111) surface. This is the structure of the lowest total energy, whereas the 2×1 reconstruction we have discussed above is one of a higher total energy. In Fig. 26.12 we show the results of supercell calculations for this system using a density-functional method. All atomic coordinates were optimized by moving the atoms according to the forces. These results are some of the computationally most demanding studies that have been achieved with parameter-free methods.

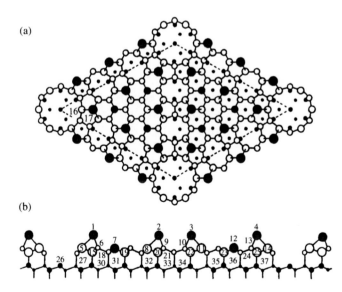

Figure 26.12 The 7×7 reconstruction of the Si(111) surface: (a) shows a view from above the surface, whereas (b) shows one from the side. Reproduced with permission from Brommer *et al.* (1992). Copyright 1992 by the American Physical Society

26.5 ADSORBANTS AND CATALYSIS

The existence of the dangling bonds may have other consequences than that of leading to surface reconstructions. Due to these bonds the surface may be highly reactive and either easily be covered with other materials or take active part in chemical reactions at the surface. Above, in Fig. 26.3, we saw an example of an Si(111) surface completely covered with either one or two layers of K atoms.

Figure 26.13 shows another example where the structural changes of a Si(111) surface are monitored as a Cl_2 molecule approaches the surface. In some cases (e.g., the three upper ones), the Cl_2 molecule dissociates into two Cl atoms that form bonds to the surface atoms at two different places, whereas in other cases (the two lower ones) the Cl_2 molecule stays more or less intact on the surface.

In this case we see that the surface leads to changes in the chemical bonds of the adsorbed molecules. This means that the surface can be used in breaking or creating chemical bonds. Thereby, molecules that in the gas phase will not react with each other, may do so on the surface, i.e., the surface becomes a catalyst.

There are other cases where the interaction of a surface with other systems is interesting. Thus, one may use the surface as a substrate for depositing two-dimensional layers of other materials. Since the electrons will be confined to a very short distance in the direction perpendicular to the film, special, so-called quantum-confinement effects lead to unusual properties for such systems compared with those of three-dimensional crystals.

In Fig. 26.3 we saw the example of K atoms deposited on the Si(111) surface. Figure 26.14 shows the two-dimensional band structures for a monolayer of As on the Ge(111) surface. In this study, a density-functional method has been applied for a supercell structure. But in addition, the quasi-particle energies that were discussed in Section 22.4 have been calculated since these provide more accurate comparisons with experiments as also seen in the figure.

Surfaces are often studied experimentally with the help of various types of tunnelling microscopes. Then, a tip is placed very close to the surface, and the tunnelling current flowing between the tip and the surface is measured as a function of the position of the tip along the surface. It is of interest to be able to simulate the outcome of such an experiment using theoretical methods, too, and in Fig. 26.15 we show the examples of such studies using four different geometries of the tip (i.e., assuming that the tip is formed by one, two, or three atoms with specific geometries). It is seen that the outcome does depend sensitively on the tip geometry, indicating that one has to be careful in interpreting the experimental results.

26.6 FILMS

One of the basic problems from a theoretical point of view when treating sur- or interfaces is that their existence leads to a lowering of the symmetry, but that

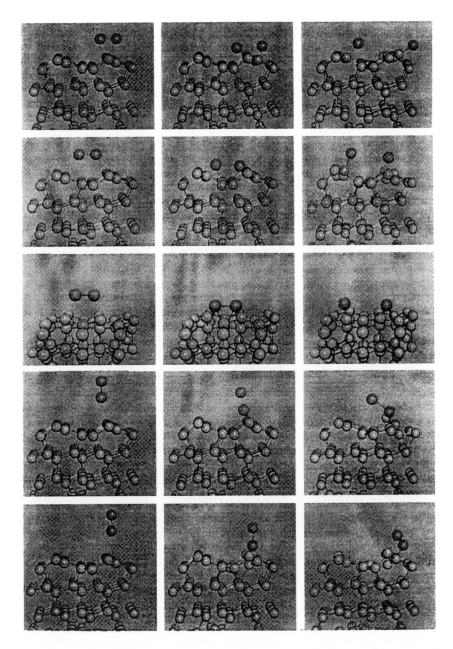

Figure 26.13 Five different scenarios (each horizontal set of three panels corresponds to one scenario) of what happens when a Cl_2 molecule (recognizable in the left panels) approaches a 2×1 reconstructed Si(111) surface. The left, middle, and right panel shows in each case the initial configuration, that just after the collision, and the final configuration, respectively. Reproduced with permission From De Vita *et al.* (1993). Copyright 1993 by the American Physical Society

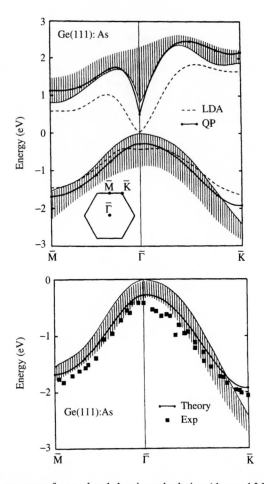

Figure 26.14 Band structures from a local-density calculation (denoted LDA) and from a quasi-particle calculation (denoted QP) for a monolayer of As atoms on the Ge(111) surface. The lower panel shows a comparison between experimental results and those of the quasi-particle calculation. The insert in the upper panel shows the two-dimensional Brillouin zone. Reproduced with permission from Hybertsen and Louie (1987). Copyright 1987 by the American Physical Society

the system remains being (approximately) infinite. We saw in Section 26.2 that through the construction of supercells one could recover the three-dimensional translational symmetry.

Another possibility is to keep the two-dimensional symmetry as that of the surface- or interface-containing system, but making the system finite in the third direction, so that it thereby becomes tractable. This corresponds to constructing systems as those of Figs. 26.1(e) and 26.1(f). Then, for the surface, one has in fact two and not one surface that may, but need not, be equivalent. If one ultimately wants to study the single surface it is desirable that the two surfaces

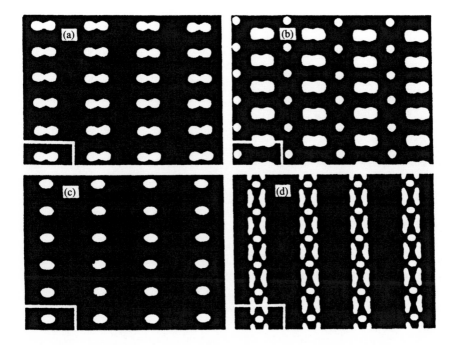

Figure 26.15 Simulated atomic-force-microscopy images of the Si(100) 2 × 1 reconstructed surface for different tips. Reproduced with permission from Abraham *et al.* (1988). Copyright 1988 by the American Physical Society

are equivalent. But consider, for instance, a GaAs crystal. One may construct surfaces where *all* surface atoms are either Ga or As atoms, and there is no need to assume that these surfaces will be equivalent. On the other hand, by constructing a system like that of Fig. 26.1(f) with both surfaces having only Ga atoms, there will be more Ga than As atoms, and, therefore, for stoichiometric reasons, one may prefer having, two different surfaces.

A further point is that in order to simulate a surface by studying a system like that of Fig. 26.1(f), the system should be made so thick that the inner parts resemble those of an infinite crystal and that the surface states of the two different interfaces do not interact with each other. Finally, for the interface, the construction of the system of Fig. 26.1(e) for the interface leads not only to one interface but also to two surfaces.

The systems of Figs. 26.1(e) and 26.1(f) are known either as slabs or as films. They are, however, not only interesting as approximate systems for surfaces or interfaces but films can exist themselves; rarely as freestanding objects, but more often deposited on some substrate. When the interactions between the substrate and the film materials are weak, it is a good approximation to consider the isolated film and study its properties without taking the effects of the substrate into account.

26.7 INTERFACES AND BAND OFFSETS

So far we have concentrated more on how one may study interfaces or surfaces theoretically and have given some few examples, but have not gone into the physical properties of the interfaces and surfaces. There are, however, a few points that deserve some extra attention, and these will be discussed here.

A first point is that we have had to assume that the two lattices of the two subsystems match. Without this assumption, the periodicity would also be destroyed parallel to the interface and the calculations would not have been possible. Often this assumption is reasonable, but there may be deviations that may change some of the conclusions that are obtained through the assumption of lattice-matching. Alternatively, often such interfaces are produced by starting out with the surface of one of the materials and gradually depositing more and more atoms of the other material. As long as the thickness of the deposited material is small, then that material may possess a structure that to a larger extent is determined by that of the first material than by that of the deposited material. That is, the deposited material may become strained.

As the second point let us consider an interface between two materials A and B. Considering each material individually we may have any of the cases of Fig. 26.16. When both materials are metals, as in Fig. 26.16(a), the Fermi level for each material separately will cut through the band regions at some point. It is very unlikely that the two Fermi energies will be identical *when* referred to a common energy zero. This means that in the general case we will have a situation like that of Fig. 26.16(a), where the Fermi level of one material is below that of the other. Upon joining the two systems, there will be a flow of electrons from the system with the higher Fermi energy to that with the lower Fermi energy (i.e., from A to B in the figure). This flow of electrons will accumulate near the interface and create a so-called dipole field that is of long range and ultimately prevent further flow.

When one of the two materials is a metal whereas the other is a semiconductor, different situations may occur. Either the Fermi energy of the metal lies in the band gap between occupied and empty bands for the semiconductor [Fig. 26.16(b)]. In that case, essentially no flow of electrons takes place. Alternatively, the Fermi energy of the metal may lie in either the conduction band [Fig. 26.16(c)] or the valence band [Fig. 26.16(d)] of the semiconductor, and, as above, an electron flow between the two systems will occur, leading to the creation of a dipole field. Such systems are known as Schottky junctions.

Finally, we may consider the case that the two materials are both semiconductors, so that one of the two situations of Figs. 26.16(e) or 26.16(f) may occur.

Here, we shall discuss the situation of Fig. 26.16(e) in some detail. Through the creation of the junction, the two Fermi levels will line up. However, since this lining-up will not move unoccupied orbitals to below the Fermi level or occupied ones to above the Fermi level, there will be no electron flow between the two materials. Neglecting the local variations in the densities of the states (i.e., the

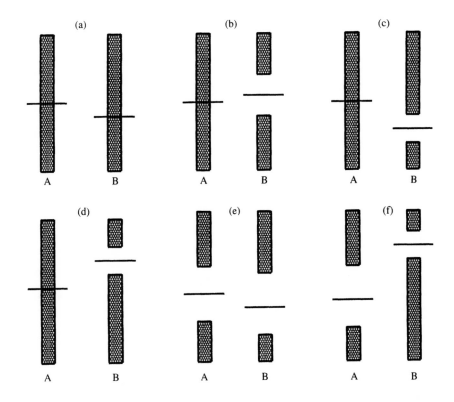

Figure 26.16 Six different examples of the energy-band regions of two materials, A and B, before joining them via an interface: (a) corresponds to two metals, (b)–(d) to one metal and one semiconductor, and (e)–(f) to two semiconductors. For each panel the crosshatched bars mark the energies of the bands (on a vertical energy scale), and the horizontal lines represent the Fermi level for each subsystem individually

fact that next to the interface the potential felt by the electrons is slightly different from that in the bulk, so that the energy levels there will be slightly different), we obtain a situation like that of Fig. 26.17. VB and CB mark the valence and conduction bands, respectively, and the common Fermi level is marked by ε_F. The top of the valence bands of the two components have different energies, as have the bottom of the conduction bands. These differences define the band offsets $\Delta\varepsilon_v$ and $\Delta\varepsilon_c$, shown in Fig. 26.17, too. They may take both negative and positive values.

By defining them according to

$$\Delta\varepsilon_v = \varepsilon_v(A) - \varepsilon_v(B) \qquad (26.16)$$

and

$$\Delta\varepsilon_c = \varepsilon_c(A) - \varepsilon_c(B), \qquad (26.17)$$

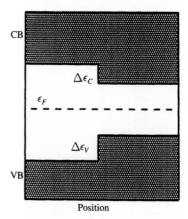

Figure 26.17 Schematic representation of the bands for a junction of two semiconductors. The dashed line mark the Fermi level, and the different band offsets are shown. The horizontal axis represents a position coordinate perpendicular to the interface

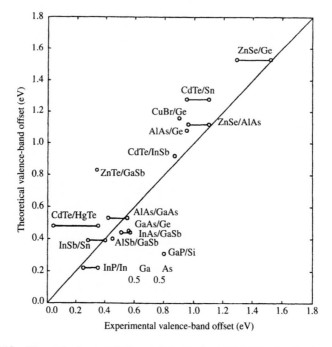

Figure 26.18 Theoretical (vertical scale) and experimental (horizontal scale) valence-band offsets for a number of semiconductor interfaces. In many cases there are more experimental results, which leads to the broader ranges of these. Reproduced with permission from Lambrecht *et al.* (1990). Copyright 1990 by the American Physical Society

where $\varepsilon_v(X)$ and $\varepsilon_c(X)$ are the energies at the top of the valence band and the bottom of the conduction band, respectively, of material X, only the difference

$$\Delta\varepsilon_c - \Delta\varepsilon_v = \varepsilon_c(A) - \varepsilon_c(B) - \varepsilon_v(A) + \varepsilon_v(B)$$
$$= E_{gap}(A) - E_{gap}(B) \qquad (26.18)$$

is determined from the properties of the pure materials without the interface. In Eq. (26.18), $E_{gap}(X)$ is the band gap of material X.

A central question is whether the band offset between two semiconductors is a unique function of the separate materials or, alternatively formulated, whether knowledge of the band offsets between materials A and B and between materials B and C can be used in determining that between A and C. This is an important question for device technologies, since the properties of the materials with such junctions (interfaces) depend very sensitively on these band offsets.

There have been some studies where a number of band offsets between different interfaces have been studied and where a 'transferability rule' as that above have been checked. So far no definite conclusions have been obtained and therefore we shall not discuss this subject further here. However, in order to illustrate the capability of the theoretical methods, we show in Fig. 26.18 experimental and theoretical valence band offsets as calculated with a density-functional method. Since density-functional methods in general predict too small band gaps (cf. Section 22), only valence band offsets are presented.

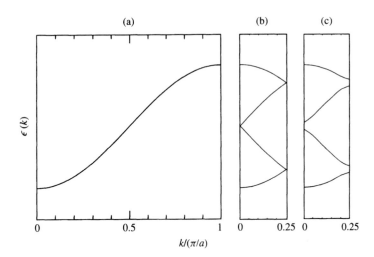

Figure 26.19 The principles behind the band folding: (a) shows a single band for a chain with one atom per unit cell; (b) shows the same band but when treating the system as having four atoms per unit cell; (c) shows how the reduction of symmetry leads to the opening of gaps

26.8 SUPERLATTICES

In Section 26.2 we discussed how the construction of supercells could make the calculation of the electronic structure of interface-containing systems tractable. To this end it was important that the supercells were made so large that the introduction of periodically repeated interfaces did not obscure the results, i.e., that the orbitals located at the different interfaces did not interfer.

In some sense one takes the absolutely opposite point of view by the construction of superlattices. These are artificially, experimentally produced periodic (or, in some cases, non-periodic; here we concentrate on the periodic ones) structures consisting of a repetition of a well-defined number (n) of layers of one material, A, followed by a well-defined number (m) of layers of another material, B. The material has accordingly the stoichiometry $A_n B_m$, and the $n + m$ layers are repeated periodically. In some cases the A or B systems consist themselves of

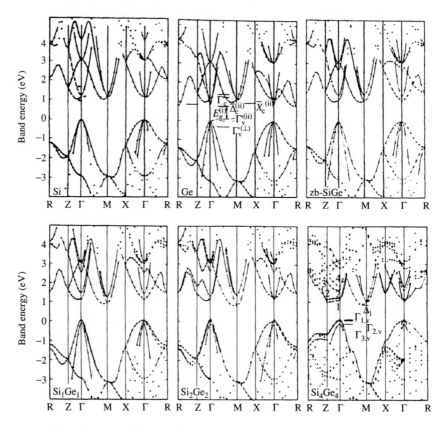

Figure 26.20 Band structures for Si_4, Ge_4, and a hypothetic zincblende SiGe structure (upper panels) together with those for Si_1Ge_1, Si_2Ge_2, and Si_4Ge_4 superlattices (lower panels). Reproduced with permission from Ciraci and Batra (1988). Copyright 1988 by the American Physical Society

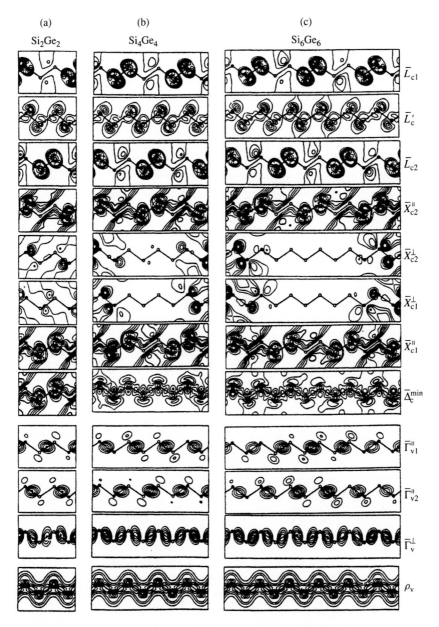

Figure 26.21 The electron density of different orbitals for Si_2Ge_2, Si_4Ge_4, and Si_6Ge_6 superlattices. Those given an index 'c' are empty, whereas those with a 'v' are occupied. The lowest panels give the total valence electron density. Si atoms are marked by black circles, Ge atoms by white circles. From Froyen *et al.* (1988). Copyright 1988 by the American Physical Society

more types of atoms; one of them could for instance be GaAs. For not too large n and m the fact that the interface states of different interfaces interact makes the electronic properties of the superlattice different from those of the individual components, A and B. By varying A and B, or the number of layers n and m, or the surfaces that form the interfaces, one can thereby to some extent manipulate the electronic properties of the superlattice, and one has entered the field of band-structure engineering.

In order to understand some of the properties of the superlattices we need to recall the principles behind band folding (cf. Section 19.3). We consider a hypothetic linear chain with one atom per unit cell. Assuming that we have only one energy band, this band may look like the one of Fig. 26.19(a). We could, however, also have treated the system as having not one but four atoms per unit cell. Thereby the first Brillouin zone would have a size of only a quarter of that of the original system, and the band structures will be obtained by folding those of Fig. 26.19(a), which leads to band structures like those of Fig. 26.19(b). There is no fundamental difference between the band structures of Fig. 26.19(a) and those of Fig. 26.19(b). The fact that the system possesses a higher symmetry than having four atoms per repeated system can in Fig. 26.19(b) be recognized through the degeneracies of the bands at $k = 0$ and $k = (\pi/4a)$. However, when the symmetry is lowered so that the system no longer has the higher symmetry, small band gaps will appear as in Fig. 26.19(c).

For the superlattices the most direct outcome of having the larger unit cell is according to the discussion above that there will be many more bands. Subsequently, when parts of the atoms are substituted periodically by others that are chemically similar to the original ones the smaller band gaps may open up. The band folding will bring orbitals that without the band folding correspond to different $\vec{k}$ vectors, to the same $\vec{k}$. This means that optical transitions that before were forbidden due to the requirement that the initial and final state should have the same $\vec{k}$ now may become allowed. However, matrix-element effects may still make them very small, and ultimately, when the larger unit cell is not dictated by a symmetry reduction but is introduced only for mathematical reasons, the forbidden transitions will remain forbidden due to vanishing transition-matrix elements.

In Fig. 26.20 we show as one example how the band structures may change when forming superlattices. In this case A and B correspond to Si and Ge, respectively, and for the sake of comparison a hypothetic zincblende SiGe structure was also considered. The increased complexity of the band structures is immediately observed, but it is also recognized how the main features of the bands for the pure materials survive for the superlattices. Finally, the electron density of various orbitals for such superlattices are shown in Fig. 26.21.

27 Non-Periodic, Extended Systems

27.1 AMORPHOUS SYSTEMS

Perfect crystalline materials are characterized by having translational symmetry in all three directions. Alternatively expressed, they possess long-range order (as well as short-range order). The introduction of impurities or surfaces destroys the translational symmetry, but the long-range order persists over major parts of the system.

The situation is different for amorphous systems. These have no long-range order and, in consequence, no translational symmetry. But in contrast to molecular systems, they are extended; i.e., approximately infinite. The lack of translational symmetry prevents the use of symmetry-adapted Bloch waves as basis functions and the introduction of a $\bar{k}$ vector.

In order to calculate the electronic properties of the amorphous systems despite these problems, one may use one out of more approximate approaches.

One approximation, which we saw exemplified in Section 20.5 for the case of amorphous selenium, is to construct a periodic structure from repeated large unit cells, where the unit cells are supposed to resemble a smaller part of the amorphous system. Thereby, one introduces some long-range order, but the hope is that by having sufficiently large unit cells the intermediate-range order (i.e., the order at longer distances than the nearest-neighbour bond lengths but at shorter distances than the size of the repeated units) is accurately described. Due to the requirement of having large unit cells this approach is computationally heavy, in particular when applying parameter-free methods.

Alternatively, one may approximate the infinite amorphous system by a finite one. Once again the system should be reasonably large in order to being able to describe the intermediate-range order. But in addition one introduces surface atoms for which dangling bonds may pose a problem (i.e., bonds towards non-existing neighbours so that the bonds are unsaturated). A common procedure against this problem is to saturate the dangling bonds through the introduction of additional atoms. Very often hydrogen atoms are placed at these dangling bonds. When applying parameter-free methods also these approaches become computationally involved.

Finally, one may abandon the idea of using parameter-free methods and instead using approximate ones like the (extended) Hückel model or related single-particle models (these are often called tight-binding models, since they assume that the Hamilton matrix elements vanish except for near neighbours) when focusing on the electronic properties. Alternatively, one may use force-field models (to be described in Chapter 30) when focusing at structural and/or dynamical properties. In both cases, the parameters that enter the models may be obtained either by fitting to experimental information or by fitting to results from parameter-free studies on some pre-selected simpler model systems.

With the simpler semiempirical models one can treat more complicated structures with a larger number of atoms (either per unit cell when considering periodically repeated units or in total when considering finite systems). It may then be important that the atoms are placed so that their structure resembles that of the amorphous system and so that the calculated properties are realistic. To this end, there exist in some cases (i.e., for some amorphous systems) artificially constructed structures that are considered good approximations to that of the infinite amorphous system.

27.2   LIQUIDS

The basic problems related to the treatment of liquids are as those of treating amorphous systems. That is, the systems are extended but contain no long-range order. Accordingly, one may apply the same approximate methods as for amorphous systems when treating liquids. We shall therefore not discuss these further here.

One example of the application of these methods was presented in Section 20.5. There, we showed examples of Car–Parrinello calculations on amorphous and liquid selenium, where repeated units consisting of 64 atoms were considered.

A special case is that of water. The bonds between the units (the water molecules) are not as strong as common covalent bonds, but are instead weaker hydrogen bonds. Hydrogen bonds have typically energies of the order of $0.1–0.5$ eV per bond. The hydrogen bonds are so weak that the properties of the individual water molecules are very similar to those of a single, isolated water molecule. On the other hand, some properties can only be understood when considering a whole sample of water molecules.

Since the hydrogen bonds are so weak, it is very difficult to describe structures containing hydrogen bonds accurately: varying the hydrogen-bond lengths by large amounts leads often to only minor changes in the total energy. Therefore, small inaccuracies (e.g., due to incomplete basis sets or to approximations when applying a density-functional scheme) may easily result in large changes in the lengths of the hydrogen bonds.

It has turned out that when using density-functional methods, the hydrogen bonds are not described correctly within a local-density approximation. The generalized-gradient approximations often perform well, and so do the hybrid methods. As a demonstration of the latter we show in Fig. 27.1 theoretically obtained pair-correlation functions for water together with the equivalent experimental results. The pair-correlation function $g_{AB}(r)$ is so defined that $g_{AB}(r)\,dr$ gives the average number of neighbours of type B to an A atom within the interval $[r, r + dr]$. The results in Fig. 27.1 were obtained by using the Car–Parrinello method for a system containing repeated units of 32 water molecules and using a hybrid density functional.

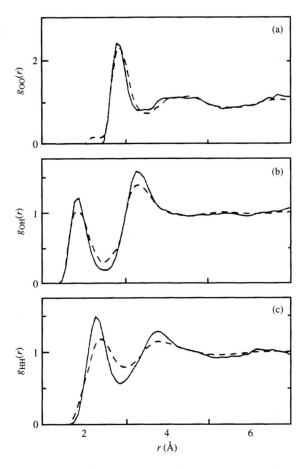

Figure 27.1 Pair correlation functions for liquid water obtained theoretically (solid curves) and compared with experiment (dashed curves). Reproduced with permission from Sprik *et al.* (1996)

A further reason for being interested in water as a liquid is due to its very common role as a solvent. The discussion of this will, however, be postponed until Chapter 32.

27.3 QUASICRYSTALS

In contrast to amorphous or liquid systems, quasicrystals are deterministic. This means that the positions of the individual atoms are well defined. On the other hand, the quasicrystals do not possess any long-range order (their construction

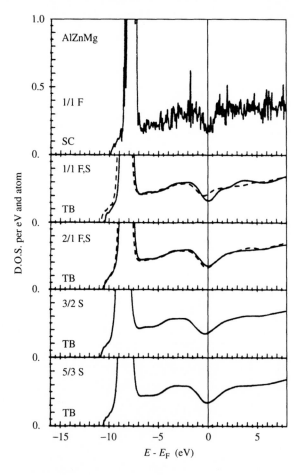

Figure 27.2 Electronic density of states relative to the Fermi level for different periodic approximations to the quasicrystals. These contain 162, 688, 2920, and 12 380 atoms per unit cell (the four lowest panels). The uppermost panel has been obtained using the standard method for the system of 162 atoms per unit cell, whereas the others have been obtained by the so-called recursion method. Reproduced with permission from Hafner and Krajčí (1992). Copyright 1992 by the American Physical Society

may, however, be considered projections onto the ordinary three-dimensional space of periodic structures in higher dimensions, but this is beyond the scope of the present manuscript). Therefore, theoretical studies of their properties are made complicated due to exactly the same facts that made those of amorphous and liquid systems complicated.

The most common approach for studying their properties is to consider so-called periodic approximants. These are periodic crystalline materials whose unit cell can be made gradually larger, whereby the material systematically approaches the real quasicrystalline material.

In Fig. 27.2 we show the calculated density of states from such a study. Systems with up to 12 380 atoms per unit cell were considered, which was possible only by using the so-called recursion method (Heine *et al.*, 1980). This method, whose detail will not be described here, is explicitly constructed for calculating the diagonal parts of the Green's function, $G(\vec{x}, \vec{x}, E)$ [i.e., not the off-diagonal parts $G(\vec{x}, \vec{x}', E)$] for large systems.

One of the interesting questions about quasicrystals is whether the fact that they are neither random nor periodic leads to special physical properties. In order to study this question in detail, it has become popular to study one-dimensional systems with the same characteristics; e.g., the Fibonacci or the Thue–Morse systems. Studies of their properties are often performed using the transfer matrices described in Section 25.4.

27.4 ALLOYS

Amorphous or liquid systems may consist of just one type of atom, but the structure is such that the system does not contain any long-range order, i.e., the translational symmetry is lacking. For an alloy, on the other hand, the atoms may sit (approximately) at regular positions, but these are occupied more or less randomly by more different types of atoms. In the simplest case we have two types of atoms and arrive at a situation like that of Fig. 27.3. We shall here restrict ourselves to the case of a two-component alloy, i.e., a system of the type $A_x B_{1-x}$. This is exactly the type of system we have shown in Fig. 27.3.

In calculating the electronic properties of such a system, the main problem is that the potential from the nuclei (and eventually also the core electrons) is not translationally symmetric. The potential from the electron–electron interactions is translationally symmetric. On the other hand, the exchange-correlation potential within a density-functional method may not be translationally invariant due to local variations in the electron density. But let us for the moment assume that these variations can be incorporated into some effective atom-centred potentials, so that the total potential can be written as

$$V(\vec{r}) = V_{\text{periodic}}(\vec{r}) + \sum_{\vec{R}} \left[\left(\frac{1}{2} + \sigma_{\vec{R}} \right) V_A(\vec{r} - \vec{R}) + \left(\frac{1}{2} - \sigma_{\vec{R}} \right) V_B(\vec{r} - \vec{R}) \right],$$

$$(27.1)$$

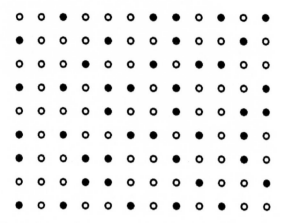

Figure 27.3   Schematic representation of a two-dimensional alloy with two types of atoms

where

$$\sigma_{\vec{R}} = \begin{cases} \frac{1}{2} & \text{if A atom at } \vec{R} \\ -\frac{1}{2} & \text{if B atom at } \vec{R} \end{cases} \tag{27.2}$$

and the $\vec{R}$ summation in Eq. (27.1) is over all atomic positions. V_A and V_B in Eq. (27.1) are the effective atomic potentials that only depend on the type of atom. Finally, V_{periodic} is the periodic part of the potential. In Eq. (27.1) we neglect variations in the atomic potentials due to differences in their surroundings, for instance that V_A may be different for an A atom that is completely surrounded by other A atoms compared with an A atom surrounded by only B atoms.

Let us now restrict ourselves to density-functional theory. Then the electronic properties of the alloy are calculated by solving the Kohn–Sham equations

$$\left[-\tfrac{1}{2} \nabla^2 + V(\vec{r}) \right] \psi_i(\vec{r}) = \varepsilon_i \psi_i(\vec{r}). \tag{27.3}$$

For the Hartree–Fock method the equations will be slightly different, but the main principles will remain unchanged.

Solving Eq. (27.3) is very complicated due to the lack of periodicity. The simplest approximation one can think of is to construct some fictive periodic crystal with an effective periodic potential like that of Fig. 27.4. This is the case for the so-called *virtual-crystal approximation*. Here, the potential of Eq. (27.1) is replaced by

$$\tilde{V}(\vec{r}) = V_{\text{periodic}}(\vec{r}) + \sum_{\vec{R}} \tilde{V}_{AB}(\vec{r} - \vec{R}) \tag{27.4}$$

where

$$\tilde{V}_{AB}(\vec{r} - \vec{R}) = x \cdot V_A(\vec{r} - \vec{R}) + (1 - x) \cdot V_B(\vec{r} - \vec{R}) \tag{27.5}$$

for the $A_x B_{1-x}$ alloy. Furthermore, the total number of electrons per unit cell is

$$\tilde{n}_{AB} = x \cdot n_A + (1 - x) \cdot n_B \tag{27.6}$$

with n_A and n_B being the number of electrons per A and B atom, respectively.

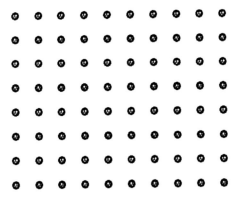

Figure 27.4 The virtual-crystal approximation

Figure 27.5 A periodic A_3B structure with A and B being the black and white circles, respectively

The idea behind this approach is that the electrons *on the average* feel an effective potential like that of Eq. (27.5), when the alloy is considered very large (infinite) and absolutely random. Therefore, this approach may be useful when considering random alloys, but one would expect it to fail for other cases when the alloy is not absolutely random. For example, for $x = 0.75$ we can construct the periodic structure A_3B of Fig. 27.5. Its electronic properties will most likely not be described accurately within the virtual-crystal potential for the $A_{0.75}B_{0.25}$ alloy.

A related approximation is the so-called *coherent-potential approximation (CPA)*. To this end we consider the Green's function that is related to Eq. (27.3), i.e.,

$$\left[-\tfrac{1}{2}\nabla^2 + V(\vec{r}) - \varepsilon\right] G(\vec{r}, \vec{r}', \varepsilon) = -\delta(\vec{r} - \vec{r}'). \tag{27.7}$$

For an infinite, periodic crystal the Green's function satisfies

$$G(\vec{r} + \vec{R}, \vec{r}' + \vec{R}, \varepsilon) = G(\vec{r}, \vec{r}', \varepsilon), \qquad (27.8)$$

where $\vec{R}$ is an arbitrary vector that belongs to the set of translation vectors that map the system onto itself. As we saw in Chapter 17, a number of fundamental quantities can be calculated directly from the Green's function (most notably the electron density and the density of states).

In the coherent-potential approximation one assumes — as for the virtual crystal approximation — that there exists some periodic crystal so that the averaged Green's function for the alloy satisfies the condition (27.8). For the virtual crystal approximation this potential was constructed via Eq. (27.5). For the coherent potential approximation one uses a somewhat different potential.

If we define the free-electron Green's function G_0 through

$$\left[-\tfrac{1}{2} \nabla^2 - \varepsilon \right] G_0(\vec{r}, \vec{r}', \varepsilon) = -\delta(\vec{r} - \vec{r}'), \qquad (27.9)$$

then the Green's function G for the alloy can be determined through Dyson's equation

$$G(\vec{r}, \vec{r}', \varepsilon) = G_0(\vec{r}, \vec{r}', \varepsilon) + \int G_0(\vec{r}, \vec{r}'', \varepsilon) V(\vec{r}'') G(\vec{r}'', \vec{r}', \varepsilon) \, d\vec{r}''. \qquad (27.10)$$

Since the wavefunctions for free electrons are well known (they are plane waves), G_0 is well known, too.

We define now the so-called T matrix through

$$\int G_0(\vec{r}, \vec{r}'', \varepsilon) V(\vec{r}'') G(\vec{r}'', \vec{r}', \varepsilon) \, d\vec{r}''$$

$$= \int \int G_0(\vec{r}, \vec{r}'', \varepsilon) T(\vec{r}'', \vec{r}''', \varepsilon) G_0(\vec{r}''', \vec{r}', \varepsilon) \, d\vec{r}'' \, d\vec{r}'''. \qquad (27.11)$$

(Notice that the free-electron Green's function G_0 occurs twice on the right-hand side but only once on the left-hand side). The T matrix describes thus the modifications of the electronic structure due to the potential, i.e., the changes compared with the free-electrons system due to the atoms.

For the coherent-potential approximation we replace T by

$$\tilde{T}(\vec{r}, \vec{r}', \varepsilon) = x \cdot T_A(\vec{r}, \vec{r}', \varepsilon) + (1 - x) \cdot T_B(\vec{r}, \vec{r}', \varepsilon), \qquad (27.12)$$

where T_A and T_B are the T matrices for the pure A and B systems, respectively. There are some important differences compared with the virtual crystal approximation, but the main idea is the same: to study some periodic approximation to the alloy. One can, however, argue that the coherent-potential approximation is more accurate than the virtual-crystal approximation, but except for giving a couple of examples this will not be discussed further here.

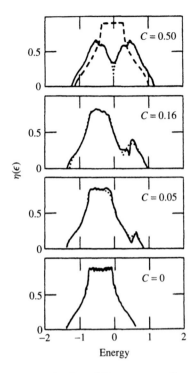

Figure 27.6 The density of states for four different cases of an alloy A_cB_{1-c} treated with a model that can be studied both exactly and with different approximations. The full curve correspond to the exact results, the dash-dotted curve to the virtual-crystal approximation, and the dotted curve to the coherent-potential approximation. Reproduced by permission of Plenum Press from Stocks and Winter (1984)

In Fig. 27.6 we show the density of states for a simple model system A_cB_{1-c} calculated both exactly and with the virtual-crystal and the coherent-potential approximation. Here, it is clearly seen that the latter is a better approximation than the former. This is confirmed in Fig. 27.7 where we show similar results for Ag_cPd_{1-c} compared with experimental results. We shall, however, not go into further discussion of this point here.

A different approach is one where statistical methods are applied for the random distribution of the two types of atoms in the alloy. The idea is that for *any* given alloy any site is occupied by either an A atom or a B atom. We can thus use the variables $\sigma_{\vec{R}}$ of Eq. (27.2) in describing the precise structure of the alloy. Subsequently we can seek to write the total energy (or any other quantity) of the alloy as a function of these variables, which actually are defined as being equivalent to spin variables.

In the simplest approximation one may use an Ising model,

$$E = \sum_{\vec{R}} \varepsilon_{\vec{R}} \sigma_{\vec{R}} + \sum_{\vec{R}_1, \vec{R}_2} J_{\vec{R}_1, \vec{R}_2} \sigma_{\vec{R}_1} \sigma_{\vec{R}_2}, \qquad (27.13)$$

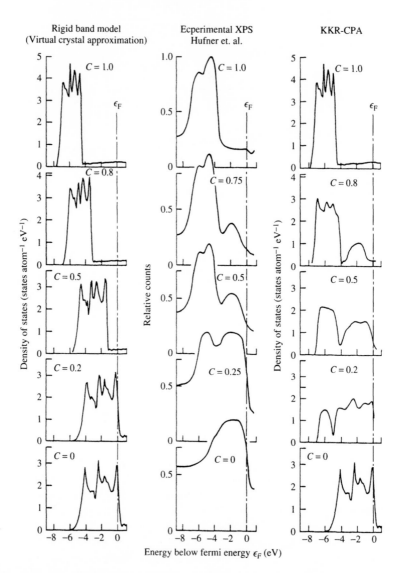

Figure 27.7 The density of states for five different cases of Ag_cPd_{1-c} alloys. The left panels show the virtual-crystal results, the middle ones experimental results, and the right ones the coherent-potential results. Reproduced by permission of Plenum Press from Stocks and Winter (1984)

where one assumes that the interaction energies $J_{\vec{R}_1, \vec{R}_2}$ are non-vanishing only for $\vec{R}_1$ and $\vec{R}_2$ being nearest neighbours. Furthermore, in that case J and ε may be assumed to be constants.

As a simple illustration of the meaning of the Hamilton operator of Eq. (27.13) we consider two nearest neighbours. Depending on whether these are both A atoms, both B atoms, or one A and one B atom, they will give a total-energy contribution of $\varepsilon + \frac{1}{4}J$, $-\varepsilon + \frac{1}{4}J$, or $-\frac{1}{4}J$. That is, the expression of Eq. (27.13) is able to discriminate between the three cases. It does, however, not include effects directly due to different surroundings of this pair of atoms.

Most often for a real system these approximations are too crude and one needs to go beyond them. Nevertheless, the general principles remain. In the general case one considers so-called cluster expansions (Sanchez et al., 1984). These are constructed as follows.

We consider a system consisting of M lattice points, $\vec{R}_1, \vec{R}_2, \ldots, \vec{R}_M$. Ultimately, M will be very large, but for the moment we consider it arbitrary. At each of these lattice points we have an atom. For our binary alloy $A_x B_{1-x}$ it may be either an A or a B atom, but in the general case we can have m different types of atoms. In order to describe what type of atom occupies a specific site, we introduce spin-like variables σ_n that are constructed from

$$\sigma_n = \begin{cases} \frac{m-1}{2} & \text{if atom type 1 at } \vec{R}_n \\ \frac{m-1}{2} - 1 & \text{if atom type 2 at } \vec{R}_n \\ \cdots & \\ -\frac{m-1}{2} & \text{if atom type } m \text{ at } \vec{R}_n. \end{cases} \qquad (27.14)$$

We stress that for our arguments we need not require that the atoms sit exactly at the lattice points but allow for relaxations as long as we still can ascribe each atom a specific site.

Any quantity will depend on the total configuration of the atoms, i.e., of the complete M-dimensional vector

$$\vec{\sigma} = (\sigma_1, \sigma_2, \ldots, \sigma_M). \qquad (27.15)$$

This means, e.g., that interchanging two atoms will change the quantity of interest. However, it seems to be reasonable to assume that the dominant contribution comes simply from the number of A and B atoms (for a binary alloy, when we consider this as for reference). However, the local properties of an A atom will depend on whether its neighbours are all B atoms or all A atoms or a mixture. Therefore, there will be a secondary effect that depends on atom pairs, i.e., for the binary alloy it will depend on the number of AA pairs, that of BB pairs, and that of AB pairs. In Fig. 27.8 we show some examples of AA and AB pairs and by looking at the figure it may be clear that a third-order effect will arise due to the fact that the different pairs have different surroundings, i.e, there will be effects that depend on the different triples of atoms. Here, not only their numbers of A and B atoms are relevant but also their precise structure (a linear ABB triple will be different from a linear BAB triple, and there will be differences depending on whether the triples are linear or bent). Continuing these arguments, we will include more and more interactions, but at some point we will assume that the effects are so small that they can be neglected. In the Ising model we have only two-body effects between nearest neighbours but in the general case we want

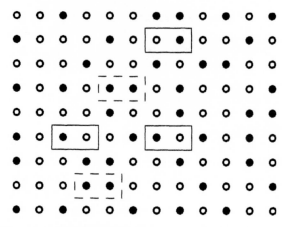

Figure 27.8   The same system as in Fig. 27.3

to go beyond those and accordingly include both more distant interactions and interactions among more sites.

For a given system of M sites where each can be occupied by one out of m different types of atoms, the quantity of interest will depend on (at most) m^M variables (corresponding to a dependence on which of the m atom types occupies each of the M sites). That is, for any set of these variables the quantity of interest, F, will take a specific value. In total,

$$F = F(\sigma_1, \sigma_2, \ldots, \sigma_M) = F(\vec{\sigma}). \qquad (27.16)$$

We seek now a complete set of orthonormal functions Π_i that depend on the spin functions of Eq. (27.14), so that

$$F = \sum_{i=1}^{m^M} \Pi_i(\vec{\sigma}) f_i. \qquad (27.17)$$

(Since we have m^M different possible values of $\vec{\sigma}$, we must have m^M different functions Π_i.) The idea is that these functions are constructed so that some of them depend only on one single spin variable, some only on two spin variables (e.g., those of two neighbouring sites), and so on. This means that the very large summation in Eq. (27.17) hopefully can be truncated after a smaller number of terms that contain the essential contributions.

The first step is to construct a set of orthogonal functions for each site. These functions depend on the spin of the site of interest. Since we have m different values of the spin variable, we can construct m different orthogonal functions. We choose the m first Chebychev polynomials $\Theta_n(\sigma_i)$, which in the case $m = 2$ are

$$\Theta_0(\sigma_i) = 1$$

$$\Theta_1(\sigma_i) = 2\sigma_i. \qquad (27.18)$$

The condition of orthogonality is explicitly

$$\frac{1}{m} \sum_{\sigma} \Theta_{n_1}(\sigma)\Theta_{n_2}(\sigma) = \delta_{n_1,n_2}, \tag{27.19}$$

where σ takes the m values of Eq. (27.14).

The next step is to use these functions in constructing functions that depend on one or more sites but that are still orthogonal. We consider accordingly a set of sites $\{n_1, n_2, \ldots, n_p\}$ and for each of the sites the polynomials $\Theta_j(\sigma_{n_i})$. We construct then the functions

$$\Pi_{n_1,n_2,\ldots,n_p;j_1,j_2,\ldots,j_p}(\sigma_{n_1}, \sigma_{n_2}, \ldots, \sigma_{n_p}) = \Theta_{j_1}(\sigma_{n_1})\ldots\Theta_{j_2}(\sigma_{n_2})\ldots\Theta_{j_p}(\sigma_{n_p}). \tag{27.20}$$

By using Eq. (27.19), one can show that these functions are orthogonal. This means that when taking any two of these functions that may be defined for different sets of sites (although the same number) and furthermore may have different polynomials for the various sites, we have

$$\sum_{\sigma_1}\sum_{\sigma_2}\cdots\sum_{\sigma_p} \Pi_{n_1,n_2,\ldots,n_p;j_1,j_2,\ldots,j_p}(\sigma_1, \sigma_2, \ldots, \sigma_p)$$

$$\times \Pi_{m_1,m_2,\ldots,m_p;k_1,k_2,\ldots,k_p}(\sigma_1, \sigma_2, \ldots, \sigma_p)$$

$$= m^p \delta_{n_1,m_1}\delta_{n_2,m_2}\cdots\delta_{n_p,m_p}\cdots\delta_{j_1,k_1}\delta_{j_2,k_2}\cdots\delta_{j_p,k_p}. \tag{27.21}$$

These functions define a complete set of functions.

Setting $p = M$ we have accordingly a complete set of functions for the complete system. Restricting now ourselves to the simple case $m = 2$ (i.e., a binary alloy), the indices j_i of the Chebychev polynomials are either 0 or 1. For those that are equal to 0, the corresponding Θ_0-function is identically equal to 1 [Eq. (27.18)] and they need therefore not be included in the product in Eq. (27.20). We end therefore up with the fact that we construct functions

$$\Pi_{n_1,n_2,\ldots,n_q}(\sigma_{n_1}, \sigma_{n_2}, \ldots, \sigma_{n_q}) = (2\sigma_{n_1}) \cdot (2\sigma_{n_2}) \cdots (2\sigma_{n_q}) \tag{27.22}$$

where the sites $n_1, n_2, \ldots, n_q$ are any set of the complete set of sites. These functions define a complete set of functions and are those we shall insert into Eq. (27.17).

Let us consider a very simple example, i.e., a linear chain of $M = 17$ sites (actually, the number $M = 17$ is not of relevance for our arguments). The sites will be numbered, $n = 1, 2, \ldots, 17$. Let us, as one example, focus on the two sites (1,4). For this pair we define according to Eq. (27.22) the function

$$\Pi_{1,4} = (2\sigma_1) \cdot (2\sigma_4) \tag{27.23}$$

which equals $+1$ if the two sites are occupied by the same type of atoms and otherwise -1. Similarly, we can also study the single sites 1 and 4. For these we define the two functions

$$\Pi_1 = (2\sigma_1)$$

$$\Pi_4 = (2\sigma_4). \qquad (27.24)$$

whose values equal $+1$ or -1 depending on whether the each site separately is occupied by an A or a B atom. Thus, the complete set of the three functions Π_1, Π_4, and $\Pi_{1,4}$ describes both single-site properties and inter-site interactions.

As discussed above, we will expect for physical reasons that only nearest-neighbour interactions are important. This means that we will only need to consider those functions in Eq. (27.22) where the set $\{n_1, n_2, \ldots, n_p\}$ is small and consists of sites that are close to each other. This means that in Eq. (27.17) we are only including a smaller set of the functions. This was exactly what we aimed at, and the method is the so-called *cluster variational method* (Sanchez *et al.*, 1984).

The next question is, how do we determine the coefficients in Eq. (27.17) to the functions Π. The current approach is that of considering more different periodic structures for which it is assumed that they can be described with the *same* coefficients (the so-called cluster expansion coefficients). Most often, the quantity of interest is the total energy per atom and this is then calculated for those different periodic structures using standard band-structure techniques. Subsequently, we attempt to describe these with an expansion like that of Eq. (27.17) including the same subset of terms as will subsequently be used for the alloy. Thereby, also the effects of structural relaxations can be incorporated by assuming that they will be similar for the periodic structures and for the alloy.

Above we studied a simple case of $M = 17$ sites and wrote down explicitly the Π functions for some few cases. We will expand the example and accordingly define *all* the single-site functions

$$\Pi_n = (2\sigma_n) \qquad (27.25)$$

as well as the nearest-neighbour functions

$$\Pi_{n,n+1} = (2\sigma_n)(2\sigma_{n+1}). \qquad (27.26)$$

Assuming that the cluster expansion coefficients for only those functions are non-vanishing and by calling them $\varepsilon/2$ and $J/4$ for the single-site and the pair functions, respectively, we arrive at

$$F = \varepsilon \sum_n \sigma_n + J \sum_n \sigma_n \sigma_{n+1} \qquad (27.27)$$

which is exactly the Ising model of Eq. (27.13). Including more cluster expansion coefficients corresponds accordingly to going beyond the Ising model.

When passing to the alloy, one needs to calculate *averaged* quantities. This means keeping the expansion coefficients in Eq. (27.17) but averaging the functions Π. Thus, for a given cluster (i.e., a given set $\{n_1, n_2, \ldots, n_p\}$; this might, e.g., be a bent triple of nearest neighbours) we consider the averaged function Π obtained by shifting the complete set of sites, $\vec{R}_{n_i} \rightarrow \vec{R}_{n_i} + \vec{R}_s$, where $\vec{R}_s$ is any lattice vector that is constant for all n_i (in the example of the bent triple, this corresponds to placing this bent triple successively at all possible positions of the system, without rotating it). The advantage is now that these averaged functions are known for a completely random alloy. Therefore, any quantity can be calculated easily from knowledge of the cluster expansion coefficients. Moreover, for periodic structures the periodicity makes the calculation of the quantity of interest straightforward.

In Fig. 27.9 we show one example of the application of this method. Here, the formation energies of different intermetallic compounds with various periodic

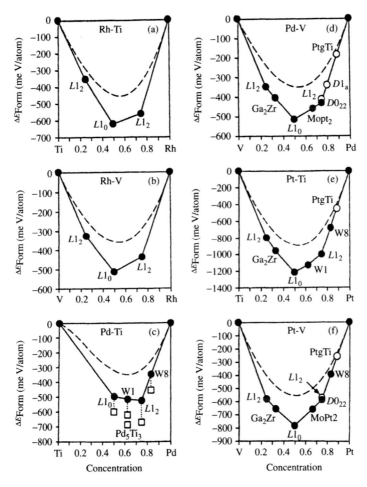

Figure 27.9 Formation energies for f.c.c. structures for different intermetallic systems. The full lines are those of ordered structures, whereas the dashed lines are those of alloys. Reproduced with permission from Wolverton *et al* (1993). Copyright 1993 by the American Physical Society

structures (marked by the different labels that describe the crystal structure) are compared with purely random alloys. In all cases, the ordered structures have lower energies than the alloy.

Also other quantities can be accessed by this method, i.e., the expansion (27.17) is not restricted to the total energy. Figures 27.10 and 27.11 give a couple of representative examples of this.

Finally, this approach is often used in performing theoretical studies of phase diagrams. This will not be discussed here but instead in Section 28.2.

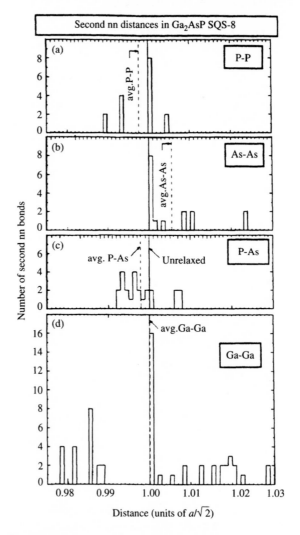

Figure 27.10 Distribution of next-nearest-neighbour distances in $GaP_{0.5}As_{0.5}$. From Wei *et al* (1990). Copyright 1990 by the American Physical Society

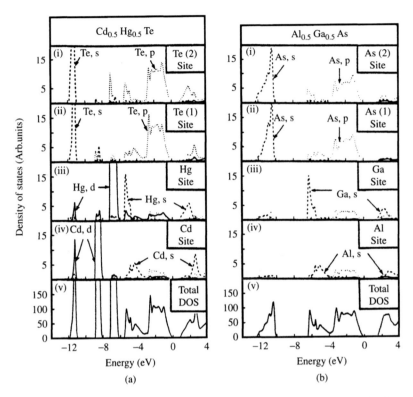

Figure 27.11 Density of states of (left part) $Cd_{0.5}Hg_{0.5}Te$ and (right part) $Al_{0.5}Ga_{0.5}As$. The method that has been applied allows for a decomposition of the total density of states (lowest panels) into different atomic and angular components as shown. From Wei et al. (1990). Copyright 1990 by the American Physical Society

27.5. ORDER-N METHODS

In particular in Sections 27.1 and 27.2 we saw that the quality of the results depends in some cases crucially on being able to treat very large systems as accurately as possible, which is often possible only with great difficulty by electronic-structure methods. Semiempirical (parametrized) models is one solution to this problem, but then the results depend critically on the quality of the model that is employed. Alternatively, one may try to make parameter-free methods work very efficiently for extended systems. In particular, the way the computational requirements depend on the size of the system of interest is a critical issue in this context, and methods that scale linearly with the number of atoms (N) are desirable in contrast to the more conventional N^3-N^7 scalings. The order-N methods are currently being developed as attempts to obtain linear scaling.

We shall here restrict ourselves to density-functional methods, but the principles are also applicable to Hartree–Fock methods.

We solve the Kohn–Sham equations

$$[-\tfrac{1}{2}\nabla^2 + V(\vec{r})]\psi_i(\vec{r}) = \varepsilon_i\psi_i(\vec{r}) \tag{27.28}$$

by expanding the solutions in a set of basis functions

$$\psi_i(\vec{r}) = \sum_j \chi_j(\vec{r})c_{ji}. \tag{27.29}$$

The density is then given as

$$\rho(\vec{r}) = \sum_i |\psi_i(\vec{r})|^2$$

$$= \sum_{j,k} \left[\sum_i c_{ij}^* c_{ik}\right] \chi_j^*(\vec{r})\chi_k(\vec{r})$$

$$\equiv \sum_{j,k} K_{jk}\chi_j^*(\vec{r})\chi_k(\vec{r}) \tag{27.30}$$

with

$$K_{jk} = \sum_i c_{ij}^* c_{ik} \tag{27.31}$$

defining a real and symmetric matrix if the basis functions are real.

We can now define the density matrix (cf. Section 14.1) analogously to the density of Eq. (27.30),

$$\rho(\vec{r}, \vec{r}') = \sum_{j,k} K_{jk}\chi_j^*(\vec{r})\chi_k(\vec{r}'). \tag{27.32}$$

If we can obtain the density matrix of Eq. (27.32), then we can calculate the total energy, the electron density, and so on. In particular by writing the total energy as a functional of the density matrix, the variational principle, when directly applied, means that we shall minimize $\frac{\delta E_{tot}}{\delta\rho(\vec{r},\vec{r}')}$ under the constraint that $\int \rho(\vec{r}, \vec{r}) \, d\vec{r} = N$, i.e., the number of electrons.

The order-N method attempts accordingly to calculate the density matrix. As a starting point it is assumed that

$$\rho(\vec{r}, \vec{r}') = 0 \quad \text{for} \quad |\vec{r} - \vec{r}'| > r_c, \tag{27.33}$$

where r_c is some cut-off distance. Experience has shown that this is often a good approximation.

In a given calculation one will attempt to determine the coefficients K_{jk} for a given basis set. This is done by requiring that the total energy is as low as possible when considering all possible variations of the coefficients. We will accordingly need to write down the total energy as a function of these coefficients and subsequently take the derivative with respect to any of them. In addition, we have to include the constraint that the density matrix leads to the correct total number of electrons. None of this will be done here, but hopefully it is obvious that it can be done very similarly to what we have done in deriving the Hartree–Fock, the Hartree–Fock–Roothaan, and the Kohn–Sham equations.

However, density matrices are idempotent. This means that

$$\rho(\vec{r}, \vec{r}\,') = \int \rho(\vec{r}, \vec{r}\,'')\rho(\vec{r}\,'', \vec{r}\,')\,\mathrm{d}\vec{r}\,''. \tag{27.34}$$

When trying to calculate a density matrix like that of Eq. (27.32), this idempotency may not be satisfied. It has turned out that it is a good idea to apply a so-called purification scheme (McWeeny, 1960). That is, for a given density matrix, which may not obey Eq. (27.34) exactly, we calculate a new one,

$$\rho(\vec{r}, \vec{r}\,') \rightarrow \tilde{\rho}(\vec{r}, \vec{r}\,') = 3 \int \rho(\vec{r}, \vec{r}\,'')\rho(\vec{r}\,'', \vec{r}\,')\,\mathrm{d}\vec{r}\,''$$

$$- 2 \int \int \rho(\vec{r}, \vec{r}\,'')\rho(\vec{r}\,'', \vec{r}\,''')\rho(\vec{r}\,''', \vec{r}\,')\,\mathrm{d}\vec{r}\,''\,\mathrm{d}\vec{r}\,'''. \tag{27.35}$$

Thereby the total energy E_{tot} becomes a functional of $\tilde{\rho}$ rather than of ρ, and it becomes non-trivial to calculate the derivatives $\frac{\partial E_{\text{tot}}}{\partial K_{jk}}$ (that are required in the minimization process).

Since the number of neighbours within a certain distance of a given atom is more or less constant for not too small systems, the calculation of the matrix elements scales linearly with the number of atoms, under the condition that

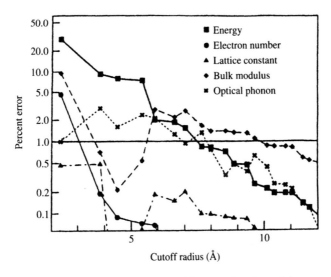

Figure 27.12 The error in % for various quantities as a function of the cut-off radius r_c for crystalline silicon. Reproduced with permission from Li et al. (1993). Copyright 1993 by the American Physical Society

Eq. (27.33) is a good approximation. The same approximation implies also that the purification (27.35) will scale linearly with the system size. By applying further, numerically specialized, techniques (see, e.g., Goedecker, 1999), one can also reformulate the problem of self-consistently determining the coefficients K (self-consistent, since the density matrix that is calculated in a given iteration will also be the one that defines the potentials) into a problem that scales only slightly more complicated than linearly with N.

The critical question is whether the approximation (27.33) is valid for not too large r_c. To address this, we show in Fig. 27.12 results of test calculations for crystalline silicon as functions of this parameter. It is seen that good accuracy is achieved for not too large values of r_c. The field is, however, still in development and, therefore, we shall not present any further results or discussions here.

28 Phase Diagrams

28.1 STRUCTURAL TRANSITIONS OF CRYSTALLINE MATERIALS

Once the total energy of a given material can be calculated it becomes possible to compare the total energy of different structural forms of the same material. In Chapter 32 we shall discuss calculated total-energy differences between different isomers of the same molecule both when being isolated in vacuum and when being dissolved in some solvent. In this subsection we shall consider the relative stability of different crystal structures.

As a first example we consider the diamond structure for C, Si, and Ge. We may calculate the total energy for this as a function of the lattice constant or, equivalently, as a function of the volume of one unit cell. This will give some curve that — hopefully — has a minimum close to the experimental value; cf. Fig. 28.1. But we may also consider other crystal structures where most likely different curves are found. And in some cases it may happen that the structure with the lowest total energy as a function of volume varies. This is, e.g., the case for Si and Ge in Fig. 28.1 but not for C in that figure. This means that the calculations can predict (or confirm) that when the volume is changed, the structure may change. Such a volume change can be produced, e.g., by applying pressure. Actually, the dashed lines in Figs. 28.1(b) and 28.1(c) (which are the common tangents to two total-energy-vs.-volume curves) have slopes that are theoretical estimates of the required pressure. This can be shown as follows.

When considering a single crystal structure, the application of pressure will change the structure so that the Gibbs free energy

$$G = E + PV - TS. \tag{28.1}$$

has its minimum. Here, E, P, V, T, and S is the total energy, the pressure, the volume, the temperature, and the entropy, respectively. At $T = 0$, the last term

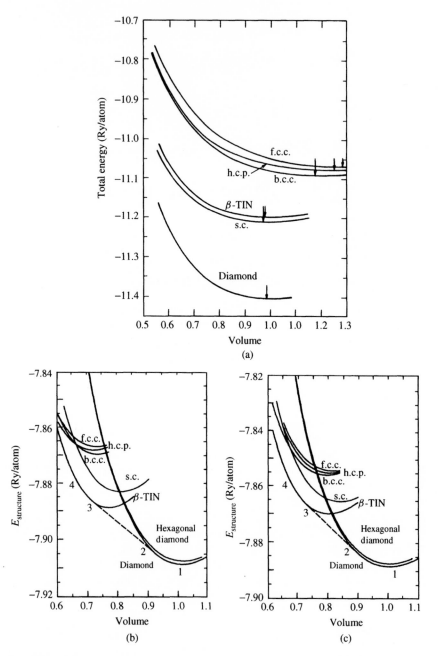

Figure 28.1 Total energy versus volume (relative to the experimental equilibrium volume) for different crystal structures for (a) C, (b) Si, and (c) Ge. Reproduced with permission from Yin and Cohen (1983, 1982). Copyright 1982, 1983 by the American Physical Society

vanishes, and by explicitly writing that E depends on the volume (cf. Fig. 28.1), we have

$$G(V) = E(V) + PV. \tag{28.2}$$

For a given pressure, the minimum is found for

$$\frac{dE(V)}{dV} = -P. \tag{28.3}$$

Thus, for a positive pressure, the minimum is found for a value where the slope of the total-energy vs. volume curve is negative. From Fig. 28.1 we see that this occurs for volumes that are smaller than those for which the minima occur, i.e., applying pressure leads to a reduction in the volume, which should be obvious.

A phase transition between two structures occurs when G of those two are identical. Since the two structures may have different volumes (cf. Fig. 28.1), we have

$$G_1(V_1) = G_2(V_2) \tag{28.4}$$

with

$$G_1(V_1) = E_1(V_1) + PV_1$$
$$G_2(V_2) = E_2(V_2) + PV_2. \tag{28.5}$$

This gives

$$E_1(V_1) - E_2(V_2) = -P \cdot (V_1 - V_2), \tag{28.6}$$

i.e., the straight line joining the two points on the total energy vs. volume curves has the slope $-P$.

Furthermore, from above we also have that for each of the two points (V_1, E_1) and (V_2, E_2) we have equations of the type (28.3), i.e.,

$$\frac{dE_1(V_1)}{dV_1} = -P$$
$$\frac{dE_2(V_2)}{dV_2} = -P. \tag{28.7}$$

Thus, $-P$ is also the slope of the two tangents of the E vs. V curves, and, accordingly, the two tangents have to lie on top of each other.

A further example is presented in Fig. 28.2. Here, we show results for solid hydrogen in different phases. The figure shows not the total energy but the enthalpy, i.e. a $P \cdot V$ term has been added to the total energy. Furthermore, since hydrogen is a very light element (there is no lighter one!), it is important to include the zero-point motion due to the vibrations, i.e., the sum $\sum_i \frac{1}{2}\omega_i$ with ω_i being the vibrational frequency of the ith vibrational mode — this will most likely be different for different structures.

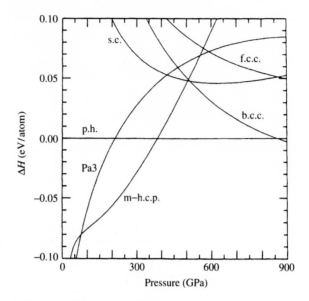

Figure 28.2 Calculated enthalpies for different structures of solid hydrogen relative to the primitive hexagonal structure. Reproduced with permission from Barbee *et al.* (1989). Copyright 1989 by the American Physical Society

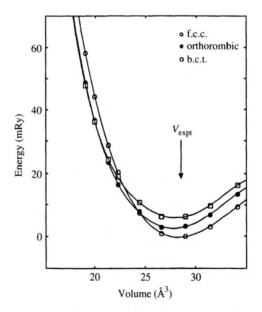

Figure 28.3 Variations in the total energy per unit cell (two atoms) for three different phases of crystalline Ce as functions of unit-cell volume. From Eriksson *et al.* (1992). Copyright 1992 by the American Physical Society

The next example, crystalline Ce, Fig. 28.3, is non-trivial since Ce has $4f$ electrons that are strongly localized but still have many properties in common with less localized valence electrons. Nevertheless, the calculations (with a density-functional method) do reproduce the correct experimental volume.

The example of Fig. 28.4 demonstrates how the calculations can predict a transition from a solid crystalline phase to a polymeric phase for which the material consists of weakly interacting chains.

Finally, in Fig. 28.5, we show results of density-functional calculations for $KNbO_3$ in an idealized cubic structure. For this, linear-response theory (see Section 21.3) has been used in calculating the full phonon dispersion curves. From Chapter 21 we know that if the dynamical matrix has negative eigenvalues (i.e., some of the vibrational or phonon frequencies become imaginary), then the structure corresponds to a saddle point for the hypersurface that describes the total energy as a function of the nuclear coordinates. The structure is accordingly unstable against a distortion as described by the imaginary-frequency vibrations. And as seen in Fig. 28.5, exactly this happens for $KNbO_3$, and ultimately the calculations can be used in explaining why $KNbO_3$ does not crystallize in the idealized cubic crystal structure.

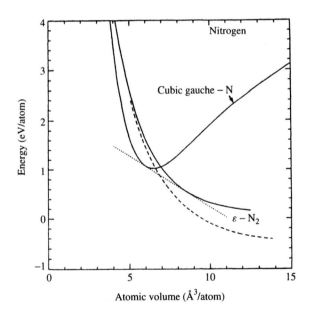

Figure 28.4 Relative total energy per atom for different phases of solid nitrogen. The 'cubic gauche' phase is a polymeric phase. From Mailhiot *et al.* (1992) The illustration was created at Lawrence Livermore National Laboratory for the US Department of Energy. Neither LLNL or the US Government makes any warranty, express or implied, or assumes any legal liability for any information, product, or process disclosed in this textbook

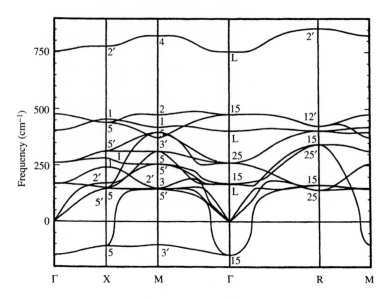

Figure 28.5 Calculated phonon dispersions of KNbO$_3$ in the ideal cubic structure. The negative frequencies are actually imaginary. Reproduced with permission from Yu and Krakauer (1995). Copyright 1995 by the American Physical Society

28.2   SEGREGATION AND PHASE SEPARATION

Until now almost all results have referred to the absolute temperature zero, $T = 0$. Only some few exceptions have been mentioned, in which molecular-dynamics calculations were assumed to be performed at a specific temperature, so that the system's properties at that temperature could be studied.

Temperature may, however, have pronounced effects. First, phase transitions where, e.g., the system changes from a solid into a liquid or where the system changes from one crystal structure to another, may occur. These transitions occur both for mono-atomic materials and for compounds. Second, for compounds, temperature may lead to the occurrence of different phases. That is, for a binary alloy A_xB_{1-x} a situation may occur where the system is a completely random alloy at one temperature but at another segregates into, e.g., A_3B and AB phases (or other phases — this depends critically on the system, the composition, and the temperature).

In Section 27.4 we saw how various properties of such alloys could be analysed with the cluster variational method, but restricted ourselves there to the $T = 0$ behaviour. In this section we shall see how these studies can be extended to $T \neq 0$.

The properties of the alloy A_xB_{1-x} (we shall here exclusively restrict ourselves to the case of binary alloys!) as a function of temperature depend first of all on the structure of it, e.g., on whether it is a purely random alloy or some phase segregation occurs. This, in turn, depends on the free energy, i.e., the total energy modified by an entropy term,

$$F = E_{tot} - T \cdot S. \tag{28.8}$$

In order to determine whether one or another structure for a given T occurs we need to determine that of the lowest free energy. In a practical calculation this means calculating the free energy for different phases and subsequently comparing them. In doing so it is important to include the boundary condition that the composition is given, i.e., that we have a mole fraction x of A atoms and a mole fraction $1 - x$ of B atoms.

When comparing the different free energies we need a common energy zero. We may take that of the isolated, pure elements, each in its ground-state phase. Then, for a given phase of the alloy, the total energy per atom, E_{tot}/N, can be written as one term by transforming the phases of the pure elements into that of the alloys,

$$xE_A^\alpha + (1 - x)E_B^\alpha, \tag{28.9}$$

where the alloy phase is labelled α, and where E_X^α is the energy per atom of transforming the ground-state phase of element X into the α phase. These energies (which, e.g., are due to differences between the crystal structures of the alloy and of the pure elements as well as in the volume per unit cell) are accessible with standard electronic-structure calculation techniques, like those whose results we have discussed in Section 28.1.

Subsequently, the two elements are brought together in the alloy. Thereby, both purely random alloys and ordered structures may occur, solely dependent on how the A and B atoms occupy the sites of the lattice in the α phase. The energies of these different cases can all be calculated by using the cluster variational method of Section 27.4. As discussed there, effects of local structure relaxations can also be incorporated without changing the formalism (although the computational demands thereby will be much larger). The cluster variation method requires the calculation of cluster interaction parameters that can be obtained by considering various periodic structures, as discussed in Section 27.4. Such studies lead to results like those shown in Fig. 28.6 for Li_cAl_{1-c}.

The major part of the temperature effects comes through the entropy term of the free energy. Entropy is far less trivial to calculate than is the total energy, and one considers therefore only those parts that are considered important.

From thermodynamics we know that entropy is a measure of order/disorder. By forming an alloy from the pure substituents the most clearly recognizable change of order is that of the more or less random distribution of atoms, of pairs of atoms, of triples of atoms, and so on. This is exactly the effect that is included in the so-called configurational entropy,

$$S = -k_B \sum_i x_i \ln x_i. \tag{28.10}$$

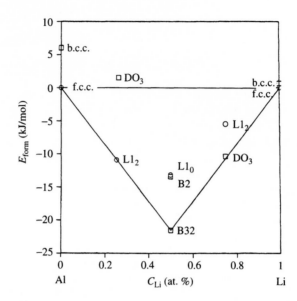

Figure 28.6 Formation energies for different ordered structures (described in the plot) of Li_cAl_{1-c}. Reproduced with permission from Sluiter *et al.* (1990). Copyright 1990 by the American Physical Society

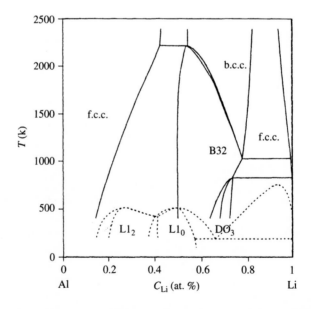

Figure 28.7 Phase diagram for different solid phases of Li_cAl_{1-c}. Reproduced with permission from Sluiter *et al.* (1990). Copyright 1990 by the American Physical Society

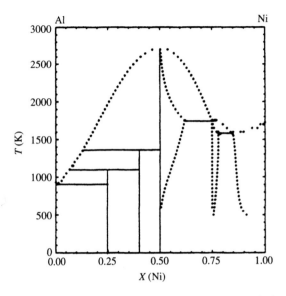

Figure 28.8 As Fig. 28.7 but for Ni_xAl_{1-x}. Reproduced by permission of Institute of Physics Publishing from Pasturel *et al.* (1992)

Here, k_B is Boltzmann's constant. Furthermore, x_i is the probability that a certain 'configuration' appears. A configuration is here a single atom, a pair of atoms, a triple of atoms, etc. For A_xB_{1-x}, one x_i is accordingly the probability of finding an A atom at a certain site (which is x), another that of finding a B atom at the site (i.e., $1 - x$), a third that of finding an AA pair at a certain pair of sites, a fourth that of finding an AB pair, a fifth that of finding a BB pair, a sixth that of finding a linear AAA triple, a seventh that of finding a bent AAA triple, and so on. For the purely random alloy these can be calculated, and also for the various ordered phases they can be calculated by carefully analysing the structure.

There may be other contributions to the entropy, e.g., due to vibrations. By including the latter one ends up with a theoretical phase diagram like that of Fig. 28.7 for Li_cAl_{1-c}. It was found that without the vibrational entropy one could not describe the phases of pure Li correctly, which is why the vibrational contribution was included. Due to its small mass and large size, Li may be an atypical example.

Considering solely configurational entropy leads to results like those of Fig. 28.8 for Ni_xAl_{1-x}.

Finally, we add that the field of determining phase diagrams from parameter-free electronic-structure calculations is not yet mature. Since 1000 K corresponds to only ~ 0.1 eV, the requirements of computational accuracy are high, and reasonable approximations can easily lead to predicted phase transitions that are more than 100 K away from the experimental values.

29   Clusters

29.1   LARGE MOLECULES

Clusters are intermediates between molecules and solids. Often, they consist of only one or two types of atoms, and the number of atoms is so large that they can not be considered 'normal' molecules, but still so small that they are not yet 'normal' solids. This can also be expressed through the statement that the number of atoms at the surface relative to the total number of atoms is non-negligible.

Clusters are often produced in the gas phase or solutions but in some sense one may also consider grains of one material within another (host) material as clusters (or colloids — in some cases these are named quantum dots), but here we shall be concerned with the former. When producing clusters or colloids in solution, their surfaces are often covered with different (often organic) molecules, so-called surfactants, that saturate the dangling bonds of the clusters or colloids.

There are two central issues that are often addressed: how do the solid-state properties develop as a function of cluster size, and which properties do they have that are different from those of the infinite crystalline counterpart?

When the number of atoms is not too small, the clusters often have structures as finite parts of the crystalline material, see Fig. 29.1. Furthermore, in that case the clusters are often close to spherically symmetric. But, as is obvious in the figure, due to the finite size of the system, it loses its periodicity, and atoms that for the infinite material would be equivalent are no longer so. This means that the computational requirements for studying the properties of the clusters become demanding as soon as the number of atoms is larger than some tens.

Nevertheless, the most direct approach for studying the properties of the clusters is to consider them as nothing but large molecules. Thereby, one applies one or more of the standard approaches for electronic-structure calculations as we have described above. In the simplest approach, one would simply take the structure as a finite part of the crystal and calculate the electronic properties

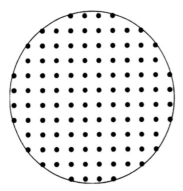

Figure 29.1 Schematic representation of a cluster as a finite, approximately spherical part of an infinite crystal. The dots represent the atoms

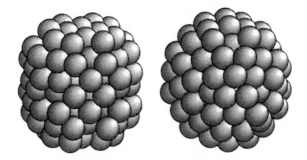

Figure 29.2 Two different structures of a cluster containing 135 atoms. Reproduced by permission of American Institute of Physics from Jennison *et al.* (1997)

for this fixed structure. More advanced studies include the fact that also the structural degrees of freedom are varied, as we saw in Section 20.5 for smaller Se_N clusters.

As a further example of this approach we show in Fig. 29.3 the density of states of clusters containing either Ru or Ag atoms with the structures of Fig. 29.2. The density of states should in principle be a set of sharp peaks at the positions of the single-particle energies, but here these peaks have been broadened in order to obtain smoother curves. The results of Fig. 29.3 have been obtained using a density-functional method.

In the example of Figs. 29.4 and 29.5 we consider smaller Na_N clusters where the structural degrees of freedom were optimized too. To this end a semiempirical method was applied. The results of Fig. 29.4 show how the clusters in fact do seek a very compact, almost spherical, shape. The results of Fig. 29.5 show, on the other hand, that the total energy as well as the single-particle energies only slowly converge to the infinite-N values, and in fact do not seem to approach such a limit for the systems that were considered in that study.

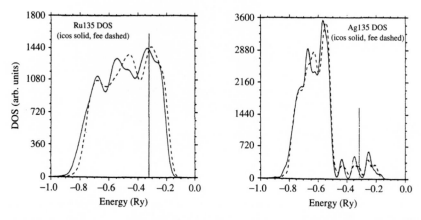

Figure 29.3 Density of states for Ru (left part) and Ag (right part) clusters with 135 atoms. The dashed and solid lines correspond to the right and left form of Fig. 29.2, respectively. Reproduced by permission of American Institute of Physics from Jennison *et al.* (1997)

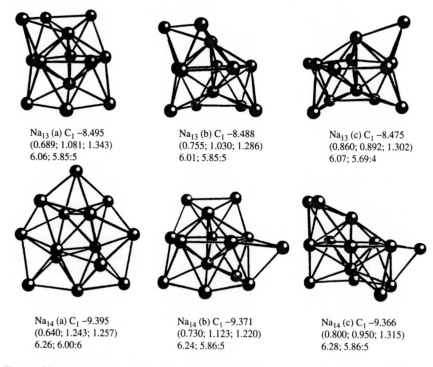

Na$_{13}$ (a) C$_1$ −8.495
(0.689; 1.081; 1.343)
6.06; 5.85:5

Na$_{13}$ (b) C$_1$ −8.488
(0.755; 1.030; 1.286)
6.01; 5.85:5

Na$_{13}$ (c) C$_1$ −8.475
(0.860; 0.892; 1.302)
6.07; 5.69:4

Na$_{14}$ (a) C$_1$ −9.395
(0.640; 1.243; 1.257)
6.26; 6.00:6

Na$_{14}$ (b) C$_1$ −9.371
(0.730; 1.123; 1.220)
6.24; 5.86:5

Na$_{14}$ (c) C$_1$ −9.366
(0.800; 0.950; 1.315)
6.28; 5.86:5

Figure 29.4 Structures of the three energetically lowest isomers of Na$_{13}$ (upper part) and Na$_{14}$ (lower part). Reproduced by permission of American Institute of Physics from Poteau and Spiegelmann (1993)

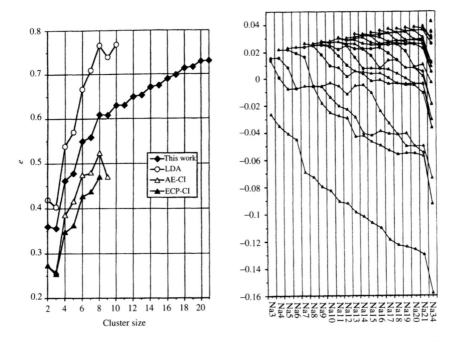

Figure 29.5 Binding energy per atom (left part) and the single-particle energies (right part) for Na$_N$ clusters as functions of N. The left-hand part includes results of different theoretical approaches. Reproduced by permission of The American Institute of Physics from Poteau and Spiegelmann (1993)

29.2 JELLIUM MODELS

Experimentally, it is possible to determine the relative abundance of the clusters as a function of N. This means that one measures the number of clusters that have a specific size. One might initially expect that this leads to a rather smooth function, but it turns out that there exist so-called magic numbers, i.e., values of N for which particularly many clusters occur. This indicates that these clusters are particularly stable, and a central question is accordingly why this is so and whether this can be described theoretically.

There may be various reasons for this behaviour and it depends partly on the type of atoms that form the clusters. In some cases, when the atoms interact very weakly, a simple packing argument can explain it, i.e., the magic numbers occur for those sizes that correspond to a close packing of hard spheres. In other cases, the bonds between the atoms are strong directional covalent bonds and detailed quantum theoretical studies are required in order to explain the occurrence of the particularly stable clusters. Finally, in some cases the valence electrons are fairly delocalized and hardly feel the precise geometrical arrangement of the nuclei but move rather in an averaged potential all over the cluster, which is, moreover, roughly spherically symmetric. This is above all the case for clusters made up of simple metals (Na, K, Rb, Cs, e.g.). Then the magic numbers can be explained as being due to the closing of electronic shells and this can be well described by the jellium model that we shall now discuss.

The idea behind the jellium model is, as mentioned above, that the electrons are so delocalized that the precise positions of the nuclei are unimportant. Instead, one focuses only on these valence electrons and assumes that everything else (i.e., nuclei and core electrons) can be smeared out to a homogeneous (charged) background density, the so-called jellium. For a spherically symmetric cluster this corresponds to the radial density shown in Fig. 29.6. The constant background density inside the cluster ρ_{jel} is determined so that it equals the density in the crystalline material, and often one introduces the so-called electron-gas parameter r_s so that

$$\frac{4\pi}{3}r_s^3 = \frac{1}{\rho_{\text{jel}}}, \tag{29.1}$$

i.e., r_s is the radius of a sphere that contains exactly one charge (see also Section 15.6). Then, R in Fig. 29.6 is given by

$$R = N^{1/3}r_s. \tag{29.2}$$

Having defined the background density we can consider either the density-functional Kohn–Sham equations

$$\left[-\tfrac{1}{2}\nabla^2 + V_{\text{ext}}(\vec{r}) + V_C(\vec{r}) + V_{\text{xc}}(\vec{r})\right]\psi_i(\vec{r}) = \varepsilon_i\psi_i(\vec{r}) \tag{29.3}$$

or the Hartree–Fock equations

$$\left[-\frac{1}{2}\nabla^2 + V_{\text{ext}}(\vec{r}) + V_C(\vec{r}) - \sum_{j=1}^{N}\hat{K}_j\right]\psi_i(\vec{r}) = \varepsilon_i\psi_i(\vec{r}). \tag{29.4}$$

In both cases, V_C is the Coulomb potential of the valence electron density (i.e., *not* that of the jellium), and V_{ext} is in the present case the Coulomb potential of the jellium, whereas it earlier in this manuscript has been that of the point-like nuclei. Finally, V_{xc} is the exchange-correlation potential in the Kohn-Sham equations,

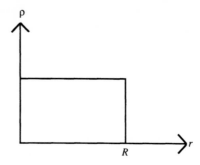

Figure 29.6 The jellium background density for a spherically symmetric system as a function of r

whereas $\hat{K}_j$ is the exchange operator for the jth (occupied) orbital within the Hartree–Fock approach.

Independently of which of the two approaches is used, the spherical symmetry allows for writing the single-particle eigenfunctions as a radial function times a harmonic function (just as for an isolated atom),

$$\psi_i(\vec{r}) = \psi_{nl}(r)Y_{lm}(\theta, \phi). \qquad (29.5)$$

Furthermore, all orbitals with the same (n, l) are energetically degenerate, and the occurrence of the magic numbers is then explained as being due to the filling

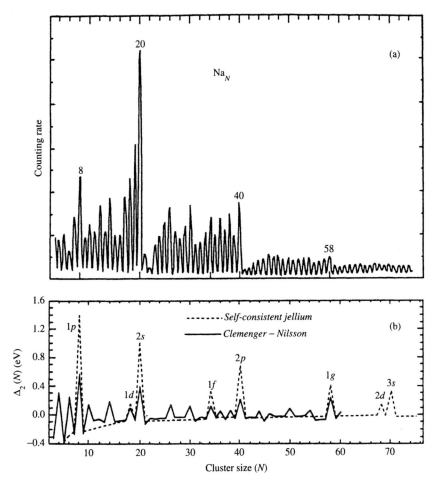

Figure 29.7 (a) Relative abundance of Na_N clusters according to experiment, and (b) the theoretical analogue. Here, the curve marked jellium represents results of self-consistent density-functional calculations with a jellium model, whereas Clemenger–Nilsson marks non-self-consistent calculations with a model potential. Reproduced with permission from de Heer (1993). Copyright 1993 by the American Physical Society

of the individual (n, l) shells. For larger systems the energies of more different (n, l) shells become so similar that the particularly stable clusters occur only when a whole group of such nearly-degenerate shells is filled.

One way of demonstrating this is to consider the total energy as a function of N. It turns out that the quantity

$$\Delta(N) = E_{\text{tot}}(N + 1) + E_{\text{tot}}(N - 1) - 2E_{\text{tot}}(N) \qquad (29.6)$$

is useful for this purpose. This has maxima (as a function of N) whenever the cluster with that number of valence electrons is particularly stable, and as seen in Fig. 29.7 the occurrence of the magic numbers can be correlated with the

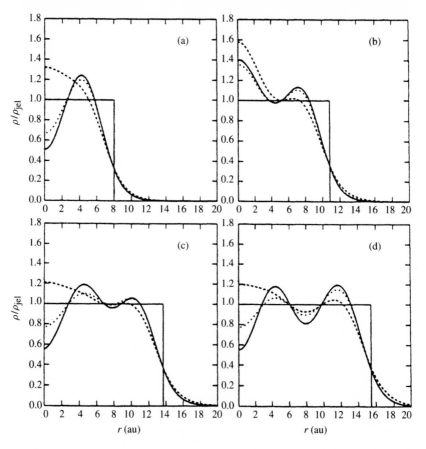

Figure 29.8 Calculated electron density for the jellium model for Na$_N$ clusters with (a) $N = 8$, (b) $N = 20$, (c) $N = 40$, and (d) $N = 58$. The rectangular profile is the jellium density, and the solid line is the density from a Hartree–Fock calculation, the dotted line is that from a density-functional calculation without correlation, and the dashed line is that from a Hartree calculation (i.e., neither exchange nor correlation effects are included). Reproduced with permission from Hansen and Nishioka (1993), *Z. Phys.* **D28**, 73, Fig. 2. Copyright 1993 Springer-Verlag

closing of electronic shells. Here, a peculiarity should be mentioned: for atoms it is customary to let $n - 1$ for any orbital ψ_{nlm} be the number of nodes in total, i.e., both radial and angular nodes are counted. For clusters, on the other hand, $n - 1$ is only the number of radial nodes. Therefore, no $1f$ function exists for atoms, but it does for clusters!

Finally, we show in Fig. 29.8 the resulting electron density from such calculations on Na_N clusters. In particular for the largest clusters we recognize an oscillatory behaviour of the electron density. These oscillations are the so-called Friedel oscillations (Friedel, 1954). Originally, they were found for an isolated single point-like impurity in an infinite, homogenous jellium, but they occur also close to surfaces of the jellium or of semi-infinite crystals.

The advantage of the jellium model is that one can treat fairly large systems, i.e., up to about 10 000 atoms, at least as long as the spherical symmetry is retained. On the other hand, the assumption of a homogeneous constant background density may be too crude an approximation. Nevertheless, it is a popular model, and many fair comparisons with experiments have been obtained with this model. Finally, it is not restricted to clusters, but has also been applied to surfaces, interfaces, etc. Friedel oscillations have also been found for these.

29.3 EMBEDDED-ATOM AND EFFECTIVE-MEDIUM METHODS

The jellium model allows for an approximate treatment of the electronic properties of large systems and not only of clusters. Its success depends on whether the electrons of the system of interest are localized or delocalized, and only in the latter case, when the precise arrangement of the nuclei is not felt strongly by the electrons, will one expect that the jellium model provides a reasonable approximation.

Also the embedded-atom (Daw and Baskes, 1984), quasiatom (Stott and Zaremba, 1980), and effective-medium (Nørskov, 1982) methods, which are closely related, work better for systems containing delocalized electrons, i.e., they are best for metals. As for the jellium model, they are not restricted to clusters but apply also to other systems. In particular large systems with low (or no) symmetry can with advantage be treated by these methods. In contrast to the jellium model, they do, however, consider the precise positions of the nuclei.

The embedded-atom and effective-medium methods are very similar and will therefore be treated as one here. Their starting point is that of an isolated atom. According to the density-functional theory one may write the total energy of this as a functional of the electron density,

$$E_{\text{tot}} = E_{\text{tot}}[\rho(\vec{r})]. \tag{29.7}$$

When this atom is placed in a host (e.g., as a part of a solid) the electron density will change,

$$\rho(\vec{r}) \rightarrow \rho(\vec{r}) + \Delta\rho(\vec{r}). \tag{29.8}$$

Assuming that we still can identify a region of space that belongs to the atom of interest, its total energy will also change due to the tails of the electron densities of the neighbouring atoms and due to electron transfers,

$$E_{tot}[\rho(\vec{r})] \rightarrow E_{tot}[\rho(\vec{r}) + \Delta\rho(\vec{r})]. \tag{29.9}$$

The atom of interest has been *embedded* in the host material.

In the general case, $\Delta\rho(\vec{r})$ is a complicated function of $\vec{r}$. However, if the electrons of the host material are fairly delocalized, $\Delta\rho(\vec{r})$ will only vary slowly in the region of the atom of interest. Then, one may assume that the additional electron density is a constant,

$$\Delta\rho(\vec{r}) = \Delta\rho_0, \tag{29.10}$$

cf. Fig. 29.9. Notice, that this density [as well as $\Delta\rho(\vec{r})$] may be positive or negative.

This approximation amounts to considering an atom embedded in a homogeneous electron gas whose electron density is given by the right-hand side of Eq. (29.10). For any system containing various types of atoms the first thing we need is thus for each type of atom (specified through its Z number) to calculate the change of its total energy when embedding it into a homogenous electron gas, i.e., the function

$$\Delta E_Z[\rho(\vec{r}), \Delta\rho_0]. \tag{29.11}$$

This function can be calculated once and for all for all the different types of atoms (Z) as well as for all the different relevant values of the density of the homogeneous electron gas ($\Delta\rho_0$).

Let us now consider a real system consisting of a number of atoms. For each atom we will assume that a certain part of space belongs to that atom. We will moreover assume that this region is spherical. When the atoms are brought together, electrons will start flowing back and forth between the atoms. We will then assume that the electrons that flow into the region of a certain atom produces a constant additional density inside that atomic region. The energy cost related to this is described by the function ΔE_Z of Eq. (29.11).

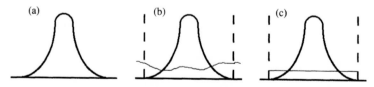

Figure 29.9   (a) The electron density of an isolated atom; (b) as (a) but with the additional tails from the surrounding atoms; (c) as (b) but with the approximation that the additional tails produce a constant electron density. The vertical dashed lines mark the boundaries of the atom

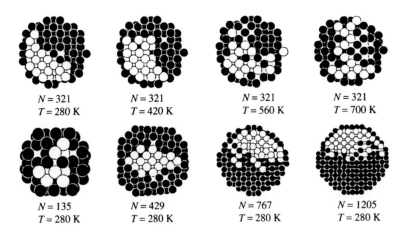

$$N = 321 \qquad N = 321 \qquad N = 321 \qquad N = 321$$
$$T = 280\ \text{K} \qquad T = 420\ \text{K} \qquad T = 560\ \text{K} \qquad T = 700\ \text{K}$$

$$N = 135 \qquad N = 429 \qquad N = 767 \qquad N = 1205$$
$$T = 280\ \text{K} \qquad T = 280\ \text{K} \qquad T = 280\ \text{K} \qquad T = 280\ \text{K}$$

Figure 29.10 Eight clusters of different sizes (given by N) and temperatures containing Ag atoms (black circles) and Cu atoms (white circles). The plots show slices through the centre. Reproduced by permission of Institute of Physics Publishing from Christensen *et al.* (1995)

In addition the atomic regions will no longer be neutral due to the electron flow. This will lead to an additional Coulomb energy

$$\frac{1}{2} \sum_{k \neq l} \frac{\Delta n_k \Delta n_l}{|\vec{R}_k - \vec{R}_l|}, \tag{29.12}$$

where Δn_k is the extra number of electrons in the sphere of atom k, that is placed at $\vec{R}_k$. For a given structure, $\{\vec{R}_k\}$, the electron transfers will adjust themselves so that the lowest total energy occurs, with the additional constraint of overall charge neutrality.

These terms comprise the first-order approximations of the embedded-atom or effective-medium methods.

There are many effects that, with this approach, are treated only approximately but that may be included more exactly through extra terms in the total-energy expression. One of the simplest ways is to replace the potentials in Eq. (29.12) by more general pair potentials,

$$\frac{1}{2} \sum_{k \neq l} \phi_{kl}(|\vec{R}_k - \vec{R}_l|), \tag{29.13}$$

where ϕ_{kl} is some function that depends on the distance between the atoms k and l and is chosen appropriately (e.g., so that the method reproduces certain physical observables accurately).

This approach often gives accurate information that can be obtained only with great difficulty by more elaborate methods. Nevertheless, it suffers from some deficiencies. First of all, there are no directional bonds, i.e., no short-range effects.

Secondly, this approach focuses on the total energy and the total electron density, so orbital effects are also abandoned. Third, the Coulomb energy neglects the fact that, for metallic systems the electrons respond to charges by screening them, so that the expression of Eq. (29.12) may be replaced by screened potentials (i.e., $|\vec{R}_k - \vec{R}_l|^{-1}$ is replaced by either an exponential or an inverse power larger than 1). All these effects can be included in various approximate ways leading to a method that can treat more and more situations and systems. We shall, however, not discuss it further here. Only the basic principle will be stressed: the method neglects the details of the electron-density variations when an atom is incorporated into a host, and various effects due to atom–atom interactions are treated only approximately.

The effective-medium and embedded-atom methods are so simple that fairly large and complex systems can be treated by using them. Moreover, since the positions of the nuclei are specified it is also possible to optimize structure, i.e., to calculate the structure with the lowest total energy. This has been done for a number of different systems, including clean surfaces and surfaces with adsorbants, interfaces, alloys, and nanostructures. Here, we shall illustrate the method through an example of the structure of two-component clusters. These are clusters consisting of two different types of atoms, in the present case Ag and Cu. Figure 29.10 shows the calculated structures for clusters with in total 321–1205 atoms at different temperatures.

30 Macromolecules

30.1 FORCE FIELDS

Force-field methods are empirical methods that are intended to be able to provide an accurate description of structural (and in some cases also vibrational, relative-stability, etc.) properties of specific classes of systems. The fundamental principle behind them is that these properties are primarily dictated by nearest-neighbour bonds. These bonds are constructed from localized electrons and an A−B bond is to some extent independent of which molecule or solid it is a part of (although there may be some differences between single, double, and triple bonds, e.g.). Thus, an A−B bond length is largely a constant for a very large class of materials, and the variations are often small and can be ascribed to bond-angle effects (i.e., effects that are due to triples of atoms). Further, higher-order effects that depend on even more atoms may also be present, but their role is most likely small. In total, we assume therefore that the total energy of any molecule can be written as

$$E_{\text{tot}} = \sum_{k,l} V_{k,l}(\vec{R}_k, \vec{R}_l) + \sum_{k,l,m} V_{k,l,m}(\vec{R}_k, \vec{R}_l, \vec{R}_m)$$
$$+ \sum_{k,l,m,n} V_{k,l,m,n}(\vec{R}_k, \vec{R}_l, \vec{R}_m, \vec{R}_n) + \cdots. \qquad (30.1)$$

Here, the first term is the two-body term, the second one is the three-body term, and so on. Furthermore, $\vec{R}_p$ is the position of the pth nucleus, and the potentials $V_{k,l,\dots}$ are assumed to depend only on the types of atoms involved as well as on their relative positions. Moreover, in most cases $V_{k,l}(\vec{R}_k, \vec{R}_l)$ depends only on the distance between the two atoms $|\vec{R}_k - \vec{R}_l|$, $V_{k,l,m}(\vec{R}_k, \vec{R}_l, \vec{R}_m)$ depends only on the three distances $|\vec{R}_k - \vec{R}_l|$, $|\vec{R}_k - \vec{R}_m|$, and $|\vec{R}_m - \vec{R}_l|$, etc. Finally, the sums in Eq. (30.1) do not include terms where one or more indices are identical. We add that the expansions (30.1) to some extent resemble the cluster expansions of Section 25.4 that we studied for alloys, although there are important differences in the systems for which the methods are developed.

The advantage with this approach is that once the potentials $V_{k,l,...}$ are known one can easily treat very large systems, also without having to assume that they are periodic or possess any other property that can be used in simplifying the calculations. Such methods are therefore common when studying structural properties of macromolecules or of other large heterogeneous systems (grains in solids, liquids, etc.). Some of the common methods include MM3 (Allinger *et al.*, 1989; Lii and Allinger, 1989), UFF (Rappé *et al.*, 1992), AMBER (Weiner and Kollman, 1981), and CHARMI (Brooks *et al.*, 1983).

Using a force field like that of Eq. (30.1) means neglecting any electronic effects. The force-field calculations do not therefore provide anything but information on structure and in some cases also on vibrations and relative stabilities. On the other hand, once the structure has been obtained, e.g., from the force fields, more sophisticated methods like parameter-free Hartree–Fock or density-functional methods can be applied in studying other properties.

The success of the force fields depends ultimately on the quality of the potentials $V_{k,l,...}$. In deriving expressions for these one is usually guided by physical and chemical intuition combined with detailed information from experimental or exact theoretical studies on small systems. Furthermore, the potentials may also differ depending on what kind of system is to be considered.

Let us consider the example of Fig. 30.1, i.e., two water molecules. If we first consider one of the water molecules separately, we may suggest using the form

$$E_{\text{tot}} = \tfrac{1}{2}k_{\text{OH}}(d_{\text{OH}_1} - d_0)^2 + \tfrac{1}{2}k_{\text{OH}}(d_{\text{OH}_2} - d_0)^2 + \tfrac{1}{2}k_{\text{HOH}}(\theta_{\text{HOH}} - \theta_0)^2. \quad (30.2)$$

Here, d_{OH_1} is the distance between the oxygen atom and one of the hydrogen atoms, whereas d_{OH_2} is the distance to the other hydrogen atom. θ_{HOH} is the H–O–H bond angle. The values d_0 and θ_0 are the equilibrium bond lengths and bond angle, respectively, and k are force constants. The four parameters of the model can be determined so that structural and vibrational properties can be described. For an improved description one may even include higher-order (anharmonic) terms in the two- and three-body terms of Eq. (30.2).

This potential is, however, inadequate when trying to apply identical potentials for all interactions of the larger system of Fig. 30.1. The potential of Eq. (30.2)

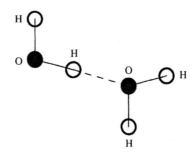

Figure 30.1 Structure of two water molecules that are connected via a hydrogen bond

diverges when the interatomic distances become large, whereas in reality one would expect the potential to approach 0 for large interatomic distances. An improved model could therefore be obtained by, e.g., requiring that the potentials of Eq. (30.2) vanish beyond certain cutoff distances, which, however, leads to non-physical discontinuities in the total energy. Alternatively, one may replace the harmonic terms with other functions that have the proper long-range behaviour. These could, e.g., include exponential functions, Morse potentials, or inverse powers of the interatomic distances. What precisely is done is a compromise between the requirements that the model should be physically realistic and that it should be sufficiently simple.

Here, we shall only mention one aspect. The water molecule has a dipole moment. This may be interpreted as originating from atomic charges, i.e., that water consists of O^{-2q} and H^q ions, where q is not necessarily an integer. Therefore, one may assume that one part of the long-range potential is a Coulomb potential from these atomic charges, i.e., that one term is

$$\frac{1}{2} \sum_{k \neq l} \frac{Q_k Q_l}{|\vec{R}_k - \vec{R}_l|}, \tag{30.3}$$

where Q_k are some effective charges placed at the nuclear positions (i.e., $+q$ and $-2q$ in the example above).

30.2 MOLECULAR MECHANICS + QUANTUM MECHANICS

Force-field methods allow only for the determination of total-energy (including structural) properties, not of electronic properties. In some cases one is, however, interested in the latter for a very large macromolecule that is too large to be treated by standard parameter-free electronic-structure calculation methods. One example is that of an ion or radical approaching a macromolecule and interacting with this at a certain (so-called active) centre. In that case it is desirable to be able to describe the electronic properties of at least a certain part of the macromolecule around the reaction centre, whereas other parts need not be described as accurately. Nevertheless, since the properties of the macromolecule near the reaction centre change, it is desirable to be able to study how the structure of the remaining parts of the macromolecule is modified due to this change. To this end one may apply methods that combine the force-field (also called molecular-mechanics) methods and the electronic-structure methods.

Consider, e.g., the example of Fig. 30.2. We have here (schematically!) a smaller molecule coming in from above that ultimately interacts with the macromolecule. We can study, quantum-mechanically, a part of the macromolecule plus the smaller molecule by, e.g., the Hartree–Fock method, i.e., by solving the

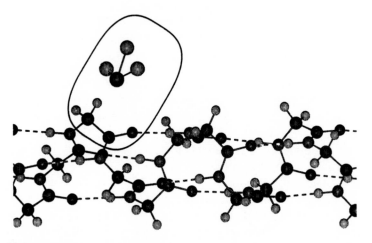

Figure 30.2 Schematic representation of a macromolecule that reacts with a smaller incoming molecule. That part inside the ellipse is treated with quantum-mechanical methods, whereas the other parts are treated using force fields

Hartree–Fock equations,

$$\left[-\frac{1}{2}\nabla^2 + V_{ext}(\vec{r}) + V_C(\vec{r}) - \sum_{j=1}^{N} \hat{K}_j\right]\psi_i(\vec{r}) = \varepsilon_i\psi_i(\vec{r}) \qquad (30.4)$$

or with a density-functional method by solving the Kohn–Sham equations

$$\left[-\frac{1}{2}\nabla^2 + V_{ext}(\vec{r}) + V_C(\vec{r}) + V_{xc}(\vec{r})\right]\psi_i(\vec{r}) = \varepsilon_i\psi_i(\vec{r}). \qquad (30.5)$$

In both cases, the external potential V_{ext} contains both a term from the nuclei of that part of the complete system that is treated quantum-mechanically (i.e., inside the ellipse in Fig. 30.2) as well as a term from the other parts of the system. The latter will be some force-field term that also acts on the electrons and that, e.g., could be the Coulomb potential from point charges as in Eq. (30.3).

Simultaneously, the force field of those parts of the system that are not treated quantum-mechanically will depend on all parts of the system, i.e., also on those parts that are treated quantum-mechanically. Once again, the latter could, e.g., be included via a point-charge model where the effective charges are then determined quantum-mechanically.

Through this approach one introduces an interplay between both parts of the macromolecule and, as a result, the complete macromolecule may respond to the approach of the smaller molecule.

One of the central problems related to the application of the MM/QM (molecular-mechanics/quantum-mechanical) hybrid methods is how to connect the two parts of the system that are treated differently. Thus, in Fig. 30.2 we see that the boundary between the two parts cuts various covalent bonds. Thus, considering the QM part as such will lead to unwanted effects due to these broken bonds. Therefore, one may include so-called link atoms (see, e.g., Bakowies and Thiel, 1996) that saturate the otherwise broken bonds. Denoting that part of the system that is treated with quantum-mechanics Q, the other part M, and the link atoms L, we end up with the following expression for the total energy:

$$E_{tot} = E_{MM}(Q + M) - E_{MM}(Q + L) + E_{QM}(Q + L), \qquad (30.6)$$

i.e., as a the molecular-mechanics energy of the complete system modified by the difference in the quantum-mechanical (QM) and molecular-mechanics (MM) energy of the Q part including the link atoms L.

Moreover, the charge distribution of the M part may lead to, e.g., long-ranged electrostatic potentials that are felt by the Q part and, accordingly, have to be included both in the MM and in the QM parts above.

31 Interactions

31.1 CHEMICAL REACTIONS

A central issue of chemistry is to understand how and why molecules change structure and, eventually, composition, both when being isolated and when interacting with other molecules. The former case corresponds to studying the reaction

$$A_1 \rightarrow A^* \rightarrow A_2, \qquad (31.1)$$

where A^* is an excited intermediate structure, whereas A_1 and A_2 are two different structures of the same molecule (i.e., two isomers).

The latter case may correspond to a reaction of the type

$$AB + C \rightarrow ABC^* \rightarrow A + BC, \qquad (31.2)$$

where ABC^* is an intermediate structure that corresponds to a so-called transition state.

Besides knowledge of the precise structures along the reaction path, first of all knowledge of the changes in the total energies is of ultimate importance. This includes knowledge of the (transition) barrier height defined as

$$E_{\text{barrier}} = \begin{cases} E_{\text{tot}}(A^*) - E_{\text{tot}}(A_1) & \text{for Eq. (31.1)} \\ E_{\text{tot}}(ABC^*) - E_{\text{tot}}(AB) - E_{\text{tot}}(C) & \text{for Eq. (31.2),} \end{cases} \qquad (31.3)$$

as well as of the energy gain (which also may be negative, i.e., a loss) of the reaction,

$$E_{\text{gain}} = \begin{cases} E_{\text{tot}}(A_1) - E_{\text{tot}}(A_2) & \text{for Eq. (31.1)} \\ E_{\text{tot}}(AB) + E_{\text{tot}}(C) - E_{\text{tot}}(A) - E_{\text{tot}}(BC) & \text{for Eq. (31.2).} \end{cases} \qquad (31.4)$$

A further issue is that of knowing whether the introduction of catalysts may lower the energy barrier, i.e., to study reactions of the type

$$M + A_1 \rightarrow M \cdot A^* \rightarrow M + A_2 \qquad (31.5)$$

or

$$M + AB + C \rightarrow M \cdot ABC^* \rightarrow M + A + BC, \qquad (31.6)$$

where M is the catalyst that takes part in the reaction but does not change either the educts or the products.

The chemical reactions involve the breaking, creation, and rearrangement of chemical bonds. If theoretical methods are to describe the chemical reactions, it is therefore important that they are able to describe the energetics involved in breaking (and, equivalently, also creating) chemical bonds.

In Tables 31.1 and 31.2 we show some examples of calculated dissociation energies as determined with a density-functional method. The results are examples of the general behaviour: the bond energies that are obtained with a local-density approximation are too high, whereas the gradient corrections bring them into close contact with experimental values. It should also be added here that Hartree–Fock bond energies usually are too low, whereas those calculated with either the CI or Møller–Plesset methods (i.e., when including correlation effects) most often are accurate.

Table 31.1 Calculated (LSD and LSD/NL) and experimental energies (in kcal/mol) for removing a carbonyl (CO) group from the listed molecules. All calculations were done with a density-functional method, where 'LSD' denotes calculations with a local-density-approximation and 'LSD/NL' denotes those that include gradient corrections. Reproduced by permission of John Wiley & Sons from Ziegler (1997)

Molecule	LSD	LSD/NL	Exp.
$Cr(CO)_6$	276	193	192, 155
$Mo(CO)_6$	226	166	170
$W(CO)_6$	249	183	192
$Fe(CO)_5$	263	192	176
$Ni(CO)_4$	192	121	104

Table 31.2 As Table 31.1 but for breaking the bonds marked by '−'. Reproduced by permission of John Wiley & Sons from Ziegler (1997)

Bond	LDA	LDA/NL	Exp.
$(CO)_5Mn\text{-}H$	324	288	284 ± 4
$(CO)_5Mn\text{-}CH_3$	322	207	192 ± 11
$(CO)_4Co\text{-}H$	332	282	280 ± 4
$(CO)_4Co\text{-}CH_3$	287	197	
$(CO)_5Mn\text{-}Mn(CO)_5$		174	171
$(CO)_4Co\text{-}Co(CO)_4$		148	61–88

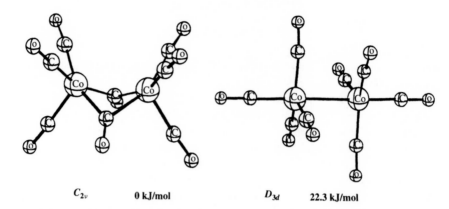

C_{2v} 0 kJ/mol D_{3d} 22.3 kJ/mol

Figure 31.1   The calculated structures of two isomers of $Co_2(CO)_8$ together with their total-energy difference. The calculations were performed using a density-functional method with a gradient-corrected approximation (GGA). Reproduced by permission of John Wiley & Sons from Ziegler (1997)

When chemical bond energies can be calculated accurately one may also calculate the structures of different isomers of the same molecule. This is examplified in Fig. 31.1 where we show the calculated structures of two isomers of $Co_2(CO)_8$.

Next we explore whether the theoretical methods can describe the energetics (i.e., the changes in the total energies) for different reactions. First, in Fig. 31.2, we show results for the isomerization of C_4H_6. It is recognized that the Møller–Plesset and the gradient-corrected density-functional calculations produce accurate results (actually a general result), but that Hartree–Fock and local-density results are less accurate.

The hydroformylation mechanism of Fig. 31.3 is a significantly more complex reaction that actually involves several steps as well as catalysts. Each of these steps has been studied with density-functional methods and here the results of one representative example will be presented briefly.

Figures 31.4 and 31.5 show accordingly the changes in the total energy and in the structure of the molecule for the first step (step b in Fig. 31.3) of the hydroformylation mechanism.

There are many other types of chemical reactions. Each of them is interesting in itself but from a general point of view they only rarely involve other features than those studied here. It is very important that all structural parameters can be optimized automatically (i.e., that forces are available). Furthermore, the transition states, being the structure of a local total-energy maximum, can be identified as having vanishing forces but in contrast to a (local or global) total-energy minimum to have one imaginary vibrational frequency (i.e., there is one negative eigenvalue of the dynamical matrix).

In some cases surfaces act as catalysts, as we saw it in Section 26.5. Then, one has to deal with the additional complexity related to having finite molecules interacting on an infinite surface, where in addition defects (e.g., extra or missing

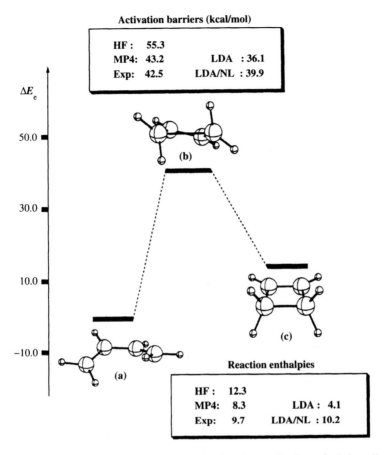

Figure 31.2 Schematic potential-energy profile for the cyclization of cis-butadiene to cyclo-butane calculated by different methods. The methods include Hartree–Fock (HF), fourth-order Møller–Plesset (MP4), local-density calculations (LDA), gradient-corrected density-functional calculations (LDA/NL), and experimental values (exp). All energies for each calculational method are given relative to that of the educt (a). Reproduced by permission of John Wiley & Sons. from Ziegler (1997)

atoms, steps, etc.) on the surface may be particularly favourable sites for the catalytic reactions. The size and the low symmetry of the complete system make theoretical studies in such cases very demanding.

31.2 HYDROGEN BONDS

Hydrogen bonds are weak bonds whose energies often are comparable with room temperature. Therefore, they can be created and broken at conditions that are normal for biological systems. Accordingly, they are extremely important for

Hydroformylation mechanism

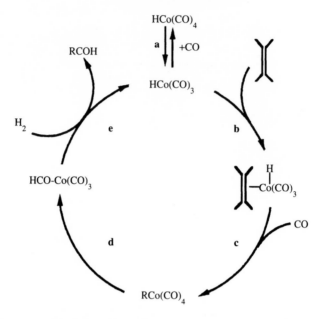

Figure 31.3 The hydroformylation reaction according to the mechanism of Heck and Breslow (1961). Reproduced by permission of John Wiley & Sons from Ziegler (1997)

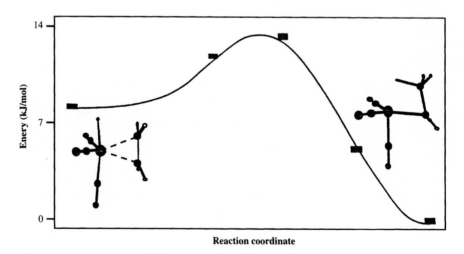

Figure 31.4 Energy profile for the step *b* of Fig. 31.3. Reproduced by permission of John Wiley & Sons from Ziegler (1997)

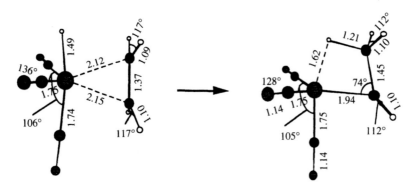

Figure 31.5 The structures a and b of Fig. 31.3. Reproduced by permission of John Wiley & Sons from Ziegler (1997)

many biochemical processes. Furthermore, both the well-known α-helical structure and the double helix are stabilized by hydrogen bonds.

The hydrogen bonds have the general form

$$A - H \cdots B, \tag{31.7}$$

i.e., the hydrogen atom forms a covalent bond to one of its two nearest neighbours, and a weaker, longer bond to the other neighbour. The $A - H - B$ bond angle is typically close to $180°$. A and B may be parts of a larger molecule, and in some cases (e.g., for water and ice), an approximately infinite network of hydrogen-bonded units may exist.

Since the hydrogen bonds are so weak, they are very difficult to treat theoretically: smaller inaccuracies may easily lead to completely wrong results. On the other hand, their importance, which was mentioned above, makes it desirable to be able to treat them accurately. It turns out, however, that these bonds are often not treated accurately with various methods, and therefore we shall here discuss results of different types of calculations on hydrogen-bonded systems.

A critical issue is whether the $A - H$ and the $H \cdots B$ distances above can be described accurately, and, whether the energy barrier for the process

$$A - H \cdots B \rightarrow A \cdots H - B \tag{31.8}$$

as well as the total-energy difference between the two structures can be calculated accurately.

In Table 31.3 we show results of different types of parameter-free calculations for the two finite molecules of Fig. 31.6. It can be seen that the density-functional calculations with a local-density approximation predict that the hydrogen atom of the hydrogen bond of malonaldehyde sits almost symmetrically between its two nearest neighbours. This is an often observed deficiency of local-density calculations on hydrogen-bonded systems: the hydrogen bonds are predicted to be much too short and strong.

Table 31.3 Structural parameters (in Å and deg.) for glyoxylic acid (upper part) and malonaldehyde (lower part). For glyoxylic acid we also give the dipole moment (in Debye). 'VWN' represents results of density-functional calculations with a local-density approximation. 'Perdew' and 'BP' are those of density-functional calculations with a gradient-corrected functional. 'HF' marks Hartree–Fock calculations, and 'MP2' marks second-order Møller-Plesset calculations. Finally, 'Exp' is experimental values. Reprinted with permission from Sim *et al.* (1992). Copyright 1992 American Chemical Society

Parameter	VWN	Perdew	BP	HF	Exp.
r(C1H2)	1.125	1.123	1.123	1.088	1.098
r(C1O1)	1.213	1.221	1.220	1.186	1.207
r(C1C2)	1.526	1.541	1.547	1.525	1.523
r(C2O3)	1.207	1.216	1.215	1.180	1.205
r(C2O2)	1.326	1.351	1.349	1.313	1.331
r(O2H1)	1.004	0.995	0.994	0.950	0.949
R(H1 $\cdots$ O1)	1.908	2.064	2.063	2.164	2.131
$\angle$(H2C1C2)	116.8	115.8	115.4	115.3	115.5
$\angle$(C2C1O1)	118.9	120.1	120.6	120.6	120.8
$\angle$(C1C2O3)	122.8	122.1	122.4	121.0	121.4
$\angle$(C1C2O2)	110.8	112.9	112.5	114.3	114.2
$\angle$(C2O2H1)	103.8	105.6	105.2	110.3	107.1
μ	2.142	1.930	1.992	2.091	1.86 ± 0.04

Parameter	VWN	Perdew	HF	MP2	Exp.
r(C1O1)	1.279	1.325	1.312	1.328	1.320
r(C1C2)	1.399	1.384	1.342	1.362	1.348
r(C2C3)	1.398	1.438	1.452	1.439	1.454
r(C3O2)	1.277	1.262	1.207	1.248	1.234
R(H1 $\cdots$ O2)	1.220	1.568	1.880	1.694	1.680
r(O1H1)	1.204	1.042	0.956	0.994	0.969
r(C1H2)	1.115	1.105	1.076	1.083	1.089
r(C2H3)	1.101	1.099	1.073	1.077	1.091
r(C3H4)	1.115	1.119	1.092	1.098	1.094
$\angle$(H1O1C1)	101.5	104.2	109.4	105.4	106.3
$\angle$(C2C3O2)	121.7	122.9	124.1	123.5	123.0

Tables 31.4–31.7 give details of a number of different types of calculations on the methane–methyl radical complex (CH_3–CH_4). In particular the variations of the total energy (Tables 31.6 and 31.7) are seen to depend strongly on the applied method, and neither the Hartree–Fock nor the local-density calculations are reliable in this respect. This indicates also that the concept of hydrogen bond is not purely that of electrostatic interactions, but that also quantum-mechanical effects (including correlation) may play a role.

31.3 SPIN–SPIN INTERACTIONS

At a few places we have discussed spin-dependent properties. In some cases, the system of interest possesses a delocalized spin density that is spread out over the complete system, but this is not always the case.

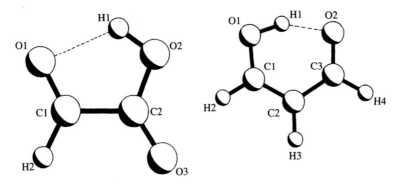

Figure 31.6 Glyoxylic acid (left part) and malonaldehyde (right part). Reprinted with permission from Sim *et al.* (1992) Copyright 1992 American Chemical Society

Table 31.4 Optimized geometries (in Å and deg.) for a methane molecule (CH_4), the methyl radical (CH_3), and the transition state (TS) for the interaction of the two systems. 'ROHF' marks Hartree–Fock calculations, 'MP2' corresponds to second-order Møller–Plesset calculations, 'LDA' to local-density calculations, 'NL' to a gradient-corrected density-functional, and the remaining entries to density-functional methods with a hybrid method that combines exact and approximate exchange interactions. Reprinted from *Chem. Phys. Lett.*, **244**, Jursic, p. 263. Copyright 1995, with permission from Elsevier Science

Method	r (CH_3)	r (CH_4)	$r1$ (TS)	$r2$ (TS)	$a1$ (TS)	$a2$ (TS)
ROHF	1.070	1.082	1.078	1.336	113.0	105.7
B3LYP	1.078	1.088	1.085	1.346	113.5	105.0
B3P86	1.078	1.088	1.085	1.339	113.6	105.0
MP2	1.074	1.085	1.083	1.323	113.4	105.2
LDA	1.092	1.101	1.100	1.334		104.7
NL	1.092	1.092	1.100	1.359		104.7

Isolated atoms from the $3d$ transition-metal or the $4f$ rare-earth series have partly filled $3d$ or $4f$ shells. These orbitals are strongly localized to the inner parts of the atoms so that, when the atoms are parts of molecular or solid compounds, these orbitals largely retain largely their atomic nature. For the isolated open-shell atoms, Hund's rule tells us that the spin tends to be maximized, i.e., the atoms have in most cases a total spin $\vec{S} \neq 0$. Also this property is essentially transferred to the molecule or solid so that it is a good approximation to assume that the spin density is localized to the inner parts of these atoms.

Table 31.5 As Table 31.4, but for the complete methane–methyl complex. The bond lengths are given in Å and the bond angles in degrees. Reprinted from *Chem. Phys. Lett.*, **244**, Jursic, p. 263. Copyright 1995, with permission from Elsevier Science

Method	$r1$	$r2$	$r3$	$r4$	$a1$	$a2$	$a3$	$a4$	$a5$
ROHF	1.084	1.083	3.400	1.072	109.4	109.5	176.9	95.3	119.3
B3LYP	1.093	1.094	3.002	1.083	109.4	109.5	176.9	91.7	120.0
B3P86	1.088	1.088	3.359	1.078	109.4	109.6	174.8	91.3	120.0
MP2	1.090	1.090	3.102	1.079	109.4	109.4	177.5	91.2	120.0
LDA			2.979						

Table 31.6 Total energies (in a.u.) for the methyl radical, methane, and the transition-state structure as well as the relative energy of the latter (in kcal/mol) in regards to separated reactants. The entries are similar to those of Table 31.4. Reprinted from *Chem. Phys. Lett.*, **244**, Jursic, p. 263. Copyright 1995, with permission from Elsevier Science

Method	CH_3	CH_4	TS	TS-reactants
ROHF	−39.57248	−40.21262	−79.72862	35.45
B3LYP	−39.85836	−40.53739	−80.37095	15.56
B3P86	−40.01690	−40.71668	−80.71156	13.82
MP2	−39.73506	−40.41079	−80.11526	19.19
LDA				2.8
NL				12.6
Exp.				14.1

Table 31.7 Total energies (a.u.) for methane–methyl radical complex and the transition-state structure, and relative energy (kcal/mol) of the transition-state structure with regards to separated reactants. The entries are similar to those of Table 31.4. Reprinted from *Chem. Phys. Lett.*, **244**, Jursic, p. 263. Copyright 1995, with permission from Elsevier Science

Method	Complex	Transition state	Relative energy
ROHF	−79.75031	−79.69400	35.33
B3LYP	−80.35742	−80.33376	14.85
B3P86	−80.73357	−80.71156	13.81
MP2	−80.00209	−79.96603	22.64
LDA			1.9
Exp.			14.1

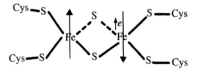

Figure 31.7 Schematic diagram of the structures and spin alignments for $Fe_2S_2(SR)_4^{3-}$ with SR being S-cysteine. Reprinted from *Coord. Chem. Rev.*, **144**, Noodleman *et al.*, Orbital interactions, electron delocalization and spin conplings in iron–sulfur clusters, p. 199. Copyright 1995 with permission from Elsevier Science

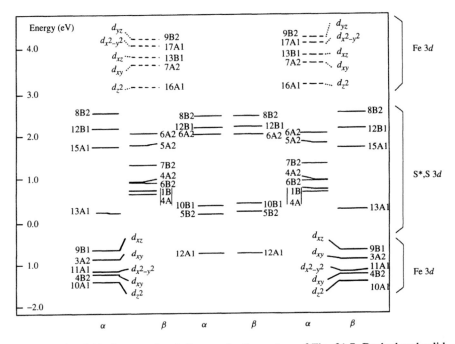

Figure 31.8 Orbital energy level diagram for the system of Fig. 31.7. Dashed and solid lines correspond to empty and filled orbitals, respectively, and the three blocks of levels (each split into the two spin components α and β) correspond to orbitals that are localized to the left part, the central part (the two S atoms), and the right part of the system of Fig. 31.7. Reprinted from *Coord. Chem. Rev.*, **144**, Noodleman *et al.*, Orbital interactions, electron delocalization and spin conflings in iron-sulfur clusters, p. 199. Copyright 1995, with permission from Elsevier Science

Consider, e.g., the system of Fig. 31.7. Except for smaller additional components, far the largest parts of the spin density originate from the two Fe atoms. These two spins may, however, interact indirectly and in some cases their interaction is well approximated through a Heisenberg Hamiltonian,

$$\hat{H}_{spin} = J\vec{S}_A \cdot \vec{S}_B, \tag{31.9}$$

where J is a coupling constant, and $\vec{S}_A$ and $\vec{S}_B$ the two spins. The total spin may take values in accordance with the angular-momentum vector addition rules,

$$S = |S_A - S_B|, |S_A - S_B| + 1, \ldots, S_A + S_B - 1, S_A + S_B. \tag{31.10}$$

The coupling between the two spins is mediated by the other electronic orbitals (in Fig. 31.7 this would mainly be through the orbitals on the central S atoms), and it is not always the case, that the Heisenberg model is a good approximation.

Also for crystalline materials or molecules with more $3d$ and/or $4f$ atoms, the coupling between the spins is very important, since these in many cases can be held responsible for the occurrence of different types of magnetism. For these, the generalization of Eq. (31.9) gives

$$\hat{H}_{\text{spin}} = \sum_{i,j} J_{ij} \vec{S}_i \cdot \vec{S}_j, \tag{31.11}$$

where the coupling constants J_{ij} often are vanishing for atoms that are not near neighbours.

Studies of the spin–spin interactions focus therefore on whether the simple models are correct, to determine the coupling constants, and to determine the parameters of additional effects, if there are any. For the system of Fig. 31.7 we show in Fig. 31.8 the orbital energy level diagram. By separating the orbitals both into different spin components and into different spatial parts we can see that the schematic picture of Fig. 31.7 is correct: the spin polarization is localized to the regions at the two terminal parts of the molecule, and it is opposite for the two parts. By mapping the results onto a model of the type (31.9) it was also found that the parameter J was of the order of some few 100 cm^{-1}, i.e., very small compared to bond energies and therefore very sensitive to approximations in the calculations.

32   Solvation

32.1   SUPERMOLECULES

A molecule in the gas phase, i.e., isolated from any other molecules, has certain properties. These properties will in most cases change when the molecule is placed in some solvent. The molecule will feel some interactions with the solvent molecules and these interactions will lead to modifications in the properties of the molecule of interest. Since many experiments and processes are carried through in the liquid phase it is desirable to be able to describe such systems by theoretical methods, too. That is, it is often not acceptable to assume that the molecular properties are unchanged from those in the gas phase.

Treating solvated molecules resembles in some sense that of treating impurities in infinite crystalline materials. In both cases, one is interested in a small, finite segment of an, in principle, infinite system. And in both cases this segment differs from the surrounding medium. On the other hand, the solvated molecules interact usually much more weakly with their surroundings than do the impurities in the crystal, i.e., no covalent bonds are formed. This means that the effects of interest are smaller, and it is often acceptable to treat them only approximately. This is important since the other important difference between the solvated molecule and the crystal impurity is that the latter system without the perturbation possesses translational periodicity and therefore can be treated with traditional band-structure methods, whereas the former system has no long-range order and can therefore be treated only with difficulty by electronic-structure methods.

The absolutely simplest approach for treating the solvated molecule resembles the cluster method for treating impurities in crystals. That is, one treats a finite system that is hopefully so large that it resembles the infinite system. For the impurity in the crystal this meant cutting covalent bonds, which led to the problems related to dangling bonds. Since the bonds between the solvent molecules are weak this problem does not occur here. Thus, with the so-called *super-molecule* approach (see, e.g., Dreyfus and Pullman 1970; Kollman and Allen,

1970) one studies the properties of an A molecule solvated in a B solvent by studying an AB_n system, where n is some integer that should not be too small. By this approach any standard, electronic-structure method for finite systems can be applied, but often the size of the system sets some limits on the practicability of the different methods.

In order to reduce the computational demands one may make explicit use of the fact that the electronic structure of the solvent molecules change only little due to the solvation, whereas that of the solute changes. Therefore, it may be a reasonable approximation to keep the electron density of the solvent molecules fixed as for the system without the solute. This is the principle behind the *frozen-density–functional-theory* method (Wesolowski and Warshel, 1993), cf. Fig. 32.1.

As the name indicates, the frozen density-functional method is based on density-functional theory. We consider a system whose electron density is split into two parts,

$$\rho(\vec{r}) = \rho_1(\vec{r}) + \rho_2(\vec{r}). \tag{32.1}$$

ρ_2 is to be the electron density of the solvent molecules, which is kept frozen as for the system without the solute (and it has accordingly to be determined in a separate calculation). On the other hand, the electron density ρ_1 is to be the

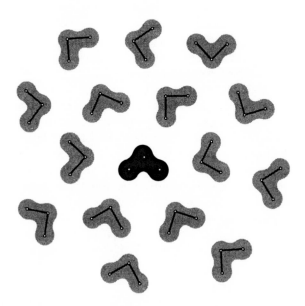

Figure 32.1   A schematic description of the frozen-density–functional-theory method, where the electron density of the solvent molecules (the lighter shaded molecules) is kept frozen, while the electron density of the solute molecule (the darker shaded molecule) is allowed to change. Reprinted with permission from Wesolowski and Warshel (1994). Copyright 1994 American Chemical Society

electron density of the solute and this is allowed to change due to the influences of the solvent molecules.

The electron density $\rho_1(\vec{r})$ is determined from the solutions to the Kohn–Sham equations,

$$\hat{h}_{\text{eff}} \psi_i(\vec{r}) = \varepsilon_i \psi_i(\vec{r}) \tag{32.2}$$

and is

$$\rho_1(\vec{r}) = \sum_{i=1}^{N_1} |\psi_i(\vec{r})|^2, \tag{32.3}$$

where

$$N_1 = \int \rho_1(\vec{r}) \, d\vec{r} \tag{32.4}$$

is the number of electrons of the solute molecule.

We need to determine the effective Hamilton operator of Eq. (32.2). This contains a kinetic-energy and a potential-energy term. For the latter we have

$$V_{\text{eff}}(\vec{r}) = V_{\text{ext}}(\vec{r}) + \int \frac{\rho_1(\vec{r}\,')}{|\vec{r} - \vec{r}\,'|} \, d\vec{r}\,' + \int \frac{\rho_2(\vec{r}\,')}{|\vec{r} - \vec{r}\,'|} \, d\vec{r}\,' + V_{\text{xc}}[\rho_1(\vec{r}) + \rho_2(\vec{r})]. \tag{32.5}$$

Here, the first term on the right-hand side is the external potential from the nuclei of the solute and the solvent, the second and third terms are the Coulombic potentials from the two electron distributions, and the last term is the exchange-correlation potential that with any local-density or generalized-gradient approximation easily can be calculated once the densities are given. Thus, these terms pose no problems.

Also the kinetic energy can be written as a functional of the electron density,

$$T[\rho(\vec{r})] = T[\rho_1(\vec{r}) + \rho_2(\vec{r})] = T_1[\rho_1(\vec{r})] + T_2[\rho_2(\vec{r})] + T_x[\rho(\vec{r})]. \tag{32.6}$$

The first and second terms are the kinetic-energy expressions for the isolated systems '1' and '2', and the last term includes all the non-additive effects that are absent in the first two.

T_2 is the kinetic energy for the solvent without the solute and can therefore be extracted from the calculations for that system. T_1 is the kinetic energy for the solute when neglecting the effects of the solvent molecules. This is therefore calculated from the standard Kohn–Sham expression

$$T_1 = \sum_{i=1}^{N_1} \langle \psi_i | -\frac{1}{2} \nabla^2 | \psi_i \rangle. \tag{32.7}$$

Finally, for T_x we may use a very approximate expression, e.g., the Thomas–Fermi expression,

$$T_x = C_{\text{TF}} \int \left\{ [\rho_1(\vec{r}) + \rho_2(\vec{r})]^{5/3} - [\rho_1(\vec{r})]^{5/3} - [\rho_2(\vec{r})]^{5/3} \right\} d\vec{r}. \tag{32.8}$$

Here, C_{TF} is a constant ($C_{\text{TF}} \simeq 2.871$).

This leads to the extra potential,

$$\frac{\delta T_x}{\delta \rho_1(\vec{r})} = \frac{5}{3} C_{TF} \{ [\rho_1(\vec{r}) + \rho_2(\vec{r})]^{2/3} - [\rho_1(\vec{r})]^{2/3} \} \tag{32.9}$$

which will be included in $\hat{h}_{eff}$. accordingly $\hat{h}_{eff}$ contains the kinetic-energy term $-\frac{1}{2}\nabla^2$, the potential of Eq. (32.5), and that of Eq. (32.9).

In total, the calculations can now proceed as follows: we calculate the electronic properties of the solvent without the solute. The electron density of this is kept frozen, and the solute is introduced. The electrons of the solute move then in the field that is the sum of the contributions from themselves and those from the solvent. The potential is thereby slightly more complicated although only slightly. For this last-mentioned system the electronic properties are calculated by solving the Kohn–Sham equations.

32.2　DIELECTRICA

A very different approach is that of treating the solvent as a dielectric continuum (Miertuš et al., 1981). To see how this works we refer to Fig. 32.2.

We assume for simplicity that the molecule of our interest, AB, has only two atoms. Furthermore, we assume that our molecule is polar, so that it effectively can be described as $A^{+q} - B^{-q}$. This molecule is shown in Fig. 32.2(a). The solvent is assumed to form a continuous medium (a dielectric continuum), Fig. 32.2(b), and when inserting the molecule into this, there will first of all be a cavity that allows for the solute to be there, as in Fig. 32.2(c). Due to the polarity of the dissolved molecule, the dielectric continuum responds by being polarized, too, as in Fig. 32.2(d). Thus, the main effects that should be incorporated when treating the solvation process are, first, the energy costs related with the creation of the cavity that will accommodate the solute and, second, the various interactions between the solute and the dielectric continuum. The latter are of electrostatic nature.

A critical issue with this approach is how to define the shape and size of the cavity. There exist various proposals but so far this question can not be considered settled. We shall here, however, not discuss it further, but nevertheless stress that the problem does exist and that it is one of the most critical aspects of this approach. Often the cavity is considered either spherical or having the shape of the solute molecule. In the latter case one may define spheres centred at each nucleus and having radii somewhat larger than the van der Waals radii of the atoms. The superposed volume of all these spheres is taken as the cavity.

Having, somehow, defined the shape and size of the cavity we need to take into account that the solvent responds to the charge distribution of the solute, which in turn responds to the electrostatic field created by the solvent. By treating the solvent as a homogenous dielectric continuum we can treat the former

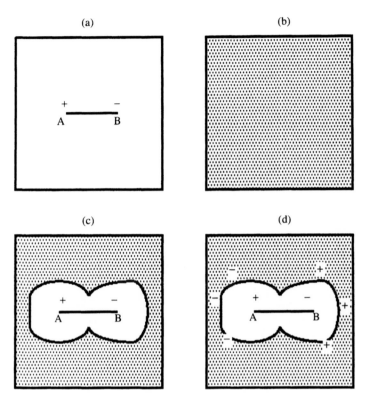

Figure 32.2 (a) The isolated A–B molecule; (b) the dielectric continuum before inserting the molecule; (c) as (b) but after the molecule has been inserted; (d) As (c) but with the response of the dielectric continuum to the charges of the molecule

using the (position- and direction-independent) polarizability χ. As discussed in Section 22.3, the polarizability describes how a system reacts to external electrostatic fields,

$$\delta\rho(\vec{r}) = \int \chi(|\vec{r} - \vec{r}'|)\delta V(\vec{r}')\,d\vec{r}'. \tag{32.10}$$

where $\delta V(\vec{r}')$ in our case is the Coulomb potential from the solute, and δ_ρ is the induced charge density of the solvent. Since χ is independent of position and direction (since the dielectric medium is homogenous), this equation becomes particularly simple.

In a practical calculation one will first define the size and shape of the cavity. Subsequently, the molecule of interest (the solute) will be inserted into the cavity and its electronic properties will be calculated using either a Hartree–Fock or a density-functional method. Here, the extra complications due to electrostatic interactions between the molecule and the surrounding medium [which may be polarized; cf. Fig. 32.2(d)] are to be included. Thereby, a self-consistent

Table 32.1 Dipole moments (in Debye) of two different isomers of three different molecules calculated by using a density-functional method with either a local-density approximation (LDA) or a gradient-corrected approximation (GGA) both in the gas phase and when being solvated in water. From Adamo and Lelj (1994)

Molecule	Gas phase LDA	Gas phase GGA	Solution LDA	Solution GGA
Formamide	3.574	3.892	4.304	4.274
Formamidic acid	1.059	1.047	1.141	1.126
2-Pyridone	4.169	4.122	5.354	5.325
2-hydroxypyridine	1.116	1.178	1.433	1.516
$[Co(NH_3)_5NO_2]^{2+}$	7.254	7.349	9.792	9.644
$[Co(NH_3)_5ONO]^{2+}$	6.672	7.085	10.360	10.198

Table 32.2 The solvation energy (E_s) (in kcal/mol) and dipole moment (μ) (in Debye) of (upper part) *syn*- and (lower part) *anti*-forms of acetic acid (Fig. 32.3) in gas phase and in aqueous solution. Relative energies of *anti*-form with respect to *syn*-form are given too (RE). LDA and GGA corresponds to a local-density and a gradient-corrected approximation, respectively, within density-functional theory. From Andzelm *et al.* (1995)

Property	Gas phase LDA	Gas phase GGA	Solution LDA	Solution GGA
μ	1.749	1.674	2.630	2.518
E_s			9.42	8.86
RE	0	0	0	0
μ	4.335	4.192	6.193	5.974
E_s			12.70	11.94
RE	5.10	4.77	1.79	1.70

procedure is required: for a given polarization of the dielectric medium, the solute will respond by having a certain electron (charge) distribution. This results, in turn, in a certain polarization of the dielectric medium, which is to be identical to the original one. As a further consequence of the redistribution of the electrons, also the structure of the solute may change.

Instead of discussing the computational details further we present some typical results from such calculations. Table 32.1 shows the dipole moments of three different molecules and for each of those molecules two different isomers are considered. Furthermore, the molecules were studied both in the gas phase and when being solvated in water. From these results it should be obvious that the solvation may have large effects on the electronic properties of the molecule.

This is further confirmed by the results of Tables 32.2 and 32.3 for the systems of Figs. 32.3 and 32.4. But the results will not be further discussed here.

Table 32.3 Geometrical parameters in gas phase (gas) and solution (sol) for neutral [III (upper part) and I (lower part)] conformers of glycine. Conformer III (gas) undergoes transformation to the zwitterion form (sol). Bond lengths are given in angstroms and bond angles in degrees. LDA and GGA have the same meaning as in Table 32.2. From Andzelm *et al.* (1995)

Parameter	Gas phase LDA	Gas phase GGA	Solution LDA	Solution GGA
O-H	1.031	1.013	1.692	1.879
N-H	1.724	1.830	1.080	1.052
N-C	1.459	1.481	1.472	1.502
C-C	1.515	1.536	1.518	1.541
C-O	1.331	1.356	1.272	1.279
C=O	1.214	1.221	1.252	1.266
H-N-C	112.6	112.1	100.2	103.4
N-C-C	110.3	111.0	106.8	108.3
C-C-O	112.2	113.1	114.9	116.3
O-H	0.994	0.989	0.997	0.992
N-H	1.031	1.028	1.033	1.029
N-C	1.436	1.461	1.441	1.464
C-C	1.495	1.523	1.490	1.516
C-O	1.348	1.374	1.336	1.360
C=O	1.216	1.223	1.224	1.230
H-N-C	108.6	108.3	108.5	108.5
N-C-C	115.4	116.0	115.0	115.8
C-C-O	112.3	111.8	112.6	112.1

Acetic acid

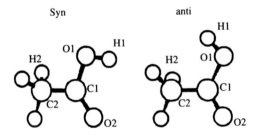

Figure 32.3 Acetic acid conformers. Reproduced by permission of American Institute of Physics from Andzelm *et al.* (1995)

32.3 POINT CHARGES

A different method for treating the effects of solvation is that of letting the solvent be represented by a set of point charges that creates an electrostatic field in which the solute is placed. This method resembles more the methods that we discussed in Chapter 30.

Glycine

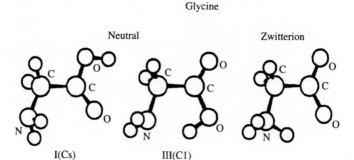

Figure 32.4 Glycine conformers. Reproduced by permission of the American Institute of Physics from Andzelm *et al.* (1995)

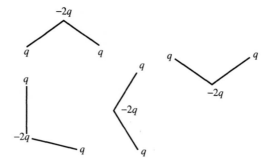

Figure 32.5 A set of point charges representing water

As a simple illustration of the ideas behind this approach we consider an aqueous solution. Each water molecule has a certain dipole moment which may be reproduced by assuming that the oxygen atom has an effective charge of $-2q$, whereas each hydrogen atom has an effective charge of $+q$. The electrostatic field generated by a single water molecule is accordingly assumed as being that of those three point charges.

Subsequently, a number of water molecules are placed at random so that both position and orientation is random, but also so that the density of water is correct. This may lead to a configuration like that of Fig. 32.5. When including a soluted molecule an additional constraint is introduced: the water molecules should not come too close to the solute. This represents a difference to the methods of the preceding subsection.

The electronic properties of the soluted molecule can now be determined using any method whereby the additional external potentials from the above distribution of point charges are included. Thereby, there is no feedback on the water from the solute, i.e., the water molecules are not allowed to respond to the presence of the solute.

Finally, in order to avoid having the selected random distribution of water molecules biased in one or another way, the calculation may be repeated for several such distributions, and, ultimately, averages are calculated for the properties of interest.

Having used water as an example we shall also add that water in fact is a complicated system to treat theoretically. Due to the hydrogen bonds that are often also are formed between the water molecules of the solvent and the solute neither a set of point charges nor a polarizable dielectric medium is able to describe accurately the details of the interactions between the solvent and the solute.

References

Abraham, F. F., Batra, I. P. and Ciraci, S. (1988) *Phys. Rev. Lett.* **60**, 1314.

Adamo, C. and Lelj, F. (1994) *Chem. Phys. Lett.* **223**, 54.

Allinger, N. L., Yuh, Y. H. and Lii, J.-H. (1989) *J. Am. Chem. Soc.* **111**, 8551.

Almbladh, C.-O. and von Barth, U. (1985) *Phys. Rev.* B **31**, 3231.

Andersen, O. K. (1975) *Phys. Rev.* B **12**, 3060.

Andersen, O. K., Jepsen, O. and Glötzel, D. (1985) in *Highlights of Condensed Matter Theory*, Ed. by F. Bassani, F. Fumi, and M. P. Tosi. North-Holland, New York.

Andzelm, J., Kölmel, C. and Klamt, A. (1995) *J. Chem. Phys.* **103**, 9312.

Bader, R. F. W (1970) *An Introduction to the Electronic Structure of Atoms and Molecules*, Clark, Irwin and Co.

Bader, R. F. W. (1990) *Atoms in Molecules: a Quantum Theory*. Clarendon, Oxford.

Bader, R. F. W. and Nguyen-Dang, T. T. (1981) *Adv. Quant. Chem.* **14**, 63.

Baird, N. C. and Dewar, M. J. S. (1969) *J. Chem. Phys.* **50**, 1262.

Bakowies, D. and Thiel, W. (1996) *J. Phys. Chem.* **100**, 10 580.

Baldereschi, A. (1973) *Phys. Rev.* B **7**, 5212.

Baraff, G. A. and Schlüter, M. (1984) *Phys. Rev.* B **30**, 3460.

Barbee, T. W. III, García, A., Cohen, M. L. and Martins, J. L. (1989) *Phys. Rev. Lett.* **62**, 1150.

Becke, A. D. (1988) *Phys. Rev.* A **38**, 3098.

Becke, A. D. (1993) *J. Chem. Phys.* **98**, 5648.

Becke, A. D. and Edgecombe, K. E. (1990) *J. Chem. Phys.* **92**, 5397.

Beeler, F. Andersen, O. K. and Scheffler, M. (1985) *Phys. Rev. Lett.* **55**, 1498.

Bethe, H. and Salpeter, E. (1957) *Quantum Mechanics of One- and Two-Electron Atoms*, Springer-Verlag, Berlin.

Beveridge, D. L. (1977) in *Semiempirical Methods of Electronic Structure Calculation*, Ed. by Segal, G. A., Plenum, New York.

Bloch, F. (1929) *Z. Phys.* **52**, 555.

Born, M. and Oppenheimer, R. (1927) *Ann. Phys.* **84**, 457.

Born, M. and von Kármán, Th. (1912) *Phys. Zeit.* **13**, 297.

Boys, S. F. and Bernardi, F. (1970) *Mol. Phys.* **19**, 553.

Brédas, J. L., Chance, R. R., Silbey, R., Nicolas, G. and Durand Ph. (1981) *J. Chem. Phys.* **75**, 255.

Breit, G. (1929) *Phys. Rev.* **34**, 553.

Breit, G. (1930) *Phys. Rev.* **36**, 383.

Breit, G. (1932) *Phys. Rev.* **39**, 616.

Brooks, B. R., Bruccoleri, R. E., Olafson, B. D., States, D. J., Swaminathan, S. and Karplus, M. J. (1983) *J. Comput. Chem.* **4**, 187.

Brommer, K. D., Needels, M., Larson, B. E. and Joannopoulos, J. D. (1992) *Phys. Rev. Lett.* **68**, 1355.

Broyden, C. G. (1965) *Math. Comput.* **19**, 577.

Bryant, G. W. and Mahan, G. D. (1978) *Phys. Rev.* B **17**, 1744.

Buda, F., Chiarotti, G. L., Car, R. and Parrinello, M. (1989) *Phys. Rev. Lett.* **63**, 294.

Car, R. and Parrinello, M. (1985) *Phys. Rev. Lett.* **55**, 2471.

Chadi, D. J. and Chang, K. J. (1988) *Phys. Rev. Lett.* **60**, 2187.

Champagne, B., Jacquemin, D. André, J.-M. and Kirtman, B. (1997) *J. Phys. Chem.* A **101**, 3158.

Chang, Ch., Pelissier, M. and Durand, Ph. (1986) *Phys. Scr.* **34**, 394.

Christensen, A., Stoltze, P. and Nørskov, J. K. (1995) *J. Phys. Condens. Matt.* **7**, 1047.

Čížek, J. (1966) *J. Chem. Phys.* **45**, 4256.

Ciraci, S. and Batra, I. P. (1987) *Phys. Rev. Lett.* **58**, 1982.

Ciraci, S. and Batra, I. P. (1988) *Phys. Rev.* B **38**, 1835.

Cowan, R. D. (1967) *Phys. Rev.* **163**, 54.

Das, G. P., Blöchl, P. Andersen, O. K., Christensen, N. E. and Gunnarsson, O. (1989) *Phys. Rev. Lett.* **63**, 1168.

Daudel, R., Leroy, G., Peeters, D. and Sana, M. (1983) *Quantum Chemistry*, John Wiley & Sons, Chichester.

Daw, M. S. and Baskes, M. I. (1984) *Phys. Rev.* B **29**, 6443.

de Heer, W. A. (1993) *Rev. Mod. Phys.* **65**, 611.

Del Bene, J. and Jaffé, H. H. (1968a) *J. Chem. Phys.* **48**, 1807.

Del Bene, J. and Jaffé. H. H. (1968b) *J. Chem. Phys.* **49**, 1221.

Del Bene, J. and Jaffé. H. H. (1969) *J. Chem. Phys.* **50**, 1126.

Delhalle, J., Delvaux, M. H., Fripiat, J. G. André, J. M. and Calais, J. L. (1988) *J. Chem. Phys.* **88**, 3141.

des Cloizeaux, J. (1964) *Phys. Rev.* **135**, A698.

De Vita, A., Štich, I., Gillan, M. J., Payne, M. C. and Clarke, L. J. (1993) *Phys. Rev. Lett.* **71**, 1276.

Dewar, M. J. S. and Klopman, G. (1967) *J. Am. Chem. Soc.* **89**, 3089.

Dewar, M. J. S., Zoebisch, E. G., Healy, E. F. and Stewart, J. J. P. (1985) *J. Am. Chem. Soc.* **107**, 3902.

Dirac, P. A. M. (1928) *Proc. Roy. Soc.* London A **117**, 610.

Douglas, M. and Kroll, N. M. (1974) *Ann. Phys.* **82**, 89.

Dreyfus, M. and Pullman, A. (1970) *Theor. Chim. Acta* **19**, 20.

Duffy, P., Chong, D. P., Casida, M. E. and Salahub, D. R. (1994) *Phys. Rev.* A **50**, 4707.

Dyson, F. J. (1949) *Phys. Rev.* **75**, 486.

Eriksson, O., Wills, J. M. and Boring, A. M. (1992) *Phys. Rev.* B **46**, 12 981.

Fässler, T. F. and Savin, A. (1997) *Chemie in unserer Zeit* **31**, 110.

Fermi, E. (1927) *Rend. Acad. Lincei* **6**, 602.

Feynman, R. P. (1939) *Phys. Rev.* **56**, 340.

Fock, V. (1930) *Z. Phys.* **61**, 126.

Foldy, L. L. and Wouthuysen, S. A. (1950) *Phys. Rev.* **78**, 29.

Ford, W. K., Duke, C. B. and Paton, A. (1982) *J. Chem. Phys.* **77**, 4564.

Friedel, J. (1954) *Adv. Phys.* **3**, 446.

Froyen, S., Wood, D. M. and Zunger, A. (1988) *Phys. Rev.* B **37**, 6893.

Fu, C.-L. and Ho, K.-M. (1983) *Phys. Rev.* B **28**, 5480.

Gáspár, R. (1954) *Acta Phys. Acad. Sci. Hung.* **3**, 263.

Ghahramani, E., Moss, D. J. and Sipe, J. E. (1990) *Phys. Rev.* B **41**, 5112.

Giannozzi, P., de Gironcoli, S., Pavone, P. and Baroni, S. (1991) *Phys. Rev.* B **43**, 7231.

Glötzel, D., Segall, B. and Andersen, O. K. (1980) *Solid State Commun.* **36**, 403.

Goedecker, S. (1999) *Rev. Mod. Phys.* **71**, 1085.

Goldstein, H. (1950) *Classical Mechanics*, Addison-Wesley, Reading.

Gombás, P. (1949) *Die statistische Theorie des Atoms und ihre Anwendungen*, Springer-Verlag, Vienna.

Grant, P. M. and Batra, I. P. (1979) *Solid State Commun.* **29**, 225.

Hafner, J. and M. Krajčí (1992) *Phys. Rev. Lett.* **68**, 2321.

Hansen, M. S. and Nishioka, H. (1993) *Z. Phys. D* **28**, 73.

Harris, J. (1984) *Phys. Rev. A* **29**, 1648.

Harrison, W. A. (1980) *Electronic Structure and the Properties of Solids*, Freeman, W. H. and Company, San Francisco.

Hartree, D. R. (1927) *Proc. Cambridge Phil. Soc.* **24**, 89, 111.

Hartree, D. R. (1928) *Proc. Cambridge Phil. Soc.* **24**, 426.

Hartree, D. R. (1929) *Proc. Cambridge Phil. Soc.* **25**, 225, 310.

Heck, R. F. and Breslow, D. S. (1961) *J. Am. Chem. Soc.* **83**, 4023.

Hedin, L. (1965) *Phys. Rev.* **139**, A796.

Hedin, L. and S. Lundqvist (1969) *Solid State Phys.* **23**, 1.

Heine, V., Haydock, R. and Kelly, M. (1980) *Solid State Phys.* **35**.

Hellmann, H. (1937) *Einführung in die Quantenchemie*, Deuticke, Leipzig.

Hess, B. A. (1986) *Phys. Rev. A* **33**, 3742.

Heully, J.-L., Lindgren, I., Lindroth, E., Lundqvist, S. and Mårtensson-Pendrill, A. M. (1986) *J. Phys. B* **19**, 2799.

Hirschfelder, J. O. and Meath, W. J. (1967) *Adv. Chem. Phys.* **12**, 3.

Hoffmann, R. (1963) *J. Chem. Phys.* **39**, 1397.

Hoffmann, R. (1964) *J. Chem. Phys.* **40**, 2047, 2474, 2480, 2745.

Hohenberg, P. and Kohn, W. (1964) *Phys. Rev.* **136**, B864.

Hohl, D. and Jones, R. O. (1991) *Phys. Rev. B* **43**, 3856.

Hohl, D., Jones, R. O., Car, R. and Parrinello, M. (1988) *J. Chem. Phys.* **89**, 6823.

Hückel, E. (1931a) *Z. Phys.* **70**, 204.

Hückel, E. (1931b) *Z. Phys.* **72**, 310.

Hückel, E. (1932) *Z. Phys.* **76**, 628.

Hughes, J. L. P. and Sipe, J. E. (1996) *Phys. Rev. B* **53**, 10 751.

Hwang, T.-S. and Wang, Y. (1998) *J. Phys. Chem. A* **102**, 3726.

Hybertsen, M. S. and Louie, S. G. (1986) *Phys. Rev. B* **34**, 5390.

Hybertsen, M. S. and Louie, S. G. (1987a) *Phys. Rev. B* **35**, 5585.

Hybertsen, M. S. and Louie, S. G. (1987b) *Phys. Rev. B* **35**, 5602.

Hybertsen, M. S. and Louie, S. G. (1987c) *Phys. Rev. Lett.* **58**, 1551.

Ishii, N. and Shimizu, T. (1995) *Chem. Phys. Lett.* **235**, 614.

Janak, J. F. (1978) *Phys. Rev. B* **18**, 7165.

Jennison, R. R., Schultz, P. A. and Sears, M. P. (1997) *J. Chem. Phys.* **106**, 1856.

Jepsen, O. and Andersen, O. K. (1971) *Solid State Commun.* **9**, 1763.

Jepsen, O., Glötzel, D. and Mackintosh, A. R. (1981) *Phys. Rev. B* **23**, 2684.

Jones, R. O. and Gunnarsson, O. (1989) *Rev. Mod. Phys.* **61**, 689.

Jursic, B. S. (1995) *Chem. Phys. Lett.* **244**, 263.

Karpfen, A. and Petkov, J. (1979) *Solid State Commun.* **29**, 251.

Klopman, G. and Evans, R. C. (1977) in *Semiempirical Methods of Electronic Structure Calculation*, Ed. by Segal, G. A., Plenum, New York.

Kohn, W. and Rostoker, N. (1954) *Phys. Rev.* **94**, 1111.

Kohn, W. and Sham, L. J. (1965) *Phys. Rev.* **140**, A1133.

Kollman, P. A. and Allen, L. C. (1970) *Theor. Chim. Acta* **18**, 399.

Koopmans, T. (1933) *Physica* **1**, 104.

Korringa, J. (1947) *Physica* **13**, 392.

Kunc, K. and Martin, R. M. (1983) in *Ab Initio Calculation of Phonon Spectra*, Ed. by J. T. Devreese, V. E. Van Doren, and P. E. Van Camp, Plenum Press, New York.

Lambrecht, W. R. L., Segall, B. and Andersen, O. K. (1990) *Phys. Rev.* B **41**, 2813.

Lee, C., Yang, W. and Parr, R. G. (1988) *Phys. Rev.* B **37**, 785.

Li, X.-P., Nunes, R. W. and Vanderbilt, D. (1993) *Phys. Rev.* B **47**, 10 891.

Lii, J.-H. and Allinger, N. L. (1989), *J. Am. Chem. Soc.* **111**, 8566, 8576.

Lindgren, I. (1971) *Int. J. Quant. Chem. Symp.* **5**, 411.

Löwdin, P.-O. (1950) *J. Chem. Phys.* **18**, 365.

Löwdin, P. O. (1955) *Phys. Rev.* **97**, 1474.

Mailhiot, C., Yang, L. H. and McMahan, A. K. (1992) *Phys. Rev.* B **46**, 14 419.

Manghi, F., Del Sole, R., Selloni, A. and Molinari, E., *Phys. Rev.* B **41**, 9935 (1990).

McWeeny, R. (1960) *Rev. Mod. Phys.* **32**, 335.

McWeeny, R. (1973) *Nature* **243**, 196.

Miertuš, S. Scrocco, E. and Tomasi, J. (1981) *Chem. Phys.* **55**, 117.

Mintmire, J. W. and White, C. T. (1983a) *Phys. Rev.* B **27**, 1447.

Mintmire, J. W. and White, C. T. (1983b) *Phys. Rev.* B **28**, 3283.

Monkhorst, H. J. and Pack, J. D. (1976) *Phys. Rev.* B **13**, 5188.

Mulliken, R. S. (1953) *J. Chem. Phys.* **23**, 1833, 1841.

Møller, C. and Plesset, M. S. (1934) *Phys. Rev.* **46**, 618.

Neumann, R. and Handy, N. C. (1995) *Chem. Phys. Lett.* **246**, 381.

Noodleman, L., Peng, C. Y., Case, D. A. and Mouesca, J.-M. (1995) *Coord. Chem. Rev.* **144**, 199.

Nørskov, J. K. (1982) *Phys. Rev.* B **26**, 2875.

Pariser, R. and Parr, R. G. (1953) *J. Chem. Phys.* **21**, 767.

Parr, R. G., Donnelly, R. A., Levy, M. and Palke, W. E. (1978) *J. Chem. Phys.* **68**, 3801.

Pasturel, A., Colinet, C., Paxton, A. T. and van Schilfgaarde, M. (1992) *J. Phys. Condens. Matt.* **4**, 945.

Pearson, R. G. (1963) *J. Am. Chem. Soc.* **85**, 3533.

Perdew, J. (1991) in *Electronic Structure of Solids '91*, Ed. by P. Ziesche and H. Eschrig, Akademie Verlag, Berlin.

Perdew, J. P. and Wang, Y. (1986) *Phys. Rev.* B **33**, 8800.

Perdew, J. P. and Zunger, A. (1981) *Phys. Rev.* B **23**, 5048.

Phillips, J. C. and Kleinman, L. (1959) *Phys. Rev.* **116**, 287.

Pople, J. A. (1953) *Trans. Faraday Soc.* **49**, 1375.

Pople, J. A., Santry, D. P. and Segal, G. A. (1965) *J. Chem. Phys.* **43**, S129.

Pople, J. A. and Segal, G. A. (1965) *J. Chem. Phys.* **43**, S136.

Pople, J. A. and Segal, G. A. (1966) *J. Chem. Phys.* **44**, 3289.

Poteau, R. and Spiegelmann, F. (1993) *J. Chem. Phys.* **98**, 6540.

Press, W. H., Teukolsky, S. A., Vetterling, W. T. and Flannery, B. P. (1992) *Numerical Recipes*, Cambridge University Press, Cambridge.

Prutton, M. (1994) *Introduction to Surface Physics*, Clarendon Press, Oxford.

Pulay, P. (1969) *Mol. Phys.* **17**, 197.

Rappé, A. K., Casewit, C. J., Colwell, K. S., Goddard, W. A. III and Skiff, W. M. (1992) *J. Am. Chem. Soc.* **114**, 10024.

Resta, R. (1994) *Rev. Mod. Phys.* **66**, 899.

Roberts, N. and Needs, R. J. (1990) *Surf. Sci.* **236**, 112.

Robertson, J. and O'Reilly, E. P. (1987) *Phys. Rev.* B **35**, 2946.

Roos, B. O., Taylor, P. R. and Siegbahn, P. E. M. (1980) *Chem. Phys.* **48**, 157.

Roothaan, C. C. J. (1951) *Rev. Mod. Phys.* **23**, 69.

Sanchez, J. M., Ducastelle, F. and Gratias, D. (1984) *Physica* A **128**, 334.

Santi, G. and Jarlborg, T. (1997) *J. Phys. Condens. Matt.* **9**, 9563.

Segal, G. A. (1967) *J. Chem. Phys.* **47**, 1876.

Seki, K., Ueno, N., Karlsson, U. O., Engelhardt, R. and Koch, E.-E. (1986) *Chem. Phys.* **105**, 247.

Sim, F., St-Amant, A., Papai, I. and Salahub, D. R. (1992) *J. Am. Chem. Soc.* **114**, 4391.

Skriver, H. L. and Rosengaard, N. M. (1992) *Phys. Rev.* B **46**, 7157.

Slater, J. C. (1937) *Phys. Rev.* **51**, 846.

Slater, J. C. (1951) *Phys. Rev.* **81**, 385.

Slater, J. C. and Koster, G. F. (1954) *Phys. Rev.* **94**, 1498.

Sluiter, M., de Fontaine, D., Guo, X. Q., Podloucky, R. and Freeman, A. J. (1990) *Phys. Rev.* B **42**, 10460.

Sprik, M., Hutter, J. and Parrinello, M. (1996) *J. Chem. Phys.* **105**, 1142.

Springborg, M. (1994a) *Solid State Commun.* **89**, 665.

Springborg, M. (1994b) *Z. Phys.* B **95**, 363.

Springborg, M. and Andersen, O. K. (1987) *J. Chem. Phys.* **87**, 7125.

Stocks, G. M. and Winter, H. (1984) in: *Electronic Structure of Complex Systems*, Ed. by P. Phariseau and W. Temmerman, Plenum Press, New York.

Stott, M. J. and Zaremba, E. (1980) *Phys. Rev.* B **22**, 1564.

Svane, A. (1987) *Phys. Rev.* B **35**, 5496.

Tarnow, E., Payne, M. C. and Joannopoulos, J. D. (1988) *Phys. Rev. Lett.* **61**, 1772.

Thomas, L. H. (1926) *Proc. Cambridge Phil. Soc.* **23**, 542.

Vanderbilt, D. and King-Smith, R. D. (1993) *Phys. Rev.* B **48**, 4442.

Van Hove, L. (1954) *Phys. Rev.* **95**, 249.

van Lenthe, E., Baerends, E. J. and Snijders, J. G. (1993) *J. Chem. Phys.* **99**, 4597.

Verlet, L. (1967) *Phys. Rev.* **159**, 98.

von Barth, U. and Hedin, L. (1972) *J. Phys.* C **5**, 1629.

Vosko, S. H., Wilk, L. and Nusair, M. (1980) *Can. J. Phys.* **58**, 1200.

Wannier, G. H. (1937) *Phys. Rev.* **52**, 191.

Wannier, G. H. (1962) *Rev. Mod. Phys.* **34**, 645.

Wei, S. and Chou, M. Y. (1992) *Phys. Rev. Lett.* **69**, 2799.

Wei, S.-H., Ferreira, L.-G., Bernard, J. E. and Zunger, A. (1990) *Phys. Rev.* B **42**, 9622.

Weiner, P. K. and Kollman, P. A. (1981) *J. Comput. Chem.* **2**, 287.

Wesolowski, T. A. and Warshel, A. (1993) *J. Phys. Chem.* **97**, 8050.

Wesolowski, T. and Warshel, A. (1994) *J. Phys. Chem.* **98**, 5183.

Whangbo, M.-H., Hoffmann, R. and Woodward, R. B. (1979) *Proc. Roy. Soc.* London A **366**, 23.

Wiberg, K. B. (1968) *J. Am. Chem. Soc.* **90**, 59.

Wolverton, C., Ceder, G., de Fontaine, D. and Dreyssé, H. (1993) *Phys. Rev.* B **48**, 7126.

Yin, M. T. and Cohen, M. L. (1982) *Phys. Rev.* B **26**, 5668.

Yin, M. T. and Cohen, M. L. (1983) *Phys. Rev. Lett.* **50**, 2006.

Young, V., Suck, S. H. and Hellmuth, E. W. (1979) *J. Appl. Phys.* **50**, 6088.

R. Yu, and Krakauer, H. (1995) *Phys. Rev. Lett.* **74**, 4067.

Ziegler, T. (1997) in *Density Functional Methods in Chemistry and Materials Science*, Ed. by M. Springborg, John Wiley, Chichester.

Index

Lightning Source UK Ltd.
Milton Keynes UK
UKOW03f1221030314

227454UK00001B/83/P